Green Synthesis in Nanomedicine and Human Health

Green Synthesis in Nanomedicine and Human Health

Edited by
Richard L. K. Glover
Daniel Nyanganyura
Maluta Steven Mufamadi
Rofhiwa Bridget Mulaudzi

CRC Press is an imprint of the
Taylor & Francis Group, an informa business

First edition published 2021
by CRC Press
6000 Broken Sound Parkway NW, Suite 300, Boca Raton, FL 33487-2742

and by CRC Press
2 Park Square, Milton Park, Abingdon, Oxon, OX14 4RN

Library of Congress Cataloging-in-Publication Data

Names: Glover, Richard L. K., editor. | Nyanganyura, Daniel, editor. | Mulaudzi, Rofhiwa Bridget, editor. | Mufamadi, Maluta Steven, editor.
Title: Green synthesis in nanomedicine and human health / Richard L. K. Glover, Daniel Nyanganyura, Rofhiwa Bridget Mulaudzi, Maluta Steven Mufamadi.
Description: First edition. | Boca Raton, FL : CRC Press, [2021] | Includes bibliographical references and index. | Summary: "Green synthesis is an emerging method for deriving nanoparticles present in natural plants for use in nanomedicine. Written by experts in the field, Green Synthesis in Nanomedicine and Human Health showcases the exciting developments of this specialty and its potential for promoting human health and wellbeing. This book gives practical information on novel preparation methods for identifying nanoparticles present in natural plants. It discusses applications of nanoparticles in combating communicable, non-communicable and vector-borne diseases. It also explores the potential for nanoparticles to combat antimicrobial resistance through improvements in treatment methods, diagnostics, and drug delivery systems. Features scientific evidence of opportunities for integrating indigenous flora into nanomedicine to develop cost-effective therapeutic and diagnostic solutions for diseases including cancer, tuberculosis, malaria and diabetes. Places green synthesis and nanomedicine in the African orthodox and traditional healthcare context. Provides policy makers with scientific evidence to inform policies for controlling or mitigating dangerous diseases. This book is essential reading for students, scientists, policymakers and practitioners of nanotechnology, and will appeal to anyone with an interest in integrating traditional African healthcare and Western medicine"-- Provided by publisher.
Identifiers: LCCN 2020042728 | ISBN 9780367710811 (paperback) | ISBN 9780367902162 (hardback) | ISBN 9781003023197 (ebook)
Subjects: LCSH: Nanomedicine. | Nanotechnology. | Biosynthesis.
Classification: LCC R857.N34 G74 2021 | DDC 610.28--dc23
LC record available at https://lccn.loc.gov/2020042728

ISBN: 9780367902162 (hbk)
ISBN: 9780367710811 (pbk)
ISBN: 9781003023197 (ebk)

Typeset in Times
by Deanta Global Publishing Services, Chennai, India

Contents

List of Figures ix
List of Tables xi
Foreword xiii
Acknowledgements xv
Editors xvii
Contributors xix

SECTION I Green Synthesis of Nanoparticles from Natural Plant Parts

General Overview 3

Maluta Steven Mufamadi, Rofhiwa Mulaudzi, Richard L.K. Glover and Daniel Nyanganyura

SECTION II Nanotechnology for Treatment of Non-Communicable Diseases

Chapter 1 Cancer Nanotheranostics: Next-Generation Early Detection and Treatment Prioritization for Cancers Using Phytonanotechnology 17

Maluta Steven Mufamadi, Marian Jiya John, Mpho Phehello Ngoepe and Palesa Rose Sekhejane

Chapter 2 Green Synthesis of Nanoparticles Using Plant Extracts: A Promising Antidiabetic Agent 31

Rofhiwa Bridget Mulaudzi, Mahwahwatse Johanna Bapela and Thilivhali Emmanuel Tshikalange

Chapter 3 Green Nanoparticles: An Alternative Therapy for Oral Candidiasis 45

Razia Z. Adam, Enas Ismail, Fanelwa Ajayi, Widadh Klein, Germana Lyimo and Ahmed A. Hussein

SECTION III Nanotechnology for Treatment of Communicable Diseases

Chapter 4 Nanotechnology and Nanomedicine to Combat Ebola Virus Disease 65

Maluta Steven Mufamadi

Chapter 5 Application of Next-Generation Plant-Derived Nanobiofabricated Drugs for the Management of Tuberculosis 81

Charles Oluwaseun Adetunji, Olugbenga Samuel Michael, Muhammad Akram, Kadiri Oseni, Ajayi Kolawole Temidayo, Osikemekha Anthony Anani, Akinola Samson Olayinka, Olerimi Samson E, Wilson Nwankwo, Iram Ghaffar and Juliana Bunmi Adetunji

SECTION IV Nanotechnology for Treatment of Vector-Borne Diseases

Chapter 6 Biogenic Nanoparticles Based Drugs Derived from Medicinal Plants: A Sustainable Panacea for the Treatment of Malaria........... 103

Charles Oluwaseun Adetunji, Olugbenga Samuel Michael, Wilson Nwankwo, Osikemekha Anthony Anani, Juliana Bunmi Adetunji, Akinola Samson Olayinka and Muhammad Akram

SECTION V Nanotechnology in Combating Antimicrobial Resistance

Chapter 7 Bioengineering of Inorganic Nanoparticle Using Plant Materials to Fight Extensively Drug-Resistant Tuberculosis 125

Mpho Phehello Ngoepe and Maluta Steven Mufamadi

Chapter 8 Recent Advances in the Utilization of Bioengineered Plant-Based Nanoparticles: A Sustainable Nanobiotechnology for the Management of Extensively Drug-Resistant Tuberculosis............... 149

Charles Oluwaseun Adetunji, Olugbenga Samuel Michael, Muhammad Akram, Kadiri Oseni, Olerimi Samson E, Osikemekha Anthony Anani, Wilson Nwankwo, Hina Anwar, Juliana Bunmi Adetunji, and Akinola Samson Olayinka

Chapter 9 Green Synthesis of Nanoparticles and Their Antimicrobial Efficacy against Drug-Resistant *Staphylococcus aureus* 167

Nonhlanhla Tlotleng, Jiya M. John, Dumisile W. Nyembe and Wells Utembe

Chapter 10 Green Metal-Based Nanoparticles Synthesized Using Medicinal Plants and Plant Phytochemicals against Multidrug-Resistant *Staphylococcus aureus* 181

Abeer Ahmed Qaed Ahmed, Lin Xiao, Tracey Jill Morton McKay and Guang Yang

SECTION VI Cross-Cutting Issues

Chapter 11 Polymer-Based Protein Delivery Systems for Loco-Regional Administration 249

Muhammad Haji Mansor, Emmanuel Garcion, Bathabile Ramalapa, Nela Buchtova, Clement Toullec, Marique Aucamp, Jean Le Bideau, François Hindré, Admire Dube, Carmen Alvarez-Lorenzo, Moreno Galleni, Christine Jérôme, and Frank Boury

Chapter 12 Nanomedicines for the Treatment of Infectious Diseases: Formulation, Delivery and Commercialization Aspects 271

Admire Dube, Boitumelo Semete-Makokotlela, Bathabile Ramalapa, Jessica Reynolds and Frank Boury

Chapter 13 Green-Synthesized Nanoparticles as Potential Sensors for Health Hazardous Compounds 291

Rachel Fanelwa Ajayi, Sphamandla Nqunqa, Yonela Mgwili, Siphokazi Tshoko, Nokwanda Ngema, Germana Lyimo, Tessia Rakgotho, Ndzumbululo Ndou,and Razia Adam

Index 315

List of Figures

Figure 1.1 Schematic diagram illustrating the mechanisms of AgNPs and AuNPs to induce cytotoxicity in a cancerous human cell 22

Figure 2.1 General methods of nanoparticle synthesis 36

Figure 3.1 HRTEM image of ZnO nanoparticles 54

Figure 3.2 Standard well diffusion test on *Candida albicans* biofilm; 100 μml ZnO-NPs (inhibition zones = 24–26 mm) 55

Figure 6.1 Management of malaria 105

Figure 6.2 Informatics and applications in drug discovery, formulation, production and treatment of malaria 111

Figure 7.1 Antimicrobial mechanism of NPs and their ions 127

Figure 7.2 Mechanism of drug resistance in *M. tuberculosis* 128

Figure 7.3 Diagram summarizing the possible mechanism of biologically mediated synthesis of nanoparticles 131

Figure 7.4 Inorganic nanoparticles ligand attachment for targeting TB-infected macrophages 137

Figure 7.5 Labelling of inorganic nanoparticles for biomedical applications 139

Figure 10.1 Typical various approaches for the synthesis and evaluation of green metal-based NPs 183

Figure 10.2 Biological (green) synthesis and the applications of metal nanoparticles in biomedical and environmental fields. (Reproduced with permission from Singh et al., 2016) 183

Figure 10.3 Historical review of MDR *S. aureus* strains and their emerging multiple resistance to drugs 187

Figure 10.4 SEM micrograph of MRSA. (a and b) Untreated control cells. (c and d) Cells treated with 25 and 50 μg/ml of AgNPs; red arrows illustrating structural deformities and irregular cell surface. (Reproduced with permission from Ansari, M.A., and Alzohairy, M.A. 2018. "One-pot facile green synthesis of silver nanoparticles using seed extract of *Phoenix dactylifera* and their bactericidal potential against MRSA". *Evidence-Based Complementary and Alternative Medicine* 2018:1–9. doi: https://doi.org/10.1155/2018/1860280) 202

Figure 10.5 HR-TEM micrograph of MRSA. (a) Untreated control cell. (b and c) Treated with 25 and 50 μg/ml AgNPs. Red arrows indicate the attachment and penetration of NPs and degradation and destruction of the outermost layers of cell wall and cytoplasmic membrane. (Reproduced with permission from Ansari, M.A., and Alzohairy, M.A. 2018. "One-pot facile green synthesis of silver nanoparticles using seed extract of *Phoenix dactylifera* and their bactericidal potential against MRSA". *Evidence-Based Complementary and Alternative Medicine* 2018:1–9. doi: https://doi.org/10.1155/2018/1860280)....203

Figure 10.6 Possible mechanism of actions for green metal-based NPs....205

Figure 11.1 Advantages of protein therapeutics for clinical applications....252

Figure 11.2 Common polymer-based systems for drug delivery applications....256

Figure 11.3 Examples of micro/nanoparticle preparation process....257

Figure 11.4 A simplified representation of a fibrous scaffold and its internal structure....261

Figure 11.5 Different modes of protein loading into a fibrous scaffold. (Adapted from J.S. Choi, K.W. Leong, H.S. Yoo, *In vivo* wound healing of diabetic ulcers using electrospun nanofibers immobilized with human epidermal growth factor (EGF), Biomaterials. 29 (2008) 587–596 and T.G. Kim, T.G. Park, Surface functionalized electrospun biodegradable. nanofibers for immobilization of bioactive molecules, Biotechnol. Prog. 22 (2006) 1108–1113)....262

Figure 12.1 Schematic diagram illustrating the intracellular locality of *M. tb* and HIV pathogens, and the various types of nanoparticles. *M. tb* is typically contained within phagosomes, while HIV is located within the nucleus. In most cases, nanoparticles will need to penetrate the cellular host/intracellular space and/or nucleus to deliver therapeutic payload....275

Figure 13.1 Structure of hydrazine....293

Figure 13.2 Structure of nitrobenzene....293

Figure 13.3 Basic structure of an electrochemical sensor....295

Figure 13.4 An illustration of the top-down and bottom-up approaches for the synthesis of nanoparticles....296

List of Tables

Table 2.1 Brief Descriptions of the Types of Diabetes 33

Table 2.2 Metal Nanoparticles Synthesis Using Different Plant Extracts with Antidiabetic Activity 38

Table 3.1 Common Risk Factors for Oral Candidiasis (Patil et al., 2015)........... 46

Table 3.2 Green-Mediated Au Nanoparticles against Different *Candida* Strains 51

Table 3.3 Recent Plant-Mediated Ag Nanoparticles against Different *Candida* Strains 52

Table 3.4 Recent Studies Using Plant-Mediated Synthesis of Zinc Oxide Nanoparticle 53

Table 3.5 Plant-Mediated Zinc Oxide Nanoparticles and *Candida* Species 53

Table 6.1 Pharmacotherapeutic Agents for Malaria Treatment 106

Table 6.2 Nanomaterial Employed for the Treatment of Malaria...................... 109

Table 7.1 Various Inorganic Nanoparticles with Antimicrobial Activity Synthesized Using Green Chemistry 133

Table 10.1 Green Synthesis of GAgNPs from Different Medicinal Plants Extracts against *S. aureus* Strains 194

Table 10.2 Green Synthesis of GAuNPs from Different Medicinal Plants Extracts against *S. aureus* Strains 197

Table 10.3 Synthesis of Different Green NPs from Variety of Medicinal Plants Extracts against *S. aureus* Strains 199

Table 10.4 Plant Phytochemicals Involved in the Formation and Stabilization of G-NPs 225

Table 12.1 List of Common Infectious Diseases, Pathogen Involved and Main Transmitting Agent 272

Table 13.1 A List of Green-Synthesized Gold Nanoparticles............................ 302

Table 13.2 A List of Green-Synthesized Silver Nanoparticles with Potential Application in Electrochemical Sensors 303

Table 13.3 Examples of Green Method Synthesized Platinum Nanoparticles with Potential Application in Electrochemical Sensors 304

List of Tables

Table 2.1 Brief Descriptions of the Types of Diabetes [illegible]

Table 3.2 Metal Nanoparticles Synthesis Using Different Plant Extracts with Antidiabetic Activity [illegible]

Table 4.1 [illegible]

Table 4.2 [illegible]

[illegible]

[illegible]

[illegible]

Table 6.2 [illegible]

Table 7.1 [illegible]

Table 9.1 [illegible]

Table [illegible] 197

Table 10.1 [illegible]

Table 10.2 [illegible]

Table 12.1 [illegible]

Table 13.1 [illegible]

Table 13.2 [illegible] Applications in Biomedical Sciences [illegible] 303

Table 13.3 Examples of Green Metal [illegible] Synthesis [illegible] Application in Biomedical Sciences [illegible]

Foreword

Nanotechnology is considered one of the most promising, cutting-edge and disruptive technological advancements. This technology can be applied in numerous industrial and biomedical sectors. Several factors, which include the versatility of nanomaterials and the relatively low production cost of nano-enabled products, make this technology particularly suitable in low-resource settings. The application of nanotechnology in water treatment, energy, agriculture and medicine is particularly relevant in countries with struggling economies. However, several factors, which include barriers to the commercialization of nanotechnology and the unknown impact of nanomaterials on the environment and humans, can impede the development of nanotechnology.

Nanomaterials can be synthesized using a top-down or bottom-up approach. The top-down approach aims to fabricate nanoscale materials by mechanical processes such as ball milling, lithography, laser and ablation. These methods of nanomaterial synthesis can be expensive and energy-intensive, and it can negatively impact the environment. The bottom-up approach, on the other hand, is based on the self-assembly of atoms into nanoscale structures and mostly involves chemical processes. These chemical approaches are not particularly eco-friendly, and there is a growing need to develop environmentally and economically friendly processes. The most common bottom-up approach for the synthesis of metallic nanomaterials is chemical reduction of the metallic cations by inorganic reducing agents to produce colloidal nanoparticles.

Green nanotechnology is a fairly new branch of nanotechnology, which aims to produce and utilize nanomaterials in a way that is safe for living organisms and their environment. It can also involve the production of nanomaterials that can solve environmental problems. It uses the principles of green chemistry, which aims to reduce or eliminate the use or generation of hazardous substances during the synthesis of chemical products. These processes also aim to develop synthetic methods, which are more energy-efficient. Aside from the fact that green nanotechnology can be beneficial to the environment, green nanomaterials produced in this fashion can also be more biocompatible and therefore more suitable for applications in nanomedicine. The impact of nanomaterials on the environment and human health is relatively unknown. The immense diversity of nanomaterials and contradictory reports on the safety of nanomaterials further exacerbate this problem. Moreover, the absence of standardized methods makes it difficult to study the impact of nanomaterials on ecosystems.

Bioinspired synthesis methods that use biological or organic reducers of metal cations have been developed to produce green nanoparticles. A number of living organisms, which include bacteria, algae, fungi and plants, have been used as sources of these biological or organic reducers to produce colloidal metal nanoparticles. The molecular interactions between microorganisms and metals, as well as between plants and metals have been well documented. Several bacterial strains are known to produce inorganic materials, including metals. The use of microorganisms and plants to develop safer nanoparticle production methods are thus not so far-fetched. Green nanotechnology aims to reduce the production of hazardous waste during the synthesis process and to develop more energy-effective methods. Bioinspired nanoparticles are considered more biocompatible than nanomaterials produced using traditional chemical synthesis methods and are therefore more suitable for biomedical applications. Bioinspired green nanoparticle synthesis using plants involves bio-reduction reactions that are catalysed by phytochemicals such as terpenoids, flavones, ketones, aldehydes, amides and carboxylic acids. These bio-reduction reactions can occur at low temperatures, which means that these reactions are more energy-efficient than traditional chemical synthesis methods. The use of plants to synthesize green nanomaterials has given rise to phytonanotechnology, which aims to synthesize nanoparticles using eco-friendly, simple, rapid and cost-effective methods.

Some of the unique physical, chemical and optical properties of nanoparticles make these materials useful for biomedical applications. Due to their small size and high surface area to volume ratio, nanoparticles are well suited as vectors to carry molecular mediators of disease therapeutics and diagnosis. Biomolecules, which include DNA, RNA and protein as well as small-molecule drugs, can be conjugated to nanoparticles for applications in therapeutics and diagnosis. The light scattering and absorption properties of plasmonic nanoparticles make these materials useful for application in the imaging and sensing.

The use of plants in African traditional healing is deeply interwoven with cultural practices and religious beliefs. Africa thus has a rich indigenous knowledge of plants which can be explored with phytonanotechnology to develop novel solutions for the treatment and diagnosis of diseases. Green nanoparticles synthesized from plants used for medicinal purposes can potentially incorporate the medicinal properties of the plant, since the same phytochemicals that are responsible for the medicinal properties of the plant can be involved in the synthesis of nanoparticles. There is thus a unique opportunity to integrate indigenous African flora into nanomedicine to develop cost-effective therapeutic and diagnostic solutions for the devastating deadly diseases such as HIV/AIDS, TB, malaria, Ebola, hypertension, heart disease and diabetes that plague Africa.

This book showcases bioinspired green synthesis of nanoparticles from plants as an emerging method for the development of nanomedicine to promote human health and well-being. The book describes the concepts of green synthesis and reviews the application of green nanoparticles for the treatment of non-communicable and communicable diseases such as cancer, TB and Ebola. Moreover, it highlights how green nanoparticles can be used to fight multidrug antimicrobial resistance. The book also focuses on the application of this technology to develop cost-effective diagnostic systems for early disease detection. A major stumbling block in science in general is the translation of research into commercial products, and the authors therefore also present considerations for commercialization of nanomedicines.

Prof. Mervin Meyer
Professor of Biotechnology
Director: DSI/Mintek Nanotechnology
Innovation Centre – Biolabels Node
South Africa

Acknowledgements

The idea of publishing a book on nanotechnology had been on the cards since 2012 as part of the International Science Council Regional Office for Africa (ISC ROA) book project under the then directorship of Dr Edith Madela-Mntla and spearheaded by Prof Malik Maaza of IThemba LABS to both of whom we owe much gratitude. Owing to several factors, however, it was put on ice.

Our sincere appreciation goes to Dr Rofhiwa Bridget Mulaudzi (Romukhu (Pty) Ltd, South Africa) and Dr Maluta Steven Mufamadi (Nabio Consulting (Pty) Ltd, South Africa) through whose initiative and tacit support, the idea was revived in 2018, albeit with a shift in thematic focus. We are equally indebted to them for acting as external editors and enthusiastically putting together a Concept Paper and draft chapter outline for consideration and endorsement by ISC ROA, assisting in developing a publishing proposal to the CRC Press and recommending potential reviewers, which enabled its acceptance as well as identifying potential contributors to each chapter.

We are grateful to Dr Hazel Mufhandu (North West University, South Africa), Dr Bhekumthetho Ncube (University of KwaZulu-Natal, South Africa), Prof Ram Prasad (Sun Yat-Sen University, China) and Prof Lebogang Katata-Seru (North West University, South Africa) for reviewing the book publishing proposal.

Dr Hazel Mufhandu (North West University, South Africa), Prof Lebogang Katata-Seru (North West University, South Africa), Dr Haly Holmes (University of the Western Cape, South Africa), Dr Dennis Arokoyo (Bowen University, Nigeria), Dr Saher Islam (University of Veterinary and Animal Sciences, Pakistan), Dr Lindiwe Thete (CSIR, South Africa), Prof T.I. Nkambule (University of South Africa, South Africa), Prof Chao Zhang (Sun Yat-Sen University, China), Dr Abdelhamid Elaissari (University of Lyon, France), Dr Clinton Rambanapasi (PharmaConnect-A, South Africa) and Dr Tatenda Dalu (University of Venda, South Africa) all come up for special mention for peer-reviewing the draft chapters of this book.

Last but not least, our heartfelt thanks and appreciation go to the contributors to the various sections and chapters of this book through whose knowledge production and generous transfer, green synthesis has been vividly showcased here as an emerging method for deriving nanoparticles from natural plant parts to be used in nanomedicine for promoting human health and well-being.

Dr Richard L. K. Glover and
Dr Daniel Nyanganyura
Internal Editors, ISC ROA, South Africa

Editors

Dr Richard L. K. Glover (Pr. Nat. Sci.) is a biological scientist with a bias in food microbiology. Dr Glover worked as a Research Officer at the Centre for Scientific Research into Plant Medicine at Mampong-Akwapim in Ghana from 1991 to 1996 where he set up its Microbiological Screening Laboratory for validation of herbal and traditional medicinal products. He joined the Department of Applied Biology, Faculty of Applied Sciences of the University for Development Studies (UDS), Navrongo Campus, as a lecturer/researcher in microbiology in 1996. He served as Head of Department and was Senior Lecturer when he left UDS in 2012. Dr Glover also led DANIDA-sponsored projects on Capability Building for Research in Traditional Fermented Food Processing in West Africa.

Dr Glover currently works as Programme Specialist (Biological Sciences) with the International Science Council [ISC] Regional Office for Africa (formerly International Council for Science [ICSU] Regional Office for Africa), Pretoria, South Africa. Dr Glover has participated in several international training programmes and conferences. He has over 20 peer-reviewed journal publications as well as several conference proceedings to his credit. He is the Regional Programme Officer (RPO) and member of the Steering Committee of the International Network for Government Science Advice (INGSA) Africa Chapter. He is a member of the South African Council for Natural Scientific Professions (SACNASP) with Registration Number 115865.

Dr Daniel Nyanganyura is Regional Director for the International Science Council for Science (ISC) Regional Office for Africa (ROA) from February 2017 and acted in this position from June 2016 to January 2017. He holds a PhD in Atmospheric Physics (2007), an MSc in Agricultural Meteorology (1999), a BSc 4th Year Honours in Physics (1997) from the University of Zimbabwe and a Licentiate Degree in Education in the Specialty of Physics and Astronomy from Enrique Josè Varona Higher Pedagogical Institute, Havana, Cuba (1991). He was Programme Specialist for Physics, Mathematics and Engineering Sciences at the International Council for Science Regional Office for Africa (2008–2016); Air Pollution Research Scientist at the Max Planck Institute for Chemistry in Mainz, Germany (August 2007–July 2008); Physics Lecturer at the University of Zimbabwe (2000–2007) and A-Level Physics and Computer Science Teacher (1991–1998). At ISC ROA, he manages and oversees ISC scientific programmes/activities in Africa. He is a member of the South African Institute of Physics; South African Society for Atmospheric Sciences; European Geophysical Union; the Air Pollution Information Network for Africa; and International Society for Agricultural Meteorology.

Dr Maluta Steven Mufamadi is the founder and managing director of Nabio Consulting (Pty) Ltd. He has earned his PhD in (nano)pharmaceutics from Wits University, entrepreneurship short courses from University of Pretoria and Ecole Polytechnique Federale de Lausanne (EPFL), Switzerland. His interest is in the health applications of nanotechnology, entrepreneurship and commercialization. He has authored and co-authored three patents, one of which is an international patent, several peer-reviewed scientific publications and book chapters. He received the awards as international inventor from the Wits Enterprise & Wits University Innovation Forum, and for the best peer-reviewed research paper of pharmaceutics (nanomedicine) from South African Academy of Pharmaceutical Sciences. In 2017, he was supported for UNESCO-Kalinga Prize for the popularization of nanotechnology in South Africa by the South African National Commission for UNESCO. From 2014, he is member of the national nanotechnologies committee (ISO TC229) at the South African Bureau of Standards (SABS).

Dr Rofhiwa Bridget Mulaudzi is the Founder and Managing Director of Romukhu (Pty) Ltd, She holds a PhD in Ethnobotany and MSc in Botany from the University of KwaZulu-Natal. She holds an Honours degree in Biochemistry and BSc degree in Biochemistry and Microbiology from the University of Venda and is registered with the South African Council for Natural Scientific Professions (SACNASP) as a Professional Natural Scientist. Apart from university degrees, Rofhiwa completed formal certificate courses in Basic, Intermediate and Advanced Project Management at the University of South Africa and a Certificate in Bio-Entrepreneurship Training Programme at Coach Lab, the Innovation Hub, Pretoria, South Africa. Dr Mulaudzi was a Researcher in Medicinal Plants at the Agricultural Research Council. Her current research projects are focusing on the Research and Development (R&D) of both medicinal plants and indigenous vegetables as medicine for human health and well-being. Dr Mulaudzi has published peer-reviewed articles in international journals and contributed chapters to books. Her current research interests include development of nanomedicine drugs using African indigenous plants and using African leafy vegetables as source of medicine. She won the Best Researcher of the year award in Plant Science presented by the Rn. Abuthahir. S, President of World Research Council, RULA Awards 2020.

Contributors

Razia Z Adam
Faculty of Dentistry, University of the Western Cape
Cape Town, South Africa

Juliana Bunmi Adetunji
Nutritional and Toxicological Research Laboratory, Department of Biochemistry Sciences, Osun State University
Osogbo, Nigeria

Charles Oluwaseun Adetunji
Microbiology, Biotechnology and Nanotechnology Laboratory, Department of Microbiology Edo University Iyamho
Edo State, Nigeria

Abeer Ahmed Qaed Ahmed
Department of Biomedical Engineering, College of Life Science and Technology, Huazhong University of Science and Technology
Wuhan, PR China

Fanelwa Ajayi
SensorLab, Chemistry Department, University of the Western Cape
Bellville, South Africa

Muhammad Akram
Department of Eastern Medicine, Government College University Faisalabad
Punjab, Pakistan

Carmen Alvarez-Lorenzo
Departamento de Farmacologia, Farmacia y Tecnología Farmacéutica, R & D Pharma Group, Facultad de Farmacia, Universidade de Santiago de Compostela
Santiago de Compostela, Spain

Osikemekha Anthony Anani
Laboratory of Ecotoxicology and Forensic Biology, Department of Biological Science, Faculty of Science, Edo University Iyamho
Edo State, Nigeria

Hina Anwar
Department of Eastern Medicine, Government College University Faisalabad
Punjab, Pakistan

Marique Aucamp
School of Pharmacy, University of the Western Cape
Bellville, South Africa

Mahwahwatse Johanna Bapela
Department of Plant and Soil Sciences, University of Pretoria
Hatfield, South Africa

Jean Le Bideau
Université de Nantes, CNRS, Institut des Matériaux Jean Rouxel
IMN, Nantes, France

Frank Boury
CRCINA, INSERM, Université de Nantes, Université d 'Angers
Angers, France

Nela Buchtova
CRCINA, INSERM, Université de Nantes, Université d'Angers
Angers, France

Admire Dube
School of Pharmacy, University of the Western Cape
Bellville, South Africa

Moreno Galleni
Laboratory for Biological Macromolecules, Center for Protein Engineering, Institut de Chimie B6, University of Liège
Sart-tilman, Liège, Belgium

Emmanuel Garcion
CRCINA, INSERM, Université de Nantes, Université d'Angers
Angers, France

Iram Ghaffar
Department of Eastern Medicine, Government College University Faisalabad
Punjab, Pakistan

Richard Lander Kwame Glover
International Science Council Regional Office for Africa (ISC ROA)
Pretoria, South Africa

François Hindré
CRCINA, INSERM, Université de Nantes, Université d'Angers
Angers, France

Ahmed A. Hussein
Natural Product Chemistry Research Group, Department of Chemistry, Cape Peninsula University of Technology
Cape Town, South Africa

Enas Ismail
Natural Product Chemistry Research Group, Department of Chemistry, Cape Peninsula University of Technology
Cape Town, South Africa

Christine Jérôme
Center for Education and Research on Macromolecules (CERM), Université de Liège
Liège, Belgium

Marian Jiya John
Nabio Consulting (Pty) Ltd
Pretoria, South Africa

Widadh Klein
Faculty of Dentistry, University of the Western Cape,
Cape Town, South Africa

Germana Lyimo
Faculty of Dentistry, University of the Western Cape
Cape Town, South Africa

Muhammad Haji Mansor
CRCINA, INSERM, Université de Nantes, Université d'Angers
Angers, France
and
Center for Education and Research on Macromolecules
(CERM), Université de Liège
Liège, Belgium

Tracey Jill Morton McKay
Department of Environmental Sciences, School of Agriculture and Environmental Sciences, University of South Africa
Florida, Roodepoort, Johannesburg, South Africa

Yonela Mgwili
SensorLab, Chemistry Department, University of the Western Cape
Bellville, South Africa

Olugbenga Samuel Michael
Cardiometabolic Research Unit, Department of Physiology, College of Health Sciences, Bowen University, Iwo
Osun State, Nigeria

Maluta Steven Mufamadi
Nabio Consulting (Pty) Ltd
Pretoria, South Africa

Rofhiwa Bridget Mulaudzi
Romukhu (Pty) Ltd
Pretoria, South Africa

Ndzumbululo Ndou
SensorLab, Chemistry Department, University of the Western Cape
Bellville, South Africa

Mpho Phehello Ngoepe
Department of Oral Biology, University of the Witwatersrand, Johannesburg
Parktown, South Africa

Daniel Nyanganyura
International Science Council Regional Office for Africa (ISC ROA)
Pretoria, South Africa

Nokwanda Ngema
SensorLab, Chemistry Department, University of the Western Cape
Bellville, South Africa

Sphamandla Nqunqa
SensorLab, Chemistry Department, University of the Western Cape
Bellville, South Africa

Wilson Nwankwo
Informatics and CyberPhysical Systems Laboratory, Department of Computer Science, Edo University Iyamho
Auchi, Edo State, Nigeria

Dumisile W Nyembe
Ngwane Teachers College
Eswatini

Akinola Samson Olayinka
Department of Physics, Faculty of Science, Edo University Iyamho
Edo State, Nigeria

Kadiri Oseni
Department of Biochemistry, Faculty of Basic Medical Sciences, Edo University Iyamho
Edo State, Nigeria

Tessia Rakgotho
SensorLab, Chemistry Department, University of the Western Cape
Bellville, South Africa

Bathabile Ramalapa
CRCINA, INSERM, Université de Nantes, Université d'Angers
Angers, France
and
Laboratory for Biological Macromolecules, Center for Protein Engineering, Institut de Chimie B6, University of Liège,
Sart-tilman, Liège, Belgium
and
Council for Scientific and Industrial Research
Brummeria, Pretoria, South Africa

Jessica Reynolds
Department of Medicine, School of Medicine and Biomedical Sciences, University at Buffalo
Buffalo, New York, USA

Olerimi Samson E.
Department of Biochemistry, Faculty of Basic Medical Sciences, Edo University Iyamho
Edo State, Nigeria

Palesa Rose Sekhejane
Africa Institute of South Africa, Human Sciences Research Council
Pretoria, South Africa

Boitumelo Semete-Makokotlela
Council for Scientific and Industrial Research, Chemicals
Pretoria, South Africa

Ajayi Kolawole Temidayo
Section of Integrative Bioenergetics Environmental and Ecotoxicological Systems, Department of Microbiology, Faculty of Life Sciences, University of Ilorin
Ilorin, Nigeria

Nonhlanhla Tlotleng
National Institute for Occupational Health, National Health Laboratory Services
Johannesburg, South Africa

Clement Toullec
CRCINA, INSERM, Université de Nantes, Université d'Angers
Angers, France
and
Center for Education and Research on Macromolecules
(CERM), Université de Liège
Liège, Belgium
and
Université de Nantes, CNRS, Institut des Matériaux Jean
Rouxel, IMN, Nantes, France

Siphokazi Tshoko
SensorLab, Chemistry Department, University of the Western Cape
Bellville, South Africa

Thilivhali Emmanuel Tshikalange
Department of Plant and Soil Sciences, University of Pretoria
Hatfield, South Africa

Wells Utembe
Nabio Consulting (Pty) Ltd
Pretoria, South Africa

Lin Xiao
Department of Biomedical Engineering, College of Life Science and Technology, Huazhong University of Science and Technology
Wuhan, PR China

Guang Yang
Department of Biomedical Engineering, College of Life Science and Technology, Huazhong University of Science and Technology
Wuhan, PR China

Section I

Green Synthesis of Nanoparticles from Natural Plant Parts

General Overview

Maluta Steven Mufamadi, Rofhiwa Mulaudzi, Richard L.K. Glover and Daniel Nyanganyura

CONTENTS

The History of Nanotechnology 4
Green Synthesis of Nanoparticles: Concepts and Scientific Context 5
Application of Green Synthesized Nanoparticles in Human Health 7
Antimicrobial Mechanism of Action 7
Anticancer Mechanism of Action 8
Non-Communicable Diseases 8
Communicable Diseases 9
Nanomedicine Innovations: New Trend 9
Commercialization Aspects 10
References 10

Green synthesis of nanoparticles (NPs) has gained extensive attention as a reliable, sustainable and eco-friendly protocol for synthesizing a wide range of materials/nanomaterials, including metal/metal oxides nanomaterials, hybrid materials and bioinspired materials (Singh et al., 2018). It is regarded as an important tool to reduce the destructive effects associated with the traditional methods of synthesis for nanoparticles commonly utilized in the laboratory and industry. The green synthesis method is also known as biological synthesis or biogenic synthesis (Salem and Fouda, 2020). It is required to avoid the production of unwanted or harmful by-products through the build-up of reliable, sustainable and eco-friendly synthesis procedures.

In recent years, green synthesis of nanoparticles from natural plant parts and/or plant crude extracts has gained a lot of attention in both medical and pharmaceutical applications (Mukherjee et al., 2014; Mufamadi et al., 2019). There are different methods of nanoparticles synthesis such as chemical, physical and biological (i.e. green) synthesis methods. Traditionally, nanoparticles production is achieved by employing physical and chemical methods. However, both methods are very expensive, and chemical method and use of harsh toxic solvents limit their applications in clinical fields and are also harmful to both human health and the environment (Caroling et al., 2015). Green synthesis methods use various green sources such as bacteria, fungi, enzyme, algae and yeast as well as plant extracts for nanoparticles production (Kaviya et al., 2011; Mufamadi and Mulaudzi, 2019; Salem and Fouda, 2020). The advantages of using plant and/or plant crude extracts during green synthesis of nanoparticles are that they contain many compounds like proteins, polysaccharides, amino acids and secondary metabolites like flavonoids, phenolic acid, alkaloids, tannins and terpenoids that are responsible for the bioreduction, capping and stabilization of metallic ions during nanoparticle formation (Aromal and Philip, 2012; Ali et al., 2020). Plant-based synthesis is an ideal approach for nanoparticles formation because it doesn't require high temperatures, energy, pressure and harmful chemical solvents. Furthermore, plant-based synthesis of nanoparticles is environment-friendly, cost-effective and easy to scale up for large-scale synthesis of nanoparticles (Benakashani et al., 2016).

The success in green synthesis of nanoparticles from natural plant parts has opened an opportunity for low- and middle-income countries such as African countries to start to elevate their research and development (R&D) in the nanomedicine area in order to address the burden of diseases that they are currently facing today. The advantages of using plant-based synthesis of nanoparticles are that it is easy to synthesize, uses single-step process, requires low or just room temperature for nanoparticles production, is easy to scale up for large-scale synthesis of nanoparticles, is easy to maintain and is inexpensive and/or economically viable. African countries, particularly those that are in sub-Sahara Africa (SSA), are facing many burdens of non-communicable diseases (NCDs), communicable diseases (CDs) and vector-borne diseases, as well as antimicrobial resistance. In order to overcome these challenges, African countries need to start to harness science-based health innovations (e.g. Green synthesis of nanoparticles from natural plant parts) as a way to achieve affordable and efficient health and medical solutions.

The main objective of this book is to showcase green synthesis as an emerging method for deriving nanoparticles from natural plant parts to be used in nanomedicine for promoting human health and well-being. It also looks at the advancement in nanoparticles production using plant parts/plant crude extracts (i.e. green synthesis) to combat non-communicable diseases (e.g. cancer, diabetes and oral candidiasis), communicable diseases (e.g. Ebola and tuberculosis) and vector-borne diseases (e.g. malaria), as well as antimicrobial resistance (extensively drug-resistant tuberculosis and antibiotic-resistant *Staphylococcus aureus*) in the African context. Furthermore, it unearths the prospects of utilizing nanoparticles to address cross-cutting disease and drug-related issues.

THE HISTORY OF NANOTECHNOLOGY

The idea of nanotechnology was first discussed in 1959 by American physicist Richard Feynman at an American Physical Society meeting at the California Institute of Technology (Caltech), in his talk entitled "There's Plenty of Room at the Bottom". In this speech, Feynman discussed the importance "of manipulating and controlling things on a small scale" and how they could "tell us much of great interest about the strange phenomena that occur in complex situations" (Feynman, 1960). The term "nanotechnology" was first introduced by the Japanese scientist Norio Taniguchi in a 1974 paper on production technology that creates objects and features on the order of a nanometre (Taniguchi, 1974). In 1986, Eric Drexler began to promote and popularize nanotechnology through his book *Engines of Creation: The Coming Era of Nanotechnology*. The invention of scanning tunnelling microscope in 1981 by

Gerd Binnig and Heinrich Rohrer at IBM Zurich Research Laboratory allowed scientists to see materials at an atomic or molecular level (Binning and Rohrer, 1986). The National Nanotechnology Initiative (NNI) in the US Federal Nanotechnology Research and Development Program defines nanotechnology as *the understanding and control of matter at dimensions of roughly 1–100 nm, where unique phenomena enable novel application.* The term nanotechnology encompasses nanoscale science, engineering and technology and involves imaging, measuring, modelling and manipulating matter at this length scale. A nanometre is one-billionth of a metre (NNI, 2020; Bhattacharya et al., 2009). Currently, there are two unique approaches for synthesizing nanoparticles: top-down and bottom-up. Top-down approach is when bulk material is broken down or cut down until it reaches the size of nanoparticles by applying external force, e.g. mechanical milling, drilling, grinding and laser ablation, while bottom-up approach refers to self-assembly of atoms, molecule by molecule until nanosize particles are formed, using chemical or physical forces, e.g. chemical reduction and biological (green) synthesis (Iqbal et al., 2012).

Nanomedicine is the medical application of nanotechnology. It uses nanoscale materials (1–100 nm) for health innovation to improve various kinds of diseases treatment, diagnosis, prevention and monitoring (Tinkle et al., 2014). The use of nanotechnology in the development of new medicines offers an enabling tool for providing new and innovative medical solutions to address unmet medical needs (Sadikot and Rubinstein 2009; Soares et al., 2018). The advantage of using nanotechnology in medicine includes target drug delivery, i.e. the capability of delivering biological or small molecules – active pharmaceutical ingredients (APIs) –to where they will be most effective. To achieve target drug delivery, APIs encapsulated nanoparticles are surface functionalized with different receptor ligands such as protein, peptide, antibody, folic acid, receptors, polysaccharide and polynucleotide so that they can be localized at the disease-specific site, without affecting the normal cells and/or healthy tissues. Site-specific drug delivery system reduced side effects and improved therapeutic efficacy and safety of therapeutic molecules (Bazak et al., 2015). The use of nanotechnology with natural products in nanomedicine is a rapidly developing innovative field that brings multiple advantages: it improves the effectiveness of natural compounds in disease treatment and prevention and/or offers better treatment outcome (Watkins et al., 2015). Incorporation of nanoparticles increases the bioavailability, targeting and controlled-release profiles of the natural products.

GREEN SYNTHESIS OF NANOPARTICLES: CONCEPTS AND SCIENTIFIC CONTEXT

The goal of using green methods for the synthesis of nanoparticles – "green nanotechnology" – was to produce nanomaterials and nano-products that are safe for human health and are environment-friendly, an alternative method for chemical and physical synthesis methods. Both chemical and green synthesis methods utilize the concepts of bottom-up approach for nanomedicine manufacturing. Green synthesis strategy uses existing principles of green chemistry and green engineering for the production of green nanomaterials and/ or nanomedicine products (Anastas and Warner, 1998). In medicine, nanotechnologies are designed to focus on the human health, environmental impact and social and economic benefits. Moreover, nanomaterials and nano-products production are achieved without the use of toxic ingredients, at low temperatures and less energy. Green synthesis strategy is cost-effective and requires low maintenance with relative reproducibility (Kalaiarasi et al., 2010). Plant extracts or microorganisms become "chemical factories" for greener nanomaterials or nano-products. During nanoparticle synthesis, both plant extracts and microorganisms can act as reducing/capping and stabilizing agents. However, plant-based synthesis approaches for nanoparticles formation are more ideal than those employing microorganism-based synthesis. Plant-based synthesis is easy and faster to formulate, and it produces nanoparticles that are more stable compared to those that are fabricated employing microorganisms-based synthesis approaches (Roy and Das, 2015; Ali et al., 2020; Salem and Fouda, 2020). In addition, plants are easily available and have a broad variety of novel metabolites or "green sources" from plant crude extracts, such as protein, vitamins, antioxidants, amino acids, enzymes, polysaccharides and secondary metabolites (e.g. flavonoids, phenolic acid, alkaloids, tannins and terpenoids). Different green sources extracted from various plant parts such as stem, root, leaves, fruit, flower, seed, callus and peel have been used as green "bio" reductants and capping agents during green synthesis of nanoparticles (Awwad and Salem, 2012; Ahmed et al., 2015; Mufamadi and Mulaudzi, 2019; Alphandery, 2020). African scientists are also using different plants parts or extracts for green synthesis of metallic nanoparticles such as *Moringa oleifera* (Leaf), *Thevetia peruviana* (leaf), *Aloe vera* (whole plant), *Agathosma betulina* (leaves), *Combretum molle* (leaves), African *Galenia africana* (whole plant), *Hypoxis hemerocallidea* (whole plant), *Microsorum punctatum* (leaf), *Cassia fistula* (leaf), *Melia azadarach* (leaf) and many more (Chiguvare et al., 2016; Oluwaniyi et al., 2016; Elbagory et al., 2017; Moodley et al., 2018; Nate et al., 2019; Kedi et al., 2020; Naseer et al., 2020). Metallic nanoparticles include silver (Ag), gold (Au), iron (Fe), copper (Cu), zinc (Zn), palladium (Pd), platinum (Pt), lead (Pb), selenium (Se), titanium dioxide (TiO_2), zinc oxide (ZnO), iron oxide (FeO) and copper oxide (CuO) (Ramesh et al., 2014; Shah et al., 2015; Avoseh et al., 2017; Chung et al., 2017; Singh et al., 2018; Mufamadi and Mulaudzi, 2019; Verma et al., 2019; Naseer et al., 2020). The mechanism of metallic nanoparticle synthesis using plant extracts involves the reduction of the silver metal ion (Ag^+) into metallic silver nanoparticles (Ag^0 or AgNPs) with a size of 1–100 nm and different shapes (e.g. spherical, hexagonal and cuboidal). The green synthesis of nanoparticles protocol is a one-step synthesis involving the assembly of zero-valent metals or metal oxides or metal salts together with plant extract broth containing metabolites (i.e. phytochemicals) as reducing, capping and stabilizing agent (Roy and Das, 2015). The formation of metallic nanoparticles is confirmed by the colour change, e.g. AgNPs from initial reagent yellow solutions to dark brown mixture solution (Moodley et al., 2018). Dynamic light scattering (DLS), scanning electron microscopy (SEM), atomic force microscopy (AFM), transmission electron microscopy (TEM), X-ray diffraction (XRD), ultraviolet and visible spectrophotometry (UV-vis), Fourier-transform infrared spectroscopy (FTIR), low-energy ion scattering (LEIS) and energy-dispersive spectroscopy (EDS) are used for further validation and physicochemical characterization of the metallic nanoparticle (e.g. size, shape, morphology and surface properties).

APPLICATION OF GREEN SYNTHESIZED NANOPARTICLES IN HUMAN HEALTH

Eco-friendly and the unique properties of green synthesized nanoparticles from plant extracts (i.e. phytochemical) make them excellent candidates for medicine and pharmaceutics applications. The novel properties of the green synthesized nanoparticles of plant extracts include optical or fluorescent properties, thermal, biosensing, photocatalyst, photodynamics, immunotherapeutic, antiangiogenesis, antibacterial, antifungal, antiparasitic, antioxidant, antiviral and anticancer activities. The applications of green synthesized nanoparticles on human health include therapy, drug delivery, diagnostics, biosensing, theranostics (a combination of therapy and diagnosis) and bio-imaging. According to Mukherjee et al. (2014), metallic nanoparticles synthesized from plant parts are promising to offer a multifunctional biomedical system, a 4-in-1 system that comprises biocompatibility, bio-imaging and antimicrobial and anticancer activities. The presence of *Olax scandens* leaf extract acted as a bio-reducing as well as stabilizing/capping agent during the formation of green AgNPs. The red fluorescence activity of AgNPs synthesized employing methanolic extract of *Olax scandens* leaf extract inside cancer cell makes it a potential diagnostic tool for cancer disease in the future. The optical and fluorescence properties of the green synthesized AgNPs and AuNPs could make them suitable candidates for replacing radiolabelled isotopes for tumour detection in the future, and ultimately they will reduce the side effects associated with radiotherapy (Ovais et al., 2016). A recent study by Varghese et al. (2020) demonstrated that green synthesized AgNPs from neem leaf extracts have biosensing and photocatalytic properties, and this was evaluated using mancozeb (MCZ) agrofungicide and optimum surface plasmon resonance (SPR). Riley and Day (2017) demonstrated the photothermal therapy employing green synthesized AuNPs. The study showed the green synthesized AuNPs to be capable of absorbing incident photons and converting them to heat to destroy cancer cells and/or tumours.

Many studies reported antimicrobial and anticancer activities employing green synthesized metallic nanoparticles from plant extract or phytochemical extracted from different parts of plants, e.g. leaves, fruit, flower, stem and root (Awwad and Salem, 2012; Chiguvare et al., 2016; Oluwaniyi et al., 2016; Elbagory et al., 2017; Kedi et al., 2020; Naseer et al., 2020). However, the antimicrobial and anticancer mechanism of action of green synthesized nanoparticles is still not fully understood. Various mechanisms have been proposed to illustrate the activity of green synthesized nanoparticles as antimicrobial and anticancer agents:

Antimicrobial Mechanism of Action

- Disabling the respiratory of chains
- Cell membrane disruption and leakage of its cellular contents

- Binding to functional group of proteins causing protein denaturation and cell death
- Blocking the DNA of replication
- Denaturation of proteins and cell death through binding to functional groups of proteins

Anticancer Mechanism of Action

- Cytotoxicity that disturbs the cell membrane permeability and cellular internalization resulting in cell death through a cascade of events
- Inducing apoptosis through caspase-dependent and mitochondrial dependent pathways
- Generation of reactive oxygen species (ROS) that leads to induction of ROS resulting in cellular damage
- Trigger upregulation of p53 protein and caspase-3 expression through apoptosis pathway activation leading to cell death
- pH-dependent release of metal ions from green synthesized nanoparticles leading to cancerous cells death
- Inhibiting the function of vascular endothelial growth factor (VEGF)-induced angiogenesis

Green synthesized nanoparticles from plant extract shown to have the potential to enhance the diagnosis and treatment of non-communicable diseases (e.g. cancer, diabetes and oral candidiasis), communicable diseases (e.g. Ebola and tuberculosis) and vector-borne diseases (e.g. malaria) as well as antimicrobial resistance (extensively drug-resistant tuberculosis and antibiotic-resistant *Staphylococcus aureus*) are discussed in this book.

Non-Communicable Diseases

NCDs are a diverse group of chronic diseases that are not communicable and cannot be passed from person to person. NCDs are responsible for the premature mortality around the world. According to the World Health Organization (WHO), the four main risk factors for major NCDs are high consumption of alcohol, tobacco consumption, lack of physical activity and unhealthy diet (WHO, 2015). According to Marquez and Farrington (2013), by 2030, NCDs such as cancers, diabetes, cardiovascular diseases (like heart attacks and stroke) and chronic respiratory diseases (pulmonary disease and asthma) will become the leading cause of death in sub-Saharan Africa. The challenge of treating NCDs is that most of them can only be diagnosed at a late stage of affliction, making it very difficult to treat or manage, e.g. cancer diseases present at a late stage (5-years later) and/or once the tumour has already formed (Mulisya et al., 2020). Combating these NCD burdens demands immediate measures and tools that could enable early and rapid diagnosis and treatment, such as nanomedicine (Chandarana et al., 2018). Nanomedicine promises to offer better approaches towards identifying NCDs at an early stage and to overcome treatment challenges associated with them.

Therapeutic benefits of using nanomedicine in NCDs include improved drug circulation times, target delivery that enhances treatment efficacy and the ability to act as a therapeutic (Sanna et al., 2014; DiSanto et al., 2015; El-Readi and Mohammad, 2019; Lafisco et al., 2019). Moreover, nanomedicine promises to provide better diagnostic tools, medical imaging and theranostics, by the combination of therapy and diagnosis in a single application (Mukherjee et al., 2014; Chandarana et al., 2018).

Communicable Diseases

CDs or infectious diseases are communicable, which means that they can be passed from person to person either directly or indirectly by disease-causing agents, humans or animals (Nash et al., 2015). The disease-causing agents include viruses, bacteria, fungi and parasites, and account for approximately 15 million deaths worldwide (Singh et al., 2017). Africa has the highest prevalence of CDs in the world. HIV/AIDS, tuberculosis, acute respiratory infections, malaria and diarrheal diseases are on top in the list of the deadliest CDs in Africa. Four of them combined are predicated to be responsible for nearly 80% of the total infectious disease burden and claiming more than 6 million people per year (Boutayeb, 2010). Since the establishment of the sustainable development goals (SDGs), Africa has set new ambitious targets which include ending the prevalence of HIV/AIDS, TB and malaria by 2030 (Narayan and Donnenfeld, 2016). However, the rising cost of the antibiotics, antimicrobial and antiparasitic resistance and also the adverse side effects due to prolonged drug use indicate that the battle against CDs is far from over (Blecher et al., 2011). Therefore, in order to envision a future without these CDs, it will require an improved antimicrobial agent or newer and more effective therapies. The unique physical and chemical properties of nanoparticles make them an ideal candidate to improve the therapeutic effects associated with current conventional therapies and/or to end the prevalence of HIV/AIDS, TB and malaria by 2030. The roles that nanomedicine is promising to play in CDs include improved therapeutic, diagnostic and prevention/vaccination approaches (Blecher et al., 2011; Dube, 2019). In the case of CD diagnostics, nanomedicine promises to offer cheap and quick diagnostic test. When it comes to CD treatments and vaccine, nanomedicine promises to offer immunotherapy that is capable of neutralizing viruses and to improve drug circulation times, target delivery, combination therapy, pharmacokinetics and pharmacodynamics.

Nanomedicine Innovations: New Trend

The convergence of material science, biotechnology, engineering and nanomedicine is promising to transform medical technology and the pharmaceutical industry. Moreover, the improved nanomedicine is promising to offer accurate diagnoses, targeting therapies with fewer side effects, and providing better medical imaging and personalized medicine (de Smet et al., 2011; Zhang et al., 2017). A new trend in nanomedicine technology is influenced by the advancement in nanoformulation that led to a new type of nanomedicine:

- Stimuli-responsive nanomedicines that trigger drug release in an on-demand manner or influence by stimulus, e.g. pH, light, thermal, redox and magnetic and electronic fields
- Nanotheranostics (i.e. smart pill) that combines therapeutics with imaging agents into one pill
- Polytherapy (i.e. polypill), which combines multiple active pharmaceutical ingredients into one pill
- Nanosensors that can diagnose certain diseases through analysing human exhaled breath, and thus recognizing the biomarker gases related to diseases in human exhalation
- Fully autonomous DNA nanorobot that is capable of transporting vaccines and drugs to the targeted site

The development of these new types of nanomedicines was to address the issue of multidrug resistance (MDR) in current antimicrobial and anticancer agents (Fattal and Tsapis, 2014; Tangden, 2014; Kim et al., 2017; Zhang et al., 2017; Li et al., 2018; Zhou et al., 2018). In addition, it was to overcome challenges associated with conventional nanomedicine such as poor stability, poor bioavailability, drug toxicity and hurtful side effects (Fattal and Tsapis, 2014).

Commercialization Aspects

Many different nanomedicines have been developed over the years to improve the therapeutic efficacy, target drug delivery system and safety for various types of diseases (Ventola, 2012; Curley et al., 2017). According to Metselaar and Lammers (2020), more than 50 nanomedicine formulations are currently approved for clinical use by the Food and Drug Administration (FDA) and many others are currently undergoing clinical trials. However, none of those approved nanomedicine formulations are from Africa. Although nanomedicines have the potential to provide affordable and efficient medical solutions, only few African countries are prioritizing on it: South Africa, Egypt, Morocco and Nigeria (Mufamadi, 2019). African countries need to start embracing their own health innovation, research and/or local technologies and stop depending on the international market for every technology they need (Singer et al., 2008). If African scientists need to commercialize their nanomedicines, they need to find a commercialization strategy that matches their potential and sustainability, such as licensing, equity investment, strategic alliances and private consortium or cluster alliance (Mufamadi, 2019). Moreover, they need to make a substantial investment towards the infrastructure along with the increase in human capacity building and nanomedicine-based products. Furthermore, they need to find a way of attracting investment from local or international investors and big pharmaceutical companies.

REFERENCES

Ahmed, S., Ahmad, M., Swami, B. L., Ikram, S. 2015. "A review on plants extracts mediated synthesis of silver nanoparticles for antimicrobial applications: A green expertise." *Journal of Advanced Research* 7: 17–28. DOI: 10.1016/j.jare.2015.02.007.

Ali, A., Ahmed, T., Wu, W., Hossain, A., Hafeez, R., Masum, MI., Wang, Y., An, Q., Sun, G., Li, B. 2020. "Advancements in plant and microbe-based synthesis of metallic nanoparticles and their antimicrobial activity against plant pathogens." *Nanomaterials* 10: 1146. DOI: 10.3390/nano10061146.

Alphandery, E. 2020. "Natural metallic nanoparticles for application in nano-oncology." *International Journal of Molecular Sciences* 21: 4412. DOI: 10.3390/ijms21124412.

Anastas, P.T. and Warner, J.C. 1998. *Green Chemistry: Theory and Practice*, 30. Oxford: Oxford University Press.

Aromal, S.A. and Philip, D. 2012. "Green synthesis of gold nanoparticles using Trigonella foenum-graecum and its size dependent catalytic activity." *Spectrochimica Acta Part A: Molecular and Biomolecular Spectroscopy* 97: 1–5. DOI: 10.1016/j.saa.2012.05.083.

Avoseh, O.N., Oyedeji, O.O., Aremu, O., Nkeh-Chungag, B.N., Songca, S.P., Oyedeji, A.O., Sneha, M.S., Oluwafemi, O.S. 2017. "Biosynthesis of silver nanoparticles from Acacia mearnsii De Wild stem bark." *Green Chemistry Letters and Reviews* 10, no. 2: 59–68. DOI: 10.1080/17518253.2017.1287310.

Awwad, A.M. and Salem, N.M. 2012. "Green synthesis of silver nanoparticles by mulberry leaves extract." *Journal of Nanoscience and Nanotechnology* 2: 125–128. DOI: 10.5923/j.nn.20120204.06.

Bazak, R., Houri, M., El Achy, S., Kamel, S., Refaat, T. 2015. "Cancer active targeting by nanoparticles: A comprehensive review of literature." *Journal of Cancer Research and Clinical Oncology* 141, no. 5: 769–784. DOI: 10.1007/s00432-014-1767-3.

Benakashani, F., Allafchian, A.R., Jalali, S.A.H. 2016. "Biosynthesis of silver nanoparticles using Capparis spinosa L. leaf extract and their antibacterial activity." *Karbala International Journal of Modern Science* 2, no. 4: 251–258. DOI: 10.1016/j.kijoms.2016.08.004.

Bhattacharya, D., Singh, S., Satnalika, N. 2009. "Nanotechnology, big things from a tiny world: A review." *International Journal of Science and Technology* 2, no. 3: 29–38.

Binnig, G. and Rohrer, H. 1986. "Scanning tunneling microscopy". *IBM Journal of Research and Development*. 30, no. 4: 355–369.

Blecher, K., Nasir, A., Friedman, A. 2011. "The growing role of nanotechnology in combating infectious disease." *Virulence* 2, no. 5: 395–401. DOI: 10.4161/viru.2.5.17035.

Boutayeb, A. 2010. "The impact of infectious diseases on the development of Africa." In *Handbook of Disease Burdens and Quality of Life Measures*, edited by Preedy V.R., Watson R.R. New York: Springer. DOI: 10.1007/978-0-387-78665-0_66.

Caroling, G., Vinodhini, E., Ranjitham, A.M., Shanthi, P. 2015. "Biosynthesis of copper nanoparticles using aqueous Phyllanthus embilica (Gooseberry) extract- characterisation and study of antimicrobial effects." *Journal of Nanostructure in Chemistry* 1, no. 2: 53–63.

Chandarana, M., Curtis, A., Hoskins, C. 2018. "The use of nanotechnology in cardiovascular disease." *Applied Nanoscience* 8: 1607–1619. DOI: 10.1007/s13204-018-0856-z.

Chiguvare, H., Oyedeji, O.O., Matewu, R., Aremu, O., Oyemitan, I.A., Oyedeji, A.O., Nkeh-Chungag, B.N., Songca, S.P, Mohan, S. and Oluwafemi, O.S. 2016. "Synthesis of silver nanoparticles using Buchu plant extracts and their analgesic properties." *Molecules* 21, no. 6: 774. DOI: 10.3390/molecules21060774.

Chung, I.M, Rahuman A.A., Marimuthu, S., Kirthi, A.V., Anbarasan, K., Padmini, P., Rajakumar, G. 2017. "Green synthesis of copper nanoparticles using Eclipta prostrata leaves extract and their antioxidant and cytotoxic activities." *Experimental and Therapeutic Medicine* 14: 18–24. DOI: 10.3892/etm.2017.4466.

Curley, P., Liptrott, N.J., Owen, A. 2017. "Advances in nanomedicine drug delivery applications for HIV therapy." *Future Science OA* 4, no. 1: FSO230. DOI: 10.4155/fsoa-2017-0069.

De Smet, M., Heijman, E., Langereis, S., Hijnen, N.M., Grüll, H. 2011. "Magnetic resonance imaging of high intensity focused ultrasound mediated drug delivery from temperature-sensitive liposomes: An in vivo proof-of-concept study." *Journal of Controlled Release* 150, no. 1: 102–110. DOI: 10.1016/j.jconrel.2010.10.036.

DiSanto R.M., Subramanian, V., Gu, Z. 2015. "Recent advances in nanotechnology for diabetes treatment." *Wiley Interdisciplinary Reviews: Nanomedicine and Nanobiotechnology* 7, no. 4: 548–564. DOI: 10.1002/wnan.1329.

Dube, A. 2019. "Nanomedicines for infectious diseases." *Pharmaceutical Research* 36, no. 4: 63. DOI: 10.1007/s11095-019-2603-x.

Elbagory, A.M., Meyer, M., Cupido, C.N., Hussein, A.A. 2017. "Inhibition of bacteria associated with wound infection by biocompatible green synthesized gold nanoparticles From South African plant extracts." *Nanomaterials* 7, no. 12: 417. DOI: 10.3390/nano7120417.

El-Readi, M.Z. and Mohammad, A.A. 2019. "Cancer nanomedicine: A new era of successful targeted therapy." *Journal of Nanomaterials* 2019: 1–13. DOI: 10.1155/2019/4927312.

Fattal, E. and Tsapis, N. 2014. "Nanomedicine technology: Current achievements and new trends." *Clinical and Translational Imaging* 2: 77–87. DOI: 10.1007/s40336-014-0053-3.

Feynman, R.P. 1960. "There's plenty of room at the bottom." *Engineering and Science* 23: 22–36.

Iqbal, P., Preece, J.A., Mendes, P.M. 2012. "Nanotechnology: The 'top-down' and 'bottom-up' approaches." In *Supramolecular Chemistry*. Chichester, UK: Wiley. DOI: 10.1002/9780470661345.smc195.

Kalaiarasi, R., Jayallakshmi, N. Venkatachalam, P. 2010. "Phytosynthesis of nanoparticles and its applications." *Plant Cell Biotechnology and Molecular Biology*. 11, no. 1/4: 1–16.

Kaviya, S., Santhanalakshmi, J., Viswanathan, B., Muthumary, J., Srinivasan, K. 2011. "Biosynthesis of silver nanoparticles using citrus sinensis peel extract and its antibacterial activity." *Spectrochimica Acta Part A: Molecular and Biomolecular Spectroscopy* 79, no. 3: 594–598. DOI: 10.1016/j.saa.2011.03.040.

Kedi, P.B.E., Nanga, C.C., Gbambie, A.P., Deli, V., Meva, F.E., Mohamed, H.E.A., Ntoumba, A.A. et al. 2020. "Biosynthesis of silver nanoparticles from Microsorum Punctatum (L.) copel fronds extract and an in-vitro anti-inflammation study." *Journal of Nanotechnology Research* 2, no. 2: 025–041 DOI: 10.26502/jnr.2688-85210014.

Kim, S., Choi, S., Jang, J., Cho, H., Kim, I. 2017. "Innovative nanosensor for disease diagnosis." *Accounts of Chemical Research* 50, no. 7: 1587–1596. DOI: 10.1021/acs.accounts.7b00047.

Lafisco, M., Alogna, A., Miragoli, M., Catalucci, D. 2019. "Cardiovascular nanomedicine: The route ahead." *Nanomedicine* 14, no. 18: 2391–2394. DOI: 10.2217/nnm-2019-0228.

Li, S., Jiang, Q., Liu, S., Zhang, Y., Tian, Y., Song, C., Wang, J. et al. 2018. "A DNA nanorobot functions as a cancer therapeutic in response to a molecular trigger in vivo." *Nature Biotechnology* 36: 258–264. DOI: 10.1038/nbt.4071.

Marquez, P.V. and Farrington, J.L. 2013. *The Challenge of Noncommunicable Diseases and Road Traffic Injuries in Sub-Saharan Africa: An Overview.* Washington, DC: The World Bank.

Metselaar, J.M. and Lammers, T. 2020. "Challenges in nanomedicine clinical translation." *Drug Delivery and Translational Research* 10: 721–725. DOI: 10.1007/s13346-020-00740-5.

Moodley, J.S., Krishna, S.B.N., Karen Pillay, K., Govender, S., Govender, P. 2018. Green synthesis of silver nanoparticles from Moringa oleifera leaf extracts and its antimicrobial potential. *Advances in Natural Sciences: Nanoscience and Nanotechnology* 9: 1–9.

Mufamadi M.S. 2019. "From lab to market: Strategies to nanotechnology commercialization in Africa." *MRS Bulletin* 44: 421–422. DOI: 10.1557/mrs.2019.134.

Mufamadi, M.S. and Mulaudzi, R.B. 2019. “Green engineering of silver nanoparticles to combat plant and foodborne pathogens: Potential economic impact and food quality.” In *Plant Nanobionics, Nanotechnology in the Life Sciences*, edited by Prasad R., 451–476. Cham: Springer. DOI: 10.1007/978-3-030-16379-2_16.

Mufamadi, M.S., George, J., Mazibuko, Z., Tshikalange, T.E. 2019. “Cancer bionanotechnology: Biogenic synthesis of metallic nanoparticles and their pharmaceutical potency.” In *Microbial Nanobionics. Nanotechnology in the Life Sciences*, edited by Prasad R. 229–252. Switzerland: Springer. DOI: 10.1007/978-3-030-16383-9_10.

Mukherjee, S., Chowdhury, D., Kotcherlakota, R., Patra, S., Vinothkumar, B., Bhadra, M.P., Sreedhar, B., Patra, C.R. 2014. “Potential theranostics application of bio-synthesized silver nanoparticles (4-in-1 system).” *Theranostics* 4, no. 3: 316–335. DOI: 10.7150/thno.7819.

Mulisya, O., Sikakulya, F.K., Mastaki, M., Gertrude, T., Mathe, J. 2020. The challenges of managing ovarian cancer in the developing world. *Case Reports in Oncological Medicine* 2020: 1–4. DOI: 10.1155/2020/8379628.

Narayan, K. and Donnenfeld, Z. 2016. “Envisioning a healthy future Africa’s shifting burden of disease.” *African Futures* Paper 18: 1–35.

Naseer, M., Aslam, U., Khalid, B., Chem, B. 2020. “Green route to synthesize zinc oxide nanoparticles using leaf extracts of Cassia fistula and Melia azadarach and their antibacterial potential.” *Scientific Reports* 10: 9055. DOI: 10.1038/s41598-020-65949-3.

Nash, A.A., Dalziel, R.G., Fitzgerald, J.R. 2015. Chapter 1: General principles. In *Mims’ Pathogenesis of Infectious Disease*, 1–7. 6th ed. Boston: Academic Press.

Nate, Z., Moloto, M.J., Sibiya, N.P., Mubiayi, P.K., Mtunzi, F.M. 2019. “Green synthesis of silver nanoparticles using aqueous extract of Combretum molle leaves, their antibacterial, antifungal and antioxidant activity.” *International Journal of Nano and Biomaterials* 8, no. 3/4:189–203. DOI: 10.1504/IJNBM.2019.104931.

NNI (National Nanotechnology Initiative). 2020. What It Is and How It Works. 20 June https://www.nano.gov/nanotech-101/what.

Oluwaniyi, O.O., Haleemat, I., Alabi, B., Bodede, O., Ayomide H.L., Oseghale, C.O. 2016. “Biosynthesis of silver nanoparticles using aqueous leaf extract of Thevetia peruviana Juss and its antimicrobial activities.” *Applied Nanoscience* 6: 903–912. DOI: 10.1007/s13204-015-0505-8.

Ovais, M., Khalil, A.T., Raza, A., Khan, M.A., Ahmad, I., Islam, N.U., Saravanan, M., Ubaid, M.F., Ali, M., Shinwari, Z.K. 2016. “Green synthesis of silver nanoparticles via plant extracts: Beginning a new era in cancer theranostics.” *Nanomedicine* 11, no. 23: 3157–3177. DOI: 10.2217/nnm-2016-0279.

Ramesh, P., Rajendran, A., Shisundaram, M.M. 2014. “Green synthesis of zinc oxide nanoparticles using flower extract cassia.” *Journal of Nanoscience and Nanotechnology* 2: 41–45.

Riley, R.S. and Day, E.S. 2017. “Gold nanoparticle-mediated photothermal therapy: Applications and opportunities for multimodal cancer treatment.” *Wiley Interdisciplinary Reviews. Nanomedicine and Nanobiotechnology* 9, no. 4:1–25. DOI: 10.1002/wnan.1449.

Roy and Das, T.K., Plant mediated green synthesis of silver NPs: A review, *International Journal of Plant Biology & Research* 3, no. 3: 1044–1055.

Sadikot, R.T. and Rubinstein, I. 2009. “Long-acting, multi-targeted nanomedicine: Addressing unmet medical need in acute lung injury.” *Journal of Biomedical Nanotechnology* 5, no. 6: 614–619. DOI: 10.1166/jbn.2009.1078.

Salem, S.S. and Fouda, A. 2020. “Green synthesis of metallic nanoparticles and their prospective biotechnological applications: An overview.” *Biological Trace Element Research* 9, no.4: 10.1002/wnan.1449. DOI: 10.1007/s12011-020-02138-3.

Sanna, V., Pala, N., Sechi, M. 2014. "Targeted therapy using nanotechnology: Focus on cancer." *International Journal of Nanomedicine* 9: 467–483.

Shah, M., Fawcett, D., Sharma, S., Tripathy, S.K., Poinern, G.E.J. 2015. "Green synthesis of metallic nanoparticles via biological entities." *Materials* 8: 7278–7308. DOI: 10.3390/ma8115377.

Singer, A. P., Al-Bader, S., Shah, R., Simiyu, K., Wiley, R. E., Kanellis, P., Pulandiran, M., Heymann, M. 2008. "Commercializing African health research: Building life science convergence platforms." *Global Forum Update on Research for Health: Technological Innovations* 5: 143–150.

Singh, J., Dutta, T., Kim, K. Rawat, M., Samddar, P., Kuma, P. 2018. "Green' synthesis of metals and their oxide nanoparticles: Applications for environmental remediation." *Journal of Nanobiotechnology* 16: 84. DOI: 10.1186/s12951-018-0408-4.

Singh, L., Kruger, H.G., Maguire, G.E.M., Govender, T. Parboosing, R. 2017. "The role of nanotechnology in the treatment of viral infections." *Therapeutic Advances in Infectious Disease* 4: 105–131. DOI: 10.1177/2049936117713593.

Soares, S., Sousa, J., Pais, A., Vitorino, C. 2018. "Nanomedicine: Principles, properties, and regulatory issues." *Frontiers in Chemistry* 6: 360. DOI: 10.3389/fchem.2018.00360.

Tangden, T. 2014. "Combination antibiotic therapy for multidrug-resistant Gram-negative bacteria." *Upsala Journal of Medical Sciences* 119, no. 2: 149–153. DOI: 10.3109/03009734.2014.899279.

Taniguchi N. 1974. "On the basic concept of 'nanotechnology'." In Proceedings of the International Conference on Production Engineering, Tokyo, Japan, 26–29 August 1974.

Tinkle S., McNeil S.E., Mühlebach S., Bawa R., Borchard G., Barenholz Y.C., Tamarkin, L., Desai, N. (2014). Nanomedicines: Addressing the scientific and regulatory gap. *Annals of the New York Academy of Sciences* 1313: 35–56. DOI: 10.1111/nyas.12403.

Varghese, A.K., Pavai, P.T., Rugmini, R., Prasad, S.M, Kamakshi, K., Sekhar, K.C. 2020. "Green synthesized Ag nanoparticles for bio-sensing and photocatalytic applications." *ACS Omega* 5(22): 13123–13129. DOI: 10.1021/acsomega.0c01136.

Ventola C.L. 2012. The nanomedicine revolution: Part 1: Emerging concepts. *Pharmacy and Therapeutics* 37, no. 9: 512–525.

Verma, A., Gautam, S.P., Bansal, K.K., Prabhakar, N., Rosenholm, J.M. 2019. "Green Nanotechnology: Advancement in Phytoformulation Research." *Medicines* 6: 39. DOI: 10.3390/medicines6010039.

Watkins, R., Wu, L., Zhang, C., Davis, R. M., Xu, B. 2015. "Natural product-based nanomedicine: Recent advances and issues." *International Journal of Nanomedicine* 10: 6055–6074. DOI: 10.2147/IJN.S92162.

World Health Organization. Non-communicable disease factsheet, January 2015, www.who.int/mediacentre/factsheets/fs355/en/.

Zhang, M., Liu, E., Cui, Y., Huang, Y. 2017. "Nanotechnology-based combination therapy for overcoming multidrug-resistant cancer." *Cancer Biology and Medicine* 14, no. 3: 212–227. DOI: 10.20892/j.issn.2095-3941.2017.0054.

Zhou, L., Wang, H., Li, Y. 2018. "Stimuli-responsive nanomedicines for overcoming cancer multidrug resistance." *Theranostics* 8, no. 4: 1059–1074. DOI: 10.7150/thno.22679.

Section II

Nanotechnology for Treatment of Non-Communicable Diseases

1 Cancer Nanotheranostics
Next-Generation Early Detection and Treatment Prioritization for Cancers Using Phytonanotechnology

Maluta Steven Mufamadi, Marian Jiya John, Mpho Phehello Ngoepe and Palesa Rose Sekhejane

CONTENTS

1.1 Introduction 17
1.2 Phytonanotechnology: Green Synthesis of Metallic Nanoparticles 19
1.3 Phytonanotechnology: Anticancer Activity 19
1.4 Phytonanotechnology: Mechanism of Action for Cancer Treatment 20
1.5 Cancer Nanotheranostics of Phytonanotechnology 23
1.6 Cancer Nanotheranostics of Phytonanotechnology: Opportunities 24
1.7 Conclusion and Future Prospects 24
References 25

1.1 INTRODUCTION

Cancer is basically a group of diseases involving genetic changes leading to uncontrolled cell proliferation in cancerous cells and tumour formation. The numbers of cancer cases are on the rise in Sub-Saharan African (SSA) due to the increase in the adoption of westernized lifestyle, changes in diet and reduction in physical activity among the African population (Chokunonga et al., 2013; Azubuike et al., 2018). The most common cancers in SSA countries are the breast and cervical cancers in women, and prostate cancer in men, as well as Kaposi's sarcoma and non-Hodgkin lymphomas as a result of the HIV/AIDS epidemic. Most cancer cases are diagnosed at an advanced stage in Africa, which contributes to poor prognoses (Stefan, 2015; Rebbeck, 2020). Current cancer treatment includes chemotherapy, radiation, surgical removal and/or a combination of any of these treatments. However, conventional cancer treatments are limited by inability to bypass biological barriers, non-specific systemic distribution, undesirable side effects and toxicity (Wicki et al., 2015; Jena et al., 2020).

Poor drug delivery to the target site leads to damage of the healthy tissues and normal cells, rather than cancer cells. Therefore, there is urgent need to develop new and innovative technologies that could overcome the limitations associated with the current conventional cancer treatment. Such technologies should have the capability of delivering anticancer drugs to where they will be most effective and should combine both diagnosis and treatment. In addition, it should have the capability to visualize tumour margins and identify tumour cell residuals and micrometastases during or post-cancer treatment or surgical removal.

Nanomedicine is the application of nanotechnology in medicine. Nanomedicine promised an immense potential towards early cancer diagnosis, a cancer diagnosis before the tumour develops and targeted drug delivery, which provide better patient prognosis. The advantage of targeted drug delivery includes the capability of delivering anticancer drugs to target-specific cancer cells and tissues (Malam et al., 2009). Nanotechnology deals with the creation and synthesis of particles at the nanoscale and molecular level (i.e. 1–100 nm). Various nanoparticle (NP) platforms to combat cancer have been extensively researched in the past 20 years, in the advancement for cancer diagnostics and minimizing the side effects associated with conventional therapy (Mufamadi et al., 2011, 2019; Wicki et al., 2015; Choi and Han, 2018; Naeem et al., 2020). NPs are surface functionalized with ligands such as monoclonal antibodies, proteins, peptide and polysaccharide for drug localization at the cancer sites, without affecting the normal cells and/or healthy tissues (Mufamadi et al., 2013; Yoo et al., 2019; Naeem et al., 2020). In addition, due to particle size, NPs are able to pass through biological barriers such as blood–brain barrier (BBB) for brain cancer (Chakroun et al., 2017; Fisusi et al., 2018). Nanotheranostics or theranostics nanoparticles promise to offer a new approach that combines diagnostics with therapeutics of cancer using nanotechnology, the next generation of early detection, with simultaneous imaging and treatment of the cancers (Keivit and Zhang, 2011; Choi et al., 2012). Nanotheranostics approach, a combination of diagnostics with treatment, paves a new way towards clinical outcomes and personalized therapy (Kim et al., 2012). Cancer theranostics nanoparticles are made up of inorganic and organic nanomaterials that are capable of diagnosing/imaging, incorporate an imaging moiety into their design and are effective for therapeutic actions. Gold nanoparticles (AuNPs), silver nanoparticles (AgNPs), copper oxide nanoparticles (CuO-NPs), zinc oxide nanoparticles (ZnO-NPs), carbon nanotubes, fullerenes and other metal oxide nanoparticles are examples of inorganic nanoparticles; lipids-based nanoparticles and polymer-based nanoparticles are examples of organic nanoparticles (Kievit and Zhang, 2011; Choi et al., 2012; Chen et al., 2017).

Phytonanotechnology or green bionanotechnology involves metallic nanoparticle synthesis using phytochemicals extracted from plant extracts and biomolecules as a green synthesis method. The utilized metals and/or metal oxides include silver, gold, titanium, iron, zinc, copper, zinc oxide and copper oxide (Patra and Baek, 2014; Barabadi et al., 2017; Nagajyothi et al., 2017; Umar et al., 2019). This eco-friendly biosynthetic procedure promises to offer the next generation of cancer therapy that are effective and inexpensive with minimal toxicity (Aromal and Philip, 2012; Barabadi et al., 2017; Mufamadi and Mulaudzi, 2019; Mufamadi et al., 2019). The unique structural, therapeutic and optical properties of silver and gold nanomaterials make them suitable candidates for cancer nanotheranostics applications (Pietro et al., 2016). The aim of this chapter is to understand the weaknesses and strengths of green bionanotechnology in cancer theranostics. This chapter will highlight the recent research advances related to plant-mediated synthesis of metallic nanoparticles for early detection of cancer, therapy and theranostics. This chapter will also look at the opportunities and challenges of using different plant sources as enabling tools towards the production of safer metallic nanoparticles for cancer treatment, imaging and nanotheranostics.

1.2 PHYTONANOTECHNOLOGY: GREEN SYNTHESIS OF METALLIC NANOPARTICLES

Phytonanotechnology is a branch of clean technologies that utilize phytoformulations to produce nanoparticles that are safe, eco-friendly and non-toxic (Rao et al., 2016; Verma et al., 2019). It is driven by the convergence and synergy of innovative technologies, including nanotechnology and biotechnology. The word "green" refers to the use of plants and their extracts and/or microorganisms in the synthesis of metallic nanoparticles (Chauhan et al., 2012; Barabadi et al., 2017; Mufamadi and Mulaudzi, 2019; Mufamadi et al., 2019). Different plant and microbial sources have been used for the green synthesis of metallic nanoparticles such as leaves, roots, seeds, fruits, bacteria, fungi and yeast (Phanjom et al., 2012; Juraifani and Ghazwani, 2015; Moghaddam et al., 2015; Perugu et al., 2016; Menon et al., 2017; Mufamadi et al., 2019). A green synthesized metallic nanoparticle includes silver, gold, titanium, copper, iron and zinc oxides (Patra and Baek, 2014; Barabadi et al., 2017; Singh et al., 2018). Green synthesis for the formation of metallic nanoparticles employs aqueous plant extracts and microbial sources (e.g. bacteria, fungi and yeast) as reducing and stabilizing agents during the synthesis (Suna et al., 2014; Sadeghi and Gholamhoseinpoor, 2015; Mohammadlou et al., 2016). The mechanism of metallic nanoparticle synthesis using plant extracts and/or microbial sources involves the reduction and capping of metal ions to form stable nanoparticles. Tran et al. (2013) showed the reduction of pure silver ions/Ag(+) ions into Ag(0) nanoparticles (AgNPs), employing amino acids and proteins from plant extracts as reducing, capping and stabilizing agents.

1.3 PHYTONANOTECHNOLOGY: ANTICANCER ACTIVITY

Many studies have demonstrated the anticancer potential of the metallic nanoparticles synthesized using plant extracts (Ramesh and Armash, 2015; Nagajyothi et al., 2017; Hembram et al., 2018; Miri et al., 2018; Mufamadi and Mulaudzi, 2019; Umar et al., 2019). Ramesh and Armash (2015) synthesized AuNPs using *Diospyros ferrea* plant materials. Scanning electron microscope (SEM) confirmed the green AuNPs with the mean particle size in the range of 70–90 nm. Anticancer activity was examined *in vitro* in human cell lines, using 3-(4,5-dimethylthiazol-2-yl)-2,5-dipheny l-2*H*-tetrazolium bromide (MTT) viability assay. The AuNPs inhibited the growth of the cancer cells significantly, in a dose- and duration-dependent manner. A similar study by Nagajyothi et al. (2017) reported the synthesis of spherical CuO-NPs with the average size of 26.6 nm using aqueous black bean extract, *Phaseolus vulgaris*. *In vitro* anticancer results indicated that CuO-NPs induced intracellular reactive oxygen species (ROS) generation, apoptosis and reduced cervical carcinoma colony growth with minimal side effects. However, anticancer activity in human cervical cancer was in a dose-dependent manner. Al-Sheddi et al. (2018) also demonstrated the anticancer potential of green synthesized AgNPs employing extract of *Nepeta deflersiana* against human cervical cancer (HeLa) cells. Miri et al. (2018) reported the synthesis of spherically uniform in shape of AuNPs with a particle size of about 25 nm using *Prosopis farcta* extract. For anticancer study, an *in vitro* cytotoxicity effect was tested on a colon (HT-29) cancerous cell line using the MTT assay. Green synthesized AgNPs destroyed a colon cell line, making them a suitable candidate for cancer treatment. Furthermore, a recent study by Umar et al. (2019) reported green synthesis of ZnO-NPs using *Albizia lebbeck* stem bark. Zetasizer and SEM results confirmed the formation of irregular spherical ZnO-NPs with an average size of 66.25 nm with a polydisparity index of 0.262. Antiproliferative and anticancer activities of the green synthesized ZnO-NPs were investigated on metastatic breast cancer cell lines, such as human breast cancer cell lines MDA-MB 231 and MCF-7. The results showed ZnO-NPs cytotoxic effects by disruption of the cell membrane integrate and the induction of membrane blebs formation in both breast cancer cell lines in a concentration-dependent manner.

1.4 PHYTONANOTECHNOLOGY: MECHANISM OF ACTION FOR CANCER TREATMENT

Nanoparticles, with a size range of 1–100 nm, possess unique physicochemical and biological properties compared to its bulk counterparts. Traditional ways of synthesizing nanoparticles include the physical and chemical methods such as ion sputtering, solvothermal synthesis, reduction and sol–gel technique (Rath et al., 2014). Both these methods involve the use of toxic chemicals, which are hazardous to the environment. Green synthesis or biological methods that use bacteria, fungi and plant extracts are gaining popularity as they do not use harsh toxic chemicals and are eco-friendly and cost-effective. Plant-mediated synthesis is preferred since, unlike bacteria and fungi, plants do not trigger biosafety issues. Also, the synthesis of nanoparticles using plants is rapid as compared to bacteria and fungi that require longer incubation times for the reduction of metal ions. In addition, the phytochemicals present in the plants can act as both reducing and stabilizing agents in the nanoparticle synthesis process (Singh et al., 2018).

Metallic nanoparticles such as AuNPs and AgNPs are ideal candidates for cancer treatment due to their unique optical and thermal characteristics. Several studies have shown the biocompatible nature and anticancer activity of green synthesized nanoparticles. Roshni et al. (2018) used the plant extracts from *Murraya koenigii* for the synthesis of benign AgNPs. The study showed that *M. koenigii* leaves play a key role in the reduction and stabilization of silver ions to AgNPs. The AgNPs produced were spherical in shape with a size range of 80.62–100.50 nm. The AgNPs showed potent anticancer activity against the human colorectal adenocarcinoma (HT-29) cell line with an IC_{50} value of 26.05 μg/ml. Cytotoxicity was due to the free oxygen radical generation which affected the apoptotic pathway and caused an antiangiogenic effect *in vitro*. Geetha et al. (2013) studied the antileukaemic cancer activity of AuNPs synthesized from the flower extract of *Couroupita guianensis*. The *C. guianensis* plant has medicinal applications, including antibiotic, antiseptic and anti-inflammatory activity. Synthesis of AuNPs was reported to be achieved in 5 min without any stirring, heating or change in pH. The green synthesized AuNPs induced apoptosis in human leukaemia cell line (HL-60) as shown by the DNA damage observed with Comet assay. A study conducted by Elia et al. (2014) looked at the efficiency of synthesizing AuNPs using four different plant extracts: *Salvia officinalis*, *Lippia citriodora*, *Pelargonium graveolens* and *Punica granatum*. Antioxidants that are most commonly found in all four plant extracts were used for the synthesis. The synthesized AuNPs were of similar size distribution, with excellent biocompatibility and stability for over three weeks. Wang et al. (2019) studied the green synthesis method of AuNPs from the aqueous extract of *Scutellaria barbata* and its potential anticancer activity against pancreatic cancer cell line (PANC-1). According to the Chinese, the *S. barbata* is the miracle root for the preservation of life and it has shown potent anticancer activity against breast cancer, colorectal cancer, hepatocarcinoma, skin cancer, lung cancer and ovarian cancer. AuNPs synthesized from the aqueous extract of *S. barbata* showed increased production of intracellular ROS in PANC-1 cells at the dose of 25 and 50 μg/ml as compared to the control cells. The study also confirmed that the synthesized AuNPs from *S. barbata* promoted apoptosis of the cancer cells by increasing the apoptosis-related protein expressions of Bax and β-actin proteins. Another study conducted by Liu et al. (2019) also showed that AuNPs, synthesized using traditional medicinal herb *Curcuma wenyujin*, induced apoptosis in the renal cancer cell line, A498, through activation of the proapoptotic proteins Bid and Bax and inactivation of anti-apoptotic proteins Bcl-2 and Bcl-xl.

The literature has shown that plant extracts are able to produce nanoparticles of specific size and shape, which aid in the large-scale production of these nanoparticles. It is also important to take into consideration the source of plant extract used for therapy purposes as they may exhibit different cytotoxic potential as evident from the research conducted by Shawkey et al. (2013). In their study, they synthesized AgNPs using the aqueous extract of fruits, seeds, leaves and roots of *Citrullus colocynthis*. The synthesized AgNPs were spherical with an average size of 7–19 nm. The anticancer activity was studied on four different human cancer cell lines: the colon (HCT-116), breast (MCF-7), liver (Hep-G2) and intestines (Caco2). The cytotoxic response varied depending on the plant part source and cancer cell line. For example, AgNPs synthesized from fruits showed cytotoxicity against MCF-7 cell lines, whereas the AgNPs synthesized from leaves, roots and seeds did not exhibit any cytotoxic response against the MCF-7 cell line. Similarly, AgNPs synthesized from root extracts showed cytotoxicity in HCT-116 and Hep-G2 with little cytotoxicity against Caco2 and MCF-7 cell lines. The cytotoxicity of AgNPs is induced by disturbing the cell membrane permeability and cellular internalization resulting in cell death through a cascade of events. A study of the exact mechanism of AgNPs inhibiting the signalling pathways will be a major breakthrough in the field of nanomedicine. Limited literature is available on *in vivo* work conducted using green synthesized nanoparticles on cancer cells. He et al. (2016) studied the anticancer activity of green synthesized AgNPs (using longan peel extracts) against human lung cancer cells (H1299), both *in vitro* and *in vivo*. In this study, they showed that spherical AgNPs with a size range of 8–22 nm were able to induce apoptosis in H1299 cells by suppressing the expression of anti-apoptotic bcl-2 family proteins. *In vivo* studies showed that AgNPs could slow down the growth of lung cancers in the mice, showing promising application in early-stage chemotherapy treatment. A size-controlled targeting strategy is recommended for effective cancer treatment using AgNPs, with smaller size showing a stronger penetration ability and greater toxicity. The efficacies of the potential anticancer drugs are determined by their ability to prolong lifespan and decrease tumour volume and viable tumour count. An *in vivo* study by Sririam et al. (2010) showed that there was a marked decrease in the number of cancer cells in AgNP-treated tumour mice, with minimal effect on the control cells. The study investigated the efficacy of bacteria-mediated synthesis of AgNPs as an anticancer agent using Dalton's lymphoma ascites (DLA) cell lines both *in vitro* and *in vivo*. They found that AgNPs showed cytotoxic activity against DLA cell lines through activation of caspase-3 enzymes, leading to apoptosis (Fig. 1.1).

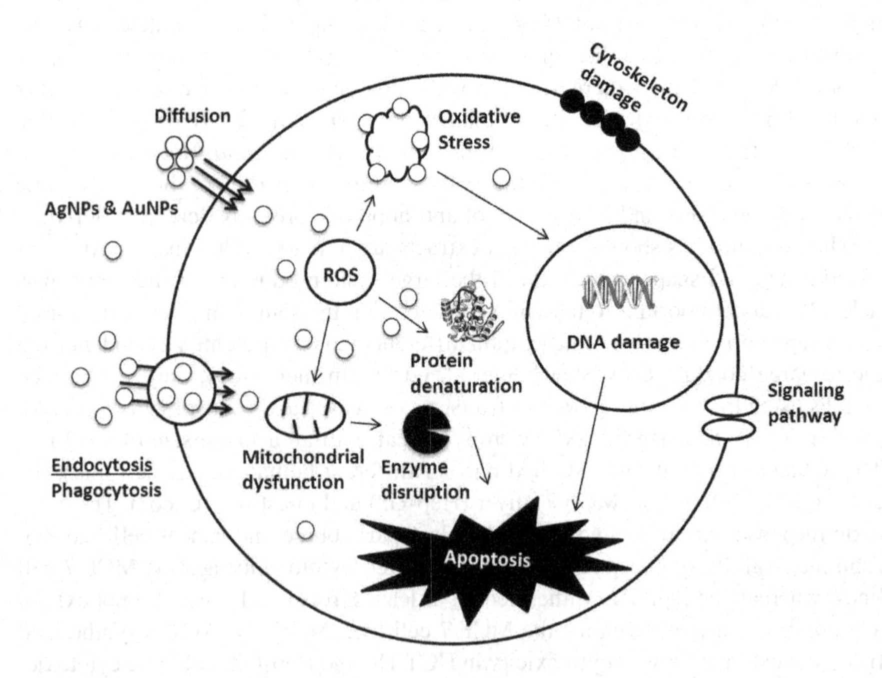

FIGURE 1.1 Schematic diagram illustrating the mechanisms of AgNPs and AuNPs to induce cytotoxicity in a cancerous human cell.

1.5 CANCER NANOTHERANOSTICS OF PHYTONANOTECHNOLOGY

Cancer nanotheranostics is a promising technology to combine bio-imaging and therapy of cancer through the use of nanotechnology and biotechnology tools (Keivit and Zhang, 2011; Ovais et al., 2016; Barabadi et al., 2017). During nanotheranostics, a single nanomedicine or theranostic nanoparticle carries both therapeutics, to improve cancer treatment, and imaging tool, to improve the cancer diagnosis approach. In addition, it has the ability to monitor cancer treatment and diagnostics in real time (Kievit and Zhang, 2011). Moreover, these green synthesized metallic nanoparticles are biocompatible to human cell lines and possess optical or fluorescent properties (Ovais et al., 2016). The unique structural, therapeutic and optical properties of AgNPs and AuNPs make them suitable candidates for cancer nanotheranostics applications (Pietro et al., 2016). AuNPs have strong surface plasmon resonance which makes it a suitable candidate for theranostics and optical imaging for early detection of cancer (Boisselier and Astruc, 2009). They can interact with light, and be detected using multiple modalities in cancer imaging, such as light scattering based imaging, photoacoustic imaging and X-ray or computed tomography (CT) based imaging (Huang et al., 2006; Aydogan et al., 2010; Jokerst et al., 2012; Liu et al., 2017). For photothermal therapy, AuNPs can absorb incident photons and convert them to heat to destroy cancer cells and/or tumours (Riley and Day, 2017). For photodynamic therapy, AuNPs could generate singlet oxygen ($^{1}O_2$) and/or ROS to kill cancer cells upon either pulse or continuous wave laser irradiation (Krpetic et al., 2010; Pasparakis, 2013). AuNPs in a nanotheranostics approach can combine both bio-imaging and phototherapy (Gao et al., 2014; Guo et al., 2017).

Many studies have demonstrated cancer nanotheranostics employing green synthesized metallic nanoparticles, nanoparticles synthesized using plant extracts and biomolecules *in vitro* (Gao et al., 2014; Mukherjee et al., 2014; Ge et al., 2015; Ovais et al., 2016; Hameed et al., 2019). Ge et al. (2015) has reported green or biological synthesis of fluorescent Au/Ce nanocluster/nanoparticles employing glutathione as a reducing and stabilizing agent during the synthesis. The results showed Au/Ce nanocluster to be highly sensitive bio-imaging agents with high fluorescence intensity. *In vitro* bio-imaging was observed in the HeLa and HepG2 cancer cell lines, and *in vivo* bio-imaging was observed in the solid tumour mouse model of cervical carcinoma. A similar study by Gao et al. (2014) showed near-infrared fluorescence imaging and anticancer activity *in vitro* in HeLa cancer cells and *in vivo* in a tumour mouse model of cervical carcinoma employing green synthesized silver nanoclusters/nanoparticles using glutathione. Mukherjee et al. (2014) demonstrated the theranostics potential of green synthesized AgNPs using *Olax scandens* leaf extract with spherical shape and particle size of 20–60 nm. The study introduced for the first time multifunctional biological activities, 4-in-1 system of green synthesized AgNPs. The green synthesized AgNPs 4-in-1 system implies (i) anticancer activity, (ii) antibacterial activity, (iii) biocompatibility/non-toxic drug delivery vehicle and (iv) diagnostic/bio-imaging system. Hameed et al. (2019) showed the theranostics potential of using green synthesized ZnO-NPs with plant extracts.

1.6 CANCER NANOTHERANOSTICS OF PHYTONANOTECHNOLOGY: OPPORTUNITIES

Nanotheranostics systems are termed multifunctional nanosystems that are capable of acting as targeted drug delivery systems, nanodiagnostics, nanosensors for early detection and target-specific destruction of cancer cells (Roma-Rodrigues et al., 2019). Green synthesized metallic nanoparticles such as AgNPs and AuNPs are easily synthesized and stable due to the presence of reducing, capping and stabilizing agents from plant extracts such as carbohydrate, proteins and polyphenols. In addition, metallic nanoparticles synthesis from biological materials or plant extracts are considered to be clean, non-toxic, biocompatible and environment-friendly compared with the physical and chemical synthesis methods involving the use of toxic chemicals, which are hazardous to the environment (Ingale and Chaudhari, 2013; Elia et al., 2014; Rath et al., 2014). The use of medicinal plants for green synthesis of metallic nanoparticles is an ideal approach for anticancer and antimicrobial activities since they can generate ROS, thus leading to apoptosis and cell death, increase in oxidative stress and cellular damages through upregulation of the p53 protein (Cragg and Newman, 2005; Sririam et al., 2010; Mei et al., 2012; Geetha et al., 2013; Gurunathan et al., 2013; Wang et al., 2019). Moreover, green synthesized AgNPs and AuNPs in cancer theranostics were reported biocompatible towards normal cell lines.

The optical properties and self-fluorescence ability of green synthesized AgNPs and AuNPs made them potential candidates to replace radiolabelled isotopes for tumour detection and ultimately reduce the side effects associated with radiotherapy (Ovais et al., 2016). In addition, because of their optical and/or fluorescence properties and the particles size of 1–100 nm, both AgNPs and AuNPs can be delivered through the BBB, and/or can also be used for bio-imaging for early diagnosis and therapy of brain cancer, at a lower cost compared to brain tumour surgical therapy (Sonali et al., 2018).

Phytochemicals in plant extracts employed in phytonanotechnology formulation or green synthesis of metallic nanoparticles as reduction, capping and stabilizing agents need in-depth research in order to understand the phytochemistry reaction and/or determine the active functional groups responsible for the production of non-toxic and biocompatible metallic nanoparticles that are efficient for cancer nanotheranostics (Ovais et al., 2016). Many studies have demonstrated the anticancer potential of metallic nanoparticles synthesized using plant extracts *in vitro* and shows non-toxic effect on normal cells, thus making them promising candidates for future cancer treatment and/or cancer nanotheranostics for personalized medicine. However, the understanding or addressing of key physiological barriers *in vivo* is the key to effectively deliver phytonanoparticles to the tumour site and for improving phytonanotechnology clinical outcomes.

1.7 CONCLUSION AND FUTURE PROSPECTS

The use of phytonanotechnology in cancer theranostics is promising to pave new ways that offer multifunctional biological activities and a 4-in-1 system such as anticancer activity, antibacterial activity, non-toxic drug delivery vehicle and bio-imaging system in a single application. Green synthesized AgNPs and AuNPs can combine both bio-imaging and phototherapies such as photodynamic and photothermal therapy that could be used to destroy cancer cells and tumours. These properties make them suitable candidates for cancer nanotheranostics. However, current phytonanotechnology or green synthesized metallic nanoparticles used for cancer nanotheranostics do not exhibit any strategy towards drug localization at the targeted cancer sites *in vivo*; therefore, one can conclude that there is still more work to be done in order to improve target delivery and clinical outcomes. Future trends in phytonanotechnology in cancer theranostics will be driven by the convergence and synergy of innovative technologies, including nanotechnology, biotechnology and photonic and artificial intelligence, and not by utilizing these technologies in isolation.

REFERENCES

Al-Sheddi, E.S., Farshori, N.N., Al-Oqail, M.M., Al-Massarani, S.M., Saquib,Q., Wahab, R., Musarrat, J., et al. 2018. "Anticancer potential of green synthesized silver nanoparticles using extract of *Nepeta deflersiana* against Human Cervical Cancer Cells (HeLA)." *Bioinorganic Chemistry and Applications*: 1–12. DOI: 10.1155/2018/9390784.

Aromal, S.A., and D. Philip. 2012. "Green synthesis of gold nanoparticles using *Trigonella foenum-graecum* and its size dependent catalytic activity." *Spectrochimica Acta Part A: Molecular and Biomolecular Spectroscopy* 97: 1–5. DOI: 10.1016/j.saa.2012.05.083.

Aydogan, B., Li, J., Rajh, T., Chaudhary, A., Chmura, S.J., Pelizzari, C., Wietholt, C., Kurtoglu, M, Redmond, P. 2010. "AuNP-DG: Deoxyglucose-labeled gold nanoparticles as X-ray computed tomography contrast agents for cancer imaging." *Molecular Imaging and Biology* 12, no. 5: 463–467. DOI: 10.1007/s11307-010-0299-8.

Azubuike, S.O., Muirhead,C., Hayes, L., McNally, R. 2018. "Rising global burden of breast cancer: The case of sub-Saharan Africa (with emphasis on Nigeria) and implications for regional development: A review." *World Journal of Surgical Oncology* 16: 63. DOI: 10.1186/s12957-018-1345-2.

Barabadi, H., Ovais, M., Shinwari, Z.K., Saravanan. M. 2017. "Anti-cancer green bionanomaterials: Present status and future prospects." *Chemistry Letters and Reviews* 10, no. 4, 285–314. DOI: 10.1080/17518253.2017.1385856.

Boisselier, E., and D. Astruc. 2009. "Gold nanoparticles in nanomedicine: Preparations, imaging, diagnostics, therapies and toxicity." *Chemical Society Reviews* 38, no. 6: 1759–1782. DOI: 10.1039/b806051g.

Chakroun, R.W., Zhang, P., Lin, R., Schiapparelli, P., Quinones-Hinojosa, A., Cui, H. 2017. "Nanotherapeutic systems for local treatment of brain tumors." *Wiley Interdisciplinary Reviews. Nanomedicine and Nanobiotechnology* 10, no. 1: 1–41. DOI: 10.1002/wnan.1479.

Chauhan, R.P.S., Gupta, C. and D. Prakash. 2012. "Methodological advancements in green nanotechnology and their applications in biological synthesis of herbal nanoparticles." *International Journal of Bioassays* 1, no. (7): 6–10.

Chen, H., Zhang, W., Zhu, G., Xie, J., Chen, X. 2017. "Rethinking cancer nanotheranostics." *Nature Reviews Materials* 2, no. 7: 1–18. DOI: 10.1038/natrevmats.2017.24.

Choi, K.Y., Liu, G., Lee, S., Chen, X. 2012. "Theranostic nanoplatforms for simultaneous cancer imaging and therapy: Current approaches and future perspectives." *Nanoscale* 4, no. 2: 330–342. doi:10.1039/c1nr11277e.

Choi, Y.H., and H.K. Han. 2018."Nanomedicines: Current status and future perspectives in aspect of drug delivery and pharmacokinetics." *Journal of Pharmaceutical Investigation* 48, no. 43–60. DOI: 10.1007/s40005-017-0370-4.

Chokunonga, E.B.M., Chirenje, Z.M., Nyakabau, A.M., Parkin, D.M. 2013. "Trends in the incidence of cancer in the black population of Harare, Zimbabwe 1991–2010." *International Journal of Cancer* 133, no. 3: 721–729. DOI: 10.1002/ijc.28063.

Cragg, G.M., and D. J. Newman. 2005. "Plants as a source of anti-cancer agents." *Journal of Ethnopharmacology* 100, no. (1–2): 72–79. DOI: 10.1016/j.jep.2005.05.011.

Elia, P., Zach, R., Hazan, S., Kolusheva, S., Porat, Z., Zeiri, Y. 2014. ''Green synthesis of gold nanoparticles using plant extracts as reducing agents.'' *International Journal of Nanomedicine* 9: 4007–4021. DOI: 10.2147/IJN.S57343.

Fisusi, F.A., Schätzlein, A.G., and I.F. Uchegbu. 2018. "Nanomedicines in the treatment of brain tumors." *Nanomedicine* 13, no. 6: 579–583. DOI: 10.2217/nnm-2017-0378.

Gao, S., Chen, D., Li, Q., Ye, J., Jiang, H., Amatore, C., Wang, X. 2014. "Near-infrared fluorescence imaging of cancer cells and tumors through specific biosynthesis of silver nanoclusters." *Scientific Reports* 4: 4384. DOI: 10.1038/srep04384.

Ge, W., Zhang, Y., Ye, J., Chen, D., Rehman, F.U., Li, Q., Chen, Y., Jiang, H., Wang, X. 2015. "Facile synthesis of fluorescent Au/Ce nanoclusters for high-sensitive bioimaging." *Journal of Nanobiotechnology* 13: 8. DOI: 10.1186/s12951-015-0071-y.

Geetha, R., Ashokkumar, T., Tamilselvan, S., Govindaraju, K., Sadiq, M., Singaravelu, G. 2013. "Green synthesis of gold nanoparticles and their anticancer activity.'' *Cancer Nanotechnology* 4: 91–98. DOI: 1007/s12645-013-0040-9.

Guo, J., Rahme, K., He, Y., Li, L., Holmes, J.D., O'Driscoll, C.M. 2017. "Gold nanoparticles enlighten the future of cancer theranostics." *International Journal of Nanomedicine* 12: 6131–6152. DOI: 10.2147/IJN.S140772.

Gurunathan, S., Han, J.W., Eppakayala, V., Jeyaraj, M., Kim, J-H. 2013. "Cytotoxicity of biologically synthesized silver nanoparticles in MDA-MB-231 human breast cancer cells." *BioMed Research International* 2013: 535796. DOI: 10.1155/2013/535796.

Hameed, S., Iqbal, J., Ali, M., Khalil, A.K., Abbasi, B.A., Numan, M., Shinwari, Z.K. 2019. Green synthesis of zinc nanoparticles through plant extracts: Establishing a novel era in cancer theranostics. *Materials Research Express* 6, no. 10: 102005. DOI: 10.1088/2053-1591.

He, Y., Zhiyun, D., Shijing, M., Yue, L., Dongli, L., Huarong, H., Sen, J., Shupeng, C., Wenjing, W., Kun, Z., Xi, Z. 2016. "Effects of green-synthesized silver nanoparticles on lung cancer cells in vitro and grown as xenograft tumors *in vivo*.'' *International Journal of Nanomedicine* 11: 1879–1887. DOI: 10.2147/IJN.S103695.

Hembram, K.C., Kumar, R., Kandha, L., Parhi, P.K., Kundu, C.N., Bindhani, B.K. 2018. "Therapeutic prospective of plant-induced silver nanoparticles: Application as antimicrobial and anticancer agent." *Artificial Cells, Nanomedicine, and Biotechnology* 46, no. (sup3): S38–S51. DOI: 10.1080/21691401.2018.1489262.

Huang, X., El-Sayed, I.H., Qian, W., El-Sayed, M.A. 2006. "Cancer cell imaging and photothermal therapy in the near-infrared region by using gold nanorods." *Journal of the American Chemical Society* 128, no. 6: 2115–2120.

Ingale. A.G. and A. N. Chaudhari. 2013. "Biogenic synthesis of nanoparticles and potential applications: An eco-friendly approach." *Journal of Nanomedicine and Nanotechnology* 4, no. 2. DOI: 10.1016/j.arabjc.2015.11.002.

Jena, L., McErlean, E. and H. McCarthy. 2020. "Delivery across the blood-brain barrier: nanomedicine for glioblastoma multiforme." *Drug Delivery and Translational Research* 10: 304–318. DOI: 10.1007/s13346-019-00679-2.

Jokerst, J.V., Cole, A.J., Van de Sompel, D., Gambhir, S.S. 2012. "Gold nanorods for ovarian cancer detection with photoacoustic imaging and resection guidance via Raman imaging in living mice." *ACS Nano* 6, no. 11: 10366–10377. DOI: 10.1021/nn304347g.

Juraifani, A.A.A.A., and A.A. Ghazwani. 2015."Biosynthesis of silver nanoparticles by Aspergillus niger, Fusarium oxysporum and Alternaria solani." *African Journal of Biotechnology* 14, no. 26: 2170–2174. DOI: 10.5897/AJB2015.14482.

Keivit, F.M and M. Zhang. 2011. "Cancer nanotheranostics: Improving imaging and therapy by targeted delivery across biological barriers." *Advanced Materials* 23, no. 36: H217–H247. DOI: 10.1002/adma.201102313.

Kim, T.H., Lee, S., Chen, X. 2012. "Nanotheranostics for personalized medicine." *Expert Review of Molecular Diagnostics* 3, no. 3: 257–269. DOI: 10.1586/erm.13.15.

Krpetić, Z., Nativo, P., Sée, V., Prior, I.A., Brust, M., Volk, M. 2010. "Inflicting controlled nonthermal damage to subcellular structures by laser-activated gold nanoparticles. *Nano Letters* 10, no. 11: 4549–4554. DOI: 10.1021/nl103142t.

Liu, W., Liu, K., Zhao, Y., Zhao, S., Luo, S., Tian, Y., et al. 2017. T1-weighted MR/CT dual-modality imaging-guided photothermal therapy using gadolinium-functionalized triangular gold nanoprism. *RSC Advances* 7, no. 26: 15702–15708. DOI: 10.1039/c7ra01101f.

Liu, R., Pei, Q., Shou, T., Zhang, W., Hu. J., Li, W. 2019. "Apoptotic effect of green synthesized gold nanoparticles from *Curcuma wenyujin* extract against human renal cell carcinoma A498 cells." *International Journal of Nanomedicine* 14: 4091–4103. DOI: 10.2147/IJN.S203222.

Malam, Y., Loizidou, M and A. M. Seifalian. 2009. "Liposomes and nanoparticles: Nanosized vehicles for drug delivery in cancer." *Trends in Pharmacological Sciences* 30, no. 11: 592–599. DOI: 10.1016/j.tips.2009.08.004.

Mei, N., Zhang, Y., Chen, Y., Guo, X., Ding, W., Ali, S.F., Biris, A.S., Rice, P., Moore, M.M., Chen, T. 2012. "Silver nanoparticle-induced mutations and oxidative stress in mouse lymphoma cells." *Environmental and Molecular Mutagenesis* 53, no. 6: 409–419. DOI: 10.1002/em.21698.

Menon, S., Rajeshkumar, S., Kumar, V.S. 2017. "A review on biogenic synthesis of gold nanoparticles, characterization, and its applications." *Resource-Efficient Technologies* 3: 516–527. DOI: 10.1016/j.reffit.2017.08.002.

Miri, A., Darroudi, M., Entezari, R., Sarani, M. 2018. "Biosynthesis of gold nanoparticles using Prosopis farcta extract and its in vitro toxicity on colon cancer cells." *Research on Chemical Intermediates* 44: 3169–3177. DOI: 10.1007/s11164-018-3299-y.

Moghaddam, A.B., Namvar, F., Moniri, M., Tahir, P.M., Azizi, S., Mohamed, R. 2015. Nanoparticles, biosynthesized by fungi and yeast: A review of their preparation, properties, and medical applications. *Molecules* 20, no. 9: 16540–16565. DOI: 10.3390/molecules200916540.

Mohammadlou, M., Maghsoudi, H., Jafarizadeh-Malmiri, H. 2016. "Review on green silver nanoparticles based on plants: Synthesis, potential applications and eco-friendly approach." *Food Research International* 23, no. 2: 446–463.

Mufamadi, M.S and R.B. Mulaudzi. 2019. "Green engineering of silver nanoparticles to combat plant and foodborne pathogens: Potential economic impact and food quality." *Plant Nanobionics, Nanotechnology in the Life Sciences*, edited by Pradad, 451–476. Springer, Cham. https://doi.org/10.1007/978-3-030-16379-2_16.

Mufamadi, M.S., Pillay, V., Choonara, Y.E., Du Toit, L.C., Modi, G., Naidoo, D., Ndesendo, V.M. 2011. "A review on composite liposomal technologies for specialized drug delivery." *Journal of Drug Delivery*: 2011: 939851. DOI: 10.1155/2011/939851.

Mufamadi, M.S., Choonara, Y.E., Kumar, P., Modi, G., Naidoo, D., van Vuuren, S., Ndesendo, V.M., Toit, L.C., Iyuke, S.E., Pillay, V. 2013. "Ligand-functionalized nanoliposomes for targeted delivery of galantamine." *International Journal of Pharmaceutic* 448, no. 1: 267–281. DOI: 10.1016/j.ijpharm.2013.03.037.

Mufamadi, M.S., George, J., Mazibuko, Z., Tshikalange, T.E. 2019. "Cancer bionanotechnology: Biogenic synthesis of metallic nanoparticles and their pharmaceutical potency." In *Microbial Nanobionics. Nanotechnology in the Life Sciences*, edited by Prasad R, 229–252. Switzerland: Springer.

Mukherjee, S., Chowdhury, D., Kotcherlakota, R., Patra, S., Vinothkumar, B., Bhadra, M.P., Sreedhar, B., Patra, C.R. 2014. "Potential theranostics application of bio-synthesized silver nanoparticles (4-in-1 system)." *Theranostics* 4, no. 3): 316–335. DOI: 10.7150/thno.7819.

Naeem, M., Awan, U.A., Subhan, F., Cao, J., Hlaing, S.P., Lee, J., Im, E., Jung, Y., Yoo, J.W. 2020. "Advances in colon-targeted nano-drug delivery systems: Challenges and solutions." *Archives of Pharmacal Research* 43, no. 1: 153–169. DOI: 10.1007/s12272-020-01219-0.

Nagajyothi, P.C., Muthuraman, P., Sreekanth, T.V.M., Kim, D.H., Shim, J. 2017. "Green synthesis: In-vitro anticancer activity of copper oxide nanoparticles against human cervical carcinoma cells." *Arabian Journal of Chemistry* 10, no. 2: 215–225. DOI: 10.1016/j.arabjc.2016.01.011.

Ovais, M., Khalil, A.T., Raza, A., Khan, M.A., Ahmad, I., Islam, N.U., Saravanan, M., Ubaid, M.F., Ali, M., Shinwari, Z.K. 2016. "Green synthesis of silver nanoparticles via plant extracts: Beginning a new era in cancer theranostics." *Nanomedicine* 11, no. 23: 3157–3177. DOI:10.2217/nnm-2016-0279.

Pasparakis, G. 2013. "Light-induced generation of singlet oxygen by naked gold nanoparticles and its implications to cancer cell phototherapy." *Small* 9, no. 24: 4130–4134. DOI: 10.1002/smll.201301365.

Patra, J.K and K. Baek. 2014. "Green nanobiotechnology: Factors affecting synthesis and characterization techniques." *Journal of Nanomaterials*: 1–19. DOI: 10.1155/2014/417305.

Perugu, S., Nagati, V., Bhanoori, M. 2016. "Green synthesis of silver nanoparticles using leaf extract of medicinally potent plant *Saraca indica*: A novel study." *Applied Nanoscience* 6: 747–753. DOI: 10.1007/s13204-015-0486-7.

Phanjom, P., Zoremi, D.E., Mazumder, J., Saha, M., Baruah, S.B. 2012. "Green synthesis of silver nanoparticles using leaf extract of *Myrica esculenta*." *International Journal of NanoScience and Nanotechnology* 3, no. 2: 73–79.

Pietro, P.D., Strano, G., Zuccarello, L., Satriano. C. 2016. "Gold and silver nanoparticles for applications in theranostics." *Current Topics in Medicinal Chemistry* 16, no. 27: 3069–3102. DOI: 10.2174/1568026616666160715163346.

Ramesh, V., and A. Armash. 2015. "Green synthesis of gold nanoparticles against pathogens and cancer cells." *International Journal of Pharmacological Research* 5, no. 10: 250–256. DOI:10.7439/ijpr.

Rao, P.V., Nallappan, D., Madhavi, K., Rahman, S., Jun Wei, L., Gan, S.H. 2016. "Phytochemicals and biogenic metallic nanoparticles as anticancer agents". *Oxidative Medicine and Cellular Longevity* 2016, no. 3685671: 1–15. DOI: 10.1155/2016/3685671.

Rath, M., Panda, S.S and K.N. Dhal. 2014. "Synthesis of silver nanoparticles from plant extract and its application in cancer treatment: A review." *International Journal of Plant, Animal and Environmental Sciences* 4, no. 3: 137–145.

Rebbeck, T.R. 2020. "Cancer in sub-Saharan Africa." *Science* 367, no. 6473: 27–28. DOI: 10.1126/science.aay4743.

Riley, R.S., and E.S. Day. 2017. "Gold nanoparticle-mediated photothermal therapy: Applications and opportunities for multimodal cancer treatment." *Wiley Interdisciplinary Reviews. Nanomedicine and Nanobiotechnology* 9, no. 4:1–25. DOI: 10.1002/wnan.1449.

Roma-Rodrigues, C., Pombo, I., Raposo, L., Pedrosa, P., Fernandes, A.R., Baptista, P.V. 2019. "Nanotheranostics targeting the tumor microenvironment." *Frontiers in Bioengineering and Biotechnology* 7, no. 197:1–18. DOI: 10.3389/fbioe.2019.00197.

Roshni, K., Younis, M., Ilakkiyapavai D, Basavaraju B, Puthamohan, V.M. 2018. "Anticancer activity of biosynthesized silver nanoparticles using *Murraya koenigii* leaf extract against HT-29 colon cancer cell line." *Journal of Cancer Science & Therapy* 10, no. 4: 072–075. DOI: 10.4172/1948-5956.1000521.

Sadeghi, B., and F. Gholamhoseinpoor. 2015. "A study on the stability and green synthesis of silver nanoparticles using *Ziziphora tenuior* (Zt) extract at room temperature." *Spectrochimica Acta Part A: Molecular and Biomolecular Spectroscopy* 134: 310–135. DOI: 10.1016/j.saa.2014.06.046.

Shawkey, A.M., Rabeh, M.A., Abdulall, A.K., Abdellatif, A.O. 2013. "Green nanotechnology: Anticancer activity of silver nanoparticles using *Citrullus colocynthis* aqueous extracts", *Advances in Life Science and Technology* 13: 60–70.

Singh, J., Dutta, T., Kim, K.H., Rawat, M., Samddar, P., Kumar, P. 2018. "Green synthesis of metals and their oxide nanoparticles: Applications for environmental remediation." *Journal of Nanobiotechnology* 16, no. 1: 84. DOI: 10.1186/s12951-018-0408-4.

Sonali, Viswanadh, M.K., Singh, R.P., Agrawal, P., Mehata, A.K., Pawde, D.M., Narendra, Roshan Sonkar, N., Muthu, M.S. 2018. Nanotheranostics: Emerging strategies for early diagnosis and therapy of brain cancer. *Nanotheranostics* 2, no. 1: 70–86. DOI: 10.7150/ntno.21638.

Sriram, M.I., Kanth, S.B.M., Kalishwaralal, K., Gurunathan, S. 2010. "Antitumor activity of silver nanoparticles in Dalton's lymphoma ascites tumor model." *International Journal of Nanomedicine* 5: 753–762. DOI: 10.2147/IJN.S11727.

Stefan, D.C. 2015. "Cancer care in Africa: An overview of resources." *Journal of Global Oncology* 1,no. 1: 30–36. DOI: 10.1200/JGO.2015.000406.

Suna, Q., Cai, X., Li, J., Zheng, M., Chen, Z., Yu, C.P. 2014. "Green synthesis of silver nanoparticles using tea leaf extract and evaluation of their stability and antibacterial activity." *Colloids and Surfaces A: Physicochemical and Engineering Aspects* 444:226–231. DOI: 10.1016/j.colsurfa.2013.12.065.

Tran, T.T.T., Vu, T.T.H., Nguyen, T.H. 2013. "Biosynthesis of silver nanoparticles using *Tithonia diversifolia* leaf extract and their antimicrobial activity." *Materials Letters* 105: 220–223. DOI: 10.1016/j.matlet.2013.04.021.

Umar, H., Kavaz, D., Rizaner, N. 2019. "Biosynthesis of zinc oxide nanoparticles using *Albizia lebbeck* stem bark, and evaluation of its antimicrobial, antioxidant, and cytotoxic activities on human breast cancer cell lines." *International Journal of Nanomedicine* 14: 87–100. DOI: 10.2147/IJN.S186888.

Verma, A., Gautam, S.P., Bansal, K.K., Prabhakar, N., Rosenholm, J.M. 2019. "Green nanotechnology: Advancement in phytoformulation research." *Medicines* 6: 39. DOI:10.3390/medicines6010039.

Wang Lei, Xu Jianwei, Yan Ye, Liu Han, Karunakaran Thiruventhan and Feng Li. 2019. "Green synthesis of gold nanoparticles from *Scutellaria barbata* and its anticancer activity in pancreatic cancer cell (PANC-1)." *Artificial Cells, Nanomedicine, and Biotechnology* 47: 1617–1627. DOI: 10.1080/21691401.2019.1594862.

Wicki, A. Witzigmann, D. Balasubramanian, V. Huwyler. J. 2015. Nanomedicine in cancer therapy: Challenges, opportunities, and clinical applications. *Journal of Controlled Release* 200: 138–157. DOI: 10.1016/j.jconrel.2014.12.030.

Yoo, J., Park, C., Yi, G., Lee, D., Koo, H. 2019. "Active targeting strategies using biological ligands for nanoparticle drug delivery systems." *Cancers* 11, no.5: 640. DOI: 10.3390/cancers11050640.

2 Green Synthesis of Nanoparticles Using Plant Extracts

A Promising Antidiabetic Agent

Rofhiwa Bridget Mulaudzi,
Mahwahwatse Johanna Bapela and
Thilivhali Emmanuel Tshikalange

CONTENTS

2.1 Introduction 31
2.2 Nanotechnology and Diabetes Management 33
2.3 Plant Material and New Perspective of Nanoparticles as an Antidiabetic Agent 34
    2.3.1 Some Antidiabetic Plant Extracts and Their Bioactive Compounds 35
    2.3.2 Green Synthesis of Nanoparticles as Potential Antidiabetic Agents 35
2.4 Conclusion 37
References 39

2.1 INTRODUCTION

Diabetes mellitus (DM) is a group of metabolic disorders in which the elevated glucose levels in the blood is observed for a long extent of time, and the prevalence of diabetes worldwide increases the high risk of heart diseases, chronic kidney diseases, stroke, foot ulcers, and damage to eyes (Ashwini and Mahalingam, 2015). Undesired effects associated with conventional antidiabetic drugs have led to the exploration of alternative remedies. Many medicinal plants are used worldwide for the treatment of diabetes, and some of them have been experimentally assessed. Medicinal plant metabolites are perceived to have lesser side effects and have good hypoglycaemic potential (Ashwini and Mahalingam, 2015; Hussain et al., 2016). Plant-based products show higher binding affinities for the specific receptor systems, and their bioactivity is often selective when compared to totally synthetic compounds. Globally, numerous plant species are documented for their ethnomedicinal use in the treatment and control of diabetes; however, only a few of them have been scientifically evaluated for their efficacy (Eddouks et al., 2002; Khan et al., 2012; Bilal et al., 2018). There exists immense anecdotal evidence on the efficacy of medicinal plants for the treatment and management of diabetes mellitus by indigenous people (Ernst, 1997; Grover et al., 2002; Aboa et al., 2008; Malviya et al., 2010; Keter and Mutiso, 2012; Moradi et al., 2018; Skalli et al., 2019).

The prevalence and severity of diabetes have invoked a resurgence of interests in bioprospecting medicinal plants for novel plant-based therapeutics with anti-hyperglycaemic activities. Many plant extracts and their isolated bioactive compounds have demonstrated significant antidiabetic activity when assessed using different types of experimental methods, and in some cases they exhibited more potency than conventional drug compounds (Kooti et al., 2016). The main active constituents of medicinal plants with antidiabetic activity include glycosides, alkaloids, polysaccharides, galactomannan gum, steroids, guanidine, carbohydrates, terpenoids, peptidoglycan, hypoglycin, glycopeptides, amino acids and inorganic ions. These phytochemicals downregulate or upregulate various metabolic cascades that directly or indirectly affect the glucose level in the human body. Patel et al. (2012) reported that anti-hyperglycaemic effects of some medicinal plants are due to their ability to restore the function of pancreatic tissues by causing an increase in insulin production. Several authors have reviewed the potential of plant secondary metabolites in suppressing and controlling blood sugar to normal levels through diverse mechanisms. However, most of the phytotherapeutic constituents have restricted bioavailability and efficacy as a result of their insolubility.

Nanotechnology is an emerging field that is presently revolutionizing medical research because of its utilization of nanomaterials, which are more biocompatible than conventional therapeutics, thereby bringing noteworthy developments in the treatment and diagnosis of diseases. A nanoparticle (NP), which is in essence a solid colloidal particle, is commonly defined as a discrete entity with at least one dimension being 100 nm (Garcia et al., 2010). Nanoparticles (NPs) demonstrate improved properties by virtue of their morphological parameters such as structure, shape, sizes (1–100 nm) and other amenable functionalities, which can be modulated for increased biocompatibility (Rafique et al., 2017; Bagyalakshmi and Haritha., 2017; Min et al., 2015). They are generally classified into organic and inorganic NPs and can be further categorized into different types based on the size, morphology and physical and chemical attributes. Such categorizations include carbon-based, ceramic, metal, semiconductor, polymeric and lipid-based nanoparticles. Organic NPs are solid particles that are made up of carbon-based compounds (mainly lipids or polymeric) and range in diameter from 10 nm to 1 μm (Kumar and Lal, 2014: Rafique et al., 2017). Inorganic NPs cover a broad range of substances such as elemental metals, metal oxides as well as metal salts. They incorporate semiconductor NPs (zinc sulphide – ZnS, zinc oxide – ZnO), metallic NPs (copper – Cu, gold – Au, aluminium – Al, silver – Ag) and magnetic NPs (copper – Co, iron –Fe, nickel – Ni). They have been given relatively more attention, and metal nanoparticles such as gold, iron, zinc, silver and nanoparticulate metal oxides have been extensively evaluated for their use as drug delivery systems in biomedical applications (Alkaladi et al., 2014).

The use of antidiabetic medicinal plants in the biosynthesis has received substantial research focus as they are considered a valuable alternative to hazardous synthetic compounds.

2.2 NANOTECHNOLOGY AND DIABETES MANAGEMENT

Diabetes mellitus has been known for more than 2,000 years and is defined as a group of metabolic disorders characterized by a complete or a relative lack of insulin (Woldu and Lenjisa, 2014). The increase in the prevalence of DM is due to three influencing factors: ethnicity, lifestyle and age. Diabetes mellitus is a life-long disease that is characterized by hyperglycaemia due to disordered metabolism of glucose (Madsen-Bouterse and Kowluru, 2008). The World Health Organization (WHO, 2016) estimates that over 170 million people worldwide are afflicted with this chronic condition, and it is projected that in the year 2030 this number will rise to over 360 million (Wild et al., 2004).

Table 2.1 shows types of diabetes and their brief description. Type II diabetes mellitus (T2DM) is more worrisome as it accounts for 90% of global diabetes mellitus cases (Badeggi et al., 2020). With T2DM, your body either resists the effects of insulin, which is a hormone that regulates the movement of sugar into your cells, or does not produce enough insulin to maintain normal glucose level (Badeggi et al., 2020). Lysy et al. (2016) reported that 425 million population of adults age between 20 and 79 suffered from diabetes. This is equivalent to 9.9% of the world's population

TABLE 2.1
Brief Descriptions of the Types of Diabetes

Type of diabetes	Brief description
Type 1	Accounts for 5–10% of all diagnosed cases, which is diagnosed earlier usually in children and youth and is characterized by the deficiency in the production of insulin. In this case, there is a destruction of islet beta cells mostly attributed to autoimmune aetiology (Madsen-Bouterse and Kowluru, 2008; Harsoliya et al., 2012)
Type 2	Has higher incidence (90–95% of the cases) and is characterized by the reduced production of insulin and/or by insulin resistance, affecting mainly muscle, liver and adipose tissue, resulting in inappropriate levels of circulating glucose (Madsen-Bouterse and Kowluru, 2008; Harsoliya et al., 2012)
Gestational diabetes	It is observed during pregnancy in a small number of women caused by interference of placental hormones interference with insulin receptor resulting in inappropriate elevated glucose levels (Madsen-Bouterse and Kowluru, 2008; Harsoliya et al., 2012)
Other specific types	Genetic disorder of β-cell function (MODY, mitochondrial DNA), genetic disorders in insulin action (lipoatrophic diabetes), exocrine pancreas diseases (pancreatitis, hemochromatosis), endocrinopathies (acromegaly, Cushing's syndrome), drug-induced (glucocorticoids, tiazidics), infections (cytomegalovirus, congenital rubeola), uncommon immunological forms (insulin receptor antibodies) and other genetic syndromes (Maraschin et al., 2010)

(Renner et al., 2020). According to an estimate by Cho et al. (2018) and Badeggi et al. (2020), if this disease is not properly managed, it would drastically increase by 48% in 28 years. Even though drugs are available on the market, they are costly and continuous administration causes side effects such as heart failure, diarrhoea, damage to the liver, dropsy, weight gain, abdominal pain, flatulence, and hyperglycaemia, necessitating the need for more potent and newer remedies (Badeggi et al., 2020; Lorenzati, et al., 2010; Ahmad et al., 2018).

To date, painless and simpler routes for insulin administration are still in demand in controlling diabetes. Conventional drug delivery systems still have several limitations, including low potency, improper ineffective dosage and limited specificity for the target (Souto et al., 2019). Nanotechnology, a field that involves nanostructures, nanoparticle design, nanomaterials and their applications in humans, is increasing in importance in diabetics research in the recent decade (Gupta, 2017). It has facilitated the development of novel glucose measurement and insulin delivery modalities, which hold the potential to dramatically improve quality of life for diabetics (DiSanto et al., 2015). Nanoparticle-based delivery systems have been proposed to overcome the enzymatic degradation in the stomach and therefore to improve permeation through the gastrointestinal tract, in order to improve oral insulin absorption (Wong et al., 2017).

Sharma et al. (2015) discussed the limitation and applications of nanoparticles in delivering insulin to the targeted organs or/and tissues. Nanotechnology is being used in non-invasive methods to engineer more effective vaccine and insulin delivery gene and cell therapies for type 1 diabetes (Veiseh et al., 2016). Veiseh et al. (2016) analysed the state of the approaches and discussed critical issues for their translation to clinical practice in managing diabetes with nanomedicine, more specifically their challenges and opportunities. Souto et al. (2019) have discussed different types of nanoparticles such as lipid nanoparticles and polymeric niosomes, dendrimers, liposomes, micelles, nano-emulsions and also drug nanosuspensions for improved delivery of different oral hypoglycaemic agents in comparison with conventional therapies. In both types of diabetes, it is necessary to ameliorate the symptoms of hyperglycaemia to reduce the rate of disease progression and its associated complications, or to manage the blood sugar level by using insulin (Souto et al., 2019; Neef and Miller, 2017; Ismail and Csóka, 2017).

2.3 PLANT MATERIAL AND NEW PERSPECTIVE OF NANOPARTICLES AS AN ANTIDIABETIC AGENT

In spite of the significant advancements made in the control and treatment of diabetes mellitus by conventional hypoglycaemic agents, there are still some limitations and side effects that are associated with synthetic antidiabetic drugs. Effective diabetes management without side effects remains a major challenge to the healthcare system. The common side effects linked to oral hypoglycaemic drugs include hypoglycaemia, weight gain, gastrointestinal disorders, lactic acid intoxication, peripheral oedema and impaired liver function (Dai et al., 2018; Liu and Yang, 2019; Ganesan and Sultan, 2020). Traditional medicines are widely used and preferred because of their perceived efficacy, relatively fewer side effects and comparatively lower costs.

2.3.1 Some Antidiabetic Plant Extracts and Their Bioactive Compounds

Antidiabetic plant extracts and their bioactive compounds target the liver, pancreas, intestine, adipose tissue and muscle, where they exert their respective modes of action. Some factors negatively affect the bioavailability of plant-based products in living systems (Samadder and Khuda-Bukhsh, 2014; Furman et al., 2020). Crude plant extracts from several plant species belonging to varied plant families, including *Aegle marmelos* L. (Rutaceae), *Balanites aegyptiaca* L., Del. (Zygophyllaceae), *Boerhavia diffusa* (Nyctaginaceae), *Camellia sinensis* L. (Theaceae), *Helicteres isora* Linn. (Sterculiaceae), *Melissa officinalis* L. (Lamiaceae), *Phaseolus vulgaris* L. (Fabaceae), *Rosmarinus officinalis* L. (Lamiaceae), *Khaya senegalensis* (Desr.) A. Juss. (Meliaceae), *Tamarindus indica* L. (Fabaceae), *Mitragyna inermis* (Willd) O Ktze. (Rubiaceae) and *Vaccinium myrtillus* L. (Ericaceae), have exhibited significant hypoglycaemic activity (Funke and Melzing, 2006; Ayodhya et al., 2010; Moradi et al., 2018; Hamza et al., 2019). Numerous antidiabetic compounds such as beta-pyrazol-1-ylalanine, epigallocatechin gallate, roseoside, cinchonain Ib, glycyrrhetinic acid, leucocyandin 3-*O*-beta-D-galactosyl cellobioside, dehydrotrametenolic acid, leucopelargonidin-3-*O*-alpha-L-rhamnoside, strictinin, pedunculagin and isostrictinin and epicatechin, demonstrated significant insulin-mimetic and antidiabetic activity with varied modes of action and some with greater efficacy than conventional hypoglycaemic agents (Saxena and Vikram, 2004; Bnouham et al., 2006; Ko et al., 2007; Qa'dan et al., 2009; Ayodhya et al., 2010; Chauhan et al., 2010; Frankish et al., 2010; Liu et al., 2020).

2.3.2 Green Synthesis of Nanoparticles as Potential Antidiabetic Agents

Green synthesis of nanoparticles has added advantage when it is combined with metal precursors to give the metal nanoparticles. The stability of green synthesis metal nanoparticles is achieved using plant extracts which act as both reducing and stabilizing agents (Ashwini and Mahalingam, 2015; Latha et al., 2015). The resources are obtained from plant materials (Saravanakumar, et al., 2017). The plant extracts employed in the normal decoction method are more suitable for the green metal nanoparticle as a capping and stabilization agent, requiring no external stabilizers (Jha and Prasad, 2010; Ashwini and Mahalingam, 2015). The reduction of metal nanoparticles with the help of plant extracts biomolecules is an eco-friendly and a low-cost approach without any side effects (Ashwini and Mahalingam, 2015; Latha et al., 2015). Nanoparticles using plant materials, algae, fungi and several useful microorganisms are in focus in the present scientific world (Patra et al., 2019). Conventional physical and chemical method is one of the beneficial green technology methods, as it provides an environmentally friendly way of synthesizing nanoparticles, with no requirement of toxic and harmful chemicals, and uses a cost-effective approach (Patra and Baek., 2014; Veerasamy et al., 2011; Nadaroğlu et al., 2017).

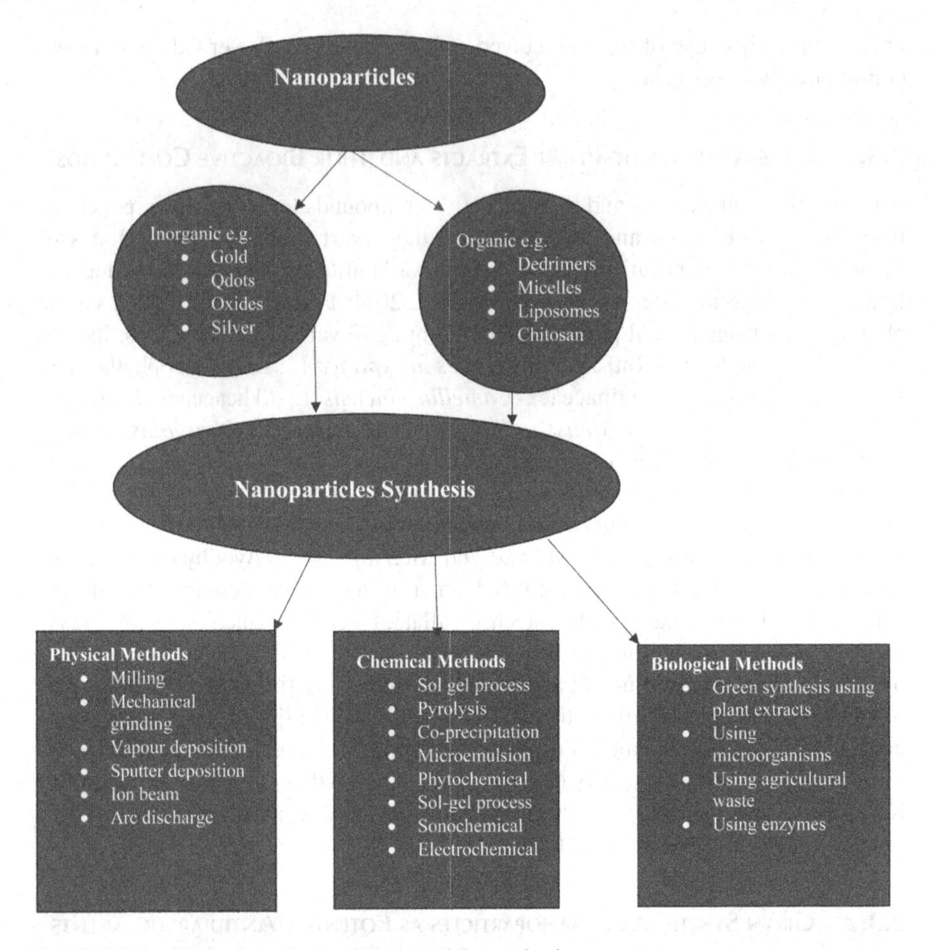

FIGURE 2.1 General methods of nanoparticle synthesis.

Figure 2.1 shows general methods of nanoparticle synthesis. Microorganisms and plant molecules such as phenolic compounds, proteins, alkaloids, enzymes, amines and pigments perform nanoparticle synthesis by reduction (Nadaroğlu et al., 2017).

Nanoparticle can limit fluctuations, reduce side effects, decrease dosage frequency and also improve patients' compliance (Souto et al., 2019). Nanoparticles are currently being explored in an attempt to achieve improved bioavailability and prolonged desired effects of antidiabetic herbal remedies in targeted organs. Studies by Samadder and Khuda-Bukhsh (2014) show that potential antidiabetic properties of nano-encapsulated forms of *Gymnema sylvestre* and *Syzygium jambolanum* have relatively more anti-hyperglycaemic effects than unencapsulated counterparts in various experimental models. According to Jamdade et al. (2019), nature has an infinite collection of therapeutic plants, which serve as a repository of bioactive principles that form the foundation to complementary and alternative medicine. Natural products are currently being explored in the plant-mediated green synthesis of nanoparticles, and the approach is developing into a new and essential branch of nanotechnology. The use of plant extracts in the production of silver nanoparticles has drawn attention due to its rapid, eco-friendly, non-pathogenic and single-step technique for the biosynthetic processes (Bagyalakshmi and Haritha, 2017).

Silver nanoparticles and *Pterocarpus marsupium* are found to be effective for antidiabetic activity (Bagyalakshmi and Haritha, 2017). The study by Prabhu et al. (2018) shows that AgNPs and leaf extract of *Pouteria sapota* have efficient antidiabetic activity in the rat model of diabetes and might have the potential for development of medical applications. Zinc oxide nanoparticles synthesized using plant extracts of *Tamarindus indica* and *Moringa oleifera* showed higher antidiabetic activity, inhibiting α-glucosidase and α-amylase, which are the important enzymes in carbohydrate metabolism (Rehana et al., 2017). Silver nanoparticles (AgNPs) using stem extracts of *Musa paradisiaca* were also effective against diabetes in the streptozotocin-induced diabetes rat model (Anbazhagan et al., 2017).

Garg et al. (2016) reported that AgNPs using *Zingiber officinale* ethanolic extract had a notable effect on blood glucose lowering in rats with STZ-induced diabetes. Some of the metal nanoparticles syntheses using extracts of different plants and antidiabetic activity are shown in Table 2.2.

Biologically synthesized AuNPs using plant extracts showed remarkable antidiabetic activity extracts of *Turbinaria conoides*, *Gymnema sylvestre*, *Sargassum swartzi*, *Chamalcostus cuspidatus*, *Hericium erinaceus*, *Sambucus nigra* and *Cassia fistula* (Badeggi et al., 2020). Studies done by Malapermal et al. (2015) and Elobeid (2016) show that gold/silver NPs of cinnamon and *Ocimum basilicum* extract lower glucose levels. Guavanoic acid, curcumin, hesperidin, diosmin and naringin, phloridzin, an antidiabetic agent found in fruits and its escin, gymnemic acid resveratrol and aglycon, are responsible for the biosynthesis of AuNPs (Badeggi et al., 2020). Tyrosine, tryptophan, chondroitin sulphate and chitosan are some of the compounds that have been used in the formation of AuNPs with antidiabetic properties (Badeggi et al., 2020).

2.4 CONCLUSION

The therapeutic value of medicinal plant created a new platform for researchers to explore the potential for herbal medicine and nanoscience as effective biomedical, and for improving existing drug delivery systems in diabetic treatment. Nanoparticles using plant extracts are the emerging novel substrates for large-scale production. The novelty of green synthesis in the field of nanotechnology is highly appreciable where the synthesis of nanoparticles has proven potential outcomes in pharmaceuticals. Therefore, this chapter provides some details of several researches that could be used for the management of antidiabetic. To date, several studies show that nanoparticles can be used for various purposes such as drug delivery at the site of disease to improve the uptake of poorly soluble drugs, targeting drugs to a specific site and increasing the drug bioavailability.

TABLE 2.2
Metal Nanoparticles Synthesis Using Different Plant Extracts with Antidiabetic Activity

Plant extracts	Antidiabetic activities of green synthesized metal nanoparticles	References
Saraca asoca (Roxb.) W.J .de Wilde (Fabaceae)	AuNP and AgNP inhibit α-amylase activity with IC_{50} values of 1.5 and 0.35 mM	Patra et al., 2018
Chamaecostus Cuspidatus (Nees & Mart.) C. Specht & D.W. Stev. (Costaceae)	1.5 mg/kg gold nanoparticles oral administration for 10 days in STZ-induced animals	Ponnanikajamideen et al., 2018
Moricandia nitens (Viv.) E.A. Durand & Barratte (Brassicaceae)	AuNP inhibit α-glucosidase with an IC_{50} value of 159.3 μg/ml	Soliman et al., 2018
Calophyllum tomentosum Wight (Calophyllaceae)	α-Amylase and α-glucosidase inhibitory assay of AgNPs results in α-glucosidase DPPIV which is greatly inhibited by Ct AgNPs compared to α-amylase	Govindappa et al., 2018
Bauhinia variegata L. (Fabaceae) flower	The IC_{50} value of AgNPs for α-amylase inhibition was 38 μg/ml	Johnson et al., 2018
Pterocarpus marsupium Roxb. (Fabaceae)	AgNPs with an IC_{50} value of 71.14% inhibit α-amylase	Bagyalakshmi and Haritha, 2017
Punica granatum L. (Lythraceae) leaf	AgNPs inhibit α-amylase and α-glucosidase with IC_{50} values of with 65.2 and 53.8 mg/ml, respectively	Saratale et al., 2017
Withania somnifera (L.) Dunal (Solanaceae)	1 mg/kg PtNP oral administration for 28 days in STZ-induced animal results in reduction in blood glucose level to normal of 117.34 ± 4.18 mg/dl	Li et al., 2017
Punica granatum L. (Punicaceae) Leaf	AgNPs antidiabetic potential exhibited effective inhibition against α-amylase and α-glucosidase with IC_{50} values of 65.2 and 53.8 μg/ml, respectively	Saratale et al., 2017
Ficus glomerata Roxb. (Moraceae)	AgNPs inhibit α-amylase and α-glucosidase with IC_{50} values of 95.28 and 97.14 μg/ml, respectively	Das et al., 2017
Ocimum basilicum L. and *O. sanctum* L. (Lamiaceae)	α-Glucosidase of the AgNPs synthesized from *O. sanctum* and *O. basilicum* indicating the activity (89.31 ± 5.319% and 79.74 ± 9.51%), respectively	Malapermal et al., 2017
Olea europaea L. (Oleaceae) leaves	2.5, 5, 10 mg/kg b.w. of NP oral administration for 60 days in alloxan-induced animals	Karthick et al., 2014
Lonicera japonica Thunb. (Caprifoliaceae) leaves	Antidiabetic ability of AgNPs was shown by the effective inhibition against α-amylase and α-glucosidase with IC_{50} values of 54.56 and 37.86 μg/ml, respectively	Balan et al., 2016

(*Continued*)

TABLE 2.2 (CONTINUED)
Metal Nanoparticles Synthesis Using Different Plant Extracts with Antidiabetic Activity

Plant extracts	Antidiabetic activities of green synthesized metal nanoparticles	References
Hibiscus sabdariffa L. (Malvaceae) leaf	Small-sized ZnO-NPs stabilized by plant metabolites had better antidiabetic effect on streptozotocin (STZ)-induced diabetic mice than that of large-sized ZnO particles	Bala et al., 2015
Heritiera fomes Buch.-Ham. (Malvaceae)	AgNPs inhibited the activity of α-amylase with an IC_{50} value of 490 mg/ml and α-glucosidase with an IC_{50} value of 385 mg/ml	Vishnu and Murugesan, 2013
Dioscorea bulbifera L. (Dioscoreaceae) tubers	CuNPs showed α-amylase with an IC_{50} value of up to 38.70 ± 1.45% and 99.09 ± 0.15% inhibition against pure α-glucosidase	Ghosh et al., 2015
Tephrosia tinctoria Pers. Fabaceae	α-Amylase and α-glucosidase inhibitory activity of AgNPs at 50 µg/ml shows 87.02 ± 1.17% and 74.34 ± 0.74%, respectively	Krishnasamy and Periyasamy, 2014
Tamarindus indica L. (Fabaceae) seed	Aqueous extract of *T. indica seed* was found to have potent antidiabetogenic activity that reduces blood sugar level in streptozotocin (STZ)-induced diabetic male rat	Maiti et al., 2004
Gnidia glauca (Fresen.) Gilg (Thymelaeaceae) leaf and stem	*G. glauca* stem synthesized CuNPs exhibited highest α-glucosidase inhibition up to 88.60±0.78% followed by *G. glauca* leaf synthesized CuNPs (86.58±3.26%)	Jamdade et al.,2019
Lonicera japonica Thunb.	α-Amylase and α-glucosidase with an IC_{50} value of 54.56 and 37.86 µg/ml, respectively	Kavitha et al., 2014

REFERENCES

Aboa, K.A., Fred-Jaiyesimib, A.A., Jaiyesimib, A.E.A. 2008. Ethnobotanical studies of medicinal plants used in the management of diabetes mellitus in South Western Nigeria. *Journal of Ethnopharmacology* 115: 67–71.

Ahmad, T., Azmi, M., Irfan, M., Moniruzzaman, M., Asghar, A., Bhattacharjee, S. 2018. Green synthesis of stabilised spherical shaped gold nanoparticles using novel aqueous *Elaeis guineensis* (Oil palm) leaves extract. *Journal of Molecular Structure* 1159: 167–173.

Alkaladi, A., Abdelazim A.M., Afifi, M. 2014. Antidiabetic activity of zinc oxide and silver nanoparticles on streptozotocin-induced diabetic rats. *International Journal of Molecular Sciences* 15: 2015–2023.

Anbazhagan, P., Murugan, K., Jaganathan, A., Sujitha, V., Samidoss, C.M., Jayashanthani, S., Amuthavalli, P., Higuchi, A., Kumar, S., Wei, H., Nicoletti, M., Canale, A., Benelli, G. 2017. Mosquitocidal, antimalarial and antidiabetic potential of *Musa paradisiaca*-synthesized silver nanoparticles: *In vivo* and *in vitro* approaches. *Journal of Cluster Science* 28: 91–107.

Ashwini, D., Mahalingam, G. 2015. Green ssynthesised metal nanoparticles, scharacterisation and its antidiabetic activities: A review. *Research Journal of Pharmacy and Technology* 13: 468–474.

Ayodhya, S., Kusum, S., Anjali, S. 2010. Hypoglycaemic activity of different extracts of various herbal plants. *International Journal of Research in Ayurveda and Pharmacy* 1: 212–224.

Badeggi, U. M., Ismail, E., Adeloye, A.O., Botha, S., Badmus, J.A., Marnewick, J.L., Cupido, C.N., Hussein. A.A. 2020. Green synthesis of gold nanoparticles capped with procyanidins from *Leucosidea sericea* as potential antidiabetic and antioxidant agents. *Biomolecules* 10: 452–476.

Bagyalakshmi, J., Haritha, H. 2017. Green synthesis and scharacterisation of silver nanoparticles using *Pterocarpus marsupium* and assessment of its *in vitro* antidiabetic activity. *American Journal of Advanced Drug Delivery* 5: 118–130.

Bala, N., Saha, S., Chakraborty, M., Maiti, M., Das, S., Basu, R., Nandy, P. 2015. Green synthesis of zinc oxide nanoparticles using *Hibiscus subdariffa* leaf extract: Effect of temperature on synthesis, anti-bacterial activity and antidiabetic activity. *RSC Advances* 5: 4993–5003.

Balan, K., Qing, W., Wang, Y., Liu, X., Palvannan, T., Wang, Y., Ma, F., Zhang, Y. 2016. Antidiabetic activity of silver nanoparticles from green synthesis using *Lonicera japonica* leaf extract. *RSC Advances* 6: 40162–40168.

Bilal, M., Iqbal, M.S., Shah, S.B., Rasheed, T., Iqbal, H.M.N. 2018. Diabetic complications and insight into antidiabetic potentialities of ethno-medicinal plants. *Recent Patents on Inflammation and Allergy Drug Discovery* 12:7–23.

Bnouham, M., Ziyyat, A., Mekhfi, H., Tahri, A., Legssyer, A. 2006. Medicinal plants with potential antidiabetic activity-a review of ten years of herbal medicine research (1990–2000). *International Journal of Diabetes and Metabolism* 14: 1–25.

Chauhan, A., Sharma, P.K., Srivastava, P., Kumar, N., Duehe, R. 2010. Plants having potential antidiabetic activity: A review. *Der Pharmacia Lettre* 2: 369–387.

Cho, N., Shaw, J.E., Karuranga, S., Huang, Y., da Rocha Fernandes, J.D., Ohlrogge, A.W., Malanda, B. 2018. IDF diabetes atlas: Global estimates of diabetes prevalence for 2017 and projections for 2045. *Diabetes Research and Clinical Practice* 138: 271–281.

Dai, X., Luo, Z.C., Zhai, L., Zhao, W.P., Huang, F. 2018. Adverse drug events associated with low-dose (10 mg) versus high-dose (25 mg) empagliflozin in patients treated for type 2 diabetes mellitus: A systematic review and meta-analysis of srandomised controlled trials. *Diabetes Therapy* 9: 753–770.

Das, M.P., Rebecca, L.J., Das, M.P. 2017. sCharacterisation of antidiabetic activity of silver nanoparticles using aqueous solution of *Ficus glomerata* (fig) gum. *International Journal of Pharma and Bio Sciences* 8:424–429.

DiSanto, R., M., Subramanian, V., Gu, Z., 2015. Recent Advances in Nanotechnology for Diabetes Treatment. Wiley interdisciplinary reviews. *Nanomedicine and nanobiotechnology* 7: 548–564.

Eddouks, M., Maghrani, M., Lemhadri, A., Ouahidi, M.L., Jouad, H. 2002. Ethnopharmacological survey of medicinal plants used for the treatment of diabetes mellitus, hypertension and cardiac diseases in the south-east region of Morocco (Tafilalet). *Journal of Ethnopharmacology* 82: 97–103.

Elobeid, M.A. 2016. Amelioration of streptozotocin induced diabetes in rats by eco-friendly composite nano-cinnamon extract. *Pakistan Journal of Zoology* 48: 645–650.

Ernst, E. 1997. Plants with hypoglycemic activity in humans. *Phytomedicine* 4: 73–78.

Frankish, N., de Sousa Menezes, F., Mills, C., Sheridan, H. 2010. Enhancement of insulin release from the beta-cell line INS-1 by an ethanolic extract of *Bauhinia variegata* and its major constituent roseoside. *Planta Medica* 76: 995–997.

Funke, I., Melzing, M.F. 2006. Traditionally used plants in diabetes therapy—phytotherapeutics as inhibitors of a-amylase activity. *Revista Brasileira de Farmacognosia* 16: 1–5.

Furman, B.L., Candasamy, M., Bhattamisra, S.K., Veettil, S.K. 2020. Reduction of blood glucose by plant extracts and their use in the treatment of diabetes mellitus; discrepancies in effectiveness between animal and human studies. *Journal of Ethnopharmacology* 247: 112264.

Ganesan, K., Sultan, S. 2020. Oral hypoglycemic medications. In StatPearls [Internet], Treasure Island, FL: StatPearls Publishing. Available from: https://www.ncbi.nlm.nih.gov/books/NBK482386/.

Garcia, M., Forbe, T., Gonzalez, E., 2010. Potential applications of nanotechnology in the agro-food sector. *Cienc e Tecnologia de Alimentos* 30: 573–581.

Garg, A., Pandey, P., Sharma, P., Shukla, A. 2016. Synthesis and scharacterisation of silver nanoparticle of ginger rhizome (*Zingiber officinale*) extract: Synthesis, scharacterisation and antidiabetic activity in streptozotocin induced diabetic rats. *European Journal of Biomedical and Pharmaceutical Sciences* 3: 605–611.

Ghosh S, More P, Nitnavare R, Jagtap S, Chippalkatti R, Derle, A., Kitture, R., Asok, A., Kale, S., Ramanamurthy, B., Bellare, J., Chopade, B.A. 2015. Antidiabetic and Antioxidant properties of copper nanoparticles synthesised by medicinal plant Dioscorea bulbifera. *Journal of Nanomedicine and Nanotechnology* S 6: 007. DOI: 10.4172/2157-7439.S6-007.

Govindappa, M., Hemashekhar, B., Arthikala, M.K., Ravishankar Rai, V., Ramachandra, Y.L. 2018. Characterisation, antibacterial, antioxidant, antidiabetic, anti-inflammatory and antityrosinase activity of green ssynthesised silver nanoparticles using *Calophyllum tomentosum* leaves extract. *Results in Physics* 9: 400–408.

Grover, J.K., Yadav, S., Vats, V., 2002. Medicinal plants of India with antidiabetic potential. *Journal of Ethnopharmacology* 81: 81–100.

Gupta, R. 2017. Diabetes treatment by nanotechnology. *Journal of Biotechnology and Biomaterials* 7: 268. DOI: 10.4172/2155-952X.1000268.

Hamza, N., Berke, B., Umar, A., Cheze, C., Gin, H., Moore, N., 2019. A review of Algerian medicinal plants used in the treatment of diabetes. *Journal of Ethnopharmacology* 23828: 111841. DOI: 10.1016/j.jep.2019.111841.

Harsoliya, M.S., Patel, V.M., Modasiya, M., Pathan, J.K., Chauhan, A., Parihar, M., Ali, M. 2012. Recent advances and applications of nanotechnology in diabetes. *International Journal of Pharmaceutical Biological Archive* 3: 255–261.

Hussain, I., Singh, N.B., Singh, A., Singh, H., Singh, S.C. 2016. Green synthesis of nanoparticles and its potential application. *Biotechnology Letters* 38: 545–560.

Ismail, R., Csóka, I. 2017. Novel strategies in the oral delivery of antidiabetic peptide drugs: Insulin, GLP 1 and its analogs. *European Journal of Pharmaceutics and Biopharmaceutics* 115: 257–267.

Jamdade, D.A., Rajpali, D., Joshi, K.A., Kitture, R., Kulkarni, A.S., Shinde, V.S., Bellare, J., Babiya, K.R., Ghosh, S. 2019. *Gnidia glauca*- and *Plumbago zeylanica*-mediated synthesis of novel copper nanoparticles as promising antidiabetic agents. *Advances in Pharmacological Sciences* 9080279. DOI: 10.1155/2019/9080279.

Jha, A.K., Prasad, K. 2010. Green synthesis of silver nanoparticles using cycas leaf. *International Journal of Green Nanotechnology: Physics and Chemistry* 1: 110–117.

Johnson, P., Krishnan, V., Loganathan, C., Govindhan, K., Raji, V., Sakayanathan, P., Vijayan, S., Sathishkumar, P., Palvannan. T. 2018. Rapid biosynthesis of *Bauhinia variegata* flower extract-mediated silver nanoparticles: An effective antioxidant scavenger and α-amylase inhibitor. *Artificial Cells, Nanomedicine, and Biotechnology* 46: 1488–1494.

Karthick, V., Kumar, V.G., Dhas, T.S., Singaravelu, G., Sadiq, A.M., Govindaraju, K. 2014. Effect of biologically ssynthesised gold nanoparticles on alloxan-induced diabetic rats-An *in vivo* approach. *Colloids Surfaces B Biointerfaces* 122: 505–511.

Kavitha, S., Dhamodharan, M., Kannan, S. 2014. Synthesis, characterisation and anti-diabetic, antibacterial activity of silver nanoparticle in terpenoid for *Andrographis paniculata*. *International Journal of Science, Engineering and Technology Research* 3: 9467–9473.

Keter, L.K., Mutiso, P.C. 2012. Ethnobotanical studies of medicinal plants used by traditional health practitioners in the management of diabetes in lower eastern province, Kenya. *Journal of Ethnopharmacology* 139: 74–80.

Khan, V., Najmi, A.K., Akhtar, M., Aqil, M., Mujeeb, M., Pillai, K.K. 2012. A pharmacological appraisal of medicinal plants with antidiabetic potential. *Journal of Pharmacy and Bioallied Sciences* 4: 27–42.

Ko, B.S., Jang, J.S., Hong, S.M., Sung, S.R., Lee, J.E., Lee, M.Y., Jeon, W.K., Park, S. 2007. Changes in components, glycyrrhizin and glycyrrhetinic acid, in raw *Glycyrrhiza uralensis* Fisch, modify insulin ssensitising and insulinotropic actions. *Bioscience, Biotechnology, and Biochemistry* 71: 1452–1461.

Kooti, W., Farokhipour, M., Asadzadeh, Z., Ashtary-Larky, D., Asadi-Samani, M. 2016. The role of medicinal plants in the treatment of diabetes: A systematic review. *Electronic Physician* 8: 1832–1842.

Krishnasamy, R., Periyasamy S. 2014. Green synthesis of silver nanoparticle using *Tephrosia tinctoria* and its antidiabetic activity. *Materials Letters* 138: 251–254.

Kumar, R., Lal, S. 2014. Synthesis of organic nanoparticles and their applications in drug delivery and food nanotechnology: A review. *Nanomaterials and Molecular Nanotechnology* 3: 4.

Latha, M., Sumathi, M., Manikandan, R., Arumugam, A., Prabhu, N.M. 2015. Biocatalytic and antibacterial svisualisation of green ssynthesised silver nanoparticles using *Hemidesmus indicus*. *Microbial Pathogenesis* 82: 43–49.

Li, Y., Zhang, J., Gu, J., Chen, S., Wang, C., Jia, W. 2017. Biosynthesis of polyphenol-stabilised nanoparticles and assessment of antidiabetic activity. *Journal of Photochemistry and Photobiology B: Biology* 169: 96–100.

Liu, R., Su, C., Xu, Y., Shang, K., Sun, K., Li, C., Lu, J. 2020. Identifying potential active components of walnut leaf that action diabetes mellitus through integration of UHPLC-Q-Orbitrap HRMS and network pharmacology analysis. *Journal of Ethnopharmacology.* 253: 1–9.

Liu, Z., Yang, B. 2019. Drug development strategy for type 2 diabetes: Targeting positive energy balances. *Current Drug Targets* 20: 879–890.

Lorenzati, B., Zucco, C., Miglietta, S., Lamberti, F., Bruno, G., 2010. Oral hypoglycemic drugs: Pathophysiological basis of their mechanism of action. *Pharmaceuticals* 3: 3005–3020.

Lysy, P.A., Corritore, E., Sokal, E.M. 2016. New insights into diabetes cell therapy. *Current Diabetes Reports* 16, 38.

Madsen-Bouterse, S.A., Kowluru, R.A. 2008. Oxidative stress and diabetic retinopathy: pathophysiological mechanisms and treatment perspectives. *Reviews in Endocrine and Metabolic Disorders* 9: 315–327.

Maiti, R., Jana, D., Das, U.B., Ghosh, D. 2004. Antidiabetic effect of aqueous extract of seed of *Tamarindus indica* in streptozotocin-induced diabetic rats. *Journal of Ethnopharmacology* 92: 85–91.

Malapermal, V., Mbatha, N., Gengan, R., Anand, K. 2015. Biosynthesis of bimetallic Au-Ag nanoparticles using *Ocimum basilicum* (L.) with antidiabetic and antimicrobial properties. *Advanced Materials Letters* 6: 1050–1057.

Malapermal, V., Botha, I., Krishna, S.B.N., Mbatha, J.N. 2017. Enhancing antidiabetic and antimicrobial performance of *Ocimum basilicum*, and *Ocimum sanctum* (L.) using silver nanoparticles. *Saudi Journal of Biological Sciences* 24: 1294–305.

Malviya, N., Jain, S., Malviya, S. 2010. Antidiabetic potential of medicinal plants. *Acta Poloniae Pharmaceutica* 67: 113–118.

Maraschin, J.F., Murussi, N. Witter, V., Silveiro, S. P. 2010. Diabetes mellitus classification. *Clinical Update* 95: 40–47.

Min, Y., Caster, J.M., Eblan, M.J., Wang, A.Z. 2015. Clinical translation of nanomedicine. *Chemical Reviews* 115: 11147–11190.

Moradi, B., Abbaszadeh, S., Shahsavari, S., Alizadeh, M., Beyranvand, F. 2018. The most useful medicinal herbs to treat diabetes. *Biomedical Research and Therapy* 5: 2538–2551.

Nadaroğlu, H., Alayli Güngör, A., İnce, S. 2017. Synthesis of nanoparticles by green synthesis method. *International Journal of Innovative Research and Reviews* 1: 6–9.

Neef, T., Miller, S.D. 2017. Tolerogenic nanoparticles to treat islet autoimmunity. *Current Diabetes Reports* 17: 84–105.

Patel, D.K., Prasad, S.K., Kumar, R., Hemalatha, S. 2012. An overview on antidiabetic medicinal plants having insulin mimetic property. *Asian Pacific Journal of Tropical Biomedicine* 2: 320–330.

Patra, J.K., Baek, K-H. 2014. Green nanobiotechnology: Factors affecting synthesis and scharacterisation techniques. *Journal of Nanomaterials*. 2014, 1–12.

Patra, J., K., Das G., Shin, H-S. 2019. Facile green biosynthesis of silver nanoparticles using *Pisum sativum* L. outer peel aqueous extract and its antidiabetic, cytotoxicity, antioxidant, and antibacterial activity. *International Journal of Nanomedicine* 14: 6679–6689.

Patra, N., Kar, D., Pal, A., Behera, A. 2018. Antibacterial, anticancer, antidiabetic and catalytic activity of bio-conjugated metal nanoparticles. *Advances in Natural Sciences: Nanoscience and Nanotechnology* 9: 1–5.

Ponnanikajamideen, M., Rajeshkumar, S., Vanaja, M., Annadurai, G. 2018. *In-vivo* antidiabetic and wound healing effect of antioxidant gold nanoparticles ssynthesised using insulin plant (*Chamaecostus cuspidatus*). *Canadian Journal of Diabetes* 43: 82–89.

Prabhu, S., Vinodhini, S., Elanchezhiyan, C., Rajeswari, D. 2018. Evaluation of antidiabetic activity of biologically ssynthesised silver nanoparticles using *Pouteria sapota* in streptozotocin-induced diabetic rats. *Journal of Diabetes* 10: 28–42.

Qa'dan, F., Verspohl, E.J., Nahrstedt, A., Petereit, F., Matalka, K.Z. 2009. Cinchonain Ib isolated from *Eriobotrya japonica* induces insulin secretion *in vitro* and *in vivo*. *Journal of Ethnopharmacology* 124: 224–227.

Rafique, M., Sadaf, I., Rafique M. S., & Tahir M. B., 2017. A review on green synthesis of silver nanoparticles and their applications, *Artificial Cells, Nanomedicine, and Biotechnology* 45:7, 1272–1291.

Rehana, D., Mahendiran, D., Kumar, R.S., Rahiman, A.K. 2017. *In vitro* antioxidant and antidiabetic activities of zinc oxide nanoparticles ssynthesised using different plant extracts. *Bioprocess and Biosystems Engineering* 40: 943–957.

Renner, S., Blutke, A., Clauss, S., Deeg, C.A., Kemter, E., Merkus, D., Wanke, R., Wolf, E. 2020. Porcine models for studying complications and organ crosstalk in diabetes mellitus. *Cell and Tissue Research* DOI: 10.1007/s00441-019-03158-9.

Rhodes, C.J. 2005. Type 2 diabetes-a matter of beta-cell life and death? *Science* 307: 380–384.

Samadder, A., Khuda-Bukhsh, A.R. 2014. Nanotechnological approaches in diabetes treatment: A new horizon. *World Journal of Translational Medicine* 3: 84–95.

Saratale, G.D., Saratale, R.G., Benelli, G., Kumar, G., Pugazhendhi, A., Kim, D.S., Shin, H.-S. 2017. Antidiabetic potential of silver nanoparticles ssynthesised with *Argyreia nervosa* leaf extract high synergistic antibacterial activity with standard antibiotics against foodborne bacteria. *Journal of Cluster Science* 28: 1709–1727.

Saratale, R.G., Shin, H.S., Kumar, G., Benelli, G., Kim, D.S., Saratale, G.D. 2017. Exploiting antidiabetic activity of silver nanoparticles ssynthesised using *Punica granatum* leaves and anticancer potential against human liver cancer cells (HepG2). *Artificial Cells, Nanomedicine, and Biotechnology* 46: 211–222.

Saravanakumar, A., Peng, M.M., Ganesh, M., Jayaprakash, J., Mohankumar, M., Jang, H.T. 2017. Low-cost and eco-friendly green synthesis of silver nanoparticles using *Prunus japonica* (Rosaceae) leaf extract and their antibacterial, antioxidant properties. *Artificial Cells, Nanomedicine and Biotechnology* 45: 1165–1171.

Saxena, A., Vikram, N.K. 2004. Role of selected Indian plants in management of type 2 diabetes: A review. *Journal of Alternative and Complementary Medicine* 10: 369–378.

Sharma G, Sharma AR, Nam J, Doss GPC, Sang-Soo Lee S-S, Chakraborty C (2015). Nanoparticle-based insulin delivery system: the next generation efficient therapy for Type 1 diabetes. *Journal of Nanobiotechnology* 13: 74.

Skalli, S., Hassikou, R., Arahou, R. 2019. An ethnobotanical survey of medicinal plants used for diabetes treatment in Rabat, Morocco. *Heliyon* 5: 1421–1424.

Soliman, N., Khalil, M., Abdel- Moaty, H., Ismael, E., Sabry, D. 2018. Anti-*Helicobacter pylori*, antidiabetic and cytotoxicity activity of sbiosynthesised gold nanoparticles using *Moricandia nitens* water extract. *Egyptian Journal of Chemistry* 61: 691–703.

Souto, E.B., Souto, S.B., Campos, J.R., Severino, P., Pashirova, T.N., Zakharova, L.Y., Silva, A.M., Durazzo, A., Lucarini, M., Izoo, A.A., Santini, A. 2019. Nanoparticles delivery systems in the treatment of diabetes complications. *Molecules* 24: 4209–4233.

Veerasamy, R., Xin, T.Z., Gunasagaran, S., Xiang, T.F.W., Yang, E.F.C., Jeyakumar, N., Dhanaraj, S.A. 2011. Biosynthesis of silver nanoparticles using Mangosteen leaf extract and evaluation of their antimicrobial activities. *Journal of Saudi Chemical Society* 15: 113–120.

Vegas, A., Veiseh, O., Gürtler, M. et al. 2016. Long-term glycemic control using polymer-encapsulated human stem cell–derived beta cells in immune-competent mice. *Nature Medicine* 22: 306–311.

Vishnu K.M., Murugesan, S., 2013. Biogenic silver nanoparticles by *Halymenia poryphyroides* and its *in vitro* antidiabetic efficacy. *Journal of Chemical and Pharmaceutical Research* 5: 1001–1008.

WHO (World Health Organization). 2016. Global Report on Diabetes 1. Diabetes Mellitus – epidemiology. 2. Diabetes Mellitus – prevention and control. 3. Diabetes, Gestational. 4. Chronic Disease. 5. Public Health. I. World Health Organization. ISBN 978 92 4 156525 7 (NLM classification: WK 810).

Wild, S., Roglic, G., Green, A., Sicree, R., King, H., 2004. Global prevalence of diabetes: Estimates for the year 2000 and projections for 2030. *Diabetes Care* 27: 1047–1053.

Woldu, M.A., Lenjisa, J.L., 2014. Nanoparticles and the new era in diabetes management. *International Journal of Basic and Clinical Pharmacology* 3: 277–284.

Wong, C.Y., Al-Salami, H., Dass, C.R. 2017. Potential of insulin nanoparticle formulations for oral delivery and diabetes treatment. *Journal of Controlled Release* 264: 247–275.

3 Green Nanoparticles
An Alternative Therapy for Oral Candidiasis

Razia Z. Adam, Enas Ismail, Fanelwa Ajayi, Widadh Klein, Germana Lyimo and Ahmed A. Hussein

CONTENTS

3.1 Oral Candidiasis 45
3.2 Management 46
3.3 Nanotechnology 47
3.4 Green Nanotechnology 47
3.5 Green Synthesis 48
  3.5.1 Sources of Green Nanoparticles 48
  3.5.2 Process of Extraction from Plants 49
3.6 Green Nanoparticles 49
3.7 Metallic NPs against *Candida* 49
3.8 Zinc Oxide NPs and *Candida* 50
3.9 Conclusion 54
References 55

3.1 ORAL CANDIDIASIS

Oral candidiasis is often associated with the very young, very old, medically compromised and/or immunocompromised (Blignaut, 2017). *Candida* is a normal oral commensal, present in 45–65% of healthy infants and 30–55% of healthy adults (Millsop and Fazel, 2016). More than 150 species of *Candida* have been described, with 95% of oral candidiasis caused by *Candida albicans* (Quindós et al., 2019).

C. albicans is a dimorphic fungal organism that grows in two different forms: yeast and hypha (Ohshima et al., 2018). *C. albicans* is usually found on the mucous membranes as yeast, hyphae and pseudohyphae when they invade tissues. Common risk factors contributing to the transition from commensal yeast to pathogenic hyphae are described in Table 3.1.

Denture stomatitis is the most common form of oral candidiasis. The colonization of the denture-based material and the strong adherence of *Candida* is an essential pathogenic factor. Although systemic antifungal agents are effective for the treatment

TABLE 3.1
Common Risk Factors for Oral Candidiasis (Patil et al., 2015)

Local factors	Systemic factors
Impaired local defence mechanisms	Impaired systemic defence mechanisms
Decreased saliva production	Primary or secondary immunodeficiency
Atrophic oral mucosa	Endocrine disorders: diabetes
Mucosal diseases	Malnutrition
Topical medications	Congenital conditions
Decreased blood supply	Broad-spectrum antibiotic therapy
Poor oral hygiene	
Dental prosthesis	
Altered or immature oral flora	

of the acute inflammation, they cannot reach a therapeutic antifungal concentration on the inner surfaces of the denture. Clinical relapse and recurrent infection make the treatment of denture stomatitis challenging (Neppelenbroek, 2016). The management of oral candidiasis is governed by four principles: early and accurate diagnosis through a detailed medical and dental history, correction of the risk factors, maintenance of proper oral hygiene of both the oral cavity and/or oral prosthesis and the appropriate use of antifungal agents (Patil et al., 2015; Quindós et al., 2019). For mild cases of oral candidiasis, topical antifungal agents such as Gentian violet and nystatin may be used (Patil et al., 2015; Fourie et al., 2016; Millsop and Fazel, 2016; Lewis and Williams, 2017; Blignaut, 2017; Quindós et al., 2019). In immunocompromised patients and those at risk of disseminated candidiasis, systemic antifungal therapies such as fluconazole and amphotericin B are recommended (Patil et al., 2015; Fourie et al., 2016; Millsop and Fazel, 2016; Lewis and Williams, 2017; Blignaut, 2017; Quindós et al., 2019).

3.2 MANAGEMENT

In denture stomatitis, incorporation of antifungal agents into a denture liner such as a tissue conditioner allows the inflamed and injured tissue to recover. However, these liners are easily degraded and colonized by microbes. The addition of nystatin, miconazole, ketoconazole, itaconazole and chlorhexidine to a tissue conditioner and a soft liner (Bueno et al., 2017) is found to be effective against *C. albicans* for 14 days after spectrophotometer analyses.

Antifungal therapy is limited to the use of azoles, echinocandins, polyenes and flucytosine (Perlin et al., 2017). The development of resistance is a consideration when treating a microbial infection. It has been found that biofilm formation on teeth and devices such as heart valves and catheters reduces the concentration of the drug by trapping it in a glucan-rich matrix polymer (Perlin et al., 2017). In yeast infections, resistance to azole antifungals has been frequently discussed and attributed to the following mechanisms: an alteration in the chemical structure of demethylase enzyme, removal of the azole from the cell by multidrug transporter pumps and compensation by other sterol synthesis enzymes in membrane biosynthesis (Lewis and Williams, 2017; Perlin et al., 2017). Echinocandins are highly effective against most *Candida* species, but the prevalence of *C. albicans* resistance has only been recorded as less than 1% (Perlin et al., 2017). However, echinocandin resistance among *C. glabrata* is now increasingly accompanied by azole resistance too (Pham et al., 2014). Resistance to polyenes, which include amphotericin B and nystatin, is unusual. However, reports of high minimum inhibitory concentration to amphotericin B has been found for *C. albicans, C. krusei, C. rugosa, C. lusitaniae and C. glabrata* (Colombo et al., 2003; Atkinson et al., 2008; Hull et al., 2012). Ultimately, there is a need for novel antifungal drugs, which address resistance, whose mechanism is more selective for fungal targets and have less host toxicity.

More recently, the addition of alternative antifungal agents such as metallic and metallic oxides nanoparticles (NPs) into tissue conditioners have been investigated against *C. albicans* (Iqbal and Zafar, 2016).

3.3 NANOTECHNOLOGY

Nanoscience focuses on the observation, study and manipulation of phenomena at the nanometre scale, while nanotechnology utilizes the unique properties of nanomaterials in the development and application of nanomaterial systems. Nanomaterials provide better functionalities in different applications, including the medical sector. Nanomaterials have unique properties due to their high surface-to-volume ratio and novel physicochemical properties such as solubility, opticality, strength, diffusivity, toxicity, magnetism and thermodynamicity (Rai et al., 2014).

NPs are the most fundamental elements of nanomaterials and different physical and chemical methods are used in their synthesis. These include processes such as photolithography, electron beam lithography, milling techniques, anodization, ion and plasma etching, chemical or electrochemical nanostructural precipitation, sol–gel processing, laser pyrolysis and chemical vapour deposition (CVD). The poor understanding of hazards associated with nanotechnology synthesis procedures has prompted researchers to move forward towards the use of green nanotechnology (Subramani et al., 2012).

3.4 GREEN NANOTECHNOLOGY

Green nanotechnology is a branch of green technology that utilizes and incorporates the main principles and concepts of green chemistry and green engineering in the design and development of green nanomaterial production methods for application in various areas (Verma et al., 2019). This has afforded researchers the opportunity to render nanotechnology less toxic throughout its life cycle and make safer and more sustainable. Green nanotechnology mainly targets reduction of waste gas emission, decreased consumption of non-renewable raw materials and increased energy efficiency. The green synthesis methods provide alternative approaches to chemical and physical methods. The green approach avoids the use of harmful and toxic chemicals as well as the need for expensive chemicals, affording an environment-friendly method of nanoparticles synthesis.

Green nanotechnology has a broad range of application, from its use in the medical field, solar cell development, biofuel and fuel cell production to various other biomedical applications. The design of nontoxic nanomaterial products has the potential to eliminate/minimize pollution and can provide a solution to existing environmental problems. It further seeks the application of nanotechnology that has a broader societal benefit by decreasing the adverse impact on health. Nanomaterials have noteworthy applications in nanobiotechnology, particularly in diagnosis, drug delivery systems (Faraji and Wipf, 2009), prostheses and implants. Advances in the medical implementations of nanotechnology have resulted in the formation of a new field called nanomedicine (Freitas, 2005). Nanomedicine includes various applications ranging from drug release with nanospheres to tissue scaffolds based on nanotechnological design that realize tissue formation, and even nanorobots for diagnostic and therapeutic purposes (Freitas, 2005). Similar to nanomedicine, incorporating nanomaterials and biotechnologies to develop nanodentistry will refine/perfect oral health through tissue engineering and nanorobots (Ramos et al., 2017).

3.5 GREEN SYNTHESIS

3.5.1 Sources of Green Nanoparticles

Green synthesis of nanoparticles can be achieved using unicellular and/or multicellular organisms such as Actinomycetes, bacteria, fungus, viruses, yeast and plants (Saratale et al., 2018). Green synthesis via natural extract seems to be a promising synthesis technique compared to other traditional physical and chemical methods, because the preparation technique of green method nanoparticles does not involve any toxic solvents or reduction agents. As a result, green synthesis of nanoparticles using microorganisms has received enormous attention in the green nanotechnology field in the past few years. Microorganisms serve as potential nanofactories for the eco-friendly and inexpensive synthesis of metallic nanoparticles like silver and gold in various shapes and forms. These different morphological forms have shown remarkable properties for use in biomedical applications as anticancer and antimicrobial compounds (Prasad, 2008).

Plants have an inherent potential for the reduction of metal precursors, capping and stabilizing agents, thereby reducing the accumulation of metals. Plant-mediated synthesis method is gaining momentum as the most promising, fast and economical route/modality in NPs synthesis techniques. Plant extract components such as sugar, flavonoid, protein, enzyme, polymer and organic acid act as reducing and capping agents and are known to take charge in the bio-induction of metal ions into nanoparticles. Plant's secondary metabolites such as flavonoids, alkaloids, phenols and terpenoids render them therapeutic (Akbar et al., 2020). These metabolites hinder virus replication without affecting the host metabolism (Hussain et al., 2017). In addition, these metabolites have also been found to be active against *C. krusei*, *C. neoformans* and *C. albicans* (Aldholmi et al., 2019). Plant extracts are obtained from different parts such as leaves, roots and fruits, and they offer numerous benefits associated with their role as natural nanoparticle factories. This is attributed to plant extracts containing a vast array of sophisticated bioactive phytochemicals (Agarwal et al., 2017; Vijayaraghavan and Ashokkumar, 2017; Akbar et al., 2020).

3.5.2 Process of Extraction from Plants

The general mechanism of green synthesis using plants is explained in three phases:

1. Activation phase: reduction of metal ions followed by nucleation of reduced atoms
2. Growth phase: spontaneous association of smaller nanoparticles into larger particles and further reduction of metal ions
3. Termination phase: determination of nanoparticle shape, size and superficial charge (Akbar et al., 2020).

The generic process may be described as follows:

1. Wash the desired plant with tap water a few times and finally with sterile distilled water.
2. Dry the plant, chop and boil in sterile deionized water.
3. The solution is then filtered to remove impurities and produces an extract.
4. A salt is added to eventually result in the formation of nanoparticles.

3.6 GREEN NANOPARTICLES

Nanoparticles have opened various fronts for the design of new materials and evaluation of their properties by modulating particle size, morphology superficial charge and distribution. The synthesis of NPs from higher plants involves the addition of water extract of the plant material to the salt. The reduction process is dependent on different factors. These include the organic functionality present on the cell wall that induces biomineralization and the reaction conditions such as pH, composition of medium, metallic salt concentration and temperature. Nanoparticle size, morphology and composition are significantly affected by these parameters. Therefore, it is essential to optimize these factors during biosynthesis to increase the efficiency of the NPs synthesis. Confirmation of the formation and properties of the NPs are obtained from different characterization techniques like UV-visible spectroscopy (UV-vis), transmission electron spectroscopy (TEM), scanning electron microscopy (SEM), energy-dispersive X-ray analysis (EDX) and X-ray diffraction spectroscopy (XRD).

3.7 METALLIC NPS AGAINST *CANDIDA*

Metallic nanoparticles, such as gold (Au) and silver (Ag), have been explored widely due to their unique antimicrobial and anticancer properties. Differing outcomes of gold nanoparticles (AuNPs) against *Candida* species have been reported (Benedec et al., 2018; Sulthana and Rajanikanth, 2018; López-Lorente et al., 2019; Zhaleh et al., 2019). The use of reducing/capping agents, varying particle size and different *Candida* spp. has showed no significant antibacterial activity for the AuNPs. However, the use of the AuNPs in conjugation with other active compounds (drugs) improved its antimicrobial/antifungal capacity (Zhang et al., 2015). The conjugation of AuNPs and the antimycotic agent amphotericin B showed synergetic activity against different *Candida* strains (Table 3.2). Also, the AuNP photosensitizer (PS) conjugate based photodynamic therapy was found to effectively deplete *C. albicans* hyphae on the superficial skin and oral mucosal *C. albicans* infection in mice (Sherwani et al., 2015).

Given all the advantages of using noble metal nanoparticles for biomedical applications, silver nanoparticles (AgNPs) have also attracted interest, as shown in Table 3.3.

In the few randomized controlled studies investigating denture acrylic modified with silver nanoparticles, all fungal species disappeared in patients with complete dentures modified with AgNPs at 0.2% concentration. However, the silver nanoparticles were chemically synthesized (Abdallah et al., 2015).

3.8 ZINC OXIDE NPS AND *CANDIDA*

Zinc oxide nanoparticles (ZnO-NPs) are inexpensive, biocompatible and relatively less toxic compared to other metal oxides NPs (Agarwal et al., 2017; Zare et al., 2017; Kalpana and Devi Rajeswari, 2018; Jiang et al., 2018). ZnO will be covered in this chapter as its unique properties make it appropriate for various biomedical applications such as anticancer, antifungal, antibacterial, antioxidant, antidiabetic, anti-inflammatory drug delivery systems and bio-imaging applications (Mishra et al., 2017; Salahuddin Siddiqi et al., 2018; Jiang et al., 2018; Kalpana and Devi Rajeswari, 2018; Akbar et al., 2020).

ZnO-NPs have been successfully manufactured from various plants or parts thereof, as well as different zinc salts (Table 3.4).

ZnO-NPs have been found to be effective against gram-positive and gram-negative bacteria (Mirzaei and Darroudi, 2017; Kalpana and Devi Rajeswari 2018; Akbar et al., 2020). However, their antifungal activity against *Candida* species is not well documented (Table 3.5).

Miri et al. (2019) demonstrated that ZnO-NPs were effective against *C. albicans*, impairing its cell membrane integrity. This may result in leakage of minerals, proteins and genetic material, eventually leading to death. The study also confirmed previous reports of its dose-dependent antifungal (Jiang et al., 2018; Miri et al., 2019). *C. albicans*' most important virulence factor is the ability to form a biofilm on almost any kind of medical device such as catheters, pacemakers, dentures, cardiac valves, etc. Chitosan-based ZnO-NPs were found to have antimicrobial and antibiofilm potential (Dhillon et al., 2014). Similarly, Jalal et al. (2018) found that green synthesized ZnO-NPs inhibited biofilm formation by up to 85%.

The mechanical properties of polymethylmethacrylate (PMMA), an acrylic resin, make it the material of choice for dentures. PMMA is easily colonized by

TABLE 3.2
Green-Mediated Au Nanoparticles against Different *Candida* Strains

Natural source	Size (nm)	MIC[a]/MIZ[b](*Candida* species)	*In vitro* or *in vivo*	Reference
Microbial exopolymers (EP)	–	Not active (*C. albicans* ATCC 10231)	*In vitro*	Scala et al., 2019
Gundelia tournefortii L.	40–45	MIC: 4.0 mg/ml (*C. albicans, C. glabrata*) MIC: 2.0 mg/ml (*C. guilliermondii, C. krusei*)	*In vitro*	Zhaleh et al., 2019
AISI 304 (Fe/Cr18/Ni10) stainless steel disc	37	MIC: ~40 mg/ml (*C. albicans*)	*In vitro*	Lopez-Lorente et al., 2019
Sodium citrate	63, 34, 17	MIC: ~20 mg/ml (*C. albicans*)	*In vitro*	Lopez-Lorente et al., 2019
Penicillium sp.	4–20	Not active (*C. albicans*)	*In vitro*	Sulthana and Rajanikanth, 2018
Origanum vulgare	36–72	Concentration not specified, inhibition zone 28 mm	*In vitro*	Benedec et al., 2018
Melon peel and peach		Not active (*C. albicans KACC30003 and KACC30066*)	*In vitro*	Patra and Baek, 2016
Conjugation 50 µg AuNPs and 5 µg amphotericin	–	MIZ: 10.0–15.5 mm *against C. albicans KACC 30003 and KACC 30062, C. glabrata KBNO6P00368, C. geochares KACC 30061* and *C. saitoana KACC 41238*	*In vitro*	Patra and Baek, 2016
Justicia glauca	~32.5	MIC: 12.5 µg/ml (*C. albicans*)	*In vitro*	Emmanuel et al., 2017
Azithromycin, Clarithromycin: AuNPs (1:1 blend)	–	MIZ: 17 mm (*C. albicans*)	*In vitro*	Emmanuel et al., 2017
Punica granatum	5–17	MIC: 310 µg/ml (*C. albicans ATCC 90028*)	*In vitro*	Lokina et al., 2014
Honey	10	MIC: 62.5 mg/ml (*C. albicans* MTCC 183)	*In vitro*	Sreelakshmi et al., 2011
	10	MIC: 250 mg/ml (*C. albicans* MTCC 10231)		Sreelakshmi et al., 2011

[a] MIC: minimum inhibitory concentration.
[b] MIZ: medium inhibition zone.

TABLE 3.3
Recent Plant-Mediated Ag Nanoparticles against Different *Candida* Strains

Natural sources (Plants)	Size (nm)	MIC[a]/MIZ[b] (*Candida* species)	Reference
Ziziphora clinopodioides Lam	5–25	*MIC:* 4±0[b] *C. albicans*, 4±0[b] *C. glabrata*, 2±0[a] *C. guilliermondii* 2±0[a] *C. krusei*	Ahmeda et al., 2020
Lantana trifolia	5–70	MIZ: 7.3–12 mm, *C. albicans*	Madivoli et al. 2020
Ganoderma lucidum	15–22	MIC: 64 μg/l *C. albicans*	Aygün et al., 2020
Aesculus hippocastanum	45–55	No effect on *C. albicans*, *C. tropicalis*, *C. krusei*	Küp et al., 2020
Thymus algeriensis	10–20	>20 mm *C. albicans*	Beldjilali et al., 2019
Syzygium cumini	10–100	MIZ: 22 mm *C. albicans*, 19 mm *C. parapsilosis*, 17 mm *C. tropicalis*	Jalal et al., 2019
Thymus algeriensis	10–20	>20 mm *C. albicans*	Beldjilali et al., 2019
Equisetum arvense L.	10–20, 40–60	*C. albicans and S. boulardii*	Miljković et al., 2019
Eichhornia crassipes	10–40	6.8 mm *Candida albicans*, 4.8 mm *Aspergillus flavus*	Mani, 2018
Dodonaea viscosa, *Hyptis suoveolens*	40–55	*C. albicans*, *C. glabrata*, *C. tropicalis*	Muthamil et al., 2018
Gymnema sylvestre	3–30	MIZ: 15.4 mm *C. albicans*, 14.2 mm *C. non-albicans*, 15.7 mm *C. tropicalis*	Netala et al., 2016
Amaranthus dubius, *Amaranthus polygonoides*, *Alternanthera sessilis*, *Portulaca oleracea*, *Pisonia grandis*, *Kedrostis foetidissima*	6–23	MIZ: 8–12 mm *C. albicans*, 8–19 mm *S. cerevisiae*	Jannathul and Lalitha, 2015

[a] MIC: minimum inhibitory concentration.
[b] MIZ: medium inhibition zone.

microorganisms and is usually a reservoir for these microorganisms, which results in many oral diseases, one of which is denture stomatitis (Neppelenbroek, 2016). The surface roughness of a denture may further encourage colonization. Several studies have investigated the addition of metal and metal oxide nanoparticles to PMM. Cierech et al. (2016) and Kamonkhantikul et al. (2017) compared the effect of ZnO-NPs incorporated into PMMA, with and without silanized modification. Kamonkhantikul et al. (2017) used chemically synthesized nanoparticles and

TABLE 3.4
Recent Studies Using Plant-Mediated Synthesis of Zinc Oxide Nanoparticle

Plants	Morphology	Size (nm)	Reference
Acalypha Fruticosa	Spherical structure	55	Vijayakumar et al., 2019
Catharanthus roseus	Hexagonal shape	50–73	Gupta et al., 2018
Tifolium pratense	Hexagonal shape	60–70	Padovani et al., 2018
Camellia sinensis	Hexagonal wurtzite structure	16	Senthilkumar and Sivakumar, 2014
Moringa Oleifera	–	40–45	Mishra et al., 2017
Cuminum cyminum	Spherical/oval	7	Zare et al., 2017
Agathosma betulina	Quasi-spherical agglomerated nanoscaled particles	12–26	Thema et al., 2015
Aspalathus linearis	Non-agglomerated quasi-spherical	1–8.5	Diallo et al., 2015
Aspalathus linearis	–	23	Nethavhanani et al (2018)

reported that the silanized groups demonstrated a greater reduction in *C. albicans* colonization than the non-silanized group. ZnO green synthesized nanoparticles have shown some promising results against *C. albicans* (Lyimo et al. 2019). High-resolution transmission electron microscopy (HRTEM) and XRD confirmed hexagonal nanoparticles averaged to be 15–30 nm in size as seen in Figure 3.1.

TABLE 3.5
Plant-Mediated Zinc Oxide Nanoparticles and *Candida* Species

Natural sources (plants)	Size (nm)	MIC[a]/MIZ[b] (*Candida* species)	Reference
Capparis zeylanica	28–30	*C. albicans* MTCC 227	Nilavukkarasi et al., 2020
Acalypha fruticosa	40	0.78 μg/ml *C. albicans* MTCC	Vijayakumar et al., 2019
P. farcta	40–80	128 μg/ml (*C. albicans)*	Miri et al., 2019
Crinum latifolium	10–30	MICs: 0.25–0.5 mg/ml (*C. albicans, C. tropicalis and C. dubliniensis*)	Jalal et al., 2018
Chelidonium majus	10	MIC: 120 μM; MFC: >2,590 M *C. albicans* ATCC 10231 and clinical strains	Dobrucka et al., 2018
Zingiber officinale	23–25	MIC not recorded *C. albicans*	Janaki et al., 2015

[a] MIC: minimum inhibitory concentration.
[b] MIZ: medium inhibition zone.

FIGURE 3.1 HRTEM image of ZnO nanoparticles.

The antimicrobial activity of ZnO-NPs against *C. albicans* was tested using standard well diffusion protocol (modified Kirby–Bauer), as shown in Figure 3.2.

Maximum inhibition zones ranged between 24 and 35 mm, signifying that the green bioinspired ZnO-NPs have potential as an effective antifungal agent. The particle size ensures increased surface area, which subsequently improves the antimicrobial effectiveness and contributes to enhanced surface reactivity (Silva et al., 2019). Remarkably, antifungal activity of the nanoparticles with a shelf life over four months at room temperature was noted. Hence, this confirms that these green ZnO-NPs are capable of rendering antifungal efficacy against *C. albicans*. Similarly, Pillai et al. (2020) reported that ZnO-NPs synthesized from *Cinnamomum tamala* were active against *C. albicans*.

The exact mode of action of ZnO against microbes is not known. However, the following has been proposed: (1) oxidative stress due to the creation of reactive oxygen species (ROS); (2) the accumulation of nanoparticles on the cell membranes resulting in disorganization and (3) the release and binding of Zn^{2+} ions to the membrane (Akbar et al., 2020).

3.9 CONCLUSION

In summary, the green synthesis of metal and metal oxide nanoparticles has been widely reported. It is low cost, nontoxic and fast. The application of these nanoparticles as an antimicrobial has proven to be effective *in vitro*. However, there is no standard measurement of antifungal activity which makes comparison between studies

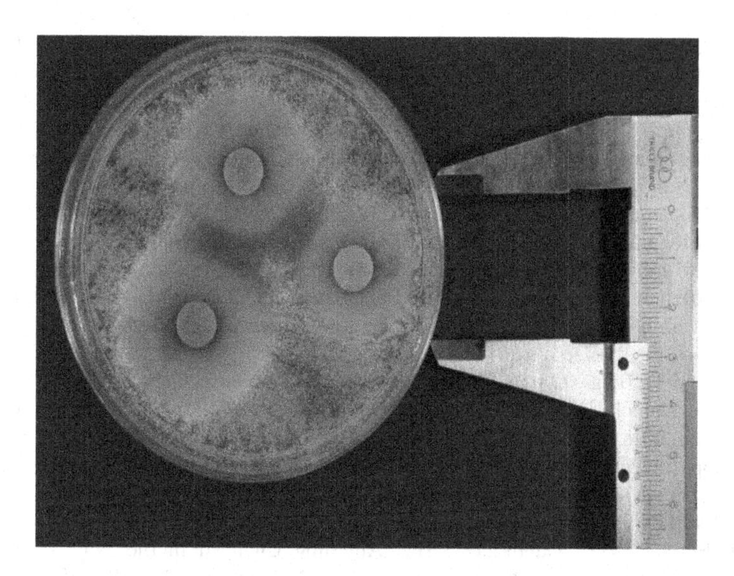

FIGURE 3.2 Standard well diffusion test on *Candida albicans* biofilm; 100 µml ZnO-NPs (inhibition zones = 24–26 mm).

difficult. The exact mechanism of action on *Candida* species and other organisms is not well understood. Further studies are required to explore the biocompatibility and the effect against *Candida* species specifically. In addition, future research should also compare the use of reference strains and clinical isolates to determine its role as therapeutic agents.

REFERENCES

Abdallah RM, RMK Emera, and A Gebreil. 2015. "Antimicrobial Activity of Silver Nanoparticles." *Egyptian Dental Journal* 61: 1039–52.

Agarwal, Happy, S. Venkat Kumar, and S. Rajeshkumar. 2017. "A Review on Green Synthesis of Zinc Oxide Nanoparticles: An Eco-Friendly Approach." *Resource-Efficient Technologies* 3, no. 4: 406–13. DOI: 10.1016/J.REFFIT.2017.03.002.

Ahmeda, Ahmad, Akram Zangeneh, Roozbeh Javad Kalbasi, Niloofar Seydi, Mohammad Mahdi Zangeneh, Sanaz Mansouri, Samaneh Goorani, and Rohallah Moradi. 2020. "Green Synthesis of Silver Nanoparticles from Aqueous Extract of Ziziphora Clinopodioides Lam and Evaluation of Their Bio-Activities under in Vitro and in Vivo Conditions." *Applied Organometallic Chemistry* 34, no. 4: 1–14. DOI: 10.1002/aoc.5358.

Akbar, Sadia, Isfahan Tauseef, Fazli Subhan, Nighat Sultana, Ibrar Khan, Umair Ahmed, and Kashif Syed Haleem. 2020. "An Overview of the Plant-Mediated Synthesis of Zinc Oxide Nanoparticles and Their Antimicrobial Potential." *Inorganic and Nano-Metal Chemistry*, 50:4,257-271, doi:10.1080/24701556.2019.1711121.

Aldholmi, Mohammed, Pascal Marchand, Isabelle Ourliac-Garnier, Patrice Le Pape, and A. Ganesan. 2019. "A Decade of Antifungal Leads from Natural Products: 2010–2019." *Pharmaceuticals* 12, no. 4: 2010–19. DOI: 10.3390/ph12040182.

Atkinson, Bradley J., Russell E. Lewis, and Dimitrios P. Kontoyiannis. 2008. "Candida Lusitaniae Fungemia in Cancer Patients: Risk Factors for Amphotericin B Failure and Outcome." *Medical Mycology* 46, no. 6, 541–46. DOI: 10.1080/13693780801968571.

Aygün, Ayşenur, Sadin Özdemir, Mehmet Gülcan, Kemal Cellat, and Fatih Şen. 2020. "Synthesis and Characterization of Reishi Mushroom-Mediated Green Synthesis of Silver Nanoparticles for the Biochemical Applications." *Journal of Pharmaceutical and Biomedical Analysis* 178. DOI: 10.1016/j.jpba.2019.112970.

Beldjilali, Mohammed, Khaled Mekhissi, Yasmina Khane, Wahiba Chaibi, Lahcène Belarbi, and Smain Bousalem. 2019. "Antibacterial and Antifungal Efficacy of Silver Nanoparticles Biosynthesized Using Leaf Extract of *Thymus algeriensis*." *Journal of Inorganic and Organometallic Polymers and Materials* 30: 2126–33. DOI: 10.1007/s10904-019-01361-3.

Benedec, Daniela, Ilioara Oniga, Flavia Cuibus, Bogdan Sevastre, Gabriela Stiufiuc, Mihaela Duma, Daniela Hanganu, Cristian Iacovita, Rares Stiufiuc, and Constantin Mihai Lucaciu. 2018. "Origanum Vulgare Mediated Green Synthesis of Biocompatible Gold Nanoparticles Simultaneously Possessing Plasmonic, Antioxidant and Antimicrobial Properties." *International Journal of Nanomedicine* 13 (February): 1041–58. DOI: 10.2147/IJN.S149819.

Blignaut, Elaine. 2017. "Candidiasis: Has Anything Changed in the Way We Manage These Patients?" *South African Dental Journal* 72, no. 8: 355–59. DOI: 10.17159/2519-0105/2017/v72no8a2.

Bueno, MG, Elen Juliana Bonassa De Sousa, Juliana Hotta, Vinícius Carvalho Porto, Vanessa Migliorini Urban, and Karin H Neppelenbroek. 2017. "Surface Properties of Temporary Soft Liners Modified by Minimum Inhibitory Concentrations of Antifungals." *Brazilian Dental Journal* 28, no. 2: 158–64. DOI: 10.1590/0103-6440201701266.

Cierech, Mariusz, Jacek Wojnarowicz, Dariusz Szmigiel, Bohdan Bączkowski, Anna Maria Grudniak, Krystyna Izabela Wolska, Witold Łojkowski, and Elżbieta Mierzwińska-Nastalska. 2016. "Preparation and Characterization of ZnO-PMMA Resin Nanocomposites for Denture Bases." *Acta of Bioengineering and Biomechanics Original Paper* 18, no. 2: 31–41. DOI: 10.5277/ABB-00232-2014-04.

Colombo, Arnaldo Lopes, Analy S. Azevedo Melo, Robert F. Crespo Rosas, Reinaldo Salomão, Marcelo Briones, Richard J. Hollis, Shawn A. Messer, and Michael A. Pfaller. 2003. "Outbreak of Candida Rugosa Candidemia: An Emerging Pathogen That May Be Refractory to Amphotericin B Therapy." *Diagnostic Microbiology and Infectious Disease* 46, no. 4: 253–57. DOI: 10.1016/S0732-8893(03)00079-8.

Dhillon, Gurpreet Singh, Surinder Kaur, and Satinder Kaur Brar. 2014. "Facile Fabrication and Characterization of Chitosan-Based Zinc Oxide Nanoparticles and Evaluation of Their Antimicrobial and Antibiofilm Activity." *International Nano Letters* 4, no. 2: 1–11. DOI: 10.1007/s40089-014-0107-6.

Diallo, A, B D Ngom, E Park, and M Maaza. 2015. "Green Synthesis of ZnO Nanoparticles by Aspalathus Linearis: Structural & Optical Properties." *Journal of Alloys and Compounds* 646: 425–30. DOI: 10.1016/j.jallcom.2015.05.242.

Dobrucka, Renata, Jolanta Dlugaszewska, and Mariusz Kaczmarek. 2018. "Cytotoxic and Antimicrobial Effects of Biosynthesized ZnO Nanoparticles Using of Chelidonium Majus Extract." *Biomedical Microdevices* 20, no. 1. DOI: 10.1007/s10544-017-0233-9.

Emmanuel, Rayappan, Muthupandian Saravanan, Muhammad Ovais, Sethuramasamy Padmavathy, Zabta Khan Shinwari, and Periyakaruppan Prakash. 2017. "Antimicrobial Efficacy of Drug Blended Biosynthesized Colloidal Gold Nanoparticles from Justicia Glauca against Oral Pathogens: A Nanoantibiotic Approach." *Microbial Pathogenesis* 113 (December): 295–302. DOI: 10.1016/j.micpath.2017.10.055.

Faraji, Amir H., and Peter Wipf. 2009. "Nanoparticles in Cellular Drug Delivery." *Bioorganic and Medicinal Chemistry*. DOI: 10.1016/j.bmc.2009.02.043.

Fourie, J, R A G Khammissa, R Ballyram, N H Wood, J Lemmer, and L Feller. 2016. "Oral Candidosis: An Update on Diagnosis, Aetiopathogenesis and Management." *South African Dental Journal* 71, no. 7: 314–18. http://www.scielo.org.za/scielo.php?pid=S0011-85162016000700007&script=sci_arttext&tlng=pt.

Freitas, Robert A. 2005. "What Is Nanomedicine?" *Nanomedicine: Nanotechnology, Biology, and Medicine* 1, no. (1): 2–9. DOI: 10.1016/j.nano.2004.11.003.

Gupta, Monika, Rajesh S. Tomar, Shuchi Kaushik, Raghvendra K. Mishra, and Divakar Sharma. 2018. "Effective Antimicrobial Activity of Green ZnO Nano Particles of Catharanthus Roseus." *Frontiers in Microbiology* 9 (september): 2030. DOI: 10.3389/fmicb.2018.02030.

Hull, Claire M., Oliver Bader, Josie E. Parker, Michael Weig, Uwe Gross, Andrew G.S. Warrilow, Diane E. Kelly, and Steven L. Kelly. 2012. "Two Clinical Isolates of Candida Glabrata Exhibiting Reduced Sensitivity to Amphotericin B Both Harbor Mutations in ERG2." *Antimicrobial Agents and Chemotherapy* 56, no. (12): 6417–21. DOI: 10.1128/AAC.01145-12.

Hussain, Wajid, Kashif Syed Haleem, Ibrar Khan, Isfahan Tauseef, Sadia Qayyum, Bilal Ahmed, and Muhammad Nasir Riaz. 2017. "Medicinal Plants: A Repository of Antiviral Metabolites." *Future Virology* 12, no. 6: 299–308. DOI: 10.2217/fvl-2016-0110.

Iqbal, Zahid, and Muhammad Sohail Zafar. 2016. "Role of Antifungal Medicaments Added to Tissue Conditioners: A Systematic Review." *Journal of Prosthodontic Research*,60 (4), 231-9 DOI: 10.1016/j.jpor.2016.03.006.

Jalal, Mohammad, Mohammad Azam Ansari, Syed Ghazanfar Ali, Haris M. Khan, and Suriya Rehman. 2018. "Anticandidal Activity of Bioinspired ZnO NPs: Effect on Growth, Cell Morphology and Key Virulence Attributes of Candida Species." *Artificial Cells, Nanomedicine and Biotechnology* 46, no. supl: 912–25. DOI: 10.1080/21691401.2018.1439837.

Jalal, Mohammad, Mohammad Azam Ansari, Mohammad A. Alzohairy, Syed Ghazanfar Ali, Haris M. Khan, Ahmad Almatroudi, and Mohammad Imran Siddiqui. 2019. "Anticandidal Activity of Biosynthesized Silver Nanoparticles: Effect on Growth, Cell Morphology, and Key Virulence Attributes of Candida Species." *International Journal of Nanomedicine* 14: 4667–79. DOI: 10.2147/IJN.S210449.

Janaki, A. Chinnammal, E. Sailatha, and S. Gunasekaran. 2015. "Synthesis, Characteristics and Antimicrobial Activity of ZnO Nanoparticles." *Spectrochimica Acta–Part A: Molecular and Biomolecular Spectroscopy* 144: 17–22. DOI: 10.1016/j.saa.2015.02.041.

Jannathul Firdhouse, M., and P. Lalitha. 2015. "Biocidal Potential of Biosynthesized Silver Nanoparticles against Fungal Threats." *Journal of Nanostructure in Chemistry* 5, no. 1: 25–33. DOI: 10.1007/s40097-014-0126-x.

Jiang, J, Jiang Pi, and Jiye Cai. 2018. "The Advancing of Zinc Oxide Nanoparticles for Biomedical Applications." *Bioinorganic Chemistry and Applications*. DOI: 10.1155/2018/1062562.

Kalpana, V. N., and V. Devi Rajeswari. 2018. "A Review on Green Synthesis, Biomedical Applications, and Toxicity Studies of ZnO NPs." *Bioinorganic Chemistry and Applications* (August): 1–12. DOI: 10.1155/2018/3569758.

Kamonkhantikul, Krid, Mansuang Arksornnukit, and Hidekazu Takahashi. 2017. "Antifungal, Optical, and Mechanical Properties of Polymethylmethacrylate Material Incorporated with Silanized Zinc Oxide Nanoparticles." *International Journal of Nanomedicine* 12 (March): 2353–60. DOI: 10.2147/IJN.S132116.

Küp, Fatma Öztürk, Seval Çoşkunçay, and Fatih Duman. 2020. "Biosynthesis of Silver Nanoparticles Using Leaf Extract of Aesculus Hippocastanum (Horse Chestnut): Evaluation of Their Antibacterial, Antioxidant and Drug Release System Activities." *Materials Science and Engineering C* 107: 110207. DOI: 10.1016/j.msec.2019.110207.

Lewis, M. A. O., and D. W. Williams. 2017. "Diagnosis and Management of Oral Candidosis." *British Dental Journal* 223, no. 9: 675–81. DOI: 10.1038/sj.bdj.2017.886.

Lokina, S., R. Suresh, K. Giribabu, A. Stephen, R. Lakshmi Sundaram, and V. Narayanan. 2014. "Spectroscopic Investigations, Antimicrobial, and Cytotoxic Activity of Green Synthesized Gold Nanoparticles." *Spectrochimica Acta Part A: Molecular and Biomolecular Spectroscopy* 129 (August): 484–90. DOI: 10.1016/j.saa.2014.03.100.

López-Lorente, Ángela Inmaculada, Soledad Cárdenas, and Zaira Isabel González-Sánchez. 2019. "Effect of Synthesis, Purification and Growth Determination Methods on the Antibacterial and Antifungal Activity of Gold Nanoparticles." *Materials Science and Engineering: C* 103 (October): 109805. DOI: 10.1016/j.msec.2019.109805.

Lyimo, Germana, Razia Adam, and Rachel Fanelwa Ajayi. 2019. "Electroanalysis of Aspalathus Linearis and Musa Paradisiaca Mediated Zinc Oxide Nanoparticles against Fungal Pathogens." Read at 70th Annual International Society of Electrochemistry Meeting held in Durban, South Africa.

Madivoli, Edwin Shigwenya, Patrick Gachoki Kareru, Anthony Ngure Gachanja, Samuel Mutuura Mugo, David Sujee Makhanu, Sammy Indire Wanakai, and Yahaya Gavamukulya. 2020. "Facile Synthesis of Silver Nanoparticles Using Lantana Trifolia Aqueous Extracts and Their Antibacterial Activity." *Journal of Inorganic and Organometallic Polymers and Material, Jan1, 1-9.* DOI: 10.1007/s10904-019-01432-5.

Mani, N. 2018. "Evaluation of Antimicrobial Activity of Silver Nanoparticle Using Eichhornia Crassipes Leaves Extract AS Prabakaran and N Mani." *Journal of Pharmacognosy and Phytochemistry* 7, no. 5: 1308–11.

Miljković, Miona, Vesna Lazić, Slađana Davidović, Ana Milivojević, Jelena Papan, Margarida M. Fernandes, Senentxu Lanceros-Mendez, S. Phillip Ahrenkiel, and Jovan M. Nedeljković. 2019. "Selective Antimicrobial Performance of Biosynthesized Silver Nanoparticles by Horsetail Extract Against E. Coli." *Journal of Inorganic and Organometallic Polymers and Materials.* Dec4, 1-10.DOI: 10.1007/s10904-019-01402-x.

Millsop, Jillian W., and Nasim Fazel. 2016. "Oral Candidiasis." *Clinics in Dermatology* 34, no. 4: 487–94. DOI: 10.1016/j.clindermatol.2016.02.022.

Miri, Abdolhossien, Nafiseh Mahdinejad, Omolbanin Ebrahimy, Mehrdad Khatami, and Mina Sarani. 2019. "Zinc Oxide Nanoparticles: Biosynthesis, Characterization, Antifungal and Cytotoxic Activity." *Materials Science and Engineering C.* 104:109981 DOI: 10.1016/j.msec.2019.109981.

Mirzaei, Hamed, and Majid Darroudi. 2017. "Zinc Oxide Nanoparticles: Biological Synthesis and Biomedical Applications." *Ceramics International* 43, no. 1: 907–14. DOI: 10.1016/J.CERAMINT.2016.10.051.

Mishra, Pawan K., Harshita Mishra, Adam Ekielski, Sushama Talegaonkar, and Bhuvaneshwar Vaidya. 2017. "Zinc Oxide Nanoparticles: A Promising Nanomaterial for Biomedical Applications." *Drug Discovery Today.*Dec 1;22(12):1825-34. DOI: 10.1016/j.drudis.2017.08.006.

Muthamil, Subramanian, Vivekanandham Amsa Devi, Boopathi Balasubramaniam, Krishnaswamy Balamurugan, and Shunmugiah Karutha Pandian. 2018. "Green Synthesized Silver Nanoparticles Demonstrating Enhanced in Vitro and in Vivo Antibiofilm Activity against Candida Spp." *Journal of Basic Microbiology* 58, no. 4: 343–57. DOI: 10.1002/jobm.201700529.

Neppelenbroek, Karin H. 2016. "Sustained Drug-Delivery System: A Promising Therapy for Denture Stomatitis?" *Journal of Applied Oral Science* 24, no. 5: 420–22. DOI: 10.1590/1678-77572016ed003.

Netala, Vasudeva Reddy, Venkata Subbaiah Kotakadi, Latha Domdi, Susmila Aparna Gaddam, Pushpalatha Bobbu, Sucharitha K. Venkata, Sukhendu Bikash Ghosh, and Vijaya Tartte. 2016. "Biogenic Silver Nanoparticles: Efficient and Effective Antifungal Agents." *Applied Nanoscience* 6, no. (4): 475–84. DOI: 10.1007/s13204-015-0463-1.

Nethavhanani, T., A Diallo, R. Madjoe, L. Kotsedi, and M. Maaza. 2018. "Synthesis of ZnO Nanoparticles by a Green Process and the Investigation of Their Physical Properties." *AIP Conference Proceedings*, 1962:1,040007 DOI: 10.1063/1.5035545.

Nilavukkarasi, M., S. Vijayakumar, and S. Prathipkumar. 2020. "Capparis Zeylanica Mediated Bio-Synthesized ZnO Nanoparticles as Antimicrobial, Photocatalytic and Anti-Cancer Applications." *Materials Science for Energy Technologies* 3 (January): 335–43. DOI: 10.1016/j.mset.2019.12.004.

Ohshima, Tomoko, Satoshi Ikawa, Katsuhisa Kitano, and Nobuko Maeda. 2018. "A Proposal of Remedies for Oral Diseases Caused by Candida: A Mini Review." *Frontiers in Microbiology,9:1522*. DOI: 10.3389/fmicb.2018.01522.

Padovani, Gislaine C., Victor P. Feitosa, Salvatore Sauro, Franklin R. Tay, Gabriela Durán, Amauri J. Paula, Nelson Durán, et al. 2018. "A Review on Green Synthesis, Biomedical Applications, and Toxicity Studies of ZnO NPs." *Bioinorganic Chemistry and Applications* 2018, no. 2: 1–12. DOI: 10.1155/2018/3569758.

Patil, Shankargouda, Roopa S. Rao, Barnali Majumdar, and Sukumaran Anil. 2015. "Clinical Appearance of Oral Candida Infection and Therapeutic Strategies." *Frontiers in Microbiology* 6 :1391. DOI: 10.3389/fmicb.2015.01391.

Patra, Jayanta Kumar, and Kwang Hyun Baek. 2016. "Comparative Study of Proteasome Inhibitory, Synergistic Antibacterial, Synergistic Anticandidal, and Antioxidant Activities of Gold Nanoparticles Biosynthesized Using Fruit Waste Materials." *International Journal of Nanomedicine* 11 (September): 4691–4705. DOI: 10.2147/IJN.S108920.

Perlin, David S, Riina Rautemaa-Richardson, and Ana Alastruey-Izquierdo. 2017. "The Global Problem of Antifungal Resistance: Prevalence, Mechanisms, and Management." *The Lancet Infectious Diseases*. 17(12):e383-92. DOI: 10.1016/S1473-3099(17)30316-X.

Pham, Cau D., Naureen Iqbal, Carol B. Bolden, Randall J. Kuykendall, Lee H. Harrison, Monica M. Farley, William Schaffner, et al. 2014. "Role of FKS Mutations in Candida Glabrata: MIC Values, Echinocandin Resistance, and Multidrug Resistance." *Antimicrobial Agents and Chemotherapy* 58, no. 8: 4690–96. DOI: 10.1128/AAC.03255-14.

Pillai, Akhilash Mohanan, Vishnu Sankar Sivasankarapillai, Abbas Rahdar, Jithu Joseph, Fardin Sadeghfar, Ronaldo Anuf A, K. Rajesh, and George Z. Kyzas. 2020. "Green Synthesis and Characterization of Zinc Oxide Nanoparticles with Antibacterial and Antifungal Activity." *Journal of Molecular Structure* 1211 (July): 128107. DOI: 10.1016/j.molstruc.2020.128107.

Prasad, S. 2008. "Nanotechnology in Medicine and Antibacterial Effect of Silver Nanoparticles." *Digest Journal of Nanomaterials and Biostructures* 3, no. 3: 115–22.

Quindós, G, S Gil-Alonso, C Marcos-Arias, E Sevillano, E Mateo, N Jauregizar, and E Eraso. 2019. "Therapeutic Tools for Oral Candidiasis: Current and New Antifungal Drugs." *Medicina Oral, Patologia Oral y Cirugia Bucal* 24, no. 2: e172–80. DOI: 10.4317/medoral.22978.

Rai, Mahendra, Kateryna Kon, Avinash Ingle, Nelson Duran, Stefania Galdiero, and Massimiliano Galdiero. 2014. "Broad-Spectrum Bioactivities of Silver Nanoparticles: The Emerging Trends and Future Prospects." *Applied Microbiology and Biotechnology*. 98(5):1951-61. DOI: 10.1007/s00253-013-5473-x.

Ramos, Ana P, Marcos A.E. Cruz, Camila B Tovani, and Pietro Ciancaglini. 2017. "Biomedical Applications of Nanotechnology." *Biophysical Reviews*. 9:2,79-89. DOI: 10.1007/s12551-016-0246-2.

Salahuddin Siddiqi, Khwaja, Aziz ur Rahman, and Azamal Husen. 2018. "Properties of Zinc Oxide Nanoparticles and Their Activity Against Microbes." *Nanoscale Research Letters* 13: 141. DOI: 10.1186/s11671-018-2532-3.

Saratale, Rijuta Ganesh, Indira Karuppusamy, Ganesh Dattatraya Saratale, Arivalagan Pugazhendhi, Gopalakrishanan Kumar, Yooheon Park, Gajanan S. Ghodake, Ram Naresh Bharagava, J. Rajesh Banu, and Han Seung Shin. 2018. "A Comprehensive Review on Green Nanomaterials Using Biological Systems: Recent Perception and Their Future Applications." *Colloids and Surfaces B: Biointerfaces*. 170:20-35. DOI: 10.1016/j.colsurfb.2018.05.045.

Scala, Angela, Anna Piperno, Alexandru Hada, Simion Astilean, Adriana Vulpoi, Giovanna Ginestra, Andreana Marino, Antonia Nostro, Vincenzo Zammuto, and Concetta Gugliandolo. 2019. "Marine Bacterial Exopolymers-Mediated Green Synthesis of Noble Metal Nanoparticles with Antimicrobial Properties." *Polymers* 11, no. 7: 1157. DOI: 10.3390/polym11071157.

Senthilkumar, S R, and T Sivakumar. 2014. "Green Tea (Camellia Sinensis) Mediated Synthesis of Zinc Oxide (ZnO) Nanoparticles and Studies on Their Antimicrobial Activities." *International Journal of Pharmacy and Pharmaceutical Sciences* 6, no. 6, 461-5 https://innovareacademics.in/journal/ijpps/Vol6Issue6/9715.pdf.

Sherwani, Mohd. Asif, Saba Tufail, Aijaz Ahmed Khan, and Mohammad Owais. 2015. "Gold Nanoparticle-Photosensitizer Conjugate Based Photodynamic Inactivation of Biofilm Producing Cells: Potential for Treatment of *C. albicans* Infection in BALB/c Mice." Edited by Joy Sturtevant. *PLOS ONE* 10, no. 7: e0131684. DOI: 10.1371/journal.pone.0131684.

Silva, Bruna Lallo da, Marina Paiva Abuçafy, Eloisa Berbel Manaia, João Augusto Oshiro Junior, Bruna Galdorfini Chiari-Andréo, Rosemeire C.L.R. Pietro, and Leila Aparecida Chiavacci. 2019. "Relationship between Structure and Antimicrobial Activity of Zinc Oxide Nanoparticles: An Overview." *International Journal of Nanomedicine*. 14:9395 DOI: 10.2147/IJN.S216204.

Sreelakshmi, Ch., K. K. R. Datta, J. S. Yadav, and B. V. Subba Reddy. 2011. "Honey Derivatized Au and Ag Nanoparticles and Evaluation of Its Antimicrobial Activity." *Journal of Nanoscience and Nanotechnology* 11, no. 8: 6995–7000. DOI: 10.1166/jnn.2011.4240.

Subramani, Karthikeyan, Waqar Ahmed, and James K. Hartsfield. 2012. *Nanobiomaterials in Clinical Dentistry. Nanobiomaterials in Clinical Dentistry.* DOI: 10.1016/C2012-0-01361-0.

Sulthana, R. Nigar, and A. Rajanikanth. 2018. "Synthesis of Gold Nanoparticles Using Penicillium Sp. and Their Antibacterial Activity against Human Pathogens." *International Journal of Current Research and Academic Review* 6, no. (7): 19–24. DOI: 10.20546/ijcrar.2018.607.003.

Thema, F T, E Manikandan, M S Dhlamini, and M Maaza. 2015. "Green Synthesis of ZnO Nanoparticles via Agathosma Betulina Natural Extract." *Materials Letters* 161: 124–27. DOI: 10.1016/j.matlet.2015.08.052.

Verma, Ajay, Surya Gautam, Kuldeep Bansal, Neeraj Prabhakar, and Jessica Rosenholm. 2019. "Green Nanotechnology: Advancement in Phytoformulation Research." *Medicines* 6, no. 1: 39. DOI: 10.3390/medicines6010039.

Vijayakumar, S., P. Arulmozhi, N. Kumar, B. Sakthivel, S. Prathip Kumar, and P.K. Praseetha. 2019. "Acalypha Fruticosa L. Leaf Extract Mediated Synthesis of ZnO Nanoparticles: Characterization and Antimicrobial Activities." *Materials Today: Proceedings* (July), 23:73-80. DOI: 10.1016/j.matpr.2019.06.660.

Vijayaraghavan, K., and T. Ashokkumar. 2017. "Plant-Mediated Biosynthesis of Metallic Nanoparticles: A Review of Literature, Factors Affecting Synthesis, Characterization Techniques and Applications." *Journal of Environmental Chemical Engineering*, 5(5):4866-83. DOI: 10.1016/j.jece.2017.09.026.

Zare, Elham, Shahram Pourseyedi, Mehrdad Khatami, and Esmaeel Darezereshki. 2017. "Simple Biosynthesis of Zinc Oxide Nanoparticles Using Nature's Source, and It's in Vitro Bio-Activity." *Journal of Molecular Structure* 1146: 96–103. DOI: 10.1016/j.molstruc.2017.05.118.

Zhaleh, Mohsen, Akram Zangeneh, Samaneh Goorani, Niloofar Seydi, Mohammad Mahdi Zangeneh, Reza Tahvilian, and Elham Pirabbasi. 2019. "In Vitro and in Vivo Evaluation of Cytotoxicity, Antioxidant, Antibacterial, Antifungal, and Cutaneous Wound Healing Properties of Gold Nanoparticles Produced via a Green Chemistry Synthesis Using Gundelia Tournefortii L. as a Capping and Reducing Agent." *Applied Organometallic Chemistry* (June),33(9): e5015. DOI: 10.1002/aoc.5015.

Zhang, Ying, Thabitha P. Shareena Dasari, Hua Deng, and Hongtao Yu. 2015. "Antimicrobial Activity of Gold Nanoparticles and Ionic Gold." *Journal of Environmental Science and Health, Part C* 33, no. 3: 286–327. DOI: 10.1080/10590501.2015.1055161.

[illegible] K., and T. [illegible]. [illegible] "[illegible] Biosynthesis of [illegible] Nanoparticles: A Review on Alternative [illegible] Synthesis, Characterization Techniques and Applications." [illegible] 5(6): [illegible]. DOI: [illegible]

Zare, Ehsan, Shadi Pourseyedi, Mehrdad Khatami, and [illegible]. [illegible]. "Simple Biosynthesis of Zinc Oxide Nanoparticles Using Nature's Source, and It's In Vitro Bio-Activity." Journal of Molecular Structure [illegible]: 96–[illegible]. DOI: 10.1016/j.molstruc.2017.05.118.

[illegible] Zangeneh, [illegible] 2019. "In Vitro and In Vivo Evaluation of [illegible] Cutaneous Wound [illegible] Nanoparticles [illegible] as [illegible]." Applied Organometallic Chemistry [illegible]

[illegible]

Section III

Nanotechnology for Treatment of Communicable Diseases

4 Nanotechnology and Nanomedicine to Combat Ebola Virus Disease

Maluta Steven Mufamadi

CONTENTS

4.1 Introduction 65
4.2 Ebola Virus Pathogenesis: Antiviral, Vaccine and Diagnostic Device Opportunities 66
4.3 Nanotechnology Platforms 67
4.4 Nanotechnology Platforms for Treatment of Ebola Virus Infection 68
    4.4.1 Lipid-Based Nanoparticles 68
    4.4.2 Polymer-Based Nanoparticles 69
    4.4.3 Inorganic Nanoparticles 70
4.5 Nanotechnology-Based Vaccines for Prevention of Ebola Virus Infection 70
4.6 Nanotechnology-Based Approach for Ebola Virus Diagnosis 72
4.7 Nanotechnology in Disinfection and Textile Applications: Future Prospects 73
4.8 Conclusion 74
References 74

4.1 INTRODUCTION

Ebola virus disease (EVD) is one of the deadliest infectious diseases in the world. It causes a severe haemorrhagic fever in humans and non-human primates with the mortality rate of 50–90%. The genus *Ebolavirus* is single-stranded, negative RNA virus that belongs to the Filoviridae family (Kuhn et al., 2010). The virus morphology consists of various structural proteins, including glycoproteins (GP), nucleoprotein (NP) and virion proteins (VPs; VP24, VP30, VP35 and VP40) which differs in function. The GP is responsible for the viral entry into host cells, while NP and VPs together with the RNA-dependent RNA polymerase are responsible for virus particle formation and replication (Huang et al., 2002, Cho and Croyle, 2013). GP, NP and VPs are potential targets for the development of EVD therapeutic agents, vaccines and diagnostic devices (Jeevan et al., 2017; Meyer et al., 2018; Bazzill et al., 2019). Since the first outbreaks in 1978 in Democratic Republic of the Congo

(DRC), many other EVD outbreaks have been reported in other sub-Saharan countries, including South Sudan, Liberia, Sierra Leone, Guinea, Nigeria, Rwanda and Uganda (WHO 2015; Engwa, 2018). The main route of transmission is through direct contact with blood and body fluids of infected persons or wild animals, and/or contact with objects such as needles, medical equipment and clothes contaminated with body fluids of infected persons from EVD (Engwa, 2018). Despite the recently licensed live-attenuated vaccine to protect adults against EVD by the Food and Drug Administration (FDA), there is still a need for the development of a safe and effective vaccine to all species of *Ebolavirus* or *Marburgvirus* to combat Ebola virus (FDA, 2019). Nanotechnology-based approaches are promising to offer better solutions for EVD prevention, diagnosis and treatment (Duan et al., 2015; Chahal et al., 2016; Fan et al., 2019). Nanotechnology is an emerging field of interdisciplinary research that cut across all other science fields, including biology, chemistry, physics, engineering and material sciences (Tarafdar et al., 2013). It deals with the creation and manipulation of the chemical and physical properties of a substance at the nanoscale and molecular level (i.e. 1–100 nm). The application of nanotechnology in medicine specifically promises to offer a unique potential for advances in the treatment, prevention and diagnosis of infectious diseases such as the EVD.

4.2 EBOLA VIRUS PATHOGENESIS: ANTIVIRAL, VACCINE AND DIAGNOSTIC DEVICE OPPORTUNITIES

Ebola virus pathogenesis is governed by interactions between the virus and host cells (Manicassamy and Rong, 2009). Infected host cells such as macrophages, dendritic cells and monocytes support virus transcription and replication which then contributes to EVD. Once you come into contact with blood or body fluids of an infected person with EVD, the virus enters the body through the mucosal membrane of host cells. The entry is governed by the interaction between virus GP and receptors on the host cell, such as T-cell immunoglobulin and mucin domain 1 (TIM-1) (Dube et al., 2010; Kondratowicz et al., 2011). During cell adhesion, the transmembrane GP allows the virus to introduce its contents into host cells and cause cell death or release of cytokines associated with inflammation, promoting haemorrhage fever and vascular leakage. In addition, the GP open reading frame of the virus gives rise to two gene products, a full-length protein (GP, with 150–170 kDa) and a non-structural small glycoprotein (SGP, with 60–70 kDa) that may alter the immune response by inhibiting neutrophil activation. The cell-surface lectins DC-SIGN and L-SIGN can also mediate GP binding to cells through viral carbohydrate determinants (Alvarez et al., 2002; Cho and Croyle, 2013). Post cell adhesion, the virus or its genome enters the cytoplasm followed by uncoating of its membrane and replication of its RNA genome. The genome is then reassembled to a matured virus and released from the infected macrophages, monocytes and dendritic cells, and begins to spread in the lymphatic system and other organs. Ebola virus endosomal escape is governed by interactions between the virus particle and host proteins such as Niemann-Pick

C1 protein. In the cytoplasm or nucleus, the ribonucleocapsid that is transcribed and the viral mRNAs are capped by the viral polymerase (King and Sharom, 2012). Ebola virus transcription and replication are governed by VP30, VP35 and the RNA-dependent RNA polymerase forming the ribonucleoprotein (RNP) complex with the viral genomic RNA. Nucleocapsid formation, viral budding, assembly and release from the host cells are governed by matrix proteins VP40 and VP24, linked to the RNP complex and the inner surface of the viral envelope (Muhlberger et al., 1999; Cho and Croyle, 2013; Hammou et al., 2016; Gordon et al., 2019). The virus particles are enclosed in a lipid bilayer envelope derived from the host cell membrane, and buds through the host endosomal sorting complexes (ESCRT) from the plasma membrane. New virions, complete virus particles released from the host cells, are governed by ESCRT machinery made up of cytosolic proteins (Gordon et al., 2019). Understanding the mechanisms of Ebola virus induced cytopathic effects and replication has facilitated the process of developing effective viral inhibitors or antivirus vaccine and smart diagnostic devices for EVD. Blocking virus entry through the virus' GP and host cell receptors or lectins, DC-SIGN interaction, preventing endosomal escape, interrupting viral RNA replication and budding and release are among the strategies used to fight EVD employing nanotechnology-based approaches (Luczkowiak et al., 2011; Duan et al., 2015; Bazzill et al., 2016; Meyer et al., 2018; Salata et al., 2019).

4.3 NANOTECHNOLOGY PLATFORMS

Many studies in the management of the EVD have tested both cellular experimental assays and *in vivo* pre-clinical platform studies (Thi et al., 2016). The studies focused on gene, drug and vaccine delivery systems against Ebola virus, thus blocking the synthesis of most disease-causing proteins or inhibitors of viral infection. Despite its powerful promises, "Naked" siRNA drug is facing many setbacks such as fast elimination in the blood circulation, poor drug localization at the targeted sites and inability of drug molecules to cross targeted cell membranes to access the cytoplasm where it functions. Appropriate delivery vehicles are therefore essential for realizing the potential of gene technology (Tam et al., 2013; Thi et al., 2015, 2016). Various nanotechnology platforms, organic and inorganic nanoparticles (NPs), promise to offer such a solution as pharmaceutical drug carriers towards the treatment of viral infections such as the Ebola virus disease (Chepurnova et al., 2003; Luczkowiak et al., 2011; Jeevan et al., 2017). Organic NPs are fabricated from proteins, carbohydrates, lipids, oil-water, polymers and other organic compounds (Luczkowiak et al., 2011; Bazzill et al., 2016; Suk et al., 2016). Inorganic NPs are fabricated from metal ions, metal oxide and quantum dot (Abo-State and Partila, 2018). However, engineered nanomaterials or NPs that have potential success in the clinic are influenced by important parameters such as fabrication strategies, e.g. physical properties or stability, drug loading and encapsulation efficiency, target delivery, drug release and toxicity (Thi et al., 2016, Fries et al., 2019). In the EVD, lipid and polymeric NPs are widely researched because of their advantage on the surface modifications with ligands to improve drug delivery specificity and therapeutic outcome.

4.4 NANOTECHNOLOGY PLATFORMS FOR TREATMENT OF EBOLA VIRUS INFECTION

4.4.1 Lipid-Based Nanoparticles

Lipid-based nanoparticle drug delivery systems such as liposome or lipoplexes have been extensively studied. Liposomes consist of a lipid bilayer that is primarily composed of amphipathic phospholipids and/or coated with polyethylene glycol (PEG) as stealth liposomes. As a drug delivery vehicle, a liposome is capable of encapsulating multiple drugs in the structure, hydrophilic drugs in the aqueous core and hydrophobic and lipophilic drugs in the bilayer interface (Mufamadi et al., 2011). Lipoplexes are cationic liposomes (positively charged) with a non-viral vector which is composed of synthetic lipid carriers that can easily bind to negatively charged DNA by charge–charge interactions, and are commonly used in gene-based delivery studies, such as siRNA technology (Hattori, 2017). Exosomes are membrane-bound extracellular vesicles (EVs) that are produced in the endosomal compartment of most eukaryotic cells (Arenaccio et al., 2019). Thi et al. (2015) showed lipid NPs encapsulation and delivery of siRNAs to target the *Zaire ebolavirus* (EBOV) Makona-infected non-human primates (NHPs) model, which closely reproduces human infection. The study reported 100% protection of the animal model using a virus isolate from a lethal challenge post treated lipid nanoparticle encapsulated siRNAs for 28 days. The treated animals had reduced viral load and plasma viremia level, and fully recovered, while the untreated control infected animals succumbed on days 8 and 9. Dunning et al. (2016) demonstrated phase 2 clinical trial studies employing TKM-130803, a lipid-based nanosystem containing siRNA directed to adult patients with severe EVD. TKM-130803 was developed by Tekmira Pharmaceuticals (currently known as the Arbutus Biopharma Corporation). The results showed that the TKM-130803 infusions were well tolerated, although they did not show improved survival of infected patients. In another study, Thi et al. (2016) reported the silencing of the *Sudan ebolavirus* (SUDV) VP35 gene of advanced *Sudan ebolavirus* infection from non-human primates employing LNP-encapsulated siRNAs. During TKM-Ebola clinical trials, 25 rhesus monkeys were challenged with a lethal dose of SUDV strains and LNP-encapsulated siRNAs post challenge, from day 1 to day 5. The outcome of this study showed 100% survival and rapid control of the SUDV replication post challenge with LNP-incorporated siRNA targeting SUDV VP35 gene. LNP-incorporated siRNA or lipid-based nanoparticles drug delivery systems showed therapeutic potential against highly lethal viral infections. Pleet et al. (2016) examined the effect of Ebola structural proteins VP40, GP, NP and virus-like particles (VLPs) in exosomes on recipient immune cells. The study showed that VP40-transfected cells packaged VP40 into exosomes has the capability of inducing apoptosis in recipient immune cells. In another study by Bazzill et al. (2016), LNP-incorporated siRNA with Ebola GP were also shown to have the capability for induction of humoral immunity against Ebola infection. Formulation of the nanoemulsion for drug delivery system is achieved mainly by colloidal dispersion of oil and water with a surfactant as a stabilizer. Nanoemulsion drug delivery systems are lipid-based formulations capable of improving the bioavailability of hydrophobic drugs (Jaiswal et al., 2015). In an earlier *in vitro* study, Chepurnova et al. (2003) investigated the antiviral activity of lipid NPs inactivation of Ebola virus using surfactant nanoemulsion. The study was tested on the *Zaire ebolavirus* strain obtained from Vero cell culture fluid and from the blood of Ebola virus infected monkeys. The study demonstrated that the nanoemulsion is capable of inactivation of the virus and also an effective disinfectant of Ebola virus in Vero cell culture.

4.4.2 Polymer-Based Nanoparticles

Similar to lipid NPs, polymer NPs are amongst the most widely explored materials for gene delivery, drug delivery system and pharmaceuticals (Bernkop-Schnürch and Dünnhaupt, 2012; Buschmann et al., 2013; Patra et al., 2018). The most common polymers used for medicine and drug delivery system or approved by the FDA include poly(ethylene glycol) (PEG), chitosan, poly(lactic-*co*-glycolic acid) (PLGA), poly(lactic acid) (PLA), poly(glutamic acid) (PGA) and polyethylenimine (PEI) (Lim et al., 2020). These polymeric materials have gained a lot of attention in NP formulations due to their favourable safety profiles with minimal toxicity, biodegradability and biocompatibility. PEG has commonly been used to surface functionalized NPs, improve stability in the bloodstream, in both polymers NPs and lipids NPs, and lower immunogenicity (Suk et al. 2016). Other polymers such as chitosan has gained interest in gene or nucleic acid delivery platforms due to its surface charge, which enables it to complex with DNA to form polymer–DNA complexes or polyplexes, and to its capability to protect DNA against nuclease degradation. The polyplexes structure is formed through the electrostatic binding force between the cationic chitosan and the anionic structure of DNA (Bravo-Anaya et al. 2016). In a study performed by Geisbert et al. (2006), they demonstrated the formation of nanopolyplexes using siRNAs and PEI. Four siRNAs were engineered to target the polymerase gene of the Zaire species of EBOV (ZEBOV). The results showed that nanopolyplexes have the capability to combat Ebola virus by protecting guinea pigs against viremia and death post exposure with a lethal challenge of ZEBOV. Luczkowiak et al. (2011) demonstrated viral inhibition using pseudosaccharide (polymannosylated ligands) functionalized dendrimers as adequate ligands to block DC-specific C-type lectin-like cell-surface receptor. Virus-like glycodendrinanoparticles have been reported to be capable of blocking Ebola viral infection using a novel quasi-equivalent nested polyvalency approach (Ribeiro-Viana et al., 2012). Munoz et al. (2016) demonstrated the use of water-soluble tridecafullerenes (also called superballs or giant globular multivalent glycofullerenes) to block cell-surface lectin receptors, as a novel strategy towards the inhibition of Ebola virus entry into cells. The infection assay showed potent inhibition of cell infection by the superballs (an artificial Ebola virus) at very low concentration, with half-maximum inhibitory concentrations in the subnanomolar range. The tridecafullerenes were decorated with 120 peripheral carbohydrate subunits, a hexakis adduct of fullerene C60 synthesis employing copper-catalysed azide-alkyne cycloaddition. The tridecafullerenes were later designed into multivalent glycosylated nanostructures to minimize the development of resistance by

Ebola virus mutations (Illescas et al., 2017). The results of this study showed tridecafullerenes as a biocompatible carbon platform of scaffold fullerene C60 connected to 12 sugars-containing fullerene units, with a total of 120 mannoses, a unique symmetrical and 3D globular structure exhibition with an outstanding antiviral activity against Ebola virus with IC_{50} values in the subnanomolar range. In another study, Jeevan et al. (2017) reported Ebola virus inhibition employing graphene nanosheets. This potential pharmacological agent, a nanoparticle-based therapy made up of graphene, was able to disrupt the Ebola virus matrix through protein VP40 and graphene interactions that cause the disruption of the virus life cycle. In a recent study, three different nanocarbon-based glycoconjugates (functionalized with either glycodendrons or glycofullerenes) were used as a virus-mimicking nanocarbon platform to inhibit Ebola virus infection (Rodríguez-Pérez et al., 2017). The multivalent carbohydrates and the nanocarbons showed efficiency in blocking DC-SIGN-mediated viral infection when used in an artificial Ebola virus infection model assay and a cellular experimental assay, respectively. The results showed that nanocarbons functionalized with glycofullerenes are strong inhibitors of Ebola viral infection.

4.4.3 Inorganic Nanoparticles

Among other nanostructures, metallic NPs are used for drug delivery system, antiviral agents, imaging and medical applications (Galdiero et al., 2011; Mufamadi et al., 2019). Metallic NPs are fabricated from metal ions, metal oxides and quantum dots employing various methods such as chemical, physical and biological methods (Abo-State and Partila, 2018). However, chemical and physical methods use toxic chemicals that are not suitable for medical applications. On the other hand, biological synthesis of metal NPs are non-toxic and environment-friendly and uses biological materials such as plants extracts and microorganisms and reducing and stabilizing agents (Monika et al., 2015; Haggag et al., 2019). Metallic NPs such as gold nanoparticles (AuNPs) and silver nanoparticles (AgNPs) are widely researched for antimicrobial or antiviral activities due to their unique physical and chemical properties and inherent inhibitory potential (Galdiero et al., 2011; Avilala and Golla, 2019). The antiviral activities of metallic NPs include interaction with the surface GP of the viral envelope or binding with viral and/or cellular factors in the host cells and preventing viral replication (Bowman et al., 2008; O'Connell, 2014; Singh et al., 2017).

4.5 NANOTECHNOLOGY-BASED VACCINES FOR PREVENTION OF EBOLA VIRUS INFECTION

Antiviral vaccine development has been the most successful in preventing the epidemic and pandemic viral infections in the past, employing either biotechnological or biomedical approaches (Graham, 2013; Yassine et al., 2015; Graham and Sullivan, 2018). The cost, safety and immunogenicity of vaccines or antiviral agents used to block viral infection are very essential for the development of an ideal vaccine (Brito et al., 2014; Rappuoli et al., 2014). Among these, nanotechnology-based vaccines have been identified as potent vaccine candidates (Cho and Croyle, 2013; Bogers et al., 2015). Some of them are already in clinical trials and/or approved by the US FDA for protective immunity against lethal Ebola virus (Cho and Croyle, 2013; Chahal et al., 2016). Various candidates of nanotechnology-based vaccines against Ebola virus infection have been explored; in this section, we will highlight the recent studies only (Chahal et al., 2016; Yang et al., 2017; Meyer et al., 2018; Bazzill et al., 2019; Fan et al., 2019). Chahal et al. (2016) developed a rapid-response, fully synthetic and adjuvant-free dendrimer NPs vaccine platform with a broad protection against Ebola virus, *Toxoplasma gondii* and influenza H1N1. The dendrimer NPs were encapsulated with multiple antigen-producing RNAs, including replicons, and was responsible for eliciting $CD8^+$ T-cell response and the production of appropriate antibody response, and generated protective immunity with a single dose. In another study, Yang et al. (2017) reported the development of DNA vaccine coated on PLGA-PLL/γPGA NPs as a highly immunogenic platform for vaccine against Ebola virus. The nanoparticle delivery system showed increases in vaccine thermostability and immunogenicity compared with free vaccine. Nanoparticles-based vaccine administered to the skin using a microneedle patch produced stronger immune responses than did intramuscularly administered NPs vaccine. Meyer et al. (2018) developed two mRNA vaccines based on the Ebola virus envelope GP formulated with lipid NPs. The results of this study demonstrated efficient delivery of mRNAs by the lipid NPs. *In vivo* studies in guinea pigs showed that modified mRNA-based vaccines are capable of eliciting protective immunity against lethal Ebola virus, by inducing Ebola virus specific IgG neutralizing antibody responses. The results also showed that a vaccination of modified mRNAs-based vaccine conferred protection with 100% survival of guinea pigs after Ebola virus infection. In a recent study, two forms of lipid-based nanoparticles loaded with recombinant Ebola virus glycoprotein (rGP) were produced by Bazzill et al. (2019). *In vivo* study in mice showed the NPs vaccine efficiently generated germinal centre B cells and polyfunctional T-cell responses, while inducing robust neutralizing antibody responses. A phase 1 clinical trial study in healthy adults showed recombinant EBOV NPs vaccine formulated with Matrix-M™ adjuvant elicited potent, robust and persistent immune response against Ebola virus infection (Fries et al., 2019). In another recent study, Fan et al. (2019) demonstrated that a recombinant protein-based vaccine elicited potent immune activation with protein antigens and protected mice against Ebola virus infection. Multilamellar vaccine particle system (MVPS) composed of lipid–hyaluronic acid multi-cross-linked hybrid NPs was used for delivery of protein antigens against Ebola virus. The MVPS were efficiently accumulated in dendritic cells and promoted antigen processing. *In vivo* study in mice immunized with the MVPS elicited long-lasting antigen specific to $CD8^+$ and $CD4^+$ T-cell immune responses, including humoral immunity. MVPS delivering Ebola virus GP achieved 80% protection rate against lethal Ebola virus infection with single-dose vaccination. The above studies suggest that the development of nanotechnology-based vaccine approaches is safe and a promising platform to generate protective immunity (or high immunogenicity) against Ebola virus infection, with just a single-dose vaccine.

4.6 NANOTECHNOLOGY-BASED APPROACH FOR EBOLA VIRUS DIAGNOSIS

The conventional methods to confirm the presence of Ebola virus include enzyme-linked immunosorbent assay (ELISA), electron microscopy and reverse-transcriptase polymerase chain reaction (RT-PCR). Although the RT-PCR method is very sensitive, it is also very expensive, requires specialized equipment and highly trained staff to work in a sophisticated laboratory environment – a Biosafety level 4 (BSL-4) containment laboratory (Butler, 2014). Therefore, there is a need to develop highly sensitive detection techniques and user-friendly and inexpensive diagnostic devices to prevent and control Ebola virus infection rates, particularly for those living in rural and remote areas. Nanotechnology-based diagnostic approaches have been reported as an enabling tool for rapid, highly sensitive and cost-effective Ebola virus diagnosis (Yanik et al., 2010; Daaboul et al., 2014; Duan et al., 2015; Yen et al., 2015; Li et al., 2017; Agrawal et al., 2018). The nanozyme strip test has been used for detecting of Ebola virus (Duan et al., 2015). Duan and colleagues engineered the nanozyme strip using Fe_3O_4 magnetic nanoparticle (MNP) as a nanozyme probe. The results showed the nanozyme strip to be capable of detecting Ebola virus GP at 1 ng/ml. The test was very simple, faster (within 30 min) and 100-fold more sensitive compared with the standard ELISA strip method. The AuNPs binding assay has been investigated for the capability of Ebola virus diagnosis. Feizpour et al. (2015) demonstrated the use of AuNPs binding assay, an optical approach for measuring the lipid contents in Ebola VLPs and viral envelope membranes such as phosphatidylserine (PS) and monosialotetrahexosylganglioside (GM1). Both PS and GM1 are examples of lipids that are known to contribute to virus attachment, uptake or mediate interactions between virus particles and host cells. This novel method was able to quantify the lipid contents in VLP and viral envelope membranes employing plasmonic NPs. A similar study, using coloured AgNPs with optical properties, has reported to be capable of detecting Ebola virus, using NPs conjugated to antibody in a lateral flow strip (Yen et al., 2015). Another example of metallic NPs, platinum NPs incorporated volumetric bar-chart chip, showed the capability for detecting Ebola virus DNA as low as 16 pM (Wang et al., 2016). The results also showed the platinum NPs to be highly sensitive, and catalysts integrated with the hybridization chain reaction. In addition, platinum NPs were capable of quantifying and visualizing detection of single base mismatches of DNA hybridization. Tsang et al. (2016) demonstrated a low-cost, rapid and ultrasensitive detection employing upconversion NPs conjugated with oligonucleotide probe and AuNPs linked to a target Ebola virus oligonucleotide. The results also showed an increased, light interaction from 523, 546 and 654 nm throughout the nanopore walls of a nanoporous alumina membrane. Li et al. (2017) demonstrated a digital triplex DNA assay based on plasmonic nanocrystals specific to Ebola virus (EV), Variola virus (VV) and *Bacillus anthracis* (BA). Metal nanocrystals encoded with AuNPs were engineered specific to Ebola virus. The results showed a high detection of multiple DNA molecules, with detection limits of 0.5–3.0 fM. In a recent study, Hu et al. (2017) developed novel multifunctional nanosphere and compared it with quantum dots and AuNPs. This dual-signal readout nanosphere comprised of fluorescence signalling for quantitative detection and calorimetric signalling for visual detection. Additionally, this assay enables naked eye detection of Ebola virus glycoprotein as low as 2 ng/ml within a period of 20 min, with a quantitative detection limit of 0.18 ng/ml. Agrawal et al. (2018) developed optical biosensors based on the spherical polymer NPs employed with scattered light biosensing techniques. In this case, nanostructures were fabricated through resembling (individual filamentous virions) two-layer substrate of the silicon wafer overlaid with the silicon oxide film employing electron beam lithography. Upon fabrication, the polymer NPs were characterized for their dimensions by scanning electron microscopy (SEM) and atomic force microscope for subsequent correlation to the signal of those from Ebola VLPs. The results showed Ebola VLPs generated using matrix protein VP40, which exhibit characteristic filamentous morphology. Lipid-based nanoparticles, liposomes have also been researched as nanocarriers towards an Ebola virus diagnostic approach. Yazan et al. (2016) developed liposome viral-like NPs encapsulated with Ebola nucleic acid, Ebola partial GP gene (700 bp) and whole GP gene (2031 bp). SEM and transmission electron microscopy (TEM) analyses revealed the formation of stable liposome NPs with a size ranging from 80 to 100 nm with zeta potential ranging from –30 and –70 mV. In addition, a PCR method was used to quantitate liposomes without extraction, using triton X-100 (as low as 0.5% per reaction) to improve the amplification.

4.7 NANOTECHNOLOGY IN DISINFECTION AND TEXTILE APPLICATIONS: FUTURE PROSPECTS

The Ebola virus disease is a highly contagious infection, and it is transmitted among humans through close and direct physical contact with body fluids of the Ebola-infected person (Engwa, 2018). Health workers, friends and family of infected people are at a high risk of being infected by this deadly pathogen due to the environmental and patient care items contamination. Therefore, disinfection of the contaminated patient care units, instruments, clothes, homes and hospital premises is the most important step to control and prevent the spread of infectious diseases such as EVD before patient isolation (Rutala and Weber, 1999; Cozad and Jones, 2003). Although there are many disinfection methods, nanomaterial-based disinfectants have been well documented due to their antimicrobial effects and unique chemical and physical properties. NPs as disinfectant agents with great virucidal effects include silver, gold, titanium and zinc (Khandelwal et al., 2014; Tarhan, 2018). A recent study showed surfactant nanoemulsion as an emerging disinfectant agent with great virucidal effect against Ebola virus (Chepurnov et al., 2003). Surfactant nanoemulsion is fabricated from detergents and vegetable oil suspended in water. Another nanotechnological approach that could assist in the prevention of the spread of Ebola virus is employed in the textile clothing industry. Functional nano-coated textiles possess many advantages, including antimicrobial activities and hydrophobic (water and stain repellent) and self-cleaning properties (Dastjerdi et al., 2012; Wu et al., 2016). Since Ebola is spread through contact with fluids, the nano-coated personal protective equipment (PPE, e.g. body overall, face mask, faceshield, safety goggle, gloves, and shoe cover) therefore could assist in preventing contamination and spreading of the Ebola virus disease during patient care. The advantage is that nano-coated PPE are capable of trapping and killing viruses immediately upon contact.

4.8 CONCLUSION

In summary, an understanding of the mechanisms of Ebola virus entry and/or adhesion processes on surface receptors of the host cells, endosomal escape, cytopathic effects, transcription, replication, budding and virion release from infected host cells has facilitated the process of vaccine, antiviral therapy and smart diagnostic device development against Ebola virus infection. In addition, it has also helped towards unpacking or providing new information about the complex Ebola virus pathogenesis and the immune response to the virus. Nanotechnology-based approaches promise to offer unique opportunities to advanced treatment, prevention and diagnosis against Ebola haemorrhagic fever. In addition, nanotechnology-based approaches, in particular lipid-based nanoparticles, promises to offer a more effective and inexpensive vaccine and antiviral therapy against EVD and nano-enabled products that are safe for human consumption. Majority of nano-enabled products that are made of lipid-based nanoparticles are already approved by the FDA or in clinical trials. Nano-disinfectants and nano-coated fabrics are emerging products that promise to offer unique opportunities towards the combat of Ebola virus. Both nano-disinfectants and nanotextile products are capable of reducing the spread of Ebola virus infection from human-to-human and/or contaminated patient care items. However, there is still a need for more work to be done in both nanotechnology-based disinfectant and textile applications in order to confirm the development of nano-enabled products that are safe for both humans and the environment.

REFERENCES

Abo-State, M.A.M., and A.M. Partila. 2018. "Production of silver nanoparticles (AgNPs) by certain bacterial strains and their characterization." *Novel Research in Microbiology Journal* 1, no. (2): 19–32. DOI: 10.21608/NRMJ.2018.5834.

Agrawal, A., Majdi, J., Clouse, K.A., Stantchev, T. 2018. "Electron-beam-lithographed nanostructures as reference materials for label-free scattered-light biosensing of Single Filoviruses." *Sensors* 18, no. 6: 23. DOI: 10.3390/s18061670.

Alvarez, C.P., Lasala, F., Carrillo, J., Muñiz, O., Corbí, A.L., Delgado, R. 2002. "C-type lectins DC-SIGN and L-SIGN mediate cellular entry by Ebola virus in cis and in trans." *Journal of Virology* 76: 6841–6844. DOI: 10.1128/jvi.76.13.6841-6844.2002.

Arenaccio. C., Chiozzini, C., Ferrantelli, F., Leone, P., Olivetta, E., Federico, M. 2019. "Exosomes in therapy: Engineering, pharmacokinetics and future applications." *Current Drug Targets* 20, no. 1: 87–95. DOI: 10.2174/1389450119666180521100409.

Avilala, J., and N. Golla. 2019. "Antibacterial and antiviral properties of silver nanoparticles synthesized by marine actinomycetes". *International Journal of Pharmaceutical Sciences Review and Research* 10, no. (3): 1223–1228. DOI: 10.13040/IJPSR.0975-8232.10(3).1223-28.

Bazzill, J.D., Cooper, C.L., Fan, Y., Bavari, S., Moon, J.J. 2016. "Lipid nanoparticles incorporated with Ebola glycoprotein for induction of humoral immunity against Ebola infection." *Journal of Immunology* 196, no. 1 Supplement: 13–76.

Bazzill, J.D., Stronsky, S.M., Kalinyak, L.C., Ochyl, L.J., Steffens, J.T., van Tongeren, S.A., Cooper, C.L., Moon, J.J. 2019. "Vaccine nanoparticles displaying recombinant Ebola virus glycoprotein for induction of potent antibody and polyfunctional T cell responses." *Nanomedicine* 18: 414–425. DOI: 10.1016/j.nano.2018.11.005.

Bernkop-Schnürch, A., and S. Dünnhaupt. 2012. "Chitosan-based drug delivery systems." *European Journal of Pharmaceutics and Biopharmaceutics* 81, no. 3: 463–469. DOI: 10.1016/j.ejpb.2012.04.007.

Bogers, W.M., Oostermeijer, H., Mooij, P., Koopman, G., Verschoor, E.J., Davis, D., et al. 2015."Potent immune responses in rhesus macaques induced by nonviral delivery of a self-amplifying RNA vaccine expressing HIV type 1 envelope with a cationic nanoemulsion." *Journal of Infectious Diseases* 211, no. 6: 947–955. DOI: 10.1093/infdis/jiu522.

Bowman, M.C., Ballard, T.E., Ackerson, C.J., Feldheim, D.L., Margolis, D.M., Melander, C. 2008. "Inhibition of HIV fusion with multivalent gold nanoparticles." *Journal of the American Chemical Society* 130n no. 22: 6896–6897. DOI: 10.1021/ja710321g.

Bravo-Anaya, L.M., Soltero, J.F., Rinaudo, M. 2016. "DNA/chitosan electrostatic complex." *International Journal of Biological Macromolecules* 88: 345–353. DOI: 10.1016/j.ijbiomac.2016.03.035.

Brito, L.A., Chan, M., Shaw, C.A., Hekele, A., Carsillo, T., Schaefer, M., Archer, J.2014. "A cationic nanoemulsion for the delivery of next-generation RNA vaccines." *Molecular Therapy* 22, no. 12: 2118–2129. DOI: 10.1038/mt.2014.133.

Buschmann, M.D., Merzouki, A., Lavertu, M., Thibault, M., Jean, M., Darras, V. 2013. "Chitosans for delivery of nucleic acids." *Advanced Drug Delivery Reviews* 65, no. 9: 1234–1270. DOI: 10.1016/j.addr.2013.07.005.

Butler, D. 2014. "Ebola experts seek to expand testing." *Nature* 516, no. 7530:154–155. DOI: 10.1038/516154a.

Chahal, J.S., Khan, O.F., Cooper, C.L., McPartlan, J.S., Tsosie, J.K., Tilley, L.D., Sidik, S.M., et al. 2016. "Dendrimer-RNA nanoparticles generate protective immunity against lethal Ebola, H1N1 influenza, and Toxoplasma gondii challenges with a single dose." *Proceedings of the National Academy of Sciences of the United States of America* 113, no. 29: E4133–E4142. DOI: 10.1073/pnas.1600299113.

Chepurnova, A.A., Bakulina, L.F., Dadaeva, A.A., Ustinova, E.N., Chepurnova, T.S., Baker, J.R. 2003. "Inactivation of Ebola virus with a surfactant nanoemulsion." *Acta Tropica* 87, no. 3: 315–332. DOI: 10.1016/s0001-706x (03)00120-7.

Choi, J.H., and M.A Croyle. 2013. "Emerging Targets and Novel Approaches to Ebola Virus Prophylaxis and Treatment." *BioDrugs* 27, no. 6: 1–30. doi:10.1007/s40259-013-0046-1.

Cozad, A., and R.D. Jones. 2003. "Disinfection and the prevention of infectious disease." *American Journal of Infection Control* 31, no. 4: 243–254. DOI: 10.1067/mic.2003.49.

Daaboul, G.G., Lopez, C.A., Chinnala, J., Goldberg, B.B., Connor, J.H., Ünlü, M.S. 2014. "Digital sensing and sizing of vesicular Stomatitis virus pseudotypes in complex media; A model for Ebola and Marburg detection." *ACS Nano* 8, no. 6: 6047–6055. DOI: 10.1021/nn501312q.

Dastjerdi, R., Montazer, M., Stegmaier, T., Moghadam, M.B. 2012. "A smart dynamic self-induced orientable multiple size nano-roughness with amphiphilic feature as a stain-repellent hydrophilic surface." *Colloids and Surfaces B: Biointerfaces* 91: 280–290. DOI: 10.1016/j.colsurfb.2011.11.015.

Duan, D., Fan, K., Zhang, D., Tan, S., Liang, M., Liu, Y., Zhang, J., et al. 2015. "Nanozyme-strip for rapid local diagnosis of Ebola." *Biosensors and Bioelectronics* 74: 134–141. DOI: 10.1016/j.bios.2015.05.025.

Dube, D., Schornberg, K.L., Shoemaker, C.J., Delos, S.E., Stantchev, T.S., Clouse, K.A., Broder, C.C., White, J.M. 2010. "Cell adhesion-dependent membrane trafficking of a binding partner for the ebolavirus glycoprotein is a determinant of viral entry." *Proceedings of the National Academy of Sciences of the United States of America* 107, no. 38: 16637–16642. DOI: 10.1073/pnas.1008509107.

Dunning, J., Sahr, F., Rojek, A., Rojek, A., Gannon, F., Carson, G., Idriss, B., Massaquoi, T., et al. 2016. "Experimental treatment of Ebola virus disease with TKM-130803: A single-arm phase 2 clinical trial." *PLoS Medicine* 13, no. 4: e1001997. DOI: 10.1371/journal.pmed.1001997.

Engwa, G.A. 2018. "Ebola virus disease: Progress so far in the management of the disease." In *Current Topics in Tropical Emerging Diseases and Travel Medicine*. London: IntechOpen. DOI: 10.5772/intechopen.79053

Fan, Y., Stronsky, S.M., Xu, Y., Steffens, J.T., van Tongeren, S.A., Erwin, A., Cooper, C.L., Moon, J.J. 2019. "Multilamellar vaccine particle elicits potent immune activation with protein antigens and protects mice against Ebola virus infection." *ACS Nano* 13, no. 10: 11087–11096. DOI: 10.1021/acsnano.9b03660.

FDA (Food and Drug Administration). 2019. "First FDA-approved vaccine for the prevention of Ebola virus disease, marking a critical milestone in public health preparedness and response". Food and Drug Administration (FDA), December 19. Accessed 04 January 2020. https://www.fda.gov/news-events/press-announcements/first-fda-approved-vaccine-prevention-ebola-virus-disease-marking-critical-milestone-public-health.

Feizpour, A., Yu, X., Akiyama, H., Miller, C.M., Edmans, E., Gummuluru, S., Reinhard, B.M. 2015. "Quantifying lipid contents in enveloped virus particles with plasmonic nanoparticles." *Small* 11, no. 13: 1592–1602. DOI: 10.1002/smll.201402184.

Fries, L., Cho, I., Krähling, V., Fehling, S.K., Strecker, T., Becker, S., Hooper, J.W., et al. 2019. "A randomized, blinded, dose-ranging trial of an Ebola virus glycoprotein (EBOV GP) nanoparticle vaccine with matrix-M™ adjuvant in healthy adults." *Journal of Infectious Diseases* 222, no. 4: jiz518. DOI: 10.1093/infdis/jiz518.

Galdiero, S., Falanga, A., Vitiello, M., Cantisani, M., Marra, V., Galdiero, M. 2011. "Silver nanoparticles as potential antiviral agents." *Molecules* 16, no. 10: 8894–8918. DOI: 10.3390/molecules16108894.

Geisbert, T.W., Hensley, L.E., Kagan, E., Yu, E.Z., Geisbert, J.B., Daddario-DiCaprio, K., Fritz, E.A., et al. 2006. "Postexposure protection of guinea pigs against a lethal Ebola virus challenge is conferred by RNA interference." *Journal of Infectious Diseases* 193, no. 12: 1650–1657. DOI: 10.1086/504267.

Gordon, T.B., Hayward, J.A., Marsh, G.A., Baker, M.L., Tachedjian, G. 2019. "Host and viral proteins modulating Ebola and Marburg virus egress." *Viruses* 11, no. 1: E25. DOI: 10.3390/v11010025.

Graham, B.S. 2013. "Advances in antiviral vaccine development." *Immunological Reviews* 255, no. 1: 230–242. DOI: 10.1111/imr.12098.

Graham, B.S., and N.J Sullivan. 2018. "Emerging viral diseases from a vaccinology perspective: Preparing for the next pandemic." *Nature Immunology* 19, no. 1: 20–28. DOI: 10.1038/s41590-017-0007-9.

Haggag, E.G., Elshamy, A.M., Rabeh, M.A., Gabr, N.M., Salem, M., Youssif, K.A., Samir, A., Bin Muhsinah, A., Alsayari, A., Abdelmohsen, U.R. 2019. "Antiviral potential of green synthesized silver nanoparticles of *Lampranthus coccineus* and *Malephora lutea*." *International Journal of Nanomedicine* 14: 6217–6229. DOI: 10.2147/IJN.S214171.

Hammou, R.A., Kasmi, Y., Khataby, K., Laasri, E.L., Boughribil, S., Ennaji, M.M. 2016. "Roles of VP35, VP40 and VP24 proteins of Ebola virus in pathogenic and replication mechanisms." In *Ebola*, 101–117. London: IntechOpen. DOI: 10.5772/63830.

Hattori, Y. 2017. "Progress in the development of lipoplex and polyplex modified with anionic polymer for efficient gene delivery." *Journal of Genetic Medicine and Gene Therapy* 1: 003–018.

Hu, J., Jiang, Y.Z., Wu, L.L., Wu, Z., Bi, Y., Wong, G., Qiu, X., Chen, J., Pang, D.W., Zhang, Z.L. 2017. "Dual-signal readout nanospheres for rapid point-of-care detection of Ebola virus glycoprotein." *Analytical Chemistry* 89, no. 24: 13105–13111. DOI: 10.1021/acs.analchem.7b02222.

Huang, Y., Xu, L., Sun, Y., Nabel, G.J. 2002. "The assembly of Ebola virus nucleocapsid requires virion-associated proteins 35 and 24 and posttranslational modification of nucleoprotein." *Molecular Cell* 10, no. 2: 307–316. DOI: 10.1016/s1097-2765(02)00588-9.

Illescas, B.M., Rojo, J., Delgado, R., Martín, N. 2017. "Multivalent glycosylated nanostructures to inhibit Ebola virus infection." *Journal of the American Chemical Society* 139, no. 17: 6018–6025. DOI: 10.1021/jacs.7b01683.

Jaiswal, M., R. Dudhe, and P.K. Sharma. 2015. "Nanoemulsion: An advanced mode of drug delivery system." *3 Biotech* 5, no. 2: 123–127. DOI: 10.1007/s13205-014-0214-0.

Jeevan, B.G.C., Pokhrel, R., Bhattarai, N., Johnson, K.A., Gerstman, B.S., Stahelin, R.V., Chapagain, P.P. 2017. "Graphene-VP40 interactions and potential disruption of the Ebola virus matrix filaments." *Biochemical and Biophysical Research Communications* 493, no. 1: 176–181. DOI: 10.1016/j.bbrc.2017.

Khandelwal N., Kaur G., Kumara N., Tiwari A. 2014. "Application of silver nanoparticles in viral inhibition: A new hope for antivirals." *Digest Journal of Nanomaterials and Biostructures* 9: 175–186.

King, G., and F.J. Sharom. 2012. "Proteins that bind and move lipids: MsbA and NPC1." *Critical Reviews in Biochemistry and Molecular Biology* 47, no. 1: 75–95. DOI: 10.3109/10409238.2011.636505.

Kondratowicz, A.S., Lennemann, N.J., Sinn, P.L., Davey, R.A., Hunt, C.L., Moller-Tank, S., Meyerholz, DK., et al. 2011. "T-cell immunoglobulin and mucin domain 1 (TIM-1) is a receptor for Zaire Ebola virus and Lake Victoria Marburg virus. *Proceedings of the National Academy of Sciences of the United States of America* 108, no. 20: 8426–8431. DOI: 10.1073/pnas.1019030108.

Kuhn, J.H., Becker, S., Ebihara, H., Geisbert, T.W., Johnson, K.M., Kawaoka, Y., Lipkin, W.I. 2010. "Proposal for a revised taxonomy of the family Filoviridae: Classification, names of taxa and viruses, and virus abbreviations." *Archives of Virology* 155, no. 12: 2083–2103. DOI: 10.1007/s00705-010-0814-x.

Li, G., Zhu, L., He, Y., Tan, H., Sun, S. 2017. "Digital triplex DNA assay based on plasmonic nanocrystals." *Analytical and Bioanalytical Chemistry* 409, no. 14: 3657–3666. DOI: 10.1007/s00216-017-0307-9.

Lim, M., Badruddoza, A.Z.M., Firdous, J., Azad, M., Mannan, A., Al-Hilal, T.A., Cho, C.S., Islam, M.A. 2020. "Engineered nanodelivery systems to improve DNA vaccine technologies." *Pharmaceutics* 12, no. 1: E30. DOI: 10.3390/pharmaceutics12010030.

Luczkowiak, J., Sattin, S., Sutkevičiūtė, I., Reina, J.J., Sánchez-Navarro, M., Thépaut, M., Martínez-Prats L., et al. 2011. "Pseudosaccharide functionalized dendrimers as potent inhibitors of DC-SIGN dependent Ebola pseudotyped viral infection." *Bioconjugate Chemistry* 22, no. 7: 1354–1365. DOI: 10.1021/bc2000403.

Manicassamy, B., and L. Rong. 2009. "Expression of ebolavirus glycoprotein on the target cells enhances viral entry." *Virology Journal* 6: 75. DOI: 10.1186/1743-422X-6-75.

Meyer, M., Huang, E., Yuzhakov, O., Ramanathan, P., Ciaramella, G., Bukreyev, A. 2018. "Modified mRNA-based vaccines elicit robust immune responses and protect guinea pigs from Ebola virus disease." *Journal of Infectious Diseases* 217, no. 3: 451–455. DOI: 10.1093/infdis/jix592.

Monika, B., Anupam, B., Madhu, S., Priyanka, K. 2015. "Green synthesis of gold and silver nanoparticles." *Research Journal of Pharmaceutical, Biological and Chemical Sciences* 6, no. 3: 1710–1716.

Mufamadi, M.S., Pillay, V., Choonara, Y.E., Du Toit, L.C., Modi, G., Naidoo, D., Ndesendo, V.M. 2011. "A review on composite liposomal technologies for specialized drug delivery." *Journal of Drug Delivery* 2011: 939851. DOI: 10.1155/2011/939851.

Mufamadi, M.S., George, J., Mazibuko, Z., Tshikalange, T.E. 2019. "Cancer bionanotechnology: Biogenic synthesis of metallic nanoparticles and their pharmaceutical potency." In *Microbial Nanobionics. Nanotechnology in the Life Sciences*, edited by Prasad R, 229–252. Switzerland: Springer.

Muhlberger, E., Weik, M., Volchkov, V.E., Klenk, H.D., Becker, S. 1999. "Comparison of the transcription and replication strategies of Marburg virus and Ebola virus by using artificial replication systems." *Journal of Virology* 73, no. 3: 2333–2342. PMC104478.

Munoz, A., Sigwalt, D., Illescas, B.M., Luczkowiak, J., Rodríguez-Pérez, L., Nierengarten,I., Holler, M., et al. 2016. "Synthesis of giant globular multivalent glycofullerenes as potent inhibitors in a model of Ebola virus infection." *Nature Chemistry* 8: 50–57. DOI: 10.1038/nchem.2387.

O'Connell J. 2014. "Attack Ebola on a nanoscale." News@Northearstern, August 08. Accessed 13 January 2020. https://news.northeastern.edu/2014/08/08/ ebolananoscale/.

Patra, J.K., Das, G., Fraceto, L.F., Campos, E.V.R., Rodriguez-Torres, M.D.P., Acosta-Torres, L.S., Diaz-Torres, L.A., et al. 2018. "Nano based drug delivery systems: Recent developments and future prospects." *Journal of Nanobiotechnology* 16, no. 1: 71. DOI: 10.1186/s12951-018-0392-8.

Pleet, M.L., Mathiesen, A., DeMarino, C., Akpamagbo, Y.A., Barclay, R.A., Schwab, A., Iordanskiy, S., et al. 2016. "Ebola VP40 in exosomes can cause immune cell dysfunction." *Frontiers in Microbiology* 7, no. 7: 1765. eCollection 2016. DOI: 10.3389/fmicb.2016.01765.

Rappuoli, R., Pizza, M., Del Giudice, G., De Gregorio. E. 2014. "Vaccines, new opportunities for a new society." *Proceedings of the National Academy of Sciences of the United States of America* 111, no. 34: 12288–12293. DOI: 10.1073/pnas.1402981111.

Ribeiro-Viana, R., Sánchez-Navarro, M., Luczkowiak, J., Koeppe, J.R., Delgado, R., Rojo, J., Davis, B.G. 2012. "Virus-like glycodendrinanoparticles displaying quasi-equivalent nested polyvalency upon glycoprotein platforms potently block viral infection." *Nature Communications* 3: 1303. DOI: 10.1038/ncomms2302.

Rodríguez-Pérez, L., Ramos-Soriano, J., Pérez-Sánchez, A., Illescas, B.M., Muñoz, A., Luczkowiak, J.,et al. 2018. "Nanocarbon-based glycoconjugates as multivalent inhibitors of Ebola Virus Infection." *Journal of the American Chemical Society* 140, no. 31:9891–9898. DOI: 10.1021/jacs.8b03847

Rutala, A., and D.J. Weber. 1999. "Infection control: The role of disinfection and sterilization." *Journal of Hospital Infection* 43: S43–S55. DOI: 10.1016/S0195-6701(99)90065-8.

Salata, C., Calistri, A., Alvisi, G., Celestino, M., Parolin, C., Palù, G. 2019. "Ebola virus entry: From molecular characterization to drug discovery." *Viruses* 19;11, no. 3: E274. DOI: 10.3390/v11030274.

Singh, L., Kruger, H.G., Maguire, G.E., Govender, T., Parboosing, R. 2017. "The role of nanotechnology in the treatment of viral infections." *Therapeutic Advances in Infectious Disease* 4, no. 4: 105–131.

Suk, J.S., Xu, Q., Kim, N., Hanes, J., Ensign, L.M. 2016. "PEGylation as a strategy for improving nanoparticle-based drug and gene delivery." *Advanced Drug Delivery Reviews* 99, no. Pt A: 28–51. DOI: 10.1016/j.addr.2015.09.012.

Tam, Y.Y.C., S Chen, and P.R. Cullis. 2013. "Advances in lipid nanoparticles for siRNA delivery." *Pharmaceutics* 5, no. 3: 498–507. DOI: 10.3390/pharmaceutics5030498.

Tarafdar, J.C., Sharm, S., and Raliya, R. 2013. "Nanotechnology: Interdisciplinary science of applications." *African Journal of Biotechnology* 12, no. 3: 219–226. DOI: 10.5897/AJB12. 2481.

Tarhan, G. 2018. "What is Importance of nanotechnology in disinfection applications: A mini review. *Advances in Biotechnology & Microbiology* 9, no. 4: 0069–0070 DOI: 10.19080/AIBM.2018.09.555766.

Thi, E.P., Mire, C.E., Lee, A.C., Geisbert, J.B., Zhou, J.Z., Agans, K.N., et al. 2015. "Lipid nanoparticle siRNA treatment of Ebola virus Makona infected nonhuman primate." *Nature* 521, no. 7552: 362–365. DOI: 10.1038/nature14442.

Thi, E.P., Lee, A.C., Geisbert, J.B., Ursic-Bedoya, R., Agans, K.N., Robbins, M., Deer, D.J. 2016. "Rescue of non-human primates from advanced Sudan ebolavirus infection with lipid encapsulated siRNA." *Nature Microbiology* 1, no. 10: 16142. DOI: 10.1038/nmicrobiol. 2016.142.

Tsang, M.K., Ye, W., Wang, G., Li, J., Yang, M., Hao, J. 2016. "Ultrasensitive detection of Ebola virus oligonucleotide based on upconversion nanoprobe: Nanoporous membrane system." *ACS Nano* 10, no. 1: 598–605. DOI: 10.1021/acsnano.5b05622.

Wang, Y., Zhu, G., Qi, W., Li, Y., Song, Y. 2016. "A versatile quantitation platform based on platinum nanoparticles incorporated volumetric bar-chart chip for highly sensitive assays." *Biosensors and Bioelectronics* 85, no. 15: 777–784. DOI: 10.1016/j.bios.2016.05.090.

World Health Organization. 2015. "Ebola situation reports." http://apps.who.int/ebola/en/ebola-situation-reports.

Wu, M., Ma, B., Pan, T., Chen, S., and Sun, J. 2016. "Silver-nanoparticlecolored cotton fabrics with tunable colors and durable antibacterial and self-healing superhydrophobic properties." *Advanced Functional Materials* 26: 569–576. DOI: 10.1002/adfm.201504197.

Yang, H.W., Ye, L., Guo, X.D., Yang, C., Compans, R.W., Prausnitz, M.R. 2017. "Ebola vaccination using a DNA vaccine coated on PLGA-PLL/γPGA nanoparticles administered using a microneedle patch." *Advanced Healthcare Materials* 6, no. 1: 1600750. DOI: 10.1002/adhm.201600750.

Yanik, A.A., Huang, M., Kamohara, O., Artar, A., Geisbert, T.W., Connor, J.H., Altug, H. 2010. "An optofluidic-nanoplasmonic biosensor for direct detection of live viruses from biological media." *Nano Letters* 10, no. 12: 4962–4969.

Yassine, H.M., Boyington, J.C., McTamney, P.M., Wei, C.J., Kanekiyo, M., Kong, W.P., Gallagher, J.R. 2015. "Hemagglutinin-stem nanoparticles generate heterosubtypic influenza protection." *Nature Medicine* 21, no. 9: 1065–1070. DOI: 10.1038/nm.3927.

Yazan, H., Kledi, X., Pavel, K., David H. 2016. "Ebola liposome viral-like nanoparticles: characterization and PCR Detection." Paper presented at the NANOCON—8th International Conference on Nanomaterials—Research and Application, Brno, Czech Republic, October 19—21.

Yen, C.W., de Puig, H., Tam, J.O., Gómez-Márquez, J., Bosch, I., Hamad-Schifferli, K., Gehrke, L. 2015. "Multicolored silver nanoparticles for multiplexed disease diagnostics: Distinguishing dengue, yellow fever, and Ebola viruses." *Lab on a Chip* 15, no. 7: 1638–1641. DOI: 10.1039/c5lc00055f.

5 Application of Next-Generation Plant-Derived Nanobiofabricated Drugs for the Management of Tuberculosis

Charles Oluwaseun Adetunji, Olugbenga Samuel Michael, Muhammad Akram, Kadiri Oseni, Ajayi Kolawole Temidayo, Osikemekha Anthony Anani, Akinola Samson Olayinka, Olerimi Samson E, Wilson Nwankwo, Iram Ghaffar and Juliana Bunmi Adetunji

CONTENTS

5.1 Introduction ... 82
5.2 History of Mtb ... 83
5.3 Transmission and Pathogenesis of Mtb ... 83
5.4 Laboratory Diagnosis of Mtb Infection ... 85
5.5 Genome Sequence of Mtb ... 86
5.6 Treatment of Mtb Infection ... 86
5.7 Mode of Action for Plant-Based Nanoparticles against *Mycobacterium tuberculosis* ... 87
  5.7.1 Nanoparticles ... 87
  5.7.2 Sugar Leakages ... 88
  5.7.3 Lipid Peroxidation ... 88
5.8 Characterization of Nanoparticles and Structural Elucidation of the Biogenic Particles ... 89
5.9 Techniques Used for the Characterization of Nanoparticles ... 90
  5.9.1 Morphological Characterizations ... 90
  5.9.2 Structural Characterizations ... 90

5.9.3 Particle Size and Surface Area Characterization 90
5.9.4 Optical Characterizations 91
5.10 Synthesis of Bioengineered Plant-Based Nanoparticles and Various Phytotoxic Metabolites 91
5.11 Management of *Mycobacterium tuberculosis* Using Nanoparticles from Different Extracts with Several Examples 93
5.12 Nanostructures Delivery Systems 94
5.12.1 Polymeric Nanoparticles (PNPs) 94
5.12.2 Protein Nanoparticles 95
5.12.3 Solid-Lipid Nanoparticles (SLNs) 95
5.13 Conclusion and Recommendations 96
References 96

5.1 INTRODUCTION

Mycobacterium tuberculosis (Mtb) is a non-motile, non-sporulating, strict aerobe, acid-fast rod that usually shows up unstained with Gram stain but, like all mycobacteria, appears stained with arylmethane dyes such as carbolfuchsin and rhodamine. It is viewed as curved bacilli microscopically with size as 1–4 μm in length and 0.3–0.6 μm in width (Sakamoto, 2012; Dunn et al., 2016). Alongside other bacteria like *Corynebacterium*, *Nocardia* and *Rhodococcus*, the genus *Mycobacterium* falls under the order Actinomycetales. As intracellular pathogenic bacteria, Mtb replicate within macrophages and monocytes, which are phagocytic cells. Many species of mycobacteria are environmental, but Mtb is an obligate parasite. In culture media, Mtb is a slow-growing species with a 12- to 24-hour cell division rate and long culture period of about 21 days on agar (Sakamoto, 2012).

Mtb is also a member of the *M. tuberculosis* complex, termed as the infectious agents of tuberculosis in specific hosts, which also includes *M. bovis, M. canetti, M. africanum, M. microti, M. caprae* and *M. pinnipedii.* There was a previous widespread assumption that Mtb evolved from *M. bovis* during the livestock domestication (Sakamoto, 2012), but the whole genome sequencing projects for the two species eventually indicated that *M. bovis* has several DNA deletions while maintaining 99.95% identity with Mtb. The sequential DNA deletions of regions of difference led to the branching off of species that make up the Mtb complex (Brosch et al., 2002). It became apparent, by these genetic analyses, that members of the Mtb complex were the clonal derivative of an old parent strain of *Mycobacterium prototuberculosis*, which is also known as *M. canetti* (Gutierrez et al., 2005).

Mycobacterium tuberculosis is the bacterial human pathogen causative of tuberculosis (TB), which is still one of the top killer infectious diseases with more than a billion mortalities in the past two centuries (Smith, 2003; Sakamoto, 2012; Paulson, 2013). TB results in more mortality worldwide than any other single pathogen, with 10.4 million new cases and close to 1.7 million deaths. Approximately, one out of every three persons of the world's population is latently infected with TB and at risk of developing active TB disease (Barberis et al., 2017; Bussi and Gutierrez, 2019).

Therefore, this chapter intends to establish the effect of nanoparticles that could serve as a nanodrug which could be used for the management of tuberculosis. Modes of action and the various types of nanodrugs derived from numerous plant extracts were also highlighted. Numerous techniques used in the characterization of these biogenic nanodrugs derived from plants were also highlighted.

5.2 HISTORY OF MTB

The origin of the genus *Mycobacterium* has been traced to over 150 million years ago (Barberis et al., 2017). The bacteria were assumed to be found initially in soil and some species evolved to live in mammals. The migration of a mycobacterial pathogen from domesticated cattle to humans and its adaptation to a new host by evolution to the closely related *Mycobacterium tuberculosis* was thought to take place within 10,000–25,000 years ago (Smith, 2003). In the Middle Age, an ailment of cervical lymph nodes termed *scrofula* was depicted as a new clinical form of TB. Afterwards, tuberculosis disease was referred to as "King's evil" in England and France as it was widely believed a royal touch could bring healing to persons affected (Barberis et al., 2017). Also "consumption" and "phthisis" both terms were used in the 17th and 18th centuries to describe the disease due to associated weight loss and progressive wasting condition, respectively, until the mid-19th century when Johann Lukas Schonlein coined the term "tuberculosis". A renowned scientist named Robert Koch isolated *M. tuberculosis* and presented his novel findings to the Society of Physiology in Berlin on 24th of March 1882, and this defined a milestone in battling TB (Barberis et al., 2017).

A major breakthrough in combatting TB took place with the development of the attenuated vaccine strain, Bacille Calmette-Guerin (BCG) by Albert Calmette and Camille Guerin through multiple passages of *Mycobacterium bovis* on ox bile and glycerol-soaked potato slices between 1906 and 1919 (Sakamoto, 2012). Although it is highly effective in preventing the childhood form of TB, BCG has been observed to be inconsistently effective in adults (Glickman and Jacobs, 2001; Sakamoto, 2012). The other landmark achievement was the advent of the antimycobacterial drugs, particularly streptomycin, isoniazid, the rifamycins and pyrazinamide, in 1944, 1952, 1957 and 1980, respectively, which led to a new era in prevention, study and treatment of TB.

In the recent years till date, TB has re-emerged as a global health problem due to factors such as the absence of a fully protective TB vaccine, the slow development of the new antimicrobial drugs, need for an improved drug regimen, the current TB–HIV epidemic and the surfacing of multidrug-resistant (MDR) and extensively drug-resistant (XDR) strains of Mtb (Glickman and Jacobs, 2001, 2012; Barberis et al., 2017).

5.3 TRANSMISSION AND PATHOGENESIS OF MTB

Tuberculosis is a disease that is almost exclusively transmitted by aerosolized droplets containing infectious Mtb (Glickman and Jacobs, 2001). Mtb or tubercle bacillus is conveyed in airborne particles called droplet nuclei of 1–5 μm in diameter. Transmission of Mtb occurs through the air and not via surface contact when a normal person inhales Mtb-bearing droplet nuclei, which were generated into the air from infected lungs of persons with pulmonary TB disease as they cough, sneeze, shout or sing (Smith 2003; CDC, 2013). The factors that determine the probability of transmission of Mtb were identified as four in number. These include the immune status of the exposed individual, infectiousness of the TB disease patient, environmental conditions and exposure to an infectious person.

Infection occurs after inhalation of droplet nuclei containing tubercle bacilli that traverse the mouth or nasal routes, upper respiratory tract and bronchi down to the alveoli of the lungs (CDC, 2013). The inhaled bacilli lodge in the terminal air spaces of the lung (the lower respiratory tract) where they are phagocytosed and later replicate within alveolar macrophages and dendritic cells (Glickman and Jacobs, 2001; Bussi and Gutierrez, 2019). In some persons, the immune system is able to clean up the infection with no treatment; while in other persons, Mtb overthrows the efforts of the macrophages at its degradation and instead multiplies within the macrophages for many weeks (Dunn et al., 2016). The organism resists the host immune response or modifies this immune response to allow the host to control bacterial replication without sterilization (Glickman and Jacobs, 2001).

The bacilli spread, as they multiply, by way of lymphatic channels or through the bloodstream to other distant sites of body tissues and organs, including but not limited to the lung apices, vertebrate, peritoneum, meninges, larynx, spleen, liver, lymph nodes and genitourinary tract (CDC, 2013; Dunn et al., 2016). Within 2–8 weeks after the initial infection, the macrophages (special immune cells) ingest the tubercle bacilli and present them to other white blood cells which destroy or encapsulate most of the bacilli, leading to the formation of a barrier shell called granuloma that keeps the bacilli contained and that under control but retain the potential for reactivation. At this point, latent tuberculosis infection (LTBI) has been established. Most patients are asymptomatic within this period and often have no radiologic evidence of TB disease, but they develop cell-mediated immunity, and tests of tuberculosis infection – the tuberculin skin test (TST) or an interferon-gamma (IFN-γ) release assays (IGRAs) – become positive (CDC, 2013; Dunn et al., 2016).

However, in some people, the tubercle bacilli reactivate, subverts the immune system and multiply, leading to progression from LTBI to TB disease. Persons with active TB disease are usually infectious, spreading the bacteria to other people. TB disease is obtainable in pulmonary or extrapulmonary sites. Pulmonary TB disease patients are characterized with cough and an abnormal chest radiograph, and may be infectious (CDC, 2013). Extrapulmonary TB disease occurs in sites besides the lungs, including the larynx, the lymph nodes, the brain, the pleura, the kidneys, the bones and even joints. The disease is a chronic wasting illness with symptoms of fever, weight loss and an open abscess or lesion in which the concentration of Mtb is high (Glickman and Jacobs, 2001).

While healthy adults infected with *M. tuberculosis* have a 5–10% probability of developing the TB disease between the first and second year after infection, children who are infected but untreated have a 40–50% chance of developing the disease within six to nine months (Dunn et al., 2016). However, beyond these years, the rate of progression from LTBI to TB disease decreases significantly with increasing age (Dunn et al., 2016). Any condition or treatment that depresses the immune system such as HIV infection, diabetes mellitus, old age, malignant disease, poor nutrition, immunosuppressive medication or any other infection increases the risk of reactivation or progression from infection to disease in both adults and children (Sakamoto, 2012; Dunn et al., 2016).

5.4 LABORATORY DIAGNOSIS OF MTB INFECTION

There are two tests available to detect if a person is infected with Mtb: the Mantoux tuberculin skin test and the interferon-gamma release assays (IGRAs). Neither of the tests is preferred over the other in terms of performance (Dunn et al., 2016). However, decisions should not be based on TST or IGRA negative reaction results alone. Additional tests and information are needed to diagnose TB disease and inform public health management decisions (CDC, 2013).

Accurate identification of *Mycobacterium* species in infections is important for correct diagnosis as misdiagnosis can result in inappropriate treatment and consequentially increased mortalities of patients (Riello et al., 2016). Culture is the WHO recommended gold standard for the diagnosis of TB disease as it is important for both definitive diagnosis and determining the phenotypic drug susceptibility testing (DST) (Dunn et al., 2016).

Differential diagnosis between mycobacterial species is typically made in positive cultures based on phenotypic and biochemical traits. Currently, the conventional diagnostic methods used are bacilloscopy and microbiological culture, but the main method for bacilli detection is the Ziehl–Neelsen-specific staining technique, though it has a very low sensitivity despite its low cost and simplicity (Riello et al., 2016). The microbiological culture of Mtb is generally used in suspected pulmonary cases and in negative bacilloscopy. Culture allows for Mtb detection by culture isolation for subsequent identification of the isolated complex; it is, however, time-consuming due to slow growth of Mtb and has limited sensitivity (Dunn et al., 2016; Riello et al., 2016).

The need for a fast and reliable diagnostic laboratory test for detection of Mtb led to the development of molecular methods for detection and identification of Mtb directly from clinical specimens and/or from culture isolation. The following are some such methods.

The real-time PCR and PCR-RFLP (restriction fragment length polymorphism)-based method: The real time has the advantages of speed of detection with minimal manipulation and higher sensitivity, but it has limitations of high cost and inability to identify many species in a single reaction. However, PCR-RFLP has proved to be fast with high specificity and is relatively cost-effective. Despite its moderate complexity, PCR-RFLP continues to be the most useful tool for the identification of mycobacterial species (Riello et al., 2016).

The use of molecular techniques should be incorporated into public health systems for urgent and correct diagnosis. In recent studies, the combination of two PCR techniques in the form of a nested PCR was employed and found to be more sensitive, specific and better employable to overcome the standardization challenge than the conventional single PCR (Khosravi et al., 2017).

5.5 GENOME SEQUENCE OF MTB

The completion and first release of the complete nucleotide sequence of the circular chromosome *M. tuberculosis* strain H37Rv occurred in 1988 (Cole et al., 1998) and has immensely improved the understanding of Mtb biology in its ramifications. The *M. tuberculosis* H37Rv genome consists of 4.4×10^6 bp with a relatively high G-C content (65%) and contains about 4,000 genes. (Glickman and Jacobs, 2001; Smith 2003; Cerezo, 2015) Other unique features of the genome include its content of the two related protein families of genes named PE (derived from proline-glutamate) and PPE (derived from proline–proline glutamate), which makes up about 9% of the genome and could be implicated in the antigenic variation of Mtb during infection. About 200 genes were also annotated as encoding enzymes for the metabolism of fatty acids, comprising 6% of the total, and the genome also contains homologs of the fatty acid β-oxidation system having 36 FadD and 36 FadE homologs, which is suggestive of Mtb's involvement in lipid synthesis and assimilation as carbon sources (Glickman and Jacobs, 2001). Since the release of this complete sequence in 1988, its functional annotation has grown, giving much information to explore Mtb and its infection control (Cerezo, 2015).

5.6 TREATMENT OF MTB INFECTION

Treatment of tuberculosis infection and disease, together with correct diagnosis, represents a cardinal point in the management and control of tuberculosis (Sotgiu et al., 2015). The major goals of treatment for TB disease are to cure the TB patient, minimize risk of death and disability and reduce transmission of *M. tuberculosis* to other persons (CDC, 2013). To ensure that these goals are met, long-term antituberculosis therapy must be adhered to so as to maintain adequate blood drug level. It is recommended that TB disease treatment must be for a minimum of 6 months. A larger part of the bacterial population is cleared during the first 8 weeks of treatment, leaving behind persistent organisms that require prolonged treatment. Non-continuity of treatment for a sufficiently long duration may allow surviving bacteria to cause the infected patient to become ill and infectious again and possibly with drug-resistant disease (CDC, 2013; Sotgiu et al., 2015).

The first experimental proof of the potential efficacy of new antituberculosis drugs was seen in 1940 when Promin was administered to samples of guinea pigs. After this, several other therapeutic discoveries (monotherapy and combination therapy) with different dosages and therapy duration were made towards successful antituberculosis therapy until the recent Bedaquiline and Delamanid drug agents were approved for drug-resistant TB (Singh and Sharma, 2017; Sotgiu et al., 2015).

Regimens towards treatment of TB disease must have multiple drugs, which are active against the bacteria. The standard of care for commencing TB disease treatment is four-drug therapy to obtain a bacteriological termination in pulmonary and extrapulmonary sites. Single-drug therapy can result in the development of bacterial resistance. In the same vein, the addition of a single drug to an anti-TB regimen that is failing can lead to increased resistance. The administration of two or more drugs to which *in vitro* susceptibility testing has been demonstrated helps prevent the emergence of Mtb that are resistant to the other drugs (CDC, 2013; Sotgiu et al., 2015). Directly observed therapy (DOT) is the preferred central management strategy for treatment of TB disease as recommended by CDC and even for treatment of LTBI provided resources are enough. Almost all the treatment regimens for drug-susceptible TB disease can be administered alternatingly by direct observation. Drug-resistant TB disease should be treated all the time with a daily regimen and under direct observation. Intermittent regimens are not recommended for treatment of multidrug-resistant (MDR) TB (CDC, 2013).

As a general principle, two separate steps are recognized in the treatment of drug-susceptible tuberculosis, which are reflected in the characteristic mechanisms of action of the combined antituberculosis drug molecules – initial (or bactericidal) phase and continuation (or sterilizing) phase. During the initial phase of treatment, highly replicating mycobacteria are killed, and, as a result, with the histological pulmonary restoration and the reduction of the inflammation process, clinical recovery is achieved. This first step is crucial as an intensive phase which allows a relevant reduction of the bacterial load and consequently the reduction of the chances of selection of drug-resistant strains. However, the second step termed continuation phase is aimed at the elimination of semidormant bacteria, whose size is significantly reduced if compared with that at the beginning of the antituberculosis therapy, thus bringing about a sterilizing activity on mycobacteria in dormancy state to ensure a low probability of emergence of drug-resistant mycobacteria. The intensive phase covers a duration of four months while the continuation phase has a duration of two months.

These principles form the generally acceptable standardized regimens recommended by the World Health Organization (Sotgiu et al., 2015). Treatment of the drug-resistant tuberculosis also has its peculiar guidelines on the prescription of an efficacious drug regimen depending on the outcome of the drug susceptibility testing. It is also necessary to ensure microbiological monitoring of the efficacy of the prescribed regimen. This can be carried out by the evaluation of the sputum smear and culture conversion, particularly at the end of the intensive and continuation treatment phases.

5.7 MODE OF ACTION FOR PLANT-BASED NANOPARTICLES AGAINST *MYCOBACTERIUM TUBERCULOSIS*

5.7.1 Nanoparticles

Nanotechnology has been employed for the production of different materials at the nanoscale. It especially holds huge promise to improve human health, and it has been forecasted to be of tremendous benefits to humanity (Saravanan et al., 2018).

Nanoparticles are a wide class of materials that include particulate substances, which have one dimension less than 100 nm at least. There are several benefits to using nano-formulations for therapeutic uses, including lowering the dose of drug given to patients, resulting in less adverse reaction and perhaps decreasing the treatment time. This is made possible via enhanced sites-specific drug-bearing nanoparticle targeting resulting in increased drug concentration at the target site of interest while reducing delivery of drugs to non-target sites (Byrne et al., 2011). Nanoparticles possess large surface area compared to their volume, and their size is similar to that of macromolecules and organelles inside the cells like proteins and DNA. Interestingly, their small size allows macrophages to ingest and phagocytize them easily since macrophages ingest smaller objects or molecules more freely than do larger forms of the same material (Navalakhe and Nandedkar, 2007). Therefore, if drugs are in the nanoparticle form, it may be beneficial in treating some diseases such as TB and cancer (Clift et al., 2008; Nasiruddin et al., 2017).

5.7.2 Sugar Leakages

One of the mechanisms of antimicrobial effect of nanoparticles is the increase in cell membrane permeability of bacteria that can make cellular molecules, especially the sugar and protein, leak out of the cell (Gurunathan et al., 2014). The effect of various applications of silver nanoparticles (AgNPs) have been seen in diagnostic biomarkers, cellular labels and drug delivery system for treatment of various diseases, especially infections and cancers. Silver nanoparticles have been extensively demonstrated to exhibit sugar leakage as one its mechanism of action. Rajesh et al. (2015) made a comment that the antimicrobial effect in using silver nanoparticles may probably be the sum of distinct mechanisms of action that include reaction of silver with thiol (SH) proteins. These altered protein groups inactivate the organism by inhibiting enzymes that are involved in the respiratory chains and thus interfere with permeability of protons and phosphate. This can generate reactive oxygen untreated control. His findings also demonstrated that silver nanoparticles as an antimicrobial involved the attachment to the cellular membrane, then enters the cytoplasm causing osmotic collapse and then release of intracellular contents that include sugar and protein. The ultimate result is cell death (Yuan et al., 2017). In a similar experiment conducted by Qayyum et al. (2017), the findings were of increased sugar and protein biomolecules after treating with silver nanoparticles, at two to three folds when compared to the sample that was untreated (control) (Qayyum et al., 2017).

5.7.3 Lipid Peroxidation

Peroxidation of lipid is known to take place commonly in the membrane of the cells and organelles due to their membrane composition. The commonest targets include unsaturated lipids and cholesterol. Common inducers of lipid peroxidation include the reactive oxygen species (ROS) such as hydroxyl radical, singlet oxygen and hydroperoxyl radical. Lipid peroxidation does not involve the damage of lipids alone, but also protein and nucleic acid. Several studies have showed that several metal oxide nanoparticles have the potential to exhibit spontaneous ROS production; this can cause the cell to enter the state of oxidative stress. If the cellular antioxidant defence is low, it can lead to the damage of cellular components of proteins, nucleic acid and lipids (Premanathan et al., 2011). The generation of the fatty acid from the breakdown of lipids can cause the activation of lipid peroxidase, which will initiate a chain reaction that will disrupt the membrane of organelles and cellular membrane. This can result in cell death (Premanathan et al., 2011).

In the study investigating zinc oxide nanoparticles toxicity of prokaryotic and eukaryotic cells, antibacterial activity was tested and found that the nanoparticles with enhanced ultrasound induced lipid peroxidation in the liposomal membrane. The products of this lipid peroxidation were conjugated dienes, lipid hydroperoxides and malondialdehydes. The lipid peroxidation was a result of a chain reaction involving oxygen and mediated by free radicals. Zinc oxide nanoparticles were showed to enhance this process (Premanathan et al., 2011).

Another study was conducted to examine the effectiveness of nanocrystalline TiO_2 with bromopyrogallol (Brp@TiO_2) in photo-generating singlet oxygen, free radicals and also their ability to photosensitize peroxidation of unsaturated lipids. The result showed a type I mechanism of lipid peroxidation, which was indicated by the formation of free radicals dependent cholesterol oxidation products (Kozinska et al., 2019).

5.8 CHARACTERIZATION OF NANOPARTICLES AND STRUCTURAL ELUCIDATION OF THE BIOGENIC PARTICLES

The common techniques used in analysing the various physical and chemical properties of nanoparticles according to Khan et al. (2019) include X-ray diffraction analysis, ultraviolet-visible (UV-vis) absorption, Fourier transform infrared (FTIR), SEM, TEM, Brunauer–Emmett–Teller (BET), particle size analysis and X-ray photoelectron spectroscopy (XPS).

X-ray diffraction (XRD) analysis is commonly used to determine crystallinity and morphology of a prepared sample. Due to variation in the amount of constituent, it can show whenever there is a change in intensity. It can therefore give the amount of metallic nature of a particle, shape and size of cell from peak position and information on the electron density in the cells (Yelil et al., 2012).

In UV-vis absorption spectroscopy, optical properties are determined by the use of absorbance spectroscopy. This absorbance is measured to determine the concentration of this solution using Beer–Lambert theory. Iron nanoparticles synthesized using suitable surface plasmon resonance with high-band intensities and peaks from *Azadirachta indica* was determined through the use of UV-vis spectroscopy at the range of 216–265 nm (Monalisa and Nayak, 2013).

FTIR spectroscopy is commonly used to determine functional groups and structural features of biological extracts with nanoparticles. It measures the light's wavelength against the intensity of infrared. The spectrum which is a reflection of dependence optical properties of nanoparticles is then calculated. Various nanoparticles, especially the green synthesized silver nanoparticles, have been analysed using FTIR spectroscopy.

SEM is based almost on the same principle as that of the optical microscope, but rather than measuring photon, it measures the electrons scattered in the sample. The SEM is used for the determination of shape, size and morphology of nanoparticles. The images formed from SEM can also be magnified to about 200,000 times because the electron wavelengths are shorter than those of photons, and thus the electron can be accelerated by an electric potential.

TEM is a technique that employs the transmission of the electron beam through an ultra-thin specimen. TEM can be used in a wide range of scientific fields, both the physical science and the biological fields, including cancer research, virology, nanotechnology, and pollution (Heera and Shanmugam, 2015)

5.9 TECHNIQUES USED FOR THE CHARACTERIZATION OF NANOPARTICLES

5.9.1 Morphological Characterizations

Most nanoparticles are influenced by their morphology. Their studies involve the use of techniques such as SEM, TEM and polarized optical microscopy (POM). Mirzadeh and Akhbari (2016) studied ZnO-NP using the SEM technique which reflects the dispersion and morphological features of ZnO-modified metal organic frameworks. The morphology of gold nanoparticles was also studied using TEM (Khlebtsov and Dykman, 2011).

5.9.2 Structural Characterizations

This provides the knowledge of the bulk properties of the particular materials, and its composition as well as nature of bonding materials. The common techniques employed for structural characterization of nanoparticles include XRD, XPS, energy-dispersive X-ray (EDX), BET and Zieta-size analyser. The most commonly used technique is the XRD because it gives adequate information on the phase and crystallinity of nanoparticles (Ullah et al., 2017).

5.9.3 Particle Size and Surface Area Characterization

Techniques employed for use of size characterization include TEM, SEM, AFM, XRD and DLS (dynamic light scattering). Though the TEM, SSEM, AFM and XRD give a better idea of nanoparticle size, the DLS quantify the size in an extremely low level (Kestens, et al., 2016; Sikora, Shard and Minelli, 2016). Also, NTA (nanoparticle tracking analysis) is very useful in a biological system such as proteins and DNA which can visualize and analyse nanoparticles with diameter that range from 10 nm to 1,000 nm. It provides a more accurate result than DSC (Filipe et al., 2010; Gross et al., 2016). BET is the best technique for determining surface area (Fagerlund, 1973).

5.9.4 Optical Characterizations

The optical characterization of nanoparticles is based on Beer–Lambert theory and the principle of basic light. The techniques commonly employed in the determination of optical characterization include UV-vis, null ellipsometer and photoluminescence. These take into consideration the absorption, luminescence, reflectance and phosphorescence (Khan et al., 2019).

5.10 SYNTHESIS OF BIOENGINEERED PLANT-BASED NANOPARTICLES AND VARIOUS PHYTOTOXIC METABOLITES

Production of nanoparticles using extract is simply by the mixture of metal salt with some solution within some minutes at room temperature. Most metal nanoparticles are produced in this manner. Some factors affect the production of nanoparticles, and these have been extensively studied; they include temperature, pH, contact time, metal concentration and nature of the extraction. Several plants have been established to possess several numerous biological active constituents with many medical benefits (Adetunji et al., 2014; Adetunji and Olaleye, 2011; Adetunji et al., 2011a,b,c,d).

Silver nanoparticle is one of the most studied nanoparticles with a wide range of antimicrobial activities, and it has been synthesized from various plants extracts. An aqueous extract of *Ficus benghalensis* leaves was used to produce silver nanoparticles, with an average size of 16 nm and was noted to be an effective bactericidal drug (Saxena, Tripathi, Zafar and Singh, 2011). The reduction of nanoparticles containing silver ion from the extract *Desmodium trifolium* as described by Ahmad et al. (2010) was achieved due to the presence of hydrogen ions and ascorbic acid. The presence of hydrogen ion and ascorbic acid in the extract caused the reduction of the silver ions. Also, Kesharwani et al. (2009) synthesized silver nanoparticles that were highly stable using the plant *Datura metel*. The reduction was achieved because the extract contained alcoholic compounds, alkaloids, polysaccharides, proteins, enzymes and amino acids (Kesharwani, Yoon, Hwang and Rai, 2009). Antibiotics use when combined with the silver nanoparticles has been shown to be very effective against multidrug-resistant microorganism. Flavonoids and trepenoid compound that are present in the methanolic extract of *Eucalyptus hybrid* (safeda) in the synthesis of silver nanoparticles has been shown to be responsible for its stabilization. The silver nanoparticles formed from seed extract of *Syzygium cumini* (jambul) was shown to have an antioxidant property due to the adsorption of antioxidant materials on the surface of the nanoparticle. Another extract of *Ocimum sanctum* leaves was shown to have high levels of ascorbic acids, making it a very potent antimicrobial activity against gram-negative and gram-positive organisms (Singhal et al., 2011).

Moreover, the crude extracts derived from the peel of banana used for the synthesis of silver nanoparticle have been shown to display a wide range of antifungal and antimicrobial activities (Bankar, Joshi, Kumar and Zinjarde, 2010). Silver nanoparticles formed from peel extract of *Citrus sinensis* at 60 °C had an average size of 10 nm, while at 25 °C the size increased to 35 nm, but both were shown to have antimicrobial activities (Kaviya, Santhanalakshmi, Viswanathan, Muthumary and Srinivasan, 2011).

Silver nanoparticles was produced from extracts of *S. cumini* leaves and seeds, due to the high concentration of polyphenolic antioxidants on the surface of the particles produced using seed extract. The nanoparticles were found to have stronger antioxidant properties (Kumar, Yadav and Yadav). Silver ions loaded within the cotton fibres using the leaf extract of *Eucalyptus citriodora* (neelagiri) and *Ficus bengalensis* (marri) plants for the synthesis were described by Ravindra et al. (2010). These particles were shown to have a broad spectrum of antimicrobial activities.

The extract of the leaf *Cassia auriculata* was used to synthesize a triangular and spherical nanoparticle within 10 min with size of about 15–25 nm, as described by Kumar et al. (2011). Also, an irregular-shaped gold nanoparticle was also produced from the extract of dried clove buds of *Syzygium aromaticum* (Raghunandan, Basavaraja, Sawle, Manjunath and Venkataraman, 2010). Parida et al. (2011) also reported that the synthesis of gold nanoparticles from the extract of *Allium cepa* had its reduction process and stabilization from flavonoids seen in the extract (Parida, Bindhani and Nayak, 2011). A dilute extract from *Phyllanthus amarus* produced triangular gold nanoparticles, and the concentrated extract was seen to produce spherical nanoparticles (Kasthuri, Kathiravan and Rajendiran, 2009). Gold nanoparticle from the extract of *Terminalia chebula* has been shown to be an effective antimicrobial agent (Edison and Sethuraman, 2012). The formation of nanorods and nanowires is dependent on pH of the reaction. Castro et al. (2011) made an effective gold nanowire by making the nanoparticles using sugar beet pulp at room temperature which latter joined to form chains and nanowire.

Reduction of silver and gold nanoparticles was achieved using the leaf extract of *Cinnamonum camphora* because it contains phytochemicals such as flavone compounds, polysaccharides, phenols and terpenoids. They were found to have profound bacteriocidal effect at concentration of 45 μg/ml (Huang et al., 2007). Gold and silver nanoparticles from the extract of *Dioscorea bulbifera* when combined with antibiotics have a synergistic antibacterial effect against most organisms (Ghosh et al., 2012). The stabilization of bimetallic silver and gold nanoparticles of silver and gold was due to the presence of reducing sugar from the leaf extract of *Azadirachta indica* (Shankar, Rai, Ahmad and Sastry, 2004). Caffeine and theophylline found in the extract of *Camellia sinensis* may contribute to the reduction of silver and gold nanoparticles (Vilchis-Nestor, Sánchez-Mendieta, Camacho-López, Gómez-Espinosa and Arenas-Alatorre, 2008). Hydroxyl and carbonyl groups present in the apiin leaves extract were responsible for gold and silver nanoparticles reduction and the size and shape of these nanoparticles could be controlled using this extract (Kasthuri, Kathiravan and Rajendiran, 2009).

Silver and copper nanoparticles synthesized from the extract of *Euphorbiaceae latex* were seen to have an excellent bactericidal activity for both gram-negative and gram-positive bacteria (Valodkar et al., 2011). Nanoparticles of titanium dioxide have been synthesized from the extract of *Catharanthus roseus*, the particles were irregular and the shape and size ranged from 25 nm to 110 nm (Velayutham et al., 2011).

Gold–palladium nanoparticle using the aqueous extract of bayberry tannin has been shown to be reduced due to the preferential treatment with the phytochemical tannin (Huang et al., 2011). Antibacterial activity of palladium nanoparticles can be synthesized from extracts of coffee, tea, chitosan, grapes, cinnamon (*C. zeylanicum*) and soybean (Nadagouda and Varma, 2008). Cinnamomum zeylanicum bark is rich in methyl chavicol, linalool and eugenol, and the compounds are said to be responsible for the bioreduction of silver and palladium ions (Sathishkumar et al., 2009).

5.11 MANAGEMENT OF *MYCOBACTERIUM TUBERCULOSIS* USING NANOPARTICLES FROM DIFFERENT EXTRACTS WITH SEVERAL EXAMPLES

Zinc oxide is a cheap, non-toxic inorganic material that is readily available (Patil and Taranath, 2016). Zinc oxide nanoparticles were synthesized from *Limonia acidissima* using the microplate alamar blue assay technique, and the activity of the nanoparticle against the mycobacterium tuberculosis was determined. It was found that the nanoparticles of zinc oxide was effective in controlling the growth of these organisms at a control growth of 12.5 μg/ml (Patil and Taranath, 2016).

In a study done by Heidary et al. (2019), the effect of antimycobacterial activity of Ag, ZnO and Ag-ZnO nanoparticles was investigated on multidrug-resistant and extremely drug-resistant tuberculosis. The method of study involved the chemical and deposition method, as described by Heidary et al. (2019). The result was seen to be occurrence of synergistic effect of both silver and zinc oxide (Ag-ZnO) nanoparticles that had a higher efficacy than the individual nanoparticles in mycobacterium tuberculosis (Heidary et al., 2019).

Copper oxide nanoparticle (CuO-NP) is another nanoparticle that has been shown to be effective against mycobacterium tuberculosis. The extraction of CuO-NP was done using the leaves of *Leucaena leucocephala*. The *in vitro* test for the effectiveness of this nanoparticle was through the Lowenstein and Jensen MIC method, as described by Aher et al. (2017).

Gold nanoparticles (AuNP) were synthesized from the extracts of *Barleria prionitis*, *Plumbago zeylaniica* and *Syzygium cumini* using the method described by Salunke et al. (2014). This was seen to be very effective against the mycobacterium tuberculosis (Salunke et al., 2014). According to the report by Singh et al. (2016) who synthesized AgNP, AuNP and bimetallic Au-AgNP, these three nanoparticles had antitubercular effect but with different efficacies. It was demonstrated that the bimetallic Au-AgNP exhibited the highest activity while the AuNP was the least. Their values of MIC were less than 2.56 μg/ml for the bimetallic Au-AgNP and that of AuNP was greater than 100 μg/ml (Singh et al., 2016).

The antimicrobial activity of titanium dioxide (TiO_2) nanoparticles as described by Vaikundamoorthy et al (2019) showed a potent effect against multidrug-resistant tuberculosis. The TiO_2-NP was synthesized from titanium(IV) oxysulfate ($TiOSO_4$) solution using sol–gel method (Vaikundamoorthy et al (2019).

Aluminium hydroxide nanoparticle has been considered as an adjuvant in the production of vaccine against mycobacterium tuberculosis, as demonstrated by Amini et al. (2016). He showed that they can raise cellular immune response as a result of the raised interferon-gamma secretion, especially in the lungs macrophages when compared to use of EsxV alone as vaccine (Amini et al., 2016).

Hart et al. (2018) was able to describe Nano-FP1 which composed of emulsified yellow carnauba and palm wax with sodium myristate (YC-NaMA) coated with a fusion protein (FP1). This consists of three antigens, i.e. *Mycobacterium tuberculous* antigen Ag85B (early expression), Acr (latent expression) and HBHA (epithelium targeting). This nano-immunization showed protection in both BCG primed and BCG naïve rats; it is thus a potent vaccine for individuals whose immune system has been compromised. Hart also demonstrated that the vaccine protection was associated with cellular immune response that is antigen-specific and the humoral immune responses can be seen at both the mucosal site and the systemic site (Hart et al., 2018).

Nanoemulsion-based nanocapsules has a lipid core that are suitable for encapsulation of the drug bedaquilin, a drug for the treatment of mycobacterium tuberculosis. This was shown in an experiment conducted by De Matteisa et al. (2017). Two bedaquilin-loaded nanocarriers – chitosan nanocapsules (CS-NC) and lipid nanocapsules (LNP) – were used in this experiment. The CS-NC-loaded bedaquilin was administered via nebulization, while LNP-loaded bedaquilin was administered via intravenous route. The *in vitro* antimicrobial drug activity against tuberculosis remained unchanged after encapsulation in both nanocarriers, and there was no cytotoxic effect on other cell lines used in the experiment (De Matteisa et al., 2017).

Pinus merkusii nanoparticle of size ranging from 10 nm to 800 nm was made using tripolyphosphate (TPP) after ethanolic extraction from the herb of *P. merkusii*, as described by Sudjarwo (2019). Herbal nanoparticles are emerging as a nanomedicine of interest due to their enhanced bioavailability. The minimal inhibitory concentration (MIC) for mycobacterium tuberculosis was 1,000 mg/ml and the minimal bactericidal concentration (MBC) was 2,000 mg/ml. This made *P. merkusii* nanoparticles a potent antimycobacterial agent that required further development (Sudjarwo, 2019).

Another very common and useful nanoparticle is the chitosan which have gained a lot of recognition in fields of biomedical engineering and nanoscience. In 2018, Wardani et al. conducted a study that aimed to synthesize chitosan nanoparticles for antimycobacterial application. The chitosan nanoparticles isolated from the shrimp shell was prepared from tripolyphosphate (TPP). The result showed MIC of 1,200 μg/ml and MBC of 2,400 μg/ml. This led to the conclusion that chitosan nanoparticles are potent against mycobacterium tuberculosis (Wardani et al., 2018).

5.12 NANOSTRUCTURES DELIVERY SYSTEMS

5.12.1 Polymeric Nanoparticles (PNPs)

There are numerous particular features of polymeric nanoparticles used for antimicrobial drug delivery. Firstly, they are manufactured with abrupt size distinction and are structurally stable. Secondly, during the formation, the properties of the particle including drug release profiles, size and zeta potentials can be specifically tuned by picking out different polymer lengths, organic solvents and surfactants. Thirdly, there are functional groups of the polymeric nanoparticles that can be chemically altered with either targeting ligands or drug moieties. Formerly, the therapeutic efficiency of polymeric nanoparticles was not satisfactory because they used to clear more rapidly by reticuloendothelial system (RES) when they were administered intravenously. This problem was swept over by the discovery of long-acting nanoparticles. For selective antimicrobial drug delivery, polymeric nanoparticles are often coated with lectin; it is a protein that attach to either simple or complex carbohydrates that are located mostly on bacterial cell walls. Polymeric nanoparticles are extensively studied by several significant pulmonary drugs because of their sustained-release properties, surface modification capability and biocompatibility. Antituberculosis drugs, anticancer drugs and antiasthmatic and pulmonary hypertension drugs all are included in pulmonary drugs. Nevertheless, in the synthesis of polymeric nanoparticles, the toxicity and biodegradability should be minutely analysed to prevent aggregation of polymer carriers due to repeated dosing.

5.12.2 Protein Nanoparticles

Proteins are special class of natural molecules that have particular functions and applications in both material and biological fields. The protein-based colloidal systems may be very effective. Protein-based nanoparticles have unique characteristics: they are easily tractable for covalent attachment of ligands and drugs and have surface adjustment. Moreover, they are easily biodegradable, metabolizable and nonantigenic. The pulmonary protein therapeutics can be administered by using protein nanoparticles. The betterment of the cellular consumption of numerous substances and probability of drug-targeting particular body organ is the most significant benefit of colloidal drug carrier systems. Consequently, unwanted toxic effects of the free drug can be averted.

5.12.3 Solid-Lipid Nanoparticles (SLNs)

SLNs mainly consist of lipids which are present in solid phase at room temperature having surfactants for emulsification. It is another antimicrobial drug delivery system. In SLNs, the diameter of nanoparticles ranges from 50 nm to 1,000 nm for the delivery of different antimicrobial drugs. In the SLN preparations, solid lipids which are used include steroids (e.g. cholesterol), fatty acids (e.g. palmitic acid, behenic acid and decanoic acid), partial glycerides (e.g. glyceryl behenate and glyceryl monostearate), triglycerides (e.g. trimyristin, tripalmitin and trilaurin) and waxes (e.g. acetyl palmitate). To stabilize lipid distribution, numerous types of surfactants are applied as emulsifiers, including phosphatidylcholine, sodium cholate, soybean lecithin, poloxamer 188 and sodium glycocholate. The distinctive procedures of SLNs preparation include high shear mixing, high-pressure homogenization (HPH), spray-drying and ultra-sonication. It has an additional benefit when compared with the particles synthesized from polymeric materials that it has much higher tolerability in the lungs. Although SLNs have less value for pulmonary delivery, when physiological lipids are used then its toxicological profile is considered much better than polymer-based systems. Dry powder formulations and aqueous suspensions of SLNs are viably used for pulmonary inhalation.

5.13 CONCLUSION AND RECOMMENDATIONS

This chapter has provided a detailed information on the application of nanotechnology for the synthesis of biogenic nanodrugs that could be used for the management of tuberculosis caused by several strains of *Mycobacterium tuberculosis*. This will go a long way in mitigating the various adverse effects experienced when synthetic drugs are applied for the management of tuberculosis diseases. There is a need to explore new and novel plants, especially from various agro-ecological areas in numerous countries so as to establish their effectiveness and efficacy. There is a need to perform several *in vitro* and *in vivo* trials with these newly synthesized nanodrugs, especially to ascertain their level of toxicity. Also, more research needs to be performed to establish the effective delivery system of numerous nanodrugs in order to avoid its degradation before the active components will reach its targeted area.

REFERENCES

Adetunji C.O. and Olaleye O.O. 2011. Phytochemical screening and antimicrobial activity of the plant extracts of *Vitellaria paradoxa* against selected microbes. *Journal of Research in Biosciences* 7, no. 1: 64–69.

Adetunji C.O., Arowora K.A, Afolayan S.S., Olaleye O.O., and Olatilewa M.O. 2011a. Evaluation of antibacterial activity of leaf extract of *Chromolaena odorata*. *Science Focus* 16. no. 1: 1–6.

Adetunji C.O., Kolawole O.M., Afolayan S.S., Olaleye O.O., Umanah J.T., and Anjorin E. (2011b). Preliminary phytochemical and antibacterial properties of *Pseudocedrela kotschyi*: A potential medicinal plant. *Journal of Research in Bioscience. African Journal of Bioscience*.4, no. 1: 47–50.

Adetunji C.O., Olaleye O.O., Adetunji J.B., Oyebanji A.O. Olaleye O.O., and Olatilewa M.O. 2011c. Studies on the antimicrobial properties and phytochemical screening of methanolic extracts of *Bambusa vulgaris* leaf. *International Journal of Biochemistry* 3, no. 1. 21–26.

Adetunji C.O., Olaleye O.O., Umanah J.T., Sanu F.T, and Nwaehujor I.U. 2011d. *In vitro* antibacterial properties and preliminary phytochemical of *Kigelia Africana*. *Journal of Research in Physical Sciences* 7, no. 1: 8–11.

Adetunji, C.O, Olatunji, O.M, Ogunkunle, A.T.J, Adetunji, J.B., and Ogundare, M.O. 2014. Antimicrobial activity of ethanolic extract of *Helianthus annuus* stem. *Sikkim Manipal University Medical Journal* 1, no. 1: 79–88.

Ahmad N, Sharma S, Alam MK, Singh VN, Shamsi SF, Mehta BR, Anjum Fatma. 2010. Rapid synthesis of silver nanoparticles using dried medicinal plant of basil *Colloids and Surfaces. B: Biointerfaces* 81 (2010): 81–86

Aher, Y., Jain, G., Patil, G., Savale, A., Ghotekar, S., Pore, D., Pansambal, S., Deshmukh, K. 2017. Biosynthesis of copper oxide nanoparticles using leaves extract of Leucaena leucocephala L. and their promising upshot against diverse pathogens. *International Journal of Molecular and Clinical Microbiology* 7(1): 776–786.

Amini Y, Moradi B, Fasihi-Ramandi M. 2017. Aluminum hydroxide nanoparticles show strong activity to stimulate Th-1 immune response against tuberculosis. *Artif Cells Nanomed Biotechnol.* 45(7):1331–1335. doi: 10.1080/21691401.2016.1233111. Epub 2016 Sep 20. PMID: 27647321.

Barberis, I., Bragazzi, N.L., Galluzzo, L., and Martini M. 2017. The history of tuberculosis: From the first historical records to the isolation of Koch's bacillus. *Journal of Preventive Medicine and Hygiene* 58: E9–E12.

Brosch, R., Gordon, S.V., Marmiesse, M., Brodin, P., Buchrieser, C., Eiglmeier, K., Garnier, T. et al. 2002. A new evolutionary scenario for the *Mycobacterium tuberculosis* complex. *Proceedings of the National Academy of Sciences of the United States of America* 99: 3684–3689.

Bussi, C. and Gutierrez, M.G. 2019. *Mycobacterium tuberculosis* infection of host cells in space and time. *FEMS Microbiology Reviews* 43, no. 4: 341–361. DOI: 10.1093/femsre/fuz006.

Byrne, J., Wang, J., Napier, M., and DeSimone, J. 2011. More effective nanomedicines through particle design. *Small* 7: 1919–1931.

Castro L, Blázquez ML, Muñoz JA, González F, García-Balboa C, Ballester A. 2011. Biosynthesis of gold nanowires using sugar beet pulp. *Process Biochem* 46:1076–82.

CDC (Centers for Disease Control and Prevention). 2013. *Core Curriculum on Tuberculosis: What the Clinician Should Know*, 19–164. 6th ed. Centers for Disease Control and Prevention National Center for HIV/AIDS, Viral Hepatitis, STD, and TB Prevention Division of Tuberculosis Elimination. (CS234269) www.cdc.gov/tb.

Clift, M., Rothen-Rutishauser, B., Brown, D.M., Duffin, R., Donaldson, K., Proudfoot, L., Guy, K., and Stone, V.. 2008. The impact of different nanoparticle surface chemistry and size on uptake and toxicity in a murine macrophage cell line. *Toxicology and Applied Pharmacology* 232: 418–427.

Cole, S.T., Brosch, R., Parkhill, J., Garnier, T., Churcher, C., Harris, D., Gordon, S.V. et al. 1998. Deciphering the biology of *Mycobacterium tuberculosis* from the complete genome sequence. *Nature* 393, no. 6685: 537–544.

De Matteis L, Jary D, Lucía A, García-Embid S, Serrano-Sevilla I, Pérez D, Ainsa JA, Navarro FP, de la Fuente JM. 2018. New active formulations against M. tuberculosis: Bedaquiline encapsulation in lipid nanoparticles and chitosan nanocapsules. *Chemical Engineering Journal* 340 (2018): 181–191

Dunn, J.J., Starke, J.R., and Revell, P.A. 2016. Laboratory diagnosis of mycobacterium tuberculosis infection and disease in children. *Journal of Clinical Microbiology* 54: 1434–1441. DOI: 10.1128/JCM.03043-15.

Edison T, Sethuraman M. 2012. Instant green synthesis of silver nanoparticles using Terminalia chebula fruit extract and evaluation of their catalytic activity on reduction of Methylene Blue. *Process Biochem* 47: 1351–1357.

Ghosh S, Patil S, Ahire M, Kitture R, Kale S, Pardesi K, et al. 2012. Synthesis of silver nanoparticles using Dioscorea bulbifera tuber extract and evaluation of its synergistic potential in combination with antimicrobial agents. *Int J Nanomed* 7: 483–496.

Glickman, M.S. and Jacobs, W.R. 2001. Microbial pathogenesis of *Mycobacterium tuberculosis*: Dawn of a discipline. *Cell* 104: 477–485.

Gross J, Sayle S, Karow AR, Bakowsky U, Garidel P. 2016. Nanoparticle tracking analysis of particle size and concentration detection in suspensions of polymer and protein samples: Influence of experimental and data evaluation parameters. *Eur. J. Pharm. Biopharm.*, 104 (2016): 30–41. 10.1016/j.ejpb.2016.04.013

Gurunathan S., Han J.W., Park J.H., Kim J.H. 2014. A green chemistry approach for synthesizing biocompatible gold nanoparticles. *Nanoscale Res. Lett.* 9: 248. doi: 10.1186/1556-276X-9-248.

Gutierrez, M.C., Brisse, S., Brosch, R., Fabre, M., Omaïs, B., Marmiesse, M., Supply, P., and Vincent, V.. 2005. Ancient origin and gene mosaicism of the progenitor of *Mycobacterium tuberculosis. PLoS Pathogens* 1: e5.

Hart P, Copland A, Diogo GR, Harris S, Spallek R, Oehlmann W, Singh M, Basile J, Rottenberg M, Paul MJ, Reljic R. 2018. Nanoparticle-Fusion Protein Complexes Protect against Mycobacterium tuberculosis Infection. *Mol Ther.* 26(3): 822–833. doi: 10.1016/j.ymthe.2017.12.016. Epub 2017 Dec 22. PMID: 29518353; PMCID: PMC5910664.

Heera, P., and Shanmugam, S. 2015. Nanoparticle Characterization and Application: An Overview. *Int.J.Curr.Microbiol.App.Sci* 4 (8): 379–386.

Heidary M, Zaker Bostanabad S, Amini SM, Jafari A, Ghalami Nobar M, Ghodousi A, Kamalzadeh M, Darban-Sarokhalil D. 2019. The Anti-Mycobacterial Activity Of Ag, ZnO, And Ag- ZnO Nanoparticles Against MDR- And XDR-*Mycobacterium tuberculosis. Infect Drug Resist.* 12: 3425–3435. doi: 10.2147/IDR.S221408. PMID: 31807033; PMCID: PMC6839584.

Huang JL, Li QB, Sun DH, Lu YH, Su YB, Yang X, et al. 2007. Biosynthesis of silver and gold nanoparticles by novel sundried Cinnamomum camphora leaf. Nanotechnology 18. http://dx.doi.org/10.1088/0957-4484/18/10/105104. [article 105104].

Huang X, Wu H, Pu S, Zhang W, Liao X, Shi B. 2011. One-step room-temperature synthesis of Au@Pd core–shell nanoparticles with tunable structure using plant tannin as reductant and stabilizer. *Green Chem* 13: 950–7.

Kasthuri J, Kathiravan K, Rajendiran N. 2009. Phyllanthin-assisted biosynthesis of silver and gold nanoparticles: a novel biological approach. *J Nanopart Res.* 11: 1075–85.

Kaviya S, Santhanalakshmi J, Viswanathan B, Muthumary J, Srinivasan K. 2011. Biosynthesis of silver nanoparticles using citrus sinensis peel extract and its antibacterial activity. *Spectrochim Acta A Mol Biomol Spectrosc* 79(3): 594–859. doi: 10.1016/j.saa.2011.03.040. Epub 2011 Mar 23. PMID: 21536485.

Kesharwani J, Yoon KY, Hwang J, Rai M. 2009. Phytofabrication of silver nanoparticles by leaf extract of Datura metel: hypothetical mechanism involved in synthesis. *J Bionanosci* 3(1):39–44

Kestens V, Roebben G, Herrmann J, Jämting A, Coleman V, Minelli C, Clifford C, De Temmerman PJ, Mast J, Junjie L, Babick F, Cölfen H, Emons H. 2016. Challenges in the size analysis of a silica nanoparticle mixture as candidate certified reference material. *J. Nanopart. Res.*, 18: 171. 10.1007/s11051-016-3474-2

Khan, I., Khan, I., and Saeed, K. 2019. Nanoparticles: Properties, applications and toxicities. *Arabian Journal of Chemistry*, 908–931.

Khlebtsov N, Dykman L. 2011. Biodistribution and toxicity of engineered gold nanoparticles: a review of in vitro and in vivo studies. *Chem. Soc. Rev.*, 40 :1647–1671. 10.1039/C0CS00018C

Khosravi, A.D., Alami, A., Meghdadi, H., and Hosseini, A.A. 2017. Identification of *Mycobacterium tuberculosis* in clinical specimens of patients suspected of having extrapulmonary tuberculosis by application of nested PCR on five different genes. *Frontiers in Cellular and Infection Microbiology* 7, no. 3. 1–7. DOI: 10.3389/fcimb.2017.00003.

Kozinska, A., Zadlo, A., Labuz, P., Broniec, A., Pabisz, P., and Sarna, T. 2019. The Ability of Functionalized Fullerenes and Surface-Modified TiO2 Nanoparticles to Photosensitize Peroxidation of Lipids in Selected Model system. *Photochemistry and Photobiology* 95: 227–236.

Kumar V, Gokavarapu S, Rajeswari A, Dhas T, Karthick V, Kapadia Z. 2011. Facile green synthesis of gold nanoparticles using leaf extract of antidiabetic potent Cassia auriculata. *Colloids Surf B Biointerfaces* 87:159–63.

Kumari A, Yadav SK, Yadav SC. 2010. Biodegradable polymeric nanoparticles based drug delivery systems. *Colloids and Surfaces B: Biointerfaces* 75:1–18. DOI 10.1016/j.colsurfb.2009.09.001.

Mirzadeh E, Akhbari K. 2016. Synthesis of nanomaterials with desirable morphologies from metal–organic frameworks for various applications. *CrystEngComm.* 18: 7410–7424. 10.1039/C6CE01076H

Monalisa Pattanayak and P.L. Nayak. 2013. Green Synthesis and Characterization of Zero Valent Iron Nanoparticles from the Leaf Extract of *Azadirachta indica (Neem). World Journal of Nano Science & Technology* 2(1): 06–09. DOI: 10.5829/idosi.wjnst.2013.2.1.21132

Nadagouda MN, Varma RS. 2008. Green synthesis of silver and palladium nanoparticles at room temperature using coffee and tea extract. *Green Chem.* 10: 859–62.

Nasiruddin, M., Neyaz, M.K., and Das, S. 2017. Nanotechnology-based approach in tuberculosis treatment. *Tuberculosis Research and Treatment* 2017: 1–12.

Navalakhe, R., and Nandedkar, T. 2007. Application of nanotechnology in biomedicine. *Indian Journal of Experimental Biology* 45: 160–165.

Patil BN, Taranath TC. 2016. *Limonia acidissima* L. leaf mediated synthesis of zinc oxide nanoparticles: A potent tool against *Mycobacterium tuberculosis. Int J Mycobacteriol* 5:197–204.

Parida UK, Bindhani BK, Nayak P. 2011. Green synthesis and characterization of gold nanoparticles using onion (Allium cepa) extract. *World J Nano Sci Eng* 1:93–98.

Paulson T. 2013. Epidemiology: A mortal foe. *Nature* 502: S2–S3.

Premanathan, M., Karthikeyan, K., Jeyasubramanian, K., and Manivannan, G. 2011. Selective toxicity of ZnO nanoparticles toward Gram-positive bacteria and cancer cells by apoptosis through lipid peroxidation. *Nanomedicine: Nanotechnology, Biology, and Medicine* 7: 184–192.

Raghunandan D, Bedre MD, Basavaraja S, Sawle B, Manjunath S, Venkataraman A. 2010. Rapid biosynthesis of irregular shaped gold nanoparticles from macerated aqueous extracellular dried clove buds (Syzygium aromaticum) solution. *Colloids Surf B Biointerfaces* 79: 235–240.

Rajesh, S., Dharanishanthi, V., and Vinoth Kanna, A. 2015. Antibacterial mechanism of biogenic silvernanoparticles of Lactobacillus acidophilus. *Journal of Experimental Nanoscience*, 1143–1152.

Ravindra S, Mohan YM, Reddy NN, Raju KM. 2010. Fabrication of antibacterial cotton fibres loaded with silver nanoparticles via "green approach". *Colloids Surf A* 367:31–40.

Riello, F.N., Brígido, R.T.S., Araújo, S., Moreira, T.A., Goulart, L.R. and Goulart, I.M.B. 2016. Diagnosis of mycobacterial infections based on acid-fast bacilli test and bacterial growth time and implications on treatment and disease outcome. *BMC Infectious Diseases* 16:142. DOI 10.1186/s12879-016-1474-6.

Salunke, G. R., Ghosh, S., Santosh Kumar, R. J., Khade, S., Vashisth, P., Kale, T., Chopade, S., Pruthi, V., Kundu, G., Bellare, J. R., and Chopade, B. A. 2014. Rapid efficient synthesis and characterization of silver, gold, and bimetallic nanoparticles from the medicinal plant Plumbago zeylanica and their application in biofilm control. *International journal of nanomedicine* 9: 2635–2653. https://doi.org/10.2147/IJN.S59834

Sathishkumar M, Sneha K, Kwak IS, Mao J, Tripathy S, Yun YS. 2009. Phyto-crystallization of palladium through reduction process using Cinnamom zeylanicum bark extract. *J Hazard Mater* 171: 400–404.

Sakamoto K. 2012. The pathology of *Mycobacterium tuberculosis* Infection. *Veterinary Pathology* 49: 423–439. DOI: 10.1177/0300985811429313. http://vet.sagepub.com.

Saravanan, M., Ramachandran, B., Hamed, B., and Giardiello, M. 2018. Barriers for the development, translation, and implementation of nanomedicine: an African perspective. *Journal Interdisciplinary Nanomedicine* 3: 106–110.

Shankar SS, Rai A, Ahmad A, Sastry M. 2004. Rapid synthesis of Au, Ag, and bimetallic Au core–Ag shell nanoparticles using Neem (Azadirachta indica) leaf broth. *J Colloid Interface Sci.* 275: 496–502.

Singh, A.D. and S.K. Sharma. 2017. Extensively drug-resistant tuberculosis: Updates in epidemiology, diagnosis, and treatment. Chapter-1 in *Pulmonology.* API textbook- Update. Pg 1-7

Singh R, Nawale L, Arkile M, et al. 2016. Phytogenic silver, gold, and bimetallic nanoparticles as novel antitubercular agents. *International Journal of Nanomedicine.* 11:1889–1897. DOI: 10.2147/ijn.s102488.

Singhal G., Bhavesh R., Kasariya K., Sharma A.R., Singh R.P. 2011. Biosynthesis of silver nanoparticles using *Ocimum sanctum* (Tulsi) leaf extract and screening its antimicrobial activity. *J. Nanopart. Res.* 13: 2981–2988.

Sikora A, Shard A.G, MinelliSize C 2016. ζ-potential measurement of silica nanoparticles in serum using tunable resistive pulse sensing. *Langmuir* 32 (2016): 2216–2224. 10.1021/acs.langmuir.5b04160

Skeiky, Y.A. and J.C. Sadoff. 2006. "Advances in tuberculosis vaccine strategies." 4, no. 6: 469.

Smith I. 2003. *Mycobacterium tuberculosis* Pathogenesis and Molecular Determinants of Virulence. *Clin. Microbiol. Rev* 16(3): 463–496. DOI: 10.1128/CMR.

Sotgiu, G., Centis, R., D'ambrosio, L., and Migliori, G.B. 2015. Tuberculosis treatment and drug regimens. *Cold Spring Harbor Perspectives in Medicine* 5: 1–12. DOI: 10.1101/cshperspect.a017822.

Sudjarwo, S. A., Eraiko, K., Sudjarwo, G. W., and Koerniasari. 2019. The potency of chitosan-*Pinus merkusii* extract nanoparticle as the antioxidant and anti-caspase 3 on lead acetate-induced nephrotoxicity in rat. *Journal of advanced pharmaceutical technology & research* 10(1): 27–32. https://doi.org/10.4103/japtr.JAPTR_306_18

Vaikundamoorthy R, Mahalingam SS, Rajaram R. 2019. Size-dependent antimycobacterial activity of titanium oxide nanoparticles against Mycobacterium tuberculosis. *Journal of Materials Chemistry B* 7(27). DOI: 10.1039/C9TB00784A

Valodkar M, Nagar PS, Jadeja RN, Thounaojam MC, Devkar RV, Thakore S. 2011. Euphorbiaceaelatex induced green synthesis of non-cytotoxic metallic nanoparticle solutions: a rational approach to antimicrobial applications. *Colloids Surf A* 384: 337–344.

Vilchis-Nestor AR, Sánchez-Mendieta V, Camacho-López MA, Gómez-Espinosa RM, Arenas-Alatorre JA. 2008. Solventless synthesis and optical properties of Au and Ag nanoparticles using Camellia sinensis extract. *Mater Lett* 62:3103–3105.

Yelil, A., Hema, M., Tamilselvi, p., and Anbarasan, R. 2012. Synthesis and characterization of SiO2 nanoparticles by sol-gel process. *Indian J. Sci.*, 1(1): 6–10.

Yuan, Y.-G., Peng , Q.-L., and Sangiliyandi, G. 2017. Effects of Silver Nanoparticles on Multiple Drug-Resistant Strains of Staphylococcus aureus and Pseudomonas aeruginosa from Mastitis-Infected Goats: An Alternative Approach for Antimicrobial Therapy. *International Journal of Molecular Science* 568–590.

Wardani G, Mahmiah, Sudjarwo SA. 2018. *In vitro* Antibacterial Activity of Chitosan Nanoparticles against *Mycobacterium tuberculosis. Pharmacog J.* 10(1): 162–166.

Section IV

Nanotechnology for Treatment of Vector-Borne Diseases

6 Biogenic Nanoparticles Based Drugs Derived from Medicinal Plants

A Sustainable Panacea for the Treatment of Malaria

Charles Oluwaseun Adetunji,
Olugbenga Samuel Michael,
Wilson Nwankwo, Osikemekha Anthony Anani,
Juliana Bunmi Adetunji, Akinola Samson Olayinka
and Muhammad Akram

CONTENTS

6.1 Introduction 103
6.2 Specific Examples of Using Nanoparticles for the Management of Malaria 106
6.3 Application of Informatics and Bioinformatics in Nanomedicine for the Management of Malaria 111
6.4 Handling Uncertainties 112
6.5 Modelling, Discovery, Characterization, Screening and Cost Reduction 113
6.6 Verification and Validation 114
6.7 Clinical Research 114
6.8 Modes of the Action of Plant-Based Biogenic Nanoparticles against Malaria Parasites 115
6.9 Conclusion and Recommendations 116
References 116

6.1 INTRODUCTION

The history of malaria is often greeted with some awe, especially amongst researchers, individuals, scientists and historians who could recollect the menace and havoc this singular disease caused to humankind. It is reported that malaria is responsible for the demise of over 50 billion humans (Winegard, 2019).

Malaria is a leading cause of illness and death in the tropics due to its high prevalence and incidence, and sadly humans of all ages have remained vulnerable to the scourge of the infection till date. Notwithstanding the tremendous efforts expended on the control and eradication of the infection by the World Health Organization (WHO) and other prominent health groups over the last three decades, the situation has remained pathetic owing to worsening healthcare delivery infrastructure in developing countries, high drug resistance and emergence of more resistant strains of the parasite.

Since the discovery of the malaria parasite in 1897 in Secunderabad, India, by Ronald Ross and subsequent discovery of several antimalarial therapeutic and pharmacological agents, malaria has remained a health concern associated with remarkable human and socio-economic losses.

The aetiology of the infection is traced to plasmodium, a monocellular eukaryotic parasite that has vertebrates (e.g. humans) and insects (Anopheles mosquito) as hosts. Five species of the parasite known to cause malaria in humans have been identified: *Plasmodium ovale, Plasmodium vivax, Plasmodium falciparum, Plasmodium knowlesi* and *Plasmodium malariae.* The singular vector for the transmission of these parasites to man is the female Anopheles mosquito.

Besides Europe, the disease affects all other continents. According to the WHO, it is estimated that 3.2 billion persons across the globe are consistently at risk of the infection, with over 1.2 billion persons infected. The fatality of malaria infection is traced to the species of the infecting parasite and their location. For instance, *P. falciparum* has remained the most prominent parasite with the highest prevalence in African regions.

Therefore, this chapter provides detailed information on the application of nanodrugs for the management of malaria diseases. Detailed information on the application of nanodrugs derived from numerous plant materials is also highlighted. The modes of action through which the biogenic nanodrugs work against numerous *Plasmodium* spp. responsible for malaria diseases are also highlighted.

According to WHO (2019), malaria claims over 400,000 lives annually, with the highest burden being on children and pregnant women. In 2018, malaria was endemic in over 31 countries, with about 228 million incidences reported. According to the report, Africa and India were the worst hit. Of all the deaths resulting from malaria, 94% occurred within Africa in 2018. In Africa, the incidence of malaria has remained increasingly high in Nigeria, Ghana and Congo.

The onset of the disease is elicited when the system (liver and blood) is invaded, leading to a febrile illness that has a cycle of about 8 hours. Susceptible strains of the parasite respond well to various antimalarial agents. However, resistance against therapeutic agents develops rapidly after some administration of such agents over time.

There are two approaches to handling the menace of malaria: protection also called prevention, aimed at deploying effective mechanisms to counter the transmission or reduce the possibility of transmitting the parasite; and treatment which may be preventive or curative. Preventive treatment is often used to boost the resistance of the human system to invasion by parasites.

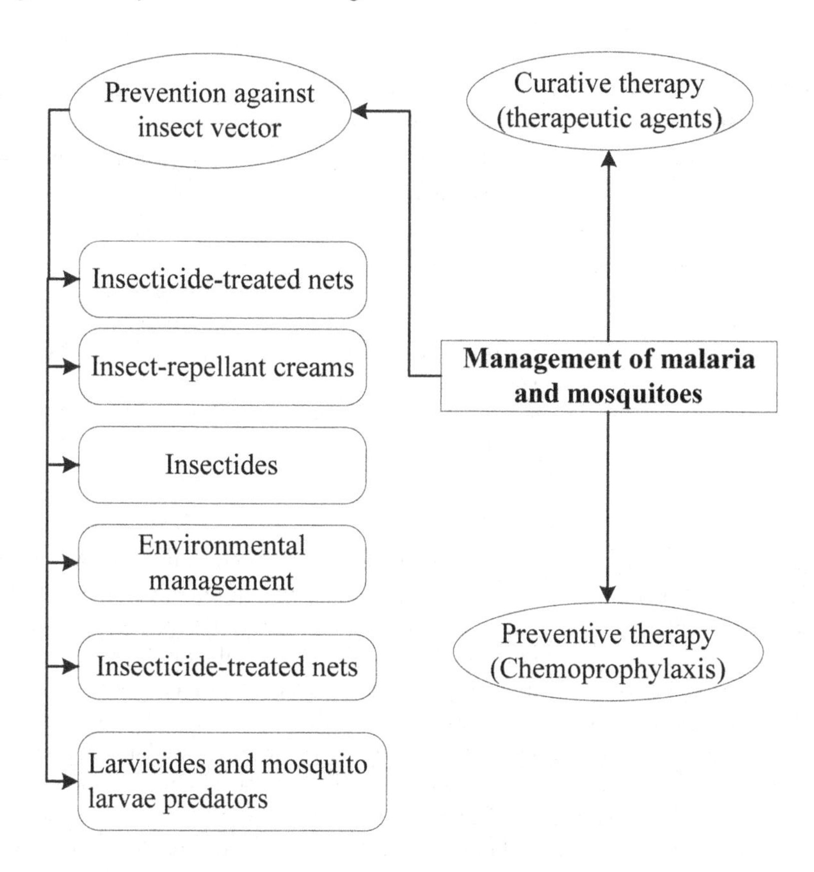

FIGURE 6.1 Management of malaria.

Figure 6.1 shows the approaches adopted to fight the malaria menace.

Table 6.1 is an articulation of the various pharmacological and therapeutic agents used in the preventive and curative treatment of malaria infection. While there are many therapeutic agents, there has not been any that could provide complete clearance of the infection.

The traditional antimalarials exhibit common challenges irrespective of the generation of the drug. The parasites soon develop resistance to the drug over a given period. Two species of *Plasmodium* have been identified as the culprit: *P. falciparum* and *P. vivax*.

Prior to the 1950s, the quinolones which are the premier antimalarials exhibited marked potency and efficacy against all strains of human malaria-causing Plasmodia. Thus, they were the first-line drugs whenever malaria is suspected. Notwithstanding the side effects of the quinolones, chemotherapeutic treatment of malaria attained a huge success. The late 1950s heralded a new development in malaria chemotherapy as drug-resistant strains of *Plasmodium falciparum* emerged in South America, Southeast Asia and Oceania. At present, no region in the world with prevalence of malaria is spared.

TABLE 6.1
Pharmacotherapeutic Agents for Malaria Treatment

S. No.	Class of therapeutics	Typical formulations	Susceptible strains	Resistant strains
1	Quinolines	Chloroquine, amodiaquine, quinine, etc.	All strains	Yes
2	Artemisinin derivatives	Artesunate	*P. Falciparum, P. malariae*	Yes
3	Antifolates	Proguanil, dapsone, pyrimethamine, sulfdoxine, etc.	All	Yes
4	Antimicrobials	Tetracyclines, e.g. doxycycline, tigecycline	*P. falciparum*	Yes

Plasmodium falciparum has developed resistance to almost all the known orthodox therapeutic agents such as mefloquine, halofantrine, quinine, sulfadoxine/pyrimethamine, artemisinin, etc. At first, it was believed that the change in trend of therapy from a single agent to multidrug combinations would add more efficacy by reason of the synergistic potentials of the agents; however, clinical trials in the last decade had seen the emergence of resistance to this combination. Consequently, it may be submitted that multidrug therapy should be re-evaluated and new therapeutic and the root cause of the resistance identified prior to the development of new agents to treat *Plasmodium falciparum* (Travassos and Laufer, 2009)

Plasmodium vivax, the second most virulent species responsible for about 47% of malaria in endemic areas, first developed resistance to chloroquine administration in 1989 in Papua New Guinea. Since then, the resistance of *P. vivax* has spread to other regions, including Southeast Asia, Africa, South America and Oceania. The discovery and isolation of artemisinin from the Chinese plant qinghaosu was greeted with much hope in the 1990s as resistance to existing pharmacological agents became a global health burden. With its rapid action and clearance rate of the drug, it was believed it would eliminate the burden created by the resistance to major antimalarials, but little was it anticipated that the WHO antimalarial guideline in 2006 wherein artemisinin combination became the most ideal chemotherapy for malaria would fail. As resistance spreads to different regions, humankind must devise new approaches towards proffering workable solutions to the menace.

6.2 SPECIFIC EXAMPLES OF USING NANOPARTICLES FOR THE MANAGEMENT OF MALARIA

The continued menace of malaria has worsened in the last decade, especially in developing countries. The foregoing is attributed to the high resistance of malarial strains to available chemotherapies (Phyo et al., 2018; Sa et al., 2019; Zhao et al.,2020; Musyoka et al., 2020; Amusengeri et al., 2020; Noisang et al., 2020). It is worth noting that no single therapeutic agent can clear the hepatic (pre-erythrocytic) and erythrocytic stages of the parasites from the body of infected persons. Oftentimes multidrug resistance also develops, making the recommended combination therapy a transaction of no value. For instance, in the early 2000s, the WHO had devised a new way of fighting malaria, especially in sub-Saharan Africa. Artemisinin combination therapeutic agents known as ACTs for short was instituted as the ideal therapeutic way. This heralded a new antimalarial trend across the globe as pharmaceutical industry recorded some boom in the production and export of ACTs to Africa. In no time, the ACTs failed and at present are facing the same fate as other older antimalarials such as quinine did in terms of potency and/or efficacy.

This sad development has led to the demand for measures that could tackle this age-old health problem. Drug resistance is a major area of concern as it is multifaceted. As posited by Tibon et al. (2020), antimalarial therapy history has shown marked improvements over the last decade from single-target-directed campaigns to multitarget therapeutic campaigns aimed at optimum clearance of parasites. As noted earlier, the failure of combination therapy demands novel approaches that could render anticipated results. According to Amusengeri et al. (2020), understanding drug resistance is prerequisite to any action geared towards discovery and development of new drugs that could be efficacious in tackling the resistance challenges posed by existing chemotherapeutic agents.

The WHO has documented some factors that promote emergence of drug-resistant parasites:

- The totality of parasites to which the antimalarial drug is intended to tackle
- Drug concentration
- Presence of some other potent antimalarials within the body system especially the blood

Majority of single and multidrug resistance have been traced to point mutation in the genes of the parasites elicited by the presence of the drug. Consequent upon mutation, binding affinity of the parasite to the drug is greatly reduced. According to WHO, the increasing drug resistance has become the greatest barrier to much efforts and many campaigns conducted to eliminate malaria.

In addition to resistance, traditional antimalarial agents have other attendant problems. These include widespread allergies, cellular toxicity, system toxicity, etc. It is reported that advancement in technology has created avenues to synthesize alternative therapies possibly from natural sources such as plants and microbes. This would promote eco-friendliness, cost-effectiveness, reduced toxicities, etc.

In the midst of the renewed widened search for possible alternatives for conventional therapeutics, herbal medicine and nanomedicine have become prospective and promising direction of not just research but also a multi-billion-dollar industry.

The history of herbs and its use in the management of healthcare problems is as old as man himself. In modern times, there has been a revitalization of herbal medical research. Thousands of phytochemicals and pharmacologically active therapeutics have been isolated for the management of several diseases, including malaria.

It has been shown that metabolites from many plants possess some good efficacy against the malarial parasite (Adeyemi et al., 2019; Ferreira et al., 2019; Manya et al., 2020; Birader et al., 2020; Herlina et al., 2019). Clinical trials have also proven that the herbal remedies may be safer and cheaper for managing malaria disease across the globe. Anand et al. (2017) noted that there are thousands of medicinal plants distributed in both temperate and tropical regions with proven medical and therapeutic uses, including the treatment of malaria. This gift of nature could be used to synthesize useful and cheap medicaments for treatment and even preventive therapies.

Recent developments in nanomedicine showcase undaunting breakthroughs that may see the exit of synthetic therapeutics in the next few years. Nanomedicine aims at synthesizing relevant nanoparticles from identified medicinal plants. Several plants have been established to possess several numerous biological active constituents with many medical benefits (Adetunji et al., 2011a–d, 2014; Adetunji and Olaleye, 2011). The opportunities presented by the global availability of abundant green plants guarantee cost-effective future medicare as nanobiotechnology would harness these materials to develop healthy and safe therapeutics that would not only surpass the potency of the traditional synthetic drugs but also align with the natural body metabolic patterns.

Five categories of nanomaterials have been recognized as effective treatment regimen for malaria. These particles can be synthesized from common green leafy medicinal plants and/or microbial sources. Research has shown that various infections could be tackled through the deployment of several nanoparticles. Infections that could be managed include helminthiasis, filariasis, tuberculosis, schistosomiasis, dengue, onchocerciasis, leishmaniasis, skin infections, malaria, etc.

In the cases of malaria, nanoparticles may be applied in two ways:

A preventive measure wherein the particular vectors are targeted and destroyed

Therapeutic agents wherein the invading parasite is targeted. The emphasis in this chapter is on antimalarial therapeutics. Table 6.2 presents the categories of no particles that could be synthesized to manage malaria.

Govindarajan and co-workers (2016) reported that *Zornia diphylla* leaf extract and silver nanoparticles (AgNP) showed dose-dependent larvicidal effect against all tested mosquito species. Therefore, *Z. diphylla* biofabricated AgNP represent a promising and eco-friendly antimalaria agent by acting against larval populations of mosquito vectors with insignificant toxic effects against other non-target organisms.

Also, Murugan et al. (2015) demonstrated that AgNP synthesized from aqueous extract of seaweeds (*Ulva lactuca*) will be a cheap, non-toxic and eco-friendly treatment against the malarial parasite. The extract of *U. lactuca* and green-synthesized AgNP possess antimalarial activity against *P. falciparum* strains 3D7 and INDO. Furthermore, *U. lactuca* extract was toxic against larval instars (I–IV) and pupae of *Anopheles stephensi* mosquitoes even if tested at low doses signifying its mosquitocidal activity. Likewise, both *U. lactuca* leaf extract and *U. lactuca*-synthesized AgNP showed higher activity against *P. falciparum* when compared to chloroquine implicating its antiplasmodial activity.

In addition, plant extract of *Pteridium aquilinum* L. and *P. aquilinum* synthesized AgNP were reported to possess mosquitocidal effects against *A. stephensi*

TABLE 6.2
Nanomaterial Employed for the Treatment of Malaria

S. No.	Nanoparticle	Source	Susceptibility
1	Liposomes	Soya bean, egg, etc.	All
2	Nucleic acid (micro-ribonucleic acid)	Chitosan (from fungi, prawns and crabs)	*P. falciparum*
3	Proteins	Gelatin	*P. falciparum*
4	Silver	*Catharanthus roseus* Linn (Panneerselvam et al., 2011), Ashoka and neem (Mishra et al., 2013), *Catharanthus roseus* Linn, etc.	
5.	Titanium	*Calotropis gigantea*, etc.	All
6.	Gold	*Polyscias scutellaria* (Yulizar et al. 2017), *Lippia citriodora*, *Pelargonium graveolens*, *Salvia officinalis*, *Punica granatum*, etc.	All

larvae and pupae, suggesting larvicidal and pupicidal toxicity. After 72 hours, 100% larval reduction was demonstrated by *Pteridium aquilinum* L. extract and *P. aquilinum* synthesized AgNP. Furthermore, *P. aquilinum* extract and AgNP led to reduced longevity and fecundity of *A. stephensi* adults. Furthermore, *P. aquilinum* leaf extract and green-synthesized AgNP showed antiplasmodial activity with higher inhibition against *P. falciparum* when compared to chloroquine. Hence, *P. aquilinum* extract might be regarded as a potential drug for malaria treatment in malaria endemic regions of the world. Essentially, *P. aquilinum* synthesized AgNP was reported to offer a cheap and valuable option in the fight against mosquito vectors (Panneerselvam et al., 2015).

Subramaniam et al. (2015) reported that AgNP synthesized using the aqueous leaf extract of *Mimusops elengi* showed larvicidal and pupicidal toxicity effects against the *A. stephensi*. It was revealed that *M. elengi* synthesized AgNP at very low dose has antimalarial effects by reducing the larval populations. Previously, *M. elengi* synthesized AgNP has been reported to have antimicrobial effects against *Klebsiella pneumoniae, Micrococcus luteus and Staphylococcus aureus* (Prakash et al. 2013). The high mortality induced by *M. elengi* synthesized AgNP against the mosquitoes during the larval stage may be because of the small size of the AgNP enabling them to go through insect's cuticle and into each cells interfering with moulting and other physiological functions (Murugan et al., 2015; Suresh et al., 2015). Furthermore, the *M. elengi* leaf extract and green-synthesized AgNP were both extremely efficient against adult mosquitoes. Therefore, mosquito adults treated with *M. elengi* synthesized AgNP usually produce offspring with distorted wings and decreased longevity probably through inhibition of the moulting process (Murugan et al., 1996).

Similarly, antiplasmodial activity of *Couroupita guianensis* synthesized gold nanoparticles (AuNPs) was demonstrated by Subramaniam et al. (2016). Malaria control is a daunting task due to the increasing number of chloroquine-resistant *Plasmodium* and pesticide-resistant *Anopheles* vectors. Therefore, newer, cheaper and safer alternative control means are necessary. *Couroupita guianensis* (Lecythidaceae), also known as a cannonball tree, is commonly utilized for the treatment of a wide spectrum of diseases in Indian traditional medicine, due to the fact that it possesses antibiotic, antifungal (Al-Dhabi et al., 2012), antidepressant (Kulkarni et al., 2011), antiseptic and analgesic (Geetha et al., 2004) effects. It is important to note that the extract of *C. guianensis* was toxic against larval instars (I–IV) and pupae of *A. stephensi*. Likewise, the green-synthesized AuNPs were highly effective against *A. stephensi* larvae and pupae, suggesting the larvicidal and pupicidal potentials of the plant extract and AuNPs. The increased mosquitocidal action of AuNPs when compared with the flower extract alone may be due to the fact that these poly-dispersed AuNPs are stable in water for some weeks, and this allows them to go through the insect cuticle and even into individual cells, where they interfere with moulting and other physiological processes (Benelli, 2016). In addition, both the *C. guianensis* flower extract and *C. guianensis* synthesized AuNPs showed enhanced antiplasmodial activity against *P. falciparum* when compared to chloroquine. Therefore, *C. guianensis* synthesized AuNPs showed significant antimalarial effects through its larvicidal and antiplasmodial effects and it may be proposed as newer and safer tools in the fight against chloroquine-resistant strain of *P. falciparum*.

The management of mosquito-borne diseases via the disruption of disease transmission by killing or preventing mosquitoes from biting humans is a critical public health concern globally and especially in the tropics (Mathew et al., 2009). Mosquito control has been largely affected by the use of conventional insecticides; however, these are not without their own problems, such as environmental degradation and the encouragement of pesticide resistance in some mosquitoes (Su and Mulla, 1998). The search for non-toxic, eco-friendly biopesticide for the control of malaria is of huge public interest. Hence, Veerakumar et al. (2013) demonstrated the larvicidal activity of *Sida acuta* aqueous extract and AgNPs synthesized using *S. acuta* plant leaf extract against the late third instar larvae of *Culex quinquefasciatus*, *A. stephensi* and *Aedes aegypti*. It is important to note that the synthesized AgNPs from *S. acuta* leaf were more toxic than crude aqueous extract of *S. acuta* leaves in these three important vector mosquito species. The AgNPs are non-toxic, affordable, biodegradable and show broad-spectrum target-specific activities against different species of vector mosquitoes. Indeed, the use of *S. acuta* synthesized silver nanoparticles can be a fast, environmentally friendly biopesticide which can serve as a novel method to develop efficient biocides for controlling the target vector mosquitoes. In the same vein, Veerakumar and Govindarajan (2014) also reported antimalarial activity of another plant-derived nanoparticle: *Feronia elephantum* synthesized silver nanoparticles. The adulticidal activity of AgNPs synthesized using *Feronia elephantum* plant leaf extract against adults of *A. stephensi, Aedes aegypti and Culex quinquefasciatus* was demonstrated by Veerakumar and Govindarajan (2014). Adult mosquitoes were treated with varying concentrations of aqueous crude extract of *F. elephantum* and synthesized AgNPs for 24 hours. *F. elephantum* treatment resulted in substantial mortality for all three important vector mosquitoes. Interestingly, the synthesized AgNPs from *F. elephantum* were extremely toxic than crude leaf aqueous extract to three important vector mosquito species. Therefore, the extracts of *F. elephantum* and green AgNPs can serve as a potential or ideal eco-friendly method for the control of the *A. stephensi, A. aegypti and C. quinquefasciatus.*

6.3 APPLICATION OF INFORMATICS AND BIOINFORMATICS IN NANOMEDICINE FOR THE MANAGEMENT OF MALARIA

Informatics contemplates a large toolset for modelling, representing, storing and disseminating information regardless of the information medium. That is, a medium may be biological, physical, chemical, artificial, etc. As a broad field and technologies, there are many subsets or rather specialties that are more refined to handle specific problems. The following subsets have been identified: bioinformatics, chemical informatics, health informatics, nanoinformatics, etc. The relationships between the various subdomains of informatics and results of utilization are shown in Fig. 6.2.

Figure 6.2 reflects an articulation of the various relationships that exist among the different strata of informatics, especially in the biomedical and chemical sciences domain. As represented in Fig. 6.2, it is deducible that bioinformatics and cheminformatics are quite complementary, whereas nanoinformatics is a hybrid or

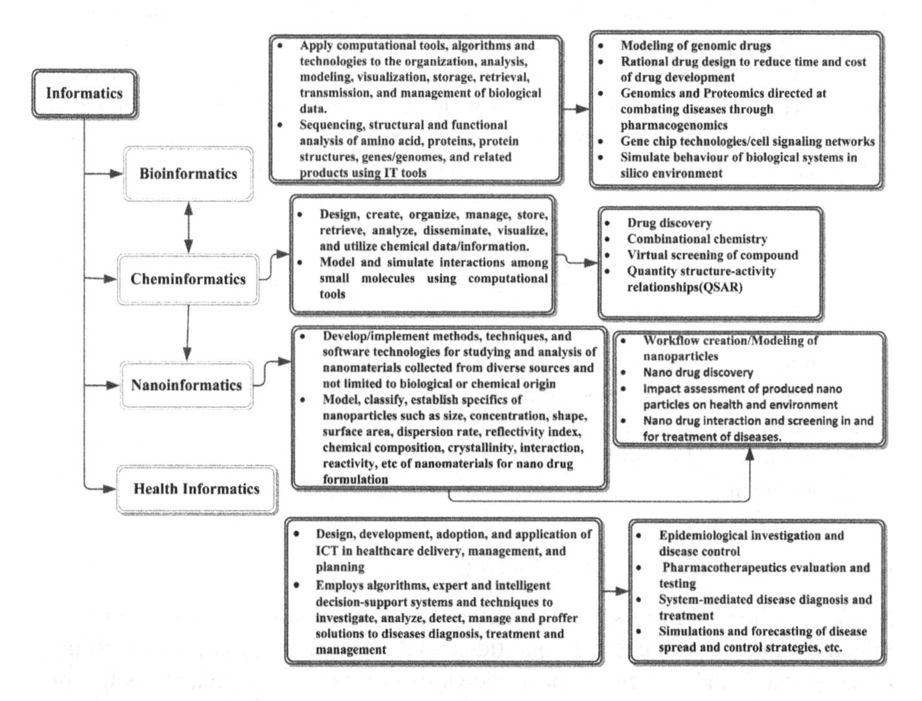

FIGURE 6.2 Informatics and applications in drug discovery, formulation, production and treatment of malaria.

somewhat convergence of the two (Moffatt, 2015), a highly differentiated component that drives nanotechnology projects. Health informatics appears to be the end point of all other variants in this context, the reason being that it explores the results from the other spheres and uses such results to manage and improve healthcare delivery systems. The scope of informatics in the diagram is also applicable from nanodrug design, discovery to formulation, nanomedicine production and use in the treatment of diseases. Note the relationships that exist and how these relationships could be explored to promote further investigation, refinement and production of nanodrugs from the raw material source such as the aforementioned medicinal plants.

Of vital interest are nanoinformatics and bioinformatics, a conglomerate of tools but with common features, that is specialized information technologies (de la Iglesia, 2013) deployed in the pursuit of nanobiotechnology research. Nanoinformatics extends beyond nanomedicine and the extent and scope of application are dependent on the nature of the nanomedicine. As nanotechnology evolves, more tools are created to help solve the information needs (Panneerselvam and Choi, 2014) that arise from it.

Management of malaria in endemic areas such as sub-Saharan Africa and India, unlike other infections and diseases with absolute prognosis, is a heinous task that requires a technology-driven toolset that can enable the professionals and researchers realize their goals. In the first place, translating research to malaria management at face value sounds odd, but the truth is that from prognosis to detection through diagnosis and/or treatment is a research process owing to the rule that at present there is no therapeutic agent that could achieve 100% curative rate in malaria attacks. Management of malaria therefore includes all activities directed towards ensuring that malaria is eradicated. It involves drug design, identification, production, clinical trials and utilization for therapeutic purposes. In all the above, informatics tools are indispensable. Informatics tools for the management of malaria fall into the following categories:

a. Modelling and design systems (used for modelling and design of antimalarial drugs). Such design includes nanostructures, etc.
b. Discovery and synthesis-support systems
c. Characterization systems
d. Validation/verification systems
e. Workflow and collaboration systems
f. Screening and impact assessment systems
g. Monitoring systems

6.4 HANDLING UNCERTAINTIES

Many nanomedicine tasks are characterized by uncertainty. Unlike some other infections and/or pathological conditions where prognosis is equivalent to diagnosis, research and experience have shown that the results of malaria control programmes are not absolute. The deliverables in such programmes are anticipated at best owing to the peculiar dynamics that may influence the outcome of the programs at some point. For instance, where nanoparticles are to be synthesized for the management of malaria or any other infection, Maojo et al. (2010) note that having respect to nature of nanoparticles wherein they exhibit diverse physico-chemical properties (shape, volume, 3D configuration, electrostatics, flexibility, etc.), which reasonably distinguish them from macromolecules, thereby significantly influencing their *in vivo–in vitro* characteristics across different cells, tissues, organs, systems and molecular environments. As a nanoparticle may be manipulated or structured to achieve some pre-specified goals based on its nanoscale characteristics and anticipated functionalization, there are more uncertainties which translate to a handicap, why? This is because these particles are structurally poly-dispersed and particulate variability exists consequent upon which different effects may be exhibited. In regard to the following, it is submitted that computationally, such tasks to a great extent follow the fuzzy logic principle consequent upon which any systematic procedures employed would need intelligent adaptive systems that are designed to handle such uncertainties. Nanoinformatics and bioinformatics integrated with artificial intelligence (AI) and machine learning (ML) systems could enhance the productivity of such programs.

Another scenario is where a drug discovery program is launched to develop new eco-friendly, safe, potent and highly efficacious drug intended to overcome the weaknesses of the available chemotherapeutic agents such a program cannot register any reasonable progress without deployment of large datasets for analysis, modelling and testing/simulation of necessary drug characteristics and component biochemical reactions. Specialized AI systems would be able to streamline and operate on these datasets. Targets would be identified including the candidate molecules. These systems would also help predict synthesis routes, cellular interaction and half-life of the prototype drug.

6.5 MODELLING, DISCOVERY, CHARACTERIZATION, SCREENING AND COST REDUCTION

Computational techniques and tools are usually applied in the design of nanodrugs for malaria. As posited by Thierry (2009), these tools may be used to design transport systems in respect of drugs and nano-devices. Other uses at this phase include stoichiometry of nanomaterial (using computational molecular approaches) and characterization.

Computational techniques have been applied for chemical simulation of the various chemical compounds which are utilized for effective development or formulation of drugs or nanodrugs (Kouvaris et al., 2012; Etheridge et al., 2014; Antreas et al., 2018).

In silico modelling of chemical structures and associated multidimensional visualizations enable and enhance virtual examination and study of interactions between compounds at different levels. It is interesting to note that in drug formulation, the desired end point is an optimized compound that yields the desirable threshold potency and efficacy. Often, the desired compound is a mix of two or more compounds which have been discovered to exhibit some synergistic effects in terms of efficacy and/or potency. Accordingly, to achieve the foregoing, multidimensional chemical modelling and characterization must be employed if optimum result is anticipated. Hence prior to extraction and modification of the different active metabolites from the medicinal plant, a classical virtual model of known metabolites could be used to understudy the new compounds as well as make predictions on their biomedical characteristics.

As opposed to time-consuming throughput screening drug chemistry studies, bioinformatics and cheminformatics (an offshoot of bioinformatics) provide an excellent platform for virtual screening. This involves the application of sophisticated computational tools to examine large *in silico* libraries of chemical compounds to discover the various compounds that exhibit interesting or desired characteristics, especially bacteriostatic or bactericidal against the targeted *Plasmodium* species. The implication is that very large chemical compounds not just restricted to plant metabolites could be understudied without conducting physical laboratory experiments. In other words, it is very much possible to discover new antimalarial drugs from virtual screening of large and diverse libraries of natural and synthetic compounds.

6.6 VERIFICATION AND VALIDATION

Another interesting application of informatics to nanodrug formulation is the use of technological tools to establish the relationships between the quantitative structure, activity and property values of chemical and biochemical compounds such as metabolites. It has been shown that the use of such techniques could lead to the accurate prediction of biochemical activity of metabolites using only their structures.

Bioinformatics plays a major role in rational drug design, thus reducing the time and cost of drug development. The foregoing is true in the synthesis of drugs as bioinformatics provides comprehensive analytical tools and huge databases for modelling of proteins, genes and related biological compounds as well as the evaluation of the interaction and behaviour, including compatibility and functionalities. In other words, it provides a very cost-effective *in silico* environment for simulating, verifying and validating the behaviour of drug and drug-related compounds in biological systems.

6.7 CLINICAL RESEARCH

Utilizing antimalarial drug formulations is very vital in any control program. Notwithstanding the category of antimalarials, clinical trials have shown that resistance to once very efficacious drugs has continued to increase. With the excellent performance of newer therapeutic agents like artemisinin groups and the ACTs, it was difficult to predict that such drugs could later be overwhelmed by resistant strains of the parasites. Today there is an overwhelming need to embark on drug discovery voyage; otherwise, it is believed that these recent class of drugs would later face the same fate chloroquine had in the 1960s.

Developing new chemotherapeutic agents for malaria is a need that cannot be shelved. The same is true for antimalarial drug improvement programmes which would complement the drug discovery programmes. Both programmes are important and should be canvassed by researchers, clinicians, health authorities, regional and international health agencies. With a collective effort, the socio-economic impact of malaria may be greatly reduced in the next few years.

Genomics, proteomics and metabolomics are domains that utilize bioinformatics apparatuses. Activities/tasks such as understanding genome sequences of the Plasmodia, especially aetiologic agents of human malaria, should be reinforced. Bioinformatics provides excellent insights and opportunities that would enable clinical researchers to identify the encoded parasite determinants in genomes.

These genomes are often the targets to which the intended or proposed drug would bind to. Thus, genome sequencing of malaria parasites would enable a thorough and precise comparative analysis to be conducted. With advanced bioinformatics technologies enhanced by AI, clinicians and researchers are provided with the necessary platform to not only discover new drugs but conduct effective tests as to the potency and/or efficacy of the discovered drug. Thus, clinical trials may be performed without actually administering the drugs on a physical patient.

6.8 MODES OF THE ACTION OF PLANT-BASED BIOGENIC NANOPARTICLES AGAINST MALARIA PARASITES

Malaria parasite (*Plasmodium* spp.) is one of the deadliest protozoans across sub-Saharan Africa, Asian and Pacific regions of the world. The main problems of the treatment of malaria parasite are the multidrug resistance it portends, which will lead to the poor efficacy of the drugs and the overload of the parasite in the blood. However, recent advances in the utilization of more effective drugs using nanoparticles (aluminium oxide, magnesium oxide, iron oxide, gold and silver) in conjunction with plant extracts (conjugates) have proven to be promising and more effective in the management of various malaria strains (Bushman et al., 2016).

Silver nanoparticles particles have been proven to be effective in the management of malaria by the restraint of the growth of the parasite *in vitro* and *in vivo* (Mishra et al., 2013; Jaganathan et al., 2016 and Murugan et al., 2016a, b) as well as plant extracts of *Artemisinin annua* and *Murraya koenigii* (Muangphrom et al., 2016; Kamarag et al., 2017). Some oxides of metal nanoparticles like CeO_2, Al_2O_3, ZrO_2, MgO and Fe_3O_4 have been known to be a good antiplasmodial against strains of malaria parasites (Inbaneso and Ravikumar, 2013), while nanoparticles of gold have proven to hinder antiplasmodial action and obstruct *in vivo* parasitemia increase (Karthik et al., 2013; Dutta et al., 2017).

The mode of action of biogenic nanoparticles with plant extracts in the management of malaria parasites is via oxidative stress, intracellular infiltration and injury, cell membrane and cell wall injury impact on the micro-*Plasmodium* strains (Dakal et al., 2016; Duran et al., 2016; Slavin et al., 2017; Roy et al., 2019). Before the conjugates (nanoparticles and plant extract(s) can exercise action on the *Plasmodium* strains, the bridge of the endoperoxide must be triggered in order to produce FRS (free radical species). Apart from the mode of action stated by Dakal et al. (2016), Duran et al. (2016), Slavin et al. (2017) and Roy et al. (2019) in nanoparticle binding with micro-plasmodium cell, Sun et al. (2015) recounted two mechanisms of actions (heme-mediated breakdown lanes and mitochondrial-activation) in which plant extracts like *Artemisia annua* and *Murraya koenigii* can be utilized in the management of *Plasmodium* spp. The first mechanism of action is via two initiation models: exposed peroxide and reductive scission models, which have been projected to both generate radicals of centred vigorous carbon (O'Neill et al., 2010). The biosynthesis of heme (Fe^{2+}) at the initial stage can provoke the breakdown of the trophozoite phase which indicate that its production can facilitate artemisinin intake by the cell (Klonis et al., 2011; Wang et al., 2015). After the heme breakdown, the HDP (heme detoxification protein) can elicit the conversion of Fe^{2+} (heme) to Fe^{3+} (hemozin) which is suitable for the *Plasmodium* parasite to strive on. But this formation of the free artemisinin-heme radical indicates an obstructive impact on sudden change (van Agtmael et al., 1999; Xie et al., 2016). The second mode of action involves cytotoxic lipid induced peroxidation through ROS (reactive oxygen species) and non-polarity of the plasmodium plasma membranes and mitochondrial (Wang et al., 2010; Mercer et al., 2011; Antoine et al., 2014; Sun et al., 2015).

6.9 CONCLUSION AND RECOMMENDATIONS

This chapter has provided a detailed insight into the application of eco-friendly nanodrugs derived from plants as sustainable natural drugs that could be used for effective management of malaria diseases. Moreover, detailed information on the application of informatics and their effectiveness in drug discovery, formulation, production and treatment of numerous types of malaria parasites were also highlighted. The mechanism of action of these nanodrugs derived from plants was discussed. Also, the relevance of modelling, discovery, characterization, screening and cost reduction of nanodrugs were also highlighted. Furthermore, the application of *in silico* modelling of chemical structures and associated multidimensional visualizations as well as application of computational techniques was also highlighted in detail. There is a need to identify new and novel biological active compound from unexploited medicinal plants with high antimalarial effectiveness that could be used in the bioengineering of these plant extracts for the synthesis of more effective nanodrugs.

REFERENCES

Adetunji, C.O. and Olaleye, O.O. 2011. Phytochemical screening and antimicrobial activity of the plant extracts of *Vitellaria paradoxa* against selected microbes. *Journal of Research in Biosciences* 7, no. 1: 64–69.

Adetunji, C.O., Arowora, K.A, Afolayan, S.S., Olaleye, O.O., and Olatilewa, M.O. 2011a. Evaluation of antibacterial activity of leaf extract of *Chromolaena odorata. Science Focus* 16, no. 1: 1–6.

Adetunji, C.O., Kolawole, O.M., Afolayan, S.S., Olaleye, O.O., Umanah, J.T., and Anjorin, E. 2011b. Preliminary phytochemical and antibacterial properties of *Pseudocedrela kotschyi*: A potential medicinal plant. *Journal of Research in Bioscience. African Journal of Bioscience* 4, no. 1: 47–50.

Adetunji, C.O., Olaleye, O.O., Adetunji, J.B., Oyebanji, A.O. Olaleye, O.O., and Olatilewa, M.O. 2011c. Studies on the antimicrobial properties and phytochemical screening of methanolic extracts of *Bambusa vulgaris* leaf. *International Journal of Biochemistry* 3(1): 21–26.

Adetunji, C.O., Olaleye, O.O., Umanah, J.T., Sanu, F.T, and Nwaehujor, I.U. 2011d. *In vitro* antibacterial properties and preliminary phytochemical of *Kigelia africana*. *Journal of Research in Physical Sciences* 7, no. 1: 8–11.

Adetunji, C.O, Olatunji, O.M, Ogunkunle, A.T.J, Adetunji, J.B., and Ogundare, M. O. 2014. Antimicrobial activity of ethanolic extract of *Helianthus annuus* stem. *Sikkim Manipal University Medical Journal* 1, no. 1: 79–88.

Adeyemi, M.M., Habila, J.D., Enemakwu, T.A., Okeniyi, S.O., and Salihu, L. 2019. Antimalarial activity of leaf extract, fractions and isolation of sterol from *Alstonia boonei*. *Tropical Journal of Natural Product Research* 3, no. 7, 221–224. DOI: 10.26538/tjnpr/v3i7.1

Al-Dhabi, N.A., Balachandran, C., Raj, M.K., Duraipandiyan, V., Muthukumar, C., Ignacimuthu, S., Khan, I.A., and Rajput, V.S. 2012. Antimicrobial, antimycobacterial and antibiofilm properties of *Couroupita guianensis* Aubl. fruit extract. *BMC Complementary and Alternative Medicine* 12: 242.

Amusengeri, A., Tata, R.B., and Bishop, Ö.T. 2020. Understanding the pyrimethamine drug resistance mechanism via combined molecular dynamics and dynamic residue network analysis. *Molecules* 25(4):904. Pages 1–24. DOI: 10.3390/molecules25040904.

Anand, K., Tiloke, C., Naidoo, P., and Chuturgoon, A.A. 2017. Phytonanotherapy for management of diabetes using green synthesis nanoparticles. *Journal of Photochemistry and Photobiology B: Biology* 173: 626–639. DOI: 10.1016/j.jphotobiol.2017.06.028.

Antoine, T., Fisher, N., Amewu, R., O'Neill, P.M., Ward, S.A., and Biagini, G.A. 2014. Rapid kill of malaria parasites by artemisinin and semi-synthetic endoperoxides involves ROS-dependent depolarization of the membrane potential. *Journal of Antimicrobial Chemotherapy* 69: 1005–1016.

Antreas, A., Georgia Melagraki, Andreas Tsoumanis, Eugenia Valsami-Jones, and Iseult Lynch. 2018. A nanoinformatics decision support tool for the virtual screening of gold nanoparticle cellular association using protein corona fingerprints. *Nanotoxicology* 12, no. 10: 1148–1165. DOI: 10.1080/17435390.2018.1504998.

Benelli G. 2016. Plant-mediated biosynthesis of nanoparticles as an emerging tool against mosquitoes of medical and veterinary importance: A review. *Parasitology Research* 115: 23–34. DOI: 10.1007/s00436-015-4800-9.

Biradar, Y.S., Bodupally, S., and Padh, H. 2020. Evaluation of antiplasmodial properties in 15 selected traditional medicinal plants from India. *Journal of Integrative Medicine* 18, no. 1: 80–85. DOI: 10.1016/j.joim.2019.11.001.

Bushman, M., Morton, L., Duah, N., Quashie, N., Abuaku, B., Koram, K.A., Dimbu, P.R. et al. 2016. Within-host competition and drug resistance in the human malaria parasite *Plasmodium falciparum*. *Proceedings of the Royal Society B: Biological Sciences* 283: 20153038.

Chellasamy Panneerselvam, Kadarkarai Murugan, Mathath Roni Al Thabiani Aziz, Udaiyan Suresh, Rajapandian Rajaganesh, Pari Madhiyazhagan, Jayapal Subramaniam et al. 2015. Fern-synthesized nanoparticles in the fight against malaria: LC/MS analysis of *Pteridium aquilinum* leaf extract and biosynthesis of silver nanoparticles with high mosquitocidal and antiplasmodial activity. *Parasitology Research* 115, no. 3: 997–1013. DOI: 10.1007/s00436-015-4828-x.

Dakal T.C., A. Kumar, R. S. Majumdar, and V. Yadav. 2016. Mechanistic basis of antimicrobial actions of silver nanoparticles. *Frontiers in Microbiology* 7: 1831.

de la Iglesia, D., Cachau, R. E., García-Remesal, M., and Maojo, V. 2013. Nanoinformatics knowledge infrastructures: Bringing efficient information management to nanomedical research. *Computational Science & Discovery* 6, no. 1: 014011. DOI: 10.1088/1749-4699/6/1/014011.

Duran N., M. Duran, M.B. de Jesus, A.B. Seabra, W.J. F´avaro, and G. Nakazato. 2016. Silver nanoparticles: A new view on mechanistic aspects on antimicrobial activity. *Nanomedicine* 12: 789–799.

Dutta, P.P., Bordoloi, M., Gogoi, K., Roy, S., Narzary, B., Bhattacharyya, D.R., Mohapatra, P.K., and Mazumder, B. 2017. Antimalarial silver and gold nanoparticles: Green synthesis, characterization and in vitro study. *Biomedicine & Pharmacotherapy* 91: 567–580.

Etheridge, M.L., Hurley, K.R., Zhang, J., Jeon, S., Ring, H.L., Hogan, C., and Bischof, J.C. 2014. Accounting for biological aggregation in heating and imaging of magnetic nanoparticles. *Technology* 2, no. 3: 214–228. DOI: 10.1142/S2339547814500198.

Ferreira, L.T., Venancio, V.P., Kawano, T., Abrão, L.C.C., Tavella, T.A., Almeida, L.D., and Costa, F.T.M. 2019. Chemical genomic profiling unveils the in vitro and in vivo antiplasmodial mechanism of açaí (*Euterpe oleracea* mart.) polyphenols. *ACS Omega* 4, no. 13: 15628–15635. DOI: 10.1021/acsomega.9b02127.

Geetha, M., Saluja, A.K., Shankar, M.B., and Mehta, R.S. 2004. Analgesic and anti-inflammatory activity of *Couroupita guianensis* Aubl. *Journal of Natural Remedies* 4: 52–55.

Govindarajan M, Rajeswary M, Muthukumaran U, Hoti SL, Khater HF, Benelli G. (2016). Single-step biosynthesis and characterization of silver nanoparticles using Zornia diphylla leaves: A potent eco-friendly tool against malaria and arbovirus vectors. *J Photochem Photobiol B*.161:482–9. doi: 10.1016/j.jphotobiol.2016.06.016.

Herlina, T., Rudiana, T., Julaeha, E., and Parubak, A.S. 2019. Flavonoids from stem bark of akway (*Drymis beccariana* gibs) and their antimalarial properties. *Journal of Physics: Conference Series* 1280, no. 2: 022010. DOI: 10.1088/1742-6596/1280/2/022010.

Inbaneson, S.J., and Ravikumar, S. 2013. In vitro antiplasmodial activity of PDDS-coated metal oxide nanoparticles against *Plasmodium falciparum*. *Applied Nanoscience* 3: 197–201.

Jaganathan, A., Murugan, K., Panneerselvam, C., Madhiyazhagan, P., Dinesh, D., Vadivalagan, C., Chandramohan, B. et al. 2016. Earthworm-mediated synthesis of silver nanoparticles: A potent tool against hepatocellular carcinoma, *Plasmodium falciparum* parasites and malaria mosquitoes. *Parasitology International* 65: 276–284.

Kaliyan Veerakumar, and Marimuthu Govindarajan. 2014. Adulticidal properties of synthesized silver nanoparticles using leaf extracts of *Feronia elephantum* (Rutaceae) against filariasis, malaria, and dengue vector mosquitoes. *Parasitology Research* 113: 4085–4096. DOI: 10.1007/s00436-014-4077-4.

Kaliyan Veerakumar, Marimuthu Govindarajan, and Mohan Rajeswary. 2013. Green synthesis of silver nanoparticles using *Sida acuta* (Malvaceae) leaf extract against *Culex quinquefasciatus*, *Anopheles stephensi*, and *Aedes aegypti* (Diptera: Culicidae). *Parasitology Research*. 112(12):4073-4085. DOI: 10.1007/s00436-013-3598-6.

Kamaraj C, Govindasamy Balasubramani, Chinnadurai Siva, Manickam Raja, Velramar Balasubramanian, Ramalingam Karthik Raja, Selvaraj Tamilselvan, Giovanni Benelli, and Pachiappan Perumal. 2017. Ag nanoparticles synthesized using b-caryophyllene isolated from *Murraya koenigii*: Antimalarial (*Plasmodium falciparum* 3D7) and anticancer activity (A549 and HeLa cell lines). *Journal of Cluster Science* 28: 1667–1684. DOI; 10.1007/s10876-017-1180-6.

Karthik, L., Kumar, G., Keswani, T., Bhattacharyya, A., Reddy, B.P., and Rao, K.B. 2013. Marine actinobacterial mediated gold nanoparticles synthesis and their antimalarial activity. *Nanomedicine: Nanotechnology, Biology and Medicine* 9: 951–960.

Klonis, N., Crespo-Ortiz, M.P., Bottova, I., Abu-Bakar, N., Kenny, S., Rosenthal, P.J., and Tilley, L. 2011. Artemisinin activity against *Plasmodium falciparum* required hemoglobin uptake and digestion. *Proceedings of the National Academy of Sciences of the United States of America* 108: 11405–11410.

Kouvaris P., Delimitis A., Zaspalis V., Papadopoulos D., Tsipas S.A., and Michailidis N. 2012. Green synthesis and characterization of silver nanoparticles produced using *Arbutus unedo* leaf extract. *Materials Letters* 76: 18–20.

Kulkarni, M., Wakade, A., Ambaye, R., and Juvekar, A. 2011. Phytochemical and pharmacological studies on the leaves of Couroupita guianensis AUBL. *Pharmacology* 3: 809–814.

MA. Laufer MK. 2009. Resistance to antimalarial drugs: molecular, pharmacological and clinical considerations. *Pediatr Res.* 65(5 Pt 2): 64R–70R. doi: 10.1203/PDR.0b013e3181a0977e.

Manya, M.H., Keymeulen, F., Ngezahayo, J., Bakari, A.S., Kalonda, M.E., Kahumba, B.J., and Lumbu, S.J. 2020. Antimalarial herbal remedies of bukavu and uvira areas in DR congo: An ethnobotanical survey. *Journal of Ethnopharmacology* 249:112422 DOI: 10.1016/j.jep.2019.112422.

Maojo, V., Martin-Sanchez, F., Kulikowski, C., Rodriguez-Paton, A., and Fritts, M. 2010. Nanoinformatics and DNA-based computing: Catalyzing nanomedicine. *Pediatric Research* 67: 481–489. DOI: 10.1203/PDR.0b013e3181d6245e.

Mathew, N., Anitha, M.G., Bala, T.S.L., Sivakumar, S.M., Narmadha, R., and Kalyanasundaram, M. 2009. Larvicidal activity of *Saraca indica*, Nyctanthes arbor-tristis, and *Clitoria ternatea* extracts against three mosquito vector species. *Parasitology Research* 104: 1017–1025.

Mercer, A.E., Copple, I.M., Maggs, J.L., O'Neill, P.M., and Park, B.K. 2011. The role of heme and the mitochondrion in the chemical and molecular mechanisms of mammalian cell death induced by the artemisinin antimalarials. *Journal of Biological Chemistry* 286, no. 2: 987–996.

Mishra, A., Kaushik, N.K., Sardar, M., and Sahal, D. 2013. Evaluation of antiplasmodial activity of green synthesized silver nanoparticles. *Colloids and Surfaces B: Biointerfaces* 111: 713–718.

Moffatt, S. 2015. Convergence of nanoinformatics nanobiotechnology and bioinformatics. *MOJ Proteomics & Bioinformatics* 2, no. 6: 193–195. DOI: 10.15406/mojpb.2015.02.0006.

Muangphrom, P., Seki, H., and Fukushima, E.O. 2016. Toshiya Muranaka1Artemisinin-based antimalarial research: application of biotechnology to the production of artemisinin, its mode of action, and the mechanism of resistance of Plasmodium parasites. *Journal of Natural Medicines* 70, no. 3: 318–334. DOI 10.1007/s11418-016-1008-y.

Murugan, K., Jeyabalan, D., Senthilkumar, N,. Babu, R., and Sivaramakrishnan, S. 1996. Antipupational effect of neem seed kernel extract against mosquito larvae of *Anopheles stephensi* (liston). *Journal of Entomological Research* 20: 137–139.

Murugan, K., Benelli, G., Panneerselvam, C., Subramaniam, J., Jeyalalitha, T., Dinesh, D., Nicoletti, M., Hwang, J.S., Suresh, U., and Madhiyazhagan, P. 2015. *Cymbopogon citratus*-synthesized gold nanoparticles boost the predation efficiency of copepod *Mesocyclops aspericornis* against malaria and dengue mosquitoes. *Experimental Parasitology* 153: 129–138.

Murugan, K., Panneerselvam, C., Samidoss, C.M., Madhiyazhagan, P., Suresh, U., Roni, M., Chandramohan, B., et al. 2016a. In vivo and in vitro effectiveness of *Azadirachta indica*-synthesized silver nanocrystals against *Plasmodium berghei* and *Plasmodium falciparum*, and their potential against malaria mosquitoes. *Research in Veterinary Science* 106: 14–22.

Murugan, K., Panneerselvam, C., Subramaniam, J., Madhiyazhagan, P., Hwang, J.S., Wang, L., Dinesh, D. et al. 2016b. Eco-friendly drugs from the marine environment: Spongeweed-synthesized silver nanoparticles are highly effective on *Plasmodium falciparum* and its vector *Anopheles stephensi*, with little non-target effects on predatory copepods. *Environmental Science and Pollution Research* 23: 16671–16685.

Musyoka, K.B., Kiiru, J.N., Aluvaala, E., Omondi, P., Chege, W.K., Judah, T., and Kimani, F.T. 2020. Prevalence of mutations in plasmodium falciparum genes associated with resistance to different antimalarial drugs in nyando, kisumu county in Kenya. *Infection, Genetics and Evolution* 78: 104121 DOI: 10.1016/j.meegid.2019.104121.

Noisang, C., Meyer, W., Sawangjaroen, N., Ellis, J., and Lee, R. 2020. Molecular detection of antimalarial drug resistance in *Plasmodium vivax* from returned travellers to NSW, Australia during 2008–2018. *Pathogens* 9(2):101 pages 1–13. DOI: 10.3390/pathogens9020101.

O'Neill, P.M., Barton, V.E., and Ward, S.A. 2010. The molecular mechanism of action of artemisinin: The debate continues. *Molecules* 15: 1705–1721.

Panneerselvam, S. and Choi, S. 2014. Nanoinformatics: Emerging databases and available tools. *International Journal of Molecular Sciences* 15, no. 5: 7158–7182. DOI: 10.3390/ijms15057158.

Panneerselvam, C., Ponarulselvam, S., and Murugan, K. 2011. Potential anti-plasmodial activity of synthesized silver nanoparticle using *Andrographis paniculata* Nees (Acanthaceae). *Archives of Applied Science Research* 3, no. 6: 208–217.

Phyo, A.P., Win, K.K., Thu, A.M., Swe, L.L., Htike, H., Beau, C., Sriprawat, K. et al. 2018. Poor response to artesunate treatment in two patients with severe malaria on the Thai–Myanmar border. *Malaria Journal* 17: 30 pages 1–5. DOI: 10.1186/s12936-018-2182-z.

Prakash, P., Gnanaprakasam, P., Emmanuel, R., Arokiyaraj, S., and Saravanan, M. 2013. Green synthesis of silver nanoparticles from leaf extract of *Mimusops elengi*, Linn. for enhanced antibacterial activity against multi drug resistant clinical isolates. *Colloids and Surfaces B: Biointerfaces* 108: 255–259.

Roy, Anupam, Onur Bulut, Sudip Some, Amit Kumar Mandal, and M. Deniz Yilmaz. 2019. Green synthesis of silver nanoparticles: Biomolecule-nanoparticle organizations targeting antimicrobial activity. *RSC Advances* 9: 2673–2702.

Sá, J.M., Kaslow, S.R., Moraes Barros, R.R., Brazeau, N.F., Parobek, C.M., Tao, D., and Wellems, T.E. 2019. *Plasmodium vivax* chloroquine resistance links to PVCRT transcription in a genetic cross. *Nature Communications* 10:4300 pages 1-10. DOI: 10.1038/s41467-019-12256-9.

Slavin Y.N., J. Asnis, U.O. H¨afeli, and H. Bach. 2017. Metal nanoparticles: Understanding the mechanisms behind antibacterial activity. *Journal of Nanobiotechnology* 15: 65.

Su, T. and Mulla, M.S. 1998. Ovicidal activity of neem products (Azadirachtin) against *Culex tarsalis* and *Culex quinquefasciatus* (Diptera:Culicidae). *Journal of the American Mosquito Control Association* 14: 204–209.

Subramaniam, J., Madhiyazhagan, P., Dinesh, D., Rajaganesh, R., Alarfaj, A.A., Nicoletti, M., Kumar, S. et al. 2015. Seaweed-synthesized silver nanoparticles: An eco-friendly tool in the fight against Plasmodium falciparum and its vector *Anopheles stephensi*? *Parasitology Research* 114, no. 11: 4087–4097. DOI: 10.1007/s00436-015-4638-1.

Subramaniam, J., Murugan, K., Panneerselvam, C., Kovendan, K., Madhiyazhagan, P., Kumar, P.M, Dinesh, D. et al. 2016. Multipurpose effectiveness of *Couroupita guianensis*-synthesized gold nanoparticles: High antiplasmodial potential, field efficacy against malaria vectors and synergy with *Aplocheilus lineatus* predators. *Environmental Science and Pollution Research* 23: 7543–7558. DOI: 10.1007/s11356-015-6007-0.

Sun, C., Li, J., Cao, Y., Long, G., and Zhou, B. 2015. Two distinct and competitive pathways confer the cellcidal actions of artemisinins. *Microbial Cell* 2: 14–25.

Suresh, U., Murugan, K., Benelli, G., Nicoletti, M., Barnard, D.R., Panneerselvam, C., Mahesh Kumar, P., Subramaniam, J., Dinesh, D., and Chandramohan, B. 2015. Tackling the growing threat of dengue: *Phyllanthus niruri*-mediated synthesis of silver nanoparticles and their mosquitocidal properties against the dengue vector *Aedes aegypti* (Diptera: Culicidae). *Parasitology Research* 114: 1551–1562.

Thierry, B. 2009. Drug nanocarriers and functional nanoparticles: Applications in cancer therapy. *Current Drug Delivery* 6, no. 4: 391–403.

Tibon, N.S., Ng, C.H., and Cheong, S.L. 2020. Current progress in antimalarial pharmacotherapy and multi-target drug discovery. *European Journal of Medicinal Chemistry* 188: 111983. Pages 1-24 DOI: 10.1016/j.ejmech.2019.111983.

van Agtmael, M.A., Eggelte, T.A., and van Boxtel, C.J. 1999. Artemisinin drugs in the treatment of malaria: From medicinal herb to registered medication. *Trends in Pharmacological Sciences* 20: 199–205.

Wang, J., Huang, L., Li, J., Fan, Q., Long, Y., Li, Y., and Zhou, B. 2010. Artemisinin directly targets malarial mitochondria through its specific mitochondrial activation. *PLoS One* 5: e9582.

Wang, J., Zhang, C.J., Chia, W.N., Loh, C.C.Y., Li, Z., Lee, Y.M., He, Y. et al. 2015. Haem-activated promiscuous targeting of artemisinin in *Plasmodium falciparum*. *Nature Communications* 6: 10111.

WHO (World Health Organization). 2019. *World Malaria Report*. Geneva: World Health Organization.

Winegrad, T.C. 2019. *The Mosquito: A Human History of Our Deadliest Predator*. Boston: Dutton.

Xie, S.C., Dogovski, C., Hanssen, E., Chiu, F., Yang, T., Crespo, M.P., Stafford, C. et al. 2016. Haemoglobin degradation underpins the sensitivity of early ring stage *Plasmodium falciparum* to artemisinins. *Journal of Cell Science* 129: 406–416.

Yulizar, Y., Utari, T., Ariyanta, H.A., and Maulina, D. 2017. Green method for synthesis of gold nanoparticles using *polyscias scutellaria* leaf extract under UV light and their catalytic activity to reduce methylene blue. *Journal of Nanomaterials* 2017: 1–6.

Zhao, L., Pi, L., Qin, Y., Lu, Y., Zeng, W., Xiang, Z., Qin, P. et al. 2020. Widespread resistance mutations to sulfadoxine-pyrimethamine in malaria parasites imported to China from Central and Western Africa. *International Journal for Parasitology: Drugs and Drug Resistance* 12: 1–6.

Section V

Nanotechnology in Combating Antimicrobial Resistance

Section V

7 Bioengineering of Inorganic Nanoparticle Using Plant Materials to Fight Extensively Drug-Resistant Tuberculosis

Mpho Phehello Ngoepe and
Maluta Steven Mufamadi

CONTENTS

7.1 Introduction 125
  7.1.1 Antimicrobial Resistance 126
  7.1.2 Extensively Drug-Resistant Tuberculosis 127
  7.1.3 Inorganic Nanotechnology 129
7.2 Bioengineering of Inorganic Nanoparticles Employing a Green Approach 130
  7.2.1 Bioengineering of Inorganic Nanoparticle for XRD-TB 132
7.3 Engineering of Functionalized Inorganic Nanoparticles 136
7.4 Engineering of Labelled Inorganic Nanoparticles 138
7.5 Mechanism of Action: Physiochemical Properties 139
7.6 Future Perspective 141
7.7 Conclusions 142
7.8 References 142

7.1 INTRODUCTION

Treatment for extensively drug-resistant tuberculosis (XDR-TB) has always been a challenge, as many patients fail to complete treatment requiring four to five active drugs over a 24-month period (Migliori et al., 2020). A large number of patients with failure to complete (treatment failure) their XDR-TB were found to have a 73% mortality rate in South Africa, where half of the deaths occurred within the first two-year time frame (Frank et al., 2019). Various treatment failure risk factors, such as TB/human immunodeficiency virus (HIV) co-infection (poor absorption of TB drugs), illicit drug use and smoking and socio-economic factors (level of education), have been observed leading to patients dying from treatment failure (Bhering et al., 2019). The era of nanotechnology targeting intracellular infections such as tuberculosis aided in overcoming treatment failure due to prolonged treatment (controlled and sustained release of drugs), high pill burden (multidrug nanoparticle drug delivery system) and low compliance (reduced side effects and favourable route of administration, e.g. inhalation) (Kaur et al., 2020). Apart from using nanoparticles (NPs) as drug carriers, inorganic nanoparticles have been shown to exert antibacterial properties, thus serving as alternative antibacterial which can overcome drug resistance associated with antibiotics (Tăbăran et al., 2020). In the present chapter, authors describe the benefits of inorganic nanoparticles synthesized from medicinal plants instead of conventional physical and chemical methods. Bio-reduction of different metallic ions by phytochemicals are highlighted and their antimicrobial activity against XDR-TB also discussed.

7.1.1 Antimicrobial Resistance

Failure to identify new antimicrobials is of major concern due to an increase in antimicrobial resistance (AMR) to antimicrobial drugs. This has made it possible to effectively control infections caused by gram-positive pathogens such as *Staphylococcus* and *Streptococcus* and *Mycobacterium tuberculosis*, respectively, since the discovery of penicillin in 1929 and streptomycin in 1943 (Brown and Wright, 2016). Due to the evolution and widespread distribution of antibiotic resistance elements in bacterial pathogens, this results in diseases that were once easily treatable becoming deadly again. To overcome this resistance, drug-resistant bacterial infections result in an increased dosage of drugs, increased toxic therapy, extended hospital stays and increased mortality. Drug resistance has been shown to occur through three mechanisms: gene acquisition, expression of resistant genes and selection of bacteria that express resistant genes (Pelgrift and Friedman, 2013). Following the acquisition of resistant genes, after exposure to the antimicrobial agent, the bacteria express genes to overcome antimicrobial activity. This leads to the selection of bacteria that express resistant genes against those susceptible to antimicrobial agents. The cause of these selective pressures has been linked to health care professionals (inappropriate prescription practices, inadequate patient education, lack of appropriate regulation) and patients' behaviour (misuse due to unnecessary and excessive, non-compliance due to missed doses or discontinuation of treatment due to side effects, taking herbal mixtures along with antimicrobial treatment) (Ayukekbong et al., 2017).

This is prevalent in cases where microbes are exposed to the drug but are not eradicated either by the microbicidal effects of the drug itself or by the microbiostatic effects (inhibit but do not kill microbes) of the drug, where the host immune system plays a key role in killing the microbes (Teixeira et al., 2018). As a result of these acquisitions of antimicrobial resistance, nanoparticles have been proposed as alternative antibiotics because they are capable of overcoming existing antibiotic resistance mechanisms by various means, such as disruption of bacterial membranes and inhibition of biofilm formation (Wang et al., 2017). The mechanism of antimicrobial action of metal oxide nanoparticles is due to (a) electrostatic interaction

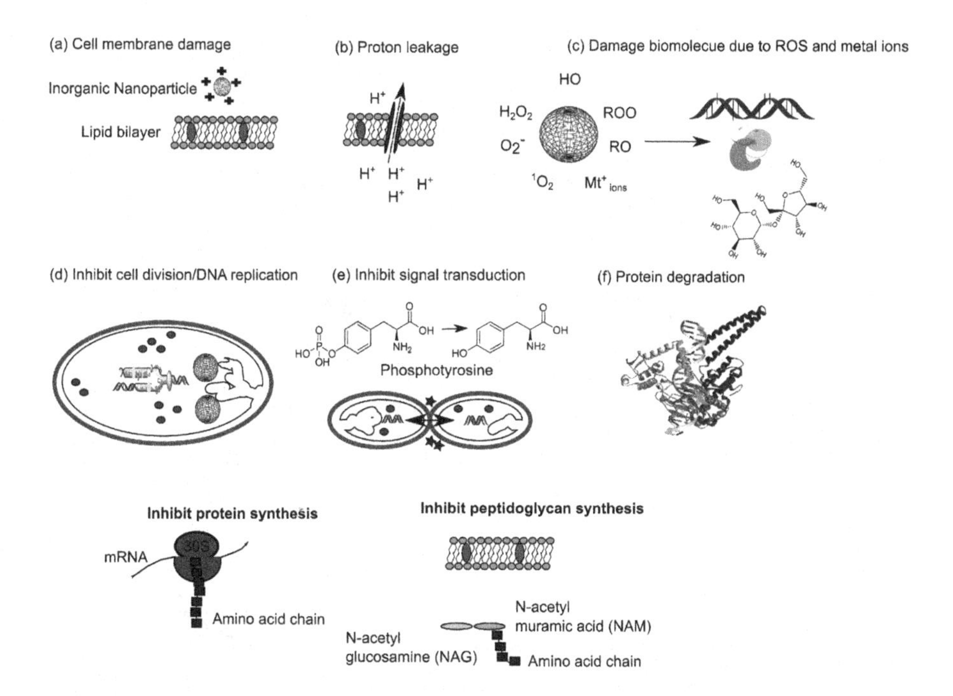

FIGURE 7.1 Antimicrobial mechanism of NPs and their ions.

between the particle and the microbe cell surface causing cell membrane damage; while cell uptake leads to (b) leakage of the proton due to disruption of the chemiosmosis process (dissipation of the proton motive force); (c) damage of organic biomolecules (carbohydrates, lipids, proteins and nucleic acids) due to the generation of reactive oxygen species (ROS) having a microbicidal effect; (d) altering cellular respiration, cell division and DNA replication due to mesosome binding; (e) inhibition of signal transduction and bacterial growth by dephosphorylation of the phosphotyrosine residue and (f) degradation of protein due to protein carbonylation leading to loss of catalytic activity of the enzyme (Fig 7.1) (Mahira et al., 2019).

7.1.2 Extensively Drug-Resistant Tuberculosis

Over the years, countries like South Africa have been gaining victory against TB and HIV. Drug-resistant TB has become a threat to the TB health care system where multidrug-resistant TB (MDR-TB) and XDR-TB infections continue to rise (Fig 7.2).

When TB is found to be resistant to one anti-tuberculosis drug, it is classified as mono-resistant TB; it is classified as poly-resistant TB when resistance is to two or more first-line anti-tuberculosis drugs (not consisting of both isoniazid and rifampicin); MDR-TB when resistance involves isoniazid and rifampicin; XDR-TB when resistance involves fluoroquinolone resistance and second-line injection (capreomycin, kanamycin and amikacin) together with isoniazid resistance and rifampicin resistance (Shah et al., 2020). As opposed to other bacteria, the resistance to

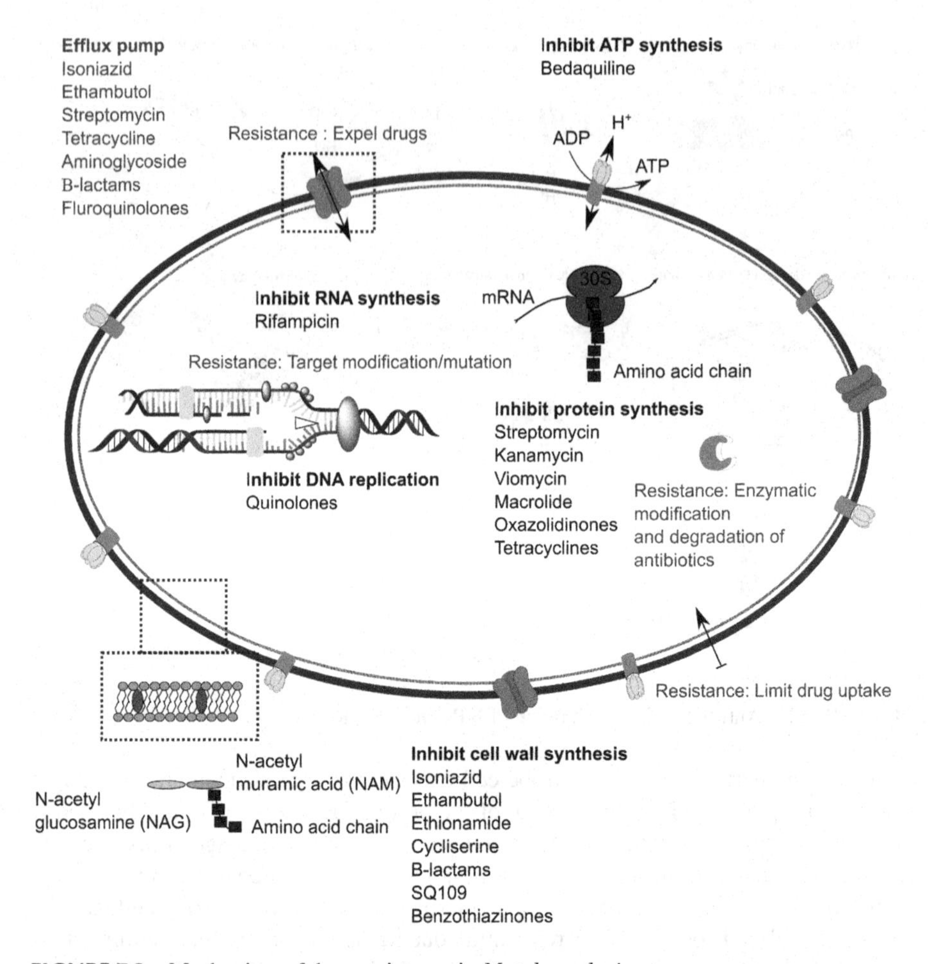

FIGURE 7.2 Mechanism of drug resistance in *M. tuberculosis.*

Mycobacterium tuberculosis arises from chromosomal mutations (acquired resistance) under the selective pressure of antibiotics instead of mutations or horizontal gene transfer mediated by phages, plasmids or transposon components. Apart from acquired drug-specific resistance, intrinsic resistance is known as well as new drugs via modification of drug target sites, drug inactivation through acetylation, drug degradation and the expulsion of drugs through efflux pumps (Nguyen, 2016). Apart from gene-directed resistance, phenotypic tolerance can lead to the bacteria able to survive drug dosing employing metabolic or physiological changes.

South Africa has one of the highest burdens of tuberculosis and drug-resistant tuberculosis in the world, whereby resistant TB strains occur through transmission rather than acquired resistance to new hosts (Shah et al., 2017). It was shown that 69–92% of XDR-TB in KwaZulu-Natal province, South Africa, can be attributed to person-to-person transmission of drug-resistant strains, rather than the acquisition of resistance in the setting of prior TB treatment. Casual contact in community settings between individuals not known to one another was a suspected mode of infection (Auld et al., 2018). For TB treatment, the estimated cost is approximately US$2,000 per patient for a standard six-month course of therapy. However, when dealing with MDR-TB (US$154,000) and XDR-TB (US$494,000), the cost can increase 25 times whilst the duration of treatment can take three times as long as for drug-susceptible TB (US$17,000) (Erck et al., 2016). Although various health systems promote Bacillus Calmette–Guérin (BCG) vaccination, various studies have shown that this does not offer complete protection from TB infection (Seaworth et al., 2014). The Directly Observed Treatment Short (DOTS) programme also faces limitations as the infrastructure required is cumbersome, labour-intensive and expensive (O'Brien and Nunn, 2001). This still causes the occurrence of the development of drug resistance in developing countries. There is a need for new drugs for TB. However, with nine of every ten new drugs failing in the human testing phase and approval screening by the FDA taking over 10–15 years, there is a need for new treatment procedures (Van Norman, 2016).

7.1.3 Inorganic Nanotechnology

Nanotechnology (submicron – <1 μm – colloidal particles) may be used to address various current limitations of TB treatment. For addressing the limitations of BCG (insufficient protection against pulmonary TB, safety concerns in children), nanoparticles can be used as carriers to deliver encapsulated vaccines (antigenic proteins) to selective sites, release them over a long period to boost the immune response and prolong the shelf life of vaccines over a wide range of temperatures (Chang and Leung, 2017; Nasiruddin et al., 2017). Citrate-coated manganese ferrite ($MnFe_2O_4$) nanoparticles fused with Mycobacterium tuberculosis fusion protein have been studied as potential nanovaccines for tuberculosis (Poon and Patel, 2020). This has given rise to nanoscale drug delivery systems (nano-DDS) whereby various materials can be used in overcoming problems associated with the increased drug resistance of infectious agents such as tuberculosis and HIV. Besides serving as carriers, the nanoparticles themselves can also act as drugs. Inorganic nanoparticles incorporating silver (Ag), gold (Au), copper (Cu), titanium (Ti), magnesium (Mg), zinc (Zn), aluminium (Al), iron (Fe) or metal oxides have been shown to have significant antimicrobial, antifungal and antiviral activity (Kumar and Anthony, 2016). Overall, metals increasingly considered to be antimicrobial agents are typically within the transition metals of the d-block (V, Ti, Cr, Co, Ni, Cu, Zn, Tb, W, Ag, Cd, Au, Hg) and a few other metals and metalloids of groups 13–16 of the periodic table (Al, Ga, Ge, As, Se, Sn, Sb, Te, Pb and Bi) (Turner, 2017). By incorporating known antibiotics into nanoparticles, the term nanoantibiotic has been introduced, which has shown improved safety and efficacy of antibiotics (Muzammil et al., 2018).

Thus, nanoantibiotics can have no immediate and intense effects, contrary to current antimicrobial agents, although potential toxicity is questionable on long-term exposures. However, caution should be taken in the formation of nanoantibiotics as studies have shown that nanoantibiotic activity is dependent on the formation of nano-pathogen physical contact, but the activity can be reduced by lipid or protein biomolecule coronas formed in physiological environments (Siemer et al., 2019). However, the use of inorganic nanoparticles has shown that they have a greater advantage as they affect multiple biological pathways found in broad species of microbes (Malekkhaiat and Malmsten, 2017). Throughout history, silver has widely been utilized for burn wound treatment, dental work, catheters and bacterial infection control, in the forms of metallic silver, silver nitrate and silver sulfadiazine (Kumar and Anthony, 2016). In the field of drug delivery, gold is used for its near-infrared (NIR) light-absorbing rangeability for imaging technology. In antimicrobial use, irradiation with focused laser pulses of the appropriate wavelengths, photo-thermal therapy (PTT), can lead to bacteria death (Boboc et al., 2017). Metal oxides such as zinc, iron and copper have also shown to affect directly where the stimulation of ROS and metal ions leads to antimicrobial activity (Boboc et al., 2017). Metal oxide nanoparticles such as titanium oxide (TiO_2), copper oxide (CuO), zinc oxide (ZnO), iron oxide (Fe_2O_3), magnesium oxide, aluminium oxide (Al_2O_3) and cerium oxide (CeO) have all been shown to have antibacterial and antibiofilm properties along with gold and silver nanoparticles (Banerjee et al., 2019).

7.2 BIOENGINEERING OF INORGANIC NANOPARTICLES EMPLOYING A GREEN APPROACH

The toxicity of the chemical reagents used in the manufacture of inorganic nanoparticles has led to the development of green technologies to address this problem. It has long been known that plants can reduce metal ions both on their surface and in various parts and tissues away from the ion penetration site. In recent years, plant extracts have been used to bio-reduce and stabilize metal ions to form nanoparticles (Fig 7.3) (Dauthal and Mukhopadhyay, 2016). These approaches provide more flexible control over the size and shape of nanoparticles (e.g. by changing medium pH and reaction temperature) and facilitate easy purification.

Plant metabolites such as sugars, terpenoids, polyphenols, alkaloids, phenolic acids and proteins play an important role in the reduction of metal ions to nanoparticles and their subsequent stability (Vijayaraghavan and Ashokkumar, 2017). It has been suggested that control over the size and morphology of nanostructures may be connected to the interaction of these biomolecules with metal ions. Flavonoids are a large group of polyphenolic compounds that comprise several classes: anthocyanins, isoflavonoids, flavonols, chalcones, flavones and flavanones, which can actively chelate and reduce metal ions into nanoparticles.

Flavonoids contain various functional groups capable of nanoparticle formation. Tautomeric transitions from the enol-form to the keto-form of flavonoids have been postulated to release a reactive hydrogen atom which can reduce the formation of metal ions into nanoparticles (Makarov et al., 2014). Various phytochemicals can be adsorbed onto the surface of a nascent nanoparticle and have chelating abilities (Adusei et al., 2019). This probably means that they are involved in the stages of initiation of nanoparticle formation (nucleation) and further aggregation, in addition to the bio-reduction stage. Glucose was also shown to be a stronger reducer than fructose for sugars because the kinetics of tautomeric shifts restrict the antioxidant

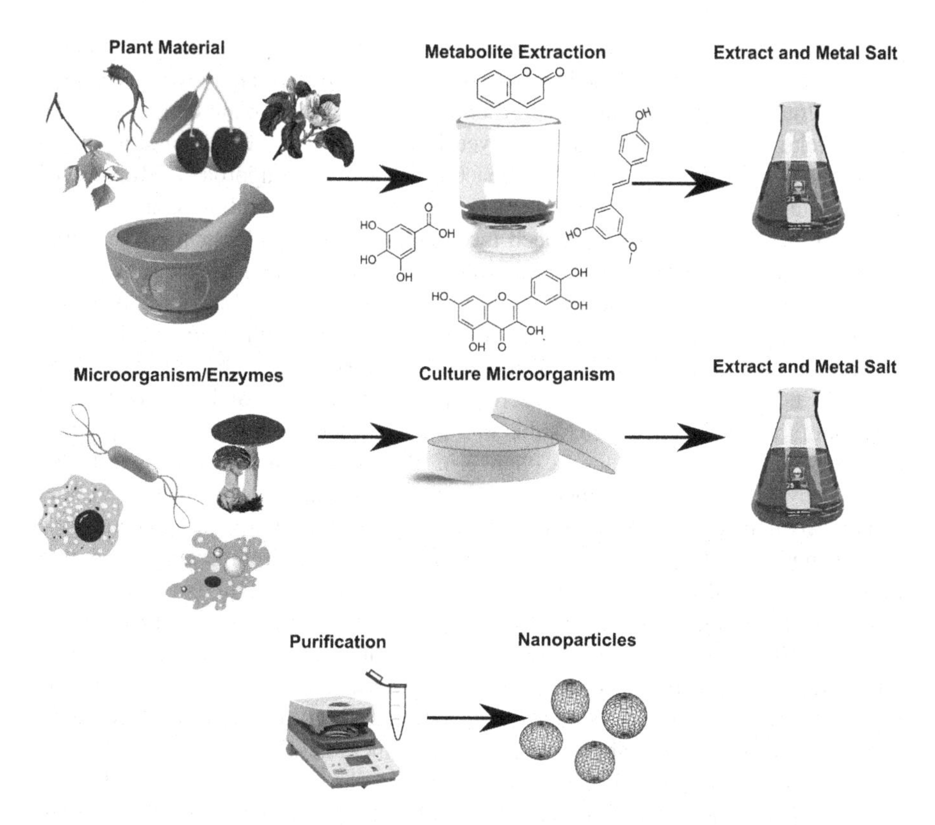

FIGURE 7.3 Diagram summarizing the possible mechanism of biologically mediated synthesis of nanoparticles.

potential for fructose (Makarov et al., 2014). Amino acids (reducing and capping agents) can bind to metal ions through the amino and carbonyl groups of the main chain or side chains, such as the carboxyl groups of aspartic and glutamic acid or a nitrogen atom of the imidazole ring of histidine (Sharma et al., 2019). By-product such as honey has also been used for the synthesis of gold, silver, carbon, platinum and palladium nanoparticles with catalytic, anticorrosive, antimicrobial and biosensing and bio-imaging ability (Balasooriya et al., 2017).

Reduction and stabilization of metal ions by a combination of biomolecules such as proteins, amino acids, enzymes, polysaccharides, alkaloids, tannins, phenols, saponins, terpenoids and vitamins are already established in medicinal plant extracts, which are environment-friendly yet chemically complex structures. Briefly, the nanoparticle synthesis protocol involves collecting part of the plant of interest from the available sites and then washing it thoroughly twice/thrice with tap water to remove both epiphytes and necrotic plants, followed by sterile distilled water to remove the associated debris if any. Clean and fresh sources are shaded for 10–15 days and then powdered with a domestic blender. Appropriate amounts of the dried powder are boiled with a specified volume of deionized distilled water (hot percolation method) for the preparation of the plant broth. The resulting infusion is then filtered thoroughly until no insoluble material has appeared in the broth (Ahmed et al., 2016a). The metal solution will then be mixed with the plant extract to make the particles. Two approaches are involved in the syntheses of inorganic nanoparticles, either from the "top-down" approach or a "down -up" approach. "Top down" approach usually includes physical (laser ablation and evaporation–condensation) synthesis, whilst "down up" approach includes either chemical (chemical reducing agents such as ascorbic acid, sodium borohydride and stabilizing agents such as surfactants to stabilize the nanoparticles) or biological (microorganisms such as plants, bacteria, fungi; biomolecules or plant extracts) techniques (Muhammad et al., 2019).

Various factors can affect the formation of nanoparticles such as reaction mixture pH, incubation temperature, reaction time, concentration and electrochemical potential of a metal ion. The pH value of a plant extract exerts a great influence on the formation of nanoparticles. A change in pH results in a charge change in the natural phytochemicals contained in an extract, which affects their ability to bind and reduce metal cations and anions in the course of nanoparticle synthesis, and this, in turn, may affect the shape, size and yield of nanoparticles. Temperature is another important factor affecting the formation of nanoparticles in plant extracts. In general, temperature elevation increases the reaction rate and efficiency of nanoparticle synthesis. Elevating the temperature increases the nucleation rate. Higher temperatures alter the interaction of phytochemicals with the nanoparticle surface, thereby inhibiting the incorporation of adjacent nanoparticles into the structure of nanoribbons. Due to the limited ability of plants to reduce metal ions, the efficiency of metal nanoparticle synthesis also depends on the electrochemical potential of an ion. The green synthesis of nanoparticles involves three phases: (1) activation phase involves the reduction of metal ions and reduced metal atoms undergo nucleation; (2) growth stage (Ostwald ripening) includes spontaneous coalescence into bigger NPs of small neighbouring NPs; (3) the termination stage in which the final form of NPs is decided(Imran Din and Rani, 2016). Table 7.1 illustrates the various inorganic nanoparticles synthesized using various metal precursors and plants.

Nanoparticles such as gallium have already been shown to inhibit the growth of virulent *Mycobacterium tuberculosis* in macrophages (Choi et al., 2017). Phytochemicals such as alkaloids (sanguiritrin, oxostephanine, micromeline, lansine, 3-formylcarbazole, 3-formyl-6-methoxycarbazole and chelerythrine), coumarins (scopoletin, umckalin, scopoletin and isobavachalcone), flavonoids/polyphenols (baicalein, catechin and epigallocatechin), terpenoids (eucalimin, (+)-bornyl piperate, bonianic acids A and bonianic acids B), quinones (1-hydroxy-benzoisochromanquinone, aloe emodin and alpha-tocopheryl quinine) and phytosteroids (ergosterol peroxide and β-sitostenone) compounds have activity against *Mycobacterium tuberculosis*, which has extensively been discussed elsewhere (Enioutina et al., 2017; Sharma and Parkash, 2017).

7.2.1 Bioengineering of Inorganic Nanoparticle for XRD-TB

The treatment success varies between TB, MDR-TB and XDR-TB at 83%, 52% and 28%, respectively (Rahman and Sarkar, 2017). This shows that more work is required

TABLE 7.1
Various Inorganic Nanoparticles with Antimicrobial Activity Synthesized Using Green Chemistry

Silver nanoparticles (Ag^+)

Feature: Work by attacking the respiratory chain and cell division that finally lead to cell death, while simultaneously releasing Ag^+ ions that enhance the antibacterial activity (Kurtjak et al., 2017).

Synthesis: The primary requirement of green synthesis of silver nanoparticles (AgNPs) is silver metal ion solution and a reducing biological agent. The major physical and chemical parameters that affect the synthesis of AgNP are reaction temperature, metal ion concentration, extract contents, pH of the reaction mixture, duration of reaction and agitation (Srikar et al., 2016).

Gold nanoparticles (Au^{3+})

Feature: In general, gold nanoparticles are considered to be inert biologically. Green-synthesized AuNPs show efficient antibacterial activity against certain strains of bacteria, especially compared to those synthesized with the chemical, due to the synergistic effect of the combination of AuNPs and extracts (Zhang et al., 2015).

Synthesis: Different ratio of gold salt and extract at different temperature and pH can be used for gold nanoparticles synthesis. The extract is simply mixed with the auric salts solution (different concentrations depending on plants parts and their species) at room temperature and their conversion into nanoparticles takes place within minutes in one-pot synthesis and eco-friendly media (Ahmed et al., 2016b).

Copper nanoparticles (Cu^{2+})

Feature: The copper oxide (Cu_2O) nanoparticles act by releasing Cu^{2+} ions, chelating of cellular enzymes and DNA damage after cell membrane penetration and interference of the biochemical pathway (Bogdanović et al., 2014).

Synthesis: The plant biomolecules (terpenoids, nimbaflavone and polyphenols) reduce Cu^{2+} ions into CuNPs and serve as capping and stabilizing agents as well (Nagar & Devra, 2018).

Palladium nanoparticles (Pd^{2+})

Feature: Palladium nanoparticles (PdNPs) synthesized via plant extract have both photocatalytic and antibacterial properties, thus making them ideal for water purification (Tahir et al., 2016b).

Synthesis: Chemical reduction and stabilization of Pd(II) by plant metabolites (antioxidants) where palladium chloride can be used as a reagent. It has been shown that reaction conditions such as pH, temperature and biomaterial dosage have no significant impacts on the shape and size of nanoparticles generated (Iravani, 2011).

Nickel/nickel oxide

Feature: Nickel (Ni) nanoparticles (release nickel ions – Ni^{2+} – inside the cell) may enter the membrane of bacterial cells and bind to phosphorous and sulfur-based functional compounds like DNA and proteins leading to cell death (oxidative stress) (Salar, 2011). The particles also have catalytic and antibacterial properties.

Synthesis: Nickel nitrate can be mixed with methanolic plant extract to produce nickel nanoparticles (NiNPs) or mixed with leaves extract where pH plays a role in the formation of nickel oxide nanoparticles (NiO-NPs) with broad-spectrum antibiotic abilities (Din et al., 2018).

Zinc oxide (ZnO)

Feature: The antimicrobial activity of zinc oxide nanoparticles (ZnO-NPs) depends on the bacterial species, where it is more versatile and more harmful towards gram-positive bacteria than gram-negative bacteria. ZnO has shown antimicrobial effects and is commonly used for food preservation, manufacturing stability and increasing the shelf life of products (Gold et al., 2018).

Synthesis: Zinc nitrate can be used for the synthesis of ZnO-NPs, by mixing with the medicinal plant extract (vitamins, flavonoids and phenolic acids) at room temperature. The reaction has to be protected from light to prevent photoinduced changes (Matinise et al., 2017).

(Continued)

TABLE 7.1 (CONTINUED)
Various Inorganic Nanoparticles with Antimicrobial Activity Synthesized Using Green Chemistry

Silver nanoparticles (Ag^{+})

Iron Oxide (Fe_3O_4)

Feature: Antimicrobial activity depends on concentration due to the release of reactive oxygen species, superoxide radicals and single oxygen created by magnetic iron oxide (Fe_3O_4-NPs). The particles were also found to have no haemolytic (cytotoxic) activity against erythrocytes (Gabrielyan et al., 2019). These particles are non-toxic, abundant, cheap, easy to produce and their magnetic properties make them applicable for extraction and biomedical application.

Synthesis: The iron nanoparticles can be prepared by mixing different ratios of plant extracts with iron chloride (Katata-Seru et al., 2018).

Cerium oxide (CeO_2)

Feature: The oxidative stress (CeO_2 consists of intrinsic oxygen defects ideal for catalytic and antioxidant application) of the components of the cell membrane of the microorganism causes microbicide activity. During this mechanism, an electron is acquired and the Ce^{4+} is transformed into Ce^{3+} on the cerium nanoparticles (CeNPs) surface (Farias et al., 2018). The nanoparticles are also described as semiconductors.

Synthesis: Plant-mediated synthesis of CeNPs has been linked to alkaloids acting as stabilizing agents whilst in fungi, enzymes, proteins and heterocyclic derivatives they are proposed to act as reducing and capping agents. The synthesis requires cerium(III) chloride salt where heat is required to form nanoparticles (Charbgoo et al., 2017).

Gallium (Ga^{3+})

Feature: Gallium (Ga^{3+}) is an iron-like metal (Fe^{3+}) that is vital for the development and metabolism of *Mycobacterium tuberculosis*. Replacing Ga^{3+} with Fe^{3+} on the active site of enzymes can make them non-functional and is a prospective strategy for novel tuberculosis treatment (Olakanmi et al., 2000).

Synthesis: Gallium nitrate can be used for the synthesis of gallium nanoparticles (GaNPs) at room temperature employing plant extract (Monika et al., 2017).

Titanium oxide (TiO_2)

Feature: Titanium nanoparticles (TiO_2-NPs) are known to have photocatalytic (disinfectant due to generation of highly reactive oxygen species – ROS) and antibacterial activity under illumination with near UV light. Furthermore, the nanoparticles are stated to have various properties such as hydrophilicity, stability, and non-toxicity and have a good economical yield (Zimbone et al., 2015).

Synthesis: Acidic solution of titanium isopropoxide solution can be mixed with plant extract for the synthesis of TiO_2 nanoparticles with hydrophilic, photocatalytic and antibacterial properties (Sankar et al., 2015).

Aluminium oxide (Al_2O_3)

Feature: Alumina nanoparticles (Al_2O_3-NPs) are thermodynamically stable across various temperature changes and have antibacterial properties against both gram-positive and gram-negative bacterial strains (Manyasree et al., 2018).

Synthesis: Aluminium nitrate can be used as a precursor by mixing with aqueous plant extracts at room temperature (Ansari et al., 2015).

(Continued)

TABLE 7.1 (CONTINUED)
Various Inorganic Nanoparticles with Antimicrobial Activity Synthesized Using Green Chemistry

Magnesium oxide (MgO)
Feature: Magnesium oxide nanoparticles (MgO-NPs) are biocompatible, stable under harsh conditions and act as antitumour and antibacterial agents (Pugazhendhi et al., 2019).
Synthesis: Magnesium nitrate can be used as a precursor in an alkaline solution containing plant extract where heat is required for the synthesis of nanoparticles (Vergheese and Vishal, 2018).

to improve treatment outcomes for XDR-TB. Treatment of biofilm-associated infections is a major challenge. The eradication of biofilm is challenging because the extracellular polymeric substances (EPS) comprising the biofilm (consisting of keto-mycolic acids) prevent the antibiotic penetration and host's immune defences into biofilms (Eldholm and Balloux, 2016). Thus, biofilm formation by pathogens appears to facilitate the survival of these pathogens in the environment and the host. Current methods to eradicate biofilms typically require excising of infected tissues combined with antibiotic therapy, invasive treatments that incur high medical costs (Wang et al., 2016a). Various agents have been designed to target the biofilm architecture, disperse the microbial cells into their more vulnerable, planktonic mode of life. These dispersing agents vary enzymes (proteases, deoxyribonucleases and glycoside hydrolases), antibiofilm peptides (human cathelicidin, LL-37) and dispersal molecules such as dispersal signals (cis-2-decanoic acid – CDA, nitric oxide), anti-matrix molecules (chitosan, D-amino acids, urea) and sequestration molecules (lactoferrin, EDTA) (Fleming and Rumbaugh, 2017). Although various penetrating agents can be conjugated to antibiotics or biocides, they can be enzymatically inactivated in biofilms.

Biofilms are critical pathogenic factors of mycobacteria, as the pathogenesis of chronic diseases are strongly associated with biofilm formation (Esteban and García-Coca, 2018). The use of biofilm-penetrating agents can make the XDR-TB treatment procedure expensive. Some studies have shown that phytochemicals are capable to act as antimycobacterial and anti-biofilm formation compounds (Bhunu et al., 2017). Thus, the use of inorganic nanoparticles can help in killing both growing and static bacteria. With advancements in drug delivery technology, the engineered particles can be utilized to target phagocytic cells infected by intracellular pathogens. A study by Jafari and colleagues (2016) has shown that mixing Ag and ZnO nanoparticles at a ratio of 8 ZnO:2 Ag, we acquired a mixture that exhibited potent antibacterial activity against Mtb and no cytotoxic effects on THP-1 cells, resulting in inhibition of both *in vitro* and *ex vivo* Mtb growth. Thus, this shows that inorganic particles have the potential to fight against XDR-TB (Jafari et al., 2016). Another study has shown the benefit of silver nanoparticles synthesized from leaf extract that possessed antibacterial and antibiofilm against *Mycobacterium tuberculosis* (Singh et al., 2016). Due to TB affecting the lungs, a study on the cytotoxicity of Ag and ZnO nanoparticles showed that Ag nanoparticles did not cause any toxic effects towards human macrophage (THP-1) and normal human lung fibroblast (MRC-5) cell lines, contrary to ZnO nanoparticles (Jafari et al., 2017).

In this chapter, various inorganic nanoparticle synthesis methods and potential applications in treating XDR-TB are discussed. These inorganic nanoparticles have a distinct advantage over conventional chemical antimicrobial agents, as their mechanism of action is also unique. As previously discussed, the limitation of the antimicrobial mechanism of chemical agents is highly dependent on the specific binding with the surface and the metabolism of agents into the microorganism. To penetrate a bacterial biofilm, nanoparticle properties such as size and surface chemistry are also discussed (Miller et al., 2015). Throughout history, these microorganisms have been evolving drug resistance over many generations with/without antimicrobial agent exposure. Apart from the synthetic procedure, there will be a discussion of how to be functionalized inorganic nanoparticles. Although various inorganic nanoparticles have been tested for antimicrobial activity due to their broad and random targeting capabilities, synthetic procedures have economic constraints (Khurana and Chudasama, 2018). The synthetic methods vary from chemical reduction, electrochemical reduction, photochemical reduction and heat evaporation.

However, these methods have several disadvantages such as the high production costs, not being environment-friendly as there is a generation of a large amount of pollution in the environment due to usage of toxic solvents and reducing agents during the production process (Yadi et al., 2018). To avoid these issues, green chemistry approaches can be employed to produce antimicrobial NPs. This chapter looks at various plant extracts utilized in NPs synthesis, with advantages such as simplicity, convenience, less energy-intensive, eco-friendly and minimize the use of unsafe materials and maximize the efficiency of the process. The phytochemicals present in the plant parts serve as natural reductants, stabilizers and capping agents of the nanoparticle. Synthesis using plant extracts generates nanoparticles of well-defined shape, structure and morphology in comparison to those obtained through the utilization of bark, tissue and whole plant. The C=O and C–O groups in the plant extract play a critical role in capping the nanoparticles (Kharissova et al., 2013).

7.3 ENGINEERING OF FUNCTIONALIZED INORGANIC NANOPARTICLES

Inorganic nanoparticles can interact with light and/or magnetic fields, thus extending their potential applications to fields such as fluorescence labelling, magnetic resonance imaging and stimulus-responsive drug delivery, which are essential for the diagnosis and treatment of disease. Appropriate design of surface ligands on these nanoparticles is necessary to facilitate their use in such applications. The surface ligands determine the physicochemical properties of the surface, such as hydrophilicity/hydrophobicity and zeta potential as well as dispersibility in solution. To facilitate cellular uptake, the nanoparticles are required to display hydrophilic moieties, which are also required to increase colloidal stability in water. The hydrophilic functional groups can include PEG, carboxylic acids, sulfonic acids, ammonium salts or zwitterions, which are essential for dispersibility in water, cellular membrane permeability, immune responses and localization *in vivo* (Fratoddi, 2017). To have surface ligands for cell membrane permeable nanoparticles, cationic ligands, such as amines or ammonium salts, can be used (endocytic cellular uptake), although their disruption of the cell membrane can create cytotoxic effects (change the cell membrane potential and intracellular concentration of calcium ions) (Kobayashi et al., 2014). A study has shown that phytochemicals-coated nanoparticles can also offer improvements in the therapeutic effects of inorganic nanoparticles, increase cellular uptake, long-term stability, increased biocompatibility (reduce inorganic nanoparticle associated genotoxicity) and bio-imaging (e.g. cinnamon-coated Au) (Cao et al., 2017).

As *M. tuberculosis* is an intracellular pathogen affecting mononuclear phagocytes (macrophages), synthesis of inorganic nanoparticles through the green synthesis method is ideal for target-directed delivery. For specific targeting against *M. tuberculosis*, macrophages are passively targeted through surface modification of nanoparticles with various ligands (Fig 7.4).

Mycobacterial cell wall mycolic acids, mannose and lactose have become one of the leading ligands for targeted drug delivery for TB management (Costa-Gouveia

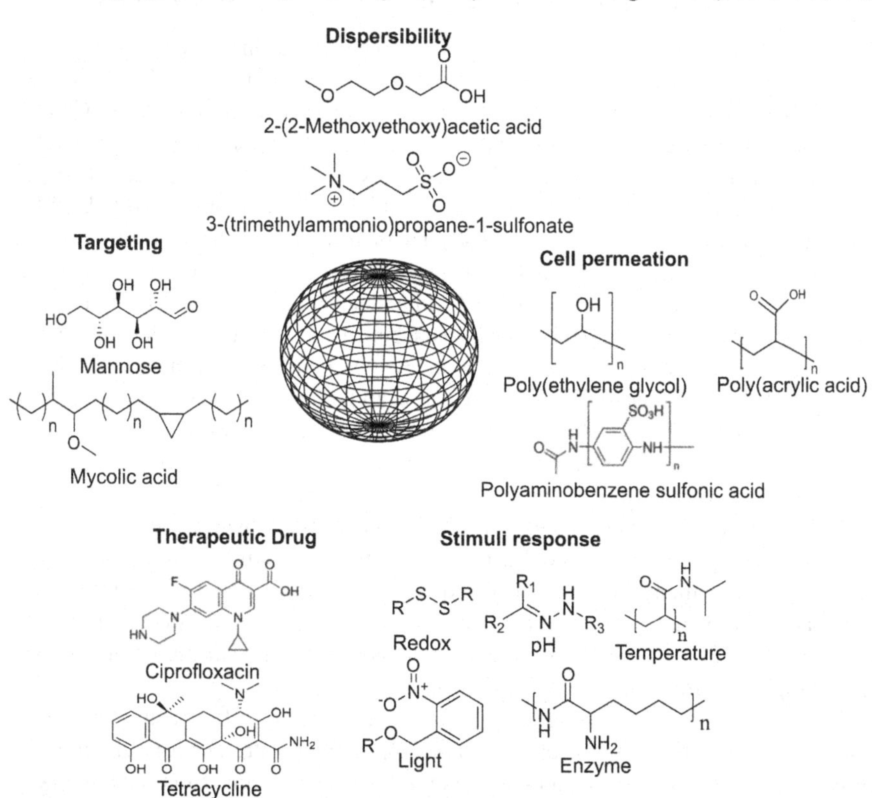

FIGURE 7.4 Inorganic nanoparticles ligand attachment for targeting TB-infected macrophages.

et al., 2017). Apart from target delivery, pharmaceutical agents can also be added to the surface of the nanoparticle. In the field of drug delivery, nanoparticles can serve as carriers of drugs either through encapsulation or surface conjugation where drug release is stimulated by an external (light, ultrasound or magnetic field) or internal stimuli (pH, redox balance or temperature changes) (Le et al., 2019). When functionalizing inorganic nanoparticles, various antibiotics such as ciprofloxacin have been conjugated to zinc oxide nanoparticles whilst ampicillin, kanamycin, streptomycin, gentamycin, neomycin have been conjugated to gold nanoparticles (AuNPs) against bacterial infections (Singh et al., 2020). A recent study has also shown the use of tetracycline as a co-reducing and stabilizing agent for the synthesis of silver and gold nanoparticles, with killing effect against both gram-negative and gram-positive tetracycline-susceptible and tetracycline-resistant bacteria (Djafari et al., 2016). By combining gold nanoparticles with a multiblock copolyester, Gajendiran and colleagues showed that multi-TB drugs (isoniazid, rifampicin and pyrazinamide) could be released over a period of 264 hours (Gajendiran et al., 2016).

7.4 ENGINEERING OF LABELLED INORGANIC NANOPARTICLES

It is of note to mention that unmodified AuNPs can be utilized in biotechnology applications as a simple and label-free colorimetric method for cadmium ions (Cd^{2+}) detection. Briefly, the unmodified AuNPs easily aggregate in a high concentration of NaCl solution, but the presence of glutathione (GSH) can prevent the salt-induced aggregation of AuNPs. When cadmium is added to these complexes with GSH as a spherical-shaped complex, it allows AuNPs to easily aggregate in high-salt conditions (Guo et al., 2014). Surface-enhanced Raman scattering (SERS) nanoparticles is another platform that does not require labelling, although molecular probe such as Victoria blue B or chelating agents can be used (Wang et al., 2016b). Gold and silver nanoparticles can be used for SERS. The advantage of this technique is that label-free SERS can provide rich spectroscopic information for detection and identification without labelling the analyte. It can also provide information about the conformation and orientation of adsorbed molecules concerning the surface. This can be utilized for identifying (in some cases also quantify) biomolecules, drug metabolites and profile spectral signatures from the molecular components of whole pathogens and cellular processes (Lane et al., 2015).

The labelling of nanoparticles can be applicable for sensing of infectious diseases employing immunoassays (plasmonic enzyme-linked immunosorbent assays – ELISAs), complementary nucleic acid sequences and direct nanoparticle interaction with bacterial species, whilst radioactive labelling of oxygen atoms in the aluminium oxide nanoparticles allows one to be able to study the *in vivo* distribution of nanoparticles (Fig 7.5) (Giner-Casares et al., 2016).

For positron emission tomography (PET), near-infrared (NIR) or magnetic resonance imaging (MRI), choice of the radioisotope (low positron energy and high $\beta+$ branching ratio), choice of a suitable nanoplatform (fluorescent emissions/magnetic resonance), radiolabelling method (irradiation/ radioactive precursors) and stability of the radiolabelled agent are essential (Chakravarty et al., 2017). Gold nanoparticles have been used for detecting cancer biomarkers (caspase-3 in myeloid leukaemia),

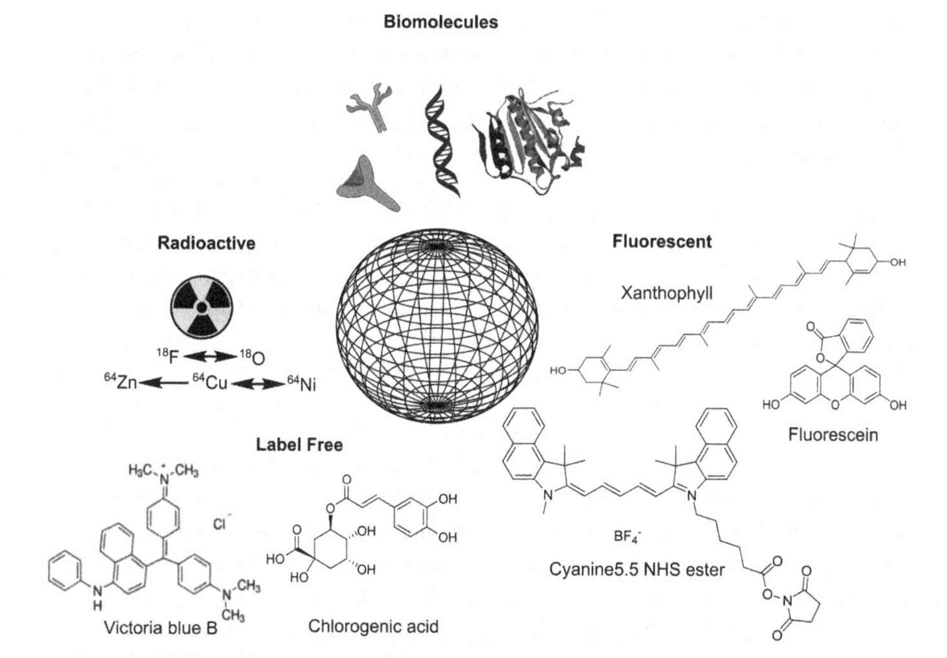

FIGURE 7.5 Labelling of inorganic nanoparticles for biomedical applications.

where the caspase-3 cleavable sequence is used to make dimers of gold nanoparticles where its cleavage leads to variation in the plasmon signal (Tajon et al., 2014). Current bio-imaging range from using superparamagnetic iron, gadolinium or manganese oxide nanoparticles for MRI to gold, bismuth and tantalum-based nanoparticles in computed tomography (CT) (Kim et al., 2018). Due to various imaging techniques requiring expensive fluorescent dyes, green synthesis of nanoparticles using plants materials can produce fluorescent, antibacterial and bio-imaging nanoparticles for live-cell imaging (Pillai et al., 2018).

To combat infectious diseases such as XDR-TB, the time-consuming polymerase chain reaction (PCR) or ELISA assays can be replaced with portable diagnostic tools. Aggregation-related colour changes of gold nanoparticles have been used for immunoassays (antibodies/nucleic acid sequences/enzyme), hemagglutinin-type assay, oligonucleotide aggregation and bacterial binding (Giner-Casares et al., 2016). The use of nanotechnology in TB diagnostics has also been applied in volatile organic compounds (VOCs)-based approach where breath test (dodecanthiol-AuNP-based sensor) or adhesive skin patch can be used for non-invasive and cheap diagnostics for TB (El-Samadony et al., 2017).

7.5 MECHANISM OF ACTION: PHYSIOCHEMICAL PROPERTIES

Several proposed mechanisms have been suggested for the inactivation of bacteria through inorganic nanoparticles: (i) metal ions interact with the thiol and phosphorus group of bacterial enzymes or DNA bases and denature them, (ii) inhibition of bacteria through respiratory mechanism, (iii) the metal ions attack on bacterial DNA and destroy them, (iv) reduction disrupts cellular transport system, (v) physical contact of metal nanoparticles disrupts bacterial cell membrane and (vi) production of highly reactive superoxide and hydroxide radicals from metal nanoparticles (Tahir et al., 2016a). The size, shape, surface charge and enhanced surface area of these nanoparticles play a pivotal role through multiple mechanisms of antibacterial activity. Small nanoparticles tend to be more toxic than large nanoparticles due to the relatively larger surface area to volume ratio as compared to larger nanoparticles. Although nanoparticles are usually synthesized in a spherical shape, other shapes such as sheets, plates, tubes, cubes, rods and triangles can also be designed. Because of the exposed planes and the oxidation levels of metals, nanotubes and nanorods are more efficient than others (Slavin et al., 2017). This is due to less stable planes requiring less energy to form oxygen vacancies.

When nanoparticles bind to the bacterial cell wall, this leads to a dissipation of the proton motive force, which in turn causes blocking of oxidative phosphorylation or forms free radicals that make the cell membrane porous which can ultimately lead to cell death. This can also lead to collapsing the membrane potential and inhibiting the ATPase activities to decrease the ATP level. However, studies have shown that gram-positive bacteria are more resistant to nanoparticle mechanisms of action, due to the protective thick peptidoglycan layer (80 versus ~8 nm of gram-negative bacteria) (Wang et al., 2017). Due to the negatively charged lipopolysaccharide molecules and a mosaic of anionic surface domains on the surface of gram-negative bacteria, this makes them more susceptible to bind to more positive ions released by nanoparticles. Upon entry into the microbe cell, the nanoparticles can start disrupting cellular mechanisms and components. In terms of destroying DNA, the nanoparticles are due to cleavage of DNA due to metal ion binding to the specific site of DNA leading to the generation of reduction/oxidation-induced DNA adducts (Bhat et al., 2019).

Activity against microbial intracellular protein is due to metal ions where they dephosphorylate peptide substrates on tyrosine residues during contact with peptides, which contribute to the inhibition of signal transduction and inhibition of bacterial growth. They are also capable of inhibiting the subunit of the ribosome from binding to transfer ribonucleic acid (tRNA). It is of note to indicate that when bacteria are accustomed to being in an environment with heavy metal stress, this makes them resistant to such metal ions. Various metal-containing products are used in medicine (e.g. calamine, silver sulfadiazine, cisplatin and melarsoprol) and agriculture, their use can be limited by acquired metal resistance (Pal et al., 2017). Resistance towards resistance is acquired through two large plasmids termed pMol-28 and pMol-30. pMol-28 confers resistance when the bacterium is exposed to Co^{2+}, Cr^{6+}, Hg^{2+} and Ni^{2+}; whereas pMol-30 is activated by Ag^{+}, Cd^{2+}, Co^{2+}, Cu^{2+}, Hg^{2+}, Pb^{2+} and Zn^{2+} (Slavin et al., 2017). Plasmids are genetic elements that can pass between bacteria; they can replicate themselves and often transfer themselves to different hosts as well. Studies have shown that bacteria are capable of buffering, efflux, storing and sequestrating metals ions when they are abundant to reduce metal intoxication through metal ion homeostasis (Chandrangsu et al., 2017). A study found issues related to silver overuse and the resulting development of bacterial resistance (Percival et al., 2005). Therefore, it is evident that like antibiotics, uncontrolled use of these metal ions can cause resistance.

7.6 FUTURE PERSPECTIVE

The use of bacteria, fungi, algae and plants biomass, as aqueous extracts or whole cells as a means to synthesize inorganic nanoparticles, appears to be a growing method that is both cost-effective and eco-friendly. This method is also allowing one to be able to efficiently control nanoparticle sizes, shapes and compositions. Although there are benefits to the biosynthesis of inorganic nanoparticles, concerns of the type of biomaterial to use in the synthesis will be crucial. Plants are easily accessible for harvesting and the required plant can be grown if need to be used in large-scale use. However, for some plant species, the duration of the plant to grow and reach maturity and the growing conditions can limit some large-scale developments. It is therefore crucial that the plant material of choice has to meet the demand for large-scale production. This issue is also applicable to the use of algae and seaweed. In terms of using microbes for biosynthesis, bioreactors can be utilized where various conditions such as temperature and pH have to be maintained for nanoparticle development. Through using current bioreactor processes used in the brewery sector, one can be able to ensure that production can be optimized to meet the demand. Thus, more work is required in looking into large-scale production of nanoparticles and the biomaterial to be used.

There is also an added benefit from the use of biomass in nanoparticle synthesis. The presence of phytochemicals also has a variety of health benefits (anti-ageing, anticancer, anti-inflammation and antioxidant) as they end on the surface of the nanoparticles. Various phytochemicals have benefits such as reducing the side effects (cytotoxicity, genotoxicity and immunotoxicity) and improving the biocompatibility (reducing oxidative stress) of the nanoparticle. More work in the identification of phytochemicals that play a role during synthesis and post-synthesis biological activity is crucial. This can help in developing eco-friendly reductant, stabilizers and capping agents for nanoparticle synthesis. In conjunction with the choice of biomass to use, once the phytochemicals have been identified, some of the plants can be saved from harvesting or can be genetically modified to produce the required phytochemicals for large-scale nanoparticle production. With the focus of medicine towards natural products growing, there are several clinical trials where inorganic nanoparticles with anti-infective properties are being evaluated (Popescu et al., 2016). Apart from the use of nanoparticles, metal-based drugs have also shown the potential to act against multidrug-resistant *M. tuberculosis* (ruthenium) strains, HIV (zinc(II), nickel(II), copper(II) and cobalt(III)) and anticancer (platinum- and ruthenium-based) activity without causing toxic effect to cell lines (Liang and Sadler, 2004; Yu et al., 2016).

7.7 CONCLUSIONS

With the number of reported new infections and death associated with XRD-TB on the rise, the need for new antimicrobial agents has been greatly affected by rapid drug resistance emergence and the long duration of drug approval to the market. Inorganic nanoparticles have been shown to have inherent antibacterial properties. By using bioengineering, these particles can further have improved biological activity due to the phytochemicals agents that are used during synthesis. The incorporation of ligands and therapeutics agents on the surface of the nanoparticles can help with site-specific targeting of TB-infected macrophages and thus help in reducing side effects associated with XRD-TB drugs. With a low chance of the bacteria developing resistance against these metals, a cocktail of the metal biological activity with known antibacterial drugs can be crucial in combating this disease. Although more work is needed (absorption, distribution, metabolism and excretion) in ensuring these metal-based nanoparticles could be applicable for human application, the nanoparticles can serve as an alternative or act in synergy with the current treatment plan for XRD-TB management.

7.8 REFERENCES

Adusei, S., Otchere, J. K., Oteng, P., Mensah, R. Q., & Tei-Mensah, E. 2019. Phytochemical analysis, antioxidant and metal chelating capacity of Tetrapleura tetraptera. *Heliyon* 5, no. 11: e02762. DOI: 10.1016/j.heliyon.2019.e02762.

Ahmed, S., Ahmad, M., Swami, B. L., & Ikram, S. 2016a. A review on plants extract mediated synthesis of silver nanoparticles for antimicrobial applications: A green expertise. *Journal of Advanced Research* 7, no. 1: 17–28. DOI: 10.1016/j.jare.2015.02.007.

Ahmed, S., Annu, Ikram, S., & Yudha S, S. 2016b. Biosynthesis of gold nanoparticles: A green approach. *Journal of Photochemistry and Photobiology B: Biology* 161: 141–153. DOI: 10.1016/j.jphotobiol.2016.04.034.

Ansari, M. A., Khan, H. M., Alzohairy, M. A., Jalal, M., Ali, S. G., & Pal, R. 2015. Green synthesis of Al_2O_3 nanoparticles and their bactericidal potential against clinical isolates of multi-drug resistant Pseudomonas aeruginosa. *Biotechnology* 31, no. 1: 153–164. DOI: 10.1007/s11274-014-1757-2.

Auld, S. C., Shah, N. S., Mathema, B., Brown, T. S., Ismail, N., Omar, S. V., Brust, J. C. M. et al. 2018. Extensively drug-resistant tuberculosis in South Africa: Genomic evidence supporting transmission in communities. *European Respiratory Journal* 52, no. 4: 1800246. DOI: 10.1183/13993003.00246-2018.

Ayukekbong, J. A., Ntemgwa, M., & Atabe, A. N. 2017. The threat of antimicrobial resistance in developing countries: causes and control strategies. *Antimicrobial Resistance & Infection Control* 6, no. 1: 47. https://doi.org/10.1186/s13756-017-0208-x.

Balasooriya, E. R., Jayasinghe, C. D., Jayawardena, U. A., Ruwanthika, R. W. D., Mendis de Silva, R., & Udagama, P. V. 2017. Honey mediated green synthesis of nanoparticles: New era of safe nanotechnology. *Journal of Nanomaterials* 2017:1–10. https://doi.org /10.1155/2017/5919836.

Banerjee, D., Shivapriya, P., Gautam, P. K., Misra, K., Sahoo, A. K., & Samanta, S. K. 2019. A review on basic biology of bacterial biofilm infections and their treatments by nanotechnology-based approaches. *Proceedings of the National Academy of Sciences, India Section B: Biological Sciences*:) 90:243–259. https://doi.org/10.1007/s40011-018-01065-7.

Bhat, I. U. H., Anwar, S. J., Subramaniam, E., & Shalla, A. H. 2019. Nanoparticles; their use as antibacterial and DNA cleaving agents. In *Nanomaterials for Healthcare, Energy and Environment*, edited by A. H. Bhat, I. Khan, M. Jawaid, F. O. Suliman, H. Al-Lawati, & S. M. Al-Kindy, 71–85. Singapore: Springer. https://doi.org/10.1007/978-981-13-9833-9_4.

Bhering, M., Duarte, R., & Kritski, A. 2019. Predictive factors for unfavourable treatment in MDR-TB and XDR-TB patients in Rio de Janeiro State, Brazil, 2000–2016. *PLOS ONE* 14, no. 11: e0218299. DOI: 10.1371/journal.pone.0218299.

Bhunu, B., Mautsa, R., & Mukanganyama, S. 2017. Inhibition of biofilm formation in Mycobacterium smegmatis by *Parinari curatellifolia* leaf extracts. *BMC Complementary and Alternative Medicine* 17, no. 1: 285. DOI: 10.1186/s12906-017-1801-5.

Boboc, M., Curti, F., Fleacă, A. M., Jianu, M. L., Roşu, A.-M., Curutiu, C., Lazar, V. et al. 2017. Chapter 14: Preparation and antimicrobial activity of inorganic nanoparticles: Promising solutions to fight antibiotic resistance. In *Nanostructures for Antimicrobial Therapy*, edited by A. Ficai & A. M. Grumezescu, 325–340. Amsterdan, The Netherlans: Elsevier. https://doi.org/10.1016/B978-0-323-46152-8.00014-7.

Bogdanović, U., Lazić, V., Vodnik, V., Budimir, M., Marković, Z., & Dimitrijević, S. 2014. Copper nanoparticles with high antimicrobial activity. *Materials Letters* 128: 75–78. DOI: 10.1016/j.matlet.2014.04.106.

Brown, E. D., & Wright, G. D. 2016. Antibacterial drug discovery in the resistance era. *Nature* 529: 336. DOI: 10.1038/nature17042.

Cao, Y., Xie, Y., Liu, L., Xiao, A., Li, Y., Zhang, C., Fang, X. et al. 2017. Influence of phytochemicals on the biocompatibility of inorganic nanoparticles: A state-of-the-art review. *Phytochemistry Reviews* 16(3): 555–563. DOI: 10.1007/s11101-017-9490-8.

Chakravarty, R., Goel, S., Dash, A., & Cai, W. 2017. Radiolabeled inorganic nanoparticles for positron emission tomography imaging of cancer: An overview. *The Quarterly Journal of Nuclear Medicine and Molecular Imaging* 61, no. 2: 181–204. DOI: 10.23736/S1824-4785.17.02969-7.

Chandrangsu, P., Rensing, C., & Helmann, J. D. 2017. Metal homeostasis and resistance in bacteria. *Nature Reviews Microbiology* 15, no. 6: 338. https://doi.org/10.1038/nrmicro.2017.15.

Chang, K. C., & Leung, C. C. 2017. BCG immunization: Efficacy, limitations, and future needs. In *Handbook of Global Tuberculosis Control*, 343–357. Springer. Boston, MA. https://doi.org/10.1007/978-1-4939-6667-7_20.

Charbgoo, F., Ahmad, M. B., & Darroudi, M. 2017. Cerium oxide nanoparticles: Green synthesis and biological applications. *International Journal of Nanomedicine* 12: 1401–1413. DOI: 10.2147/IJN.S124855.

Choi, S.-R., Britigan, B. E., Moran, D. M., & Narayanasamy, P. 2017. Gallium nanoparticles facilitate phagosome maturation and inhibit growth of virulent Mycobacterium tuberculosis in macrophages. *PLOS ONE*, 12, no. 5: e0177987–e0177987. DOI: 10.1371/journal.pone.0177987.

Costa-Gouveia, J., Aínsa, J. A., Brodin, P., & Lucía, A. 2017. How can nanoparticles contribute to antituberculosis therapy? *Drug Discovery Today* 22, no. 3: 600–607. DOI: 10.1016/j.drudis.2017.01.011.

Dauthal, P., & Mukhopadhyay, M. 2016. Noble metal nanoparticles: Plant-mediated synthesis, mechanistic aspects of synthesis, and applications. *Industrial & Engineering Chemistry Research*, 55, no. 36: 9557–9577. https://doi.org/10.1021/acs.iecr.6b00861.

Din, M. I., Nabi, A. G., Rani, A., Aihetasham, A., & Mukhtar, M. 2018. Single step green synthesis of stable nickel and nickel oxide nanoparticles from Calotropis gigantea: Catalytic and antimicrobial potentials. *Environmental Nanotechnology, Monitoring & Management*, 9: 29–36. DOI: 10.1016/j.enmm.2017.11.005.

Djafari, J., Marinho, C., Santos, T., Igrejas, G., Torres, C., Capelo, J. L., Poeta, P. et al.2016. New synthesis of gold- and silver-based nano-tetracycline composites. *ChemistryOpen* 5, no. 3: 206–212. DOI: 10.1002/open.201600016.

El-Samadony, H., Althani, A., Tageldin, M. A., & Azzazy, H. M. E. 2017. Nanodiagnostics for tuberculosis detection. *Expert Review of Molecular Diagnostics* 17, no. 5: 427–443. DOI: 10.1080/14737159.2017.1308825.

Eldholm, V., & Balloux, F. 2016. Antimicrobial resistance in mycobacterium tuberculosis: The odd one out. *Trends in Microbiology* 24, no. 8: 637–648. DOI: 10.1016/j.tim.2016.03.007.

Enioutina, E. Y., Teng, L., Fateeva, T. V., Brown, J. C. S., Job, K. M., Bortnikova, V. V., Lubov, V. et al. 2017. Phytotherapy as an alternative to conventional antimicrobials: Combating microbial resistance. *Expert Review of Clinical Pharmacology* 10, no. 11: 1203–1214. DOI:10.1080/17512433.2017.1371591.

Erck, D., Schrager, L., Piracha, S., & Manjelievskaia, J. 2016. Drug-resistant TB: Deadly, costly and in need of a vaccine. *Transactions of The Royal Society of Tropical Medicine and Hygiene* 110, no. 3: 186–191. DOI: 10.1093/trstmh/trw006.

Esteban, J., & García-Coca, M. 2018. Mycobacterium biofilms. *Frontiers in Microbiology* 8: 2651–2651. DOI: 10.3389/fmicb.2017.02651.

Farias, I. A. P., Santos, C. C. L. D., & Sampaio, F. C., 2018. Antimicrobial activity of cerium oxide nanoparticles on opportunistic microorganisms: A systematic review. *BioMed Research International* 2018: 14. DOI: 10.1155/2018/1923606.

Fleming, D., & Rumbaugh, K. P. (2017). Approaches to dispersing medical biofilms. *Microorganisms* 5, no. 2: 15. DOI: 10.3390/microorganisms5020015.

Frank, M., Adamashvili, N., Lomtadze, N., Kokhreidze, E., Avaliani, Z., Kempker, R. R., & Blumberg, H. M. 2019. Long-term follow-up reveals high post-treatment mortality among patients with extensively drug-resistant (XDR) tuberculosis in the country of Georgia. Paper presented at the Open Forum Infectious Diseases. 10.1093/ofid/ofz152.

Fratoddi, I. 2017. Hydrophobic and hydrophilic Au and Ag nanoparticles. Breakthroughs and perspectives. *Nanomaterials* 8, no. 1: 11. DOI: 10.3390/nano8010011.

Gabrielyan, L., Hovhannisyan, A., Gevorgyan, V., Ananyan, M., & Trchounian, A. 2019. Antibacterial effects of iron oxide (Fe3O4) nanoparticles: Distinguishing concentration-dependent effects with different bacterial cells growth and membrane-associated mechanisms. *Applied Microbiology and Biotechnology* 103, no. 6: 2773–2782. DOI: 10.1007/s00253-019-09653-x.

Gajendiran, M., Balashanmugam, P., Kalaichelvan, P. T., & Balasubramanian, S. 2016. Multi-drug delivery of tuberculosis drugs by π-back bonded gold nanoparticles with multiblock copolyesters. *Materials Research Express* 3, no. 6: 065401. DOI: 10.1088/2053-1591/3/6/065401.

Giner-Casares, J. J., Henriksen-Lacey, M., Coronado-Puchau, M., & Liz-Marzán, L. M. 2016. Inorganic nanoparticles for biomedicine: Where materials scientists meet medical research. *Materials Today* 19, no. 1: 19–28. DOI: 10.1016/j.mattod.2015.07.004.

Gold, K., Slay, B., Knackstedt, M., & Gaharwar, A. K. 2018. Antimicrobial activity of metal and metal-oxide based nanoparticles. *Advanced Therapeutics* 1, no. 3: 1700033. DOI: 10.1002/adtp.201700033.

Guo, Y., Zhang, Y., Shao, H., Wang, Z., Wang, X., & Jiang, X. 2014. Label-free colorimetric detection of cadmium ions in rice samples using gold nanoparticles. *Analytical Chemistry* 86, no. 17: 8530–8534. DOI: 10.1021/ac502461r.

Imran Din, M., & Rani, A. 2016. Recent advances in the synthesis and stabilization of nickel and nickel oxide nanoparticles: A green adeptness. *International Journal of Analytical Chemistry* 2016: 14. DOI: 10.1155/2016/3512145.

Iravani, S. J. 2011. Green synthesis of metal nanoparticles using plants. *Green Chemistry* 13, no. 10: 2638–2650. https://doi.org/10.1039/C1GC15386B.

Jafari, A., Mosavi, T., Mosavari, N., Majid, A., Movahedzade, F., Tebyaniyan, M., Kamalzadeh, M. et al. 2016. Mixed metal oxide nanoparticles inhibit growth of Mycobacterium tuberculosis into THP-1 cells. *International Journal of Mycobacteriology* 5 Supplement 1: S181–S183. DOI: 10.1016/j.ijmyco.2016.09.011.

Jafari, A., Kharazi, S., Mosavari, N., Movahedzadeh, F., Tebyaniyan, M., & Nodooshan, S. J. 2017. Synthesis of mixed metal oxides nano-colloidal particles and investigation of the cytotoxicity effects on the human pulmonary cell lines: A prospective approach in anti-tuberculosis inhaled nanoparticles. *Oriental Journal of Chemistry* 33, no. 3: 1529–1544. http://dx.doi.org/10.13005/ojc/330358.

Katata-Seru, L., Moremedi, T., Aremu, O. S., & Bahadur, I. (2018). Green synthesis of iron nanoparticles using Moringa oleifera extracts and their applications: Removal of nitrate from water and antibacterial activity against Escherichia coli. *Journal of Molecular Liquids* 256: 296–304. DOI: 10.1016/j.molliq.2017.11.093.

Kaur, M., Gogna, S., & Kaur, N. 2020. Nano-technological developments in tuberculosis management: An update. *Tathapi UGC CARE Journal* 19, no. 5: 313–329.

Kharissova, O. V., Dias, H. V. R., Kharisov, B. I., Pérez, B. O., & Pérez, V. M. J. 2013. The greener synthesis of nanoparticles. *Trends in Biotechnology* 31, no. 4: 240–248. DOI: 10.1016/j.tibtech.2013.01.003.

Khurana, C., & Chudasama, B. 2018. Nanoantibiotics: Strategic assets in the fight against drug-resistant superbugs. *International Journal of Nanomedicine* 13: 3–6. DOI: 10.2147/IJN.S124698.

Kim, D., Kim, J., Park, Y. I., Lee, N., & Hyeon, T. 2018. Recent development of inorganic nanoparticles for biomedical imaging. *ACS Central Science* 4, no. 3: 324–336. DOI: 10.1021/acscentsci.7b00574.

Kobayashi, K., Wei, J., Iida, R., Ijiro, K., & Niikura, K. 2014. Surface engineering of nanoparticles for therapeutic applications. *Polymer Journal* 46, no. 8: 460–468. DOI: 10.1038/pj.2014.40.

Kumar, V. V., & Anthony, S. P. 2016. Antimicrobial studies of metal and metal oxide nanoparticles. In *Surface Chemistry of Nanobiomaterials*, 265–300. Amsterdam, The Netherlands: Elsevier. https://doi.org/10.1016/B978-0-323-42861-3.00009-1.

Kurtjak, M., Aničić, N., & Vukomanovicć, M. 2017. Inorganic nanoparticles: Innovative tools for antimicrobial agents. In *Antibacterial Agents*. London: IntechOpen. http://dx.doi.org/10.5772/67904.

Lane, L. A., Qian, X., & Nie, S. 2015. SERS nanoparticles in medicine: From label-free detection to spectroscopic tagging. *Chemical Reviews* 115, no. 19: 10489–10529. DOI: 10.1021/acs.chemrev.5b00265.

Le, N. T. T., Nguyen, T. N. Q., Cao, V. D., Hoang, D. T., Ngo, V. C., Thi, H., & Thanh, T. 2019. Recent progress and advances of multi-stimuli-responsive dendrimers in drug delivery for cancer treatment. *Pharmaceutics* 11, no. 11: 591. https://dx.doi.org/10.3390%2Fpharmaceutics11110591.

Liang, X., & Sadler, P. J. 2004. Cyclam complexes and their applications in medicine. *Chemical Society Reviews* 33, no. 4: 246–266. DOI: 10.1039/b313659k.

Mahira, S., Jain, A., Khan, W., & Domb, A. J. 2019. Antimicrobial materials—An overview. In *Antimicrobial Materials for Biomedical Applications*, 1–37. Cambridge: Royal Society of Chemistry. https://doi.org/10.1039/9781788012638.

Makarov, V., Love, A., Sinitsyna, O., Makarova, S., Yaminsky, I., Taliansky, M., & Kalinina, N. J. 2014. "Green" nanotechnologies: Synthesis of metal nanoparticles using plants. *Acta Naturae* 6, no. 1: 35–44. http://www.ncbi.nlm.nih.gov/pmc/articles/pmc3999464/.

Malekkhaiat Häffner, S., & Malmsten, M. 2017. Membrane interactions and antimicrobial effects of inorganic nanoparticles. *Advances in Colloid and Interface Science* 248: 105–128. DOI: https://doi.org/10.1016/j.cis.2017.07.029.

Manyasree, D., Kiranmayi, P., & Kumar, R. 2018. Synthesis, characterization and antibacterial activity of aluminium oxide nanoparticles. *International Journal of Pharmacy and Pharmaceutical Sciences* 10, no. 1: 32–35. https://doi.org/10.22159/ijpps.2018v10i1.20636.

Matinise, N., Fuku, X. G., Kaviyarasu, K., Mayedwa, N., & Maaza, M. 2017. ZnO nanoparticles via Moringa oleifera green synthesis: Physical properties & mechanism of formation. *Applied Surface Science* 406: 339–347. DOI: 10.1016/j.apsusc.2017.01.219.

Migliori, G. B., Tiberi, S., Zumla, A., Petersen, E., Chakaya, J. M., Wejse, C., Torrico, M. M. et al. 2020. MDR/XDR-TB management of patients and contacts: Challenges facing the new decade. The 2020 clinical update by the Global Tuberculosis Network. *International Journal of Infectious Diseases*. 92: S15–S25. https://doi.org/10.1016/j.ijid.2020.01.042.

Miller, K. P., Wang, L., Benicewicz, B. C., & Decho, A. W. 2015. Inorganic nanoparticles engineered to attack bacteria. *Chemical Society Reviews* 44, no. 21: 7787–7807. DOI: 10.1039/C5CS00041F.

Monika, S., H. Ponlakshmi, S., Sundar, K., & Balakrishnan, V. 2017. Biological synthesis of gallium nanoparticles using extracts of *Andrographis paniculata. International Journal of Engineering Science, Advanced Computing and Bio-Technology* 8, no. 4: 208–222. 10.26674/ijesacbt/2017/49244.

Muhammad, W., Haleem, I., Ullah, N., Khan, M., & Kumar, N. S. 2019. Nanoparticle synthesis approaches at a glance. *Nanoscale Reports* 2: 1–10. DOI: 10.26524/nr1931.

Muzammil, S., Hayat, S., Fakhar-E-Alam, M., Aslam, B., Siddique, M., Nisar, M., Saqalein, M. et al. 2018. Nanoantibiotics: Future nanotechnologies to combat antibiotic resistance. *Frontiers in Bioscience* 10: 352–374. https://doi.org/10.2741/e827.

Nagar, N., & Devra, V. (2018). Green synthesis and characterization of copper nanoparticles using Azadirachta indica leaves. *Materials Chemistry and Physics* 213: 44–51. DOI: 10.1016/j.matchemphys.2018.04.007.

Nasiruddin, M., Neyaz, M., & Das, S. 2017. Nanotechnology-based approach in tuberculosis treatment. *Tuberculosis Research and Treatment* 2017: 1–12. https://doi.org/10.1155/2017/4920209.

Nguyen, L. 2016. Antibiotic resistance mechanisms in M. tuberculosis: An update. *Archives of Toxicology* 90, no. 7: 1585–1604. https://doi.org/10.1007/s00204-016-1727-6.

O'Brien, R. J., & Nunn, P. P. 2001. The need for new drugs against tuberculosis. 163, no. 5: 1055–1058. DOI: 10.1164/ajrccm.163.5.2007122.

Olakanmi, O., Britigan, B. E., & Schlesinger, L. S. 2000. Gallium disrupts iron metabolism of mycobacteria residing within human macrophages. *Infection and Immunity* 68, no. 10: 5619–5627. https://doi.org/10.1128/iai.68.10.5619-5627.2000.

Pal, C., Asiani, K., Arya, S., Rensing, C., Stekel, D. J., Larsson, D. J., & Hobman, J. L. 2017. Metal resistance and its association with antibiotic resistance. In *Advances in Microbial Physiology*, Vol. 70, 261–313. Amsterdam, The Netherlands: Elsevier. https://doi.org/10.1016/bs.ampbs.2017.02.001.

Pelgrift, R. Y., & Friedman, A. J. 2013. Nanotechnology as a therapeutic tool to combat microbial resistance. *Advanced Drug Delivery Reviews* 65, no. 13: 1803–1815. DOI: 10.1016/j.addr.2013.07.011.

Percival, S. L., Bowler, P. G., & Russell, D. 2005. Bacterial resistance to silver in wound care. *Journal of Hospital Infection* 60, no. 1: 1–7. DOI: 10.1016/j.jhin.2004.11.014.

Pillai, M. M., Karpagam, K. R., Begam, R., Selvakumar, R., & Bhattacharyya, A. 2018. Green synthesis of lignin based fluorescent nanocolorants for live cell imaging. *Materials Letters* 212: 78–81. DOI: 10.1016/j.matlet.2017.10.060.

Poon, C., & Patel, A. A. 2020. Organic and inorganic nanoparticle vaccines for prevention of infectious diseases. *Nano Express* 1, no. 1: 012001. DOI: 10.1088/2632-959x/ab8075.

Popescu, R. C., Andronescu, E., Oprea, A. E., & Grumezescu, A. M. 2016. Chapter 2: Toxicity of inorganic nanoparticles against prokaryotic cells. In *Nanobiomaterials in Antimicrobial Therapy*, edited by A. M. Grumezescu, 29–65. William Andrew Publishing. Norwich, NY. https://doi.org/10.1016/B978-0-323-42864-4.00002-6.

Pugazhendhi, A., Prabhu, R., Muruganantham, K., Shanmuganathan, R., & Natarajan, S. 2019. Anticancer, antimicrobial and photocatalytic activities of green synthesized magnesium oxide nanoparticles (MgONPs) using aqueous extract of Sargassum wightii. *Journal of Photochemistry and Photobiology B: Biology* 190: 86–97. DOI: 10.1016/j.jphotobiol.2018.11.014.

Rahman, M. A., & Sarkar, A. (2017). Extensively drug-resistant tuberculosis (XDR-TB): A daunting challenge to the current end TB strategy and policy recommendations. *Indian Journal of Tuberculosis* 64, no. 3: 153–160. DOI: 10.1016/j.ijtb.2017.03.006.

Salar, R. 2011. Synthesis of nickel hydroxide nanoparticles by reverse micelle method and its antimicrobial activity. *Research Journal of Chemical Sciences* 1, no. 9: 42–48.

Sankar, R., Rizwana, K., Shivashangari, K. S., & Ravikumar, V. 2015. Ultra-rapid photocatalytic activity of Azadirachta indica engineered colloidal titanium dioxide nanoparticles. *Applied Nanoscience* 5, no. 6: 731–736. DOI: 10.1007/s13204-014-0369-3.

Seaworth, B. J., Armitige, L. Y., Aronson, N. E., Hoft, D. F., Fleenor, M. E., Gardner, A. F., & Harris, D. A. 2014. Multidrug-resistant tuberculosis. Recommendations for reducing risk during travel for healthcare and humanitarian work. *Annals of the American Thoracic Society* 11, no. 3: 286–295. DOI: 10.1513/AnnalsATS.201309-312PS.

Shah, I., Poojari, V., & Meshram, H. 2020. Multi-drug resistant and extensively-drug resistant tuberculosis. *The Indian Journal of Pediatrics*, 87:833–839. https://doi.org/10.1007/s12098-020-03230-1.

Shah, N. S., Auld, S. C., Brust, J. C. M., Mathema, B., Ismail, N., Moodley, P., Mlisana, K. et al. 2017. Transmission of extensively drug-resistant tuberculosis in South Africa. *New England Journal of Medicine* 376, no. 3: 243–253. DOI: 10.1056/NEJMoa1604544.

Sharma, D., & Parkash Yadav, J. 2017. An overview of phytotherapeutic approaches for the treatment of tuberculosis. *Mini Reviews in Medicinal Chemistry* 17, no. 2: 167–183. https://doi.org/10.2174/1389557516666160505114603.

Sharma, D., Kanchi, S., & Bisetty, K. 2019. Biogenic synthesis of nanoparticles: A review. *Arabian Journal of Chemistry* 12, no. 8: 3576–3600. DOI: 10.1016/j.arabjc.2015.11.002.

Siemer, S., Westmeier, D., Barz, M., Eckrich, J., Wünsch, D., Seckert, C., Thyssen, C. et al. 2019. Biomolecule-corona formation confers resistance of bacteria to nanoparticle-induced killing: Implications for the design of improved nanoantibiotics. *Biomaterials* 192: 551–559. DOI: 10.1016/j.biomaterials.2018.11.028.

Singh, A., Gupta, A. K., & Singh, S. 2020. Molecular mechanisms of drug resistance in mycobacterium tuberculosis: Role of nanoparticles against multi-drug-resistant tuberculosis (MDR-TB). In *NanoBioMedicine*, edited by S. K. Saxena & S. M. P. Khurana, 285–314. Singapore: Springer. https://doi.org/10.1007/978-981-32-9898-9_12.

Singh, R., Nawale, L., Arkile, M., Wadhwani, S., Shedbalkar, U., Chopade, S., Sarkar, D. et al. 2016. Phytogenic silver, gold, and bimetallic nanoparticles as novel antitubercular agents. *International Journal of Nanomedicine* 11: 1889–1897. DOI: 10.2147/IJN.S102488.

Slavin, Y. N., Asnis, J., Häfeli, U. O., & Bach, H. 2017. Metal nanoparticles: Understanding the mechanisms behind antibacterial activity. *Journal of Nanobiotechnology* 15, no. 1: 65–65. DOI: 10.1186/s12951-017-0308-z.

Srikar, S. K., Giri, D. D., Pal, D. B., Mishra, P. K., & Upadhyay, S. N. 2016. Green synthesis of silver nanoparticles: A review. *Green and Sustainable Chemistry* 6, no. 1: 34–56. https://doi.org/10.1155/2015/682749.

Tăbăran, A.-F., Matea, C. T., Mocan, T., Tăbăran, A., Mihaiu, M., Iancu, C., & Mocan, L. 2020. Silver nanoparticles for the therapy of tuberculosis. *International Journal of Nanomedicine* 15: 2231–2258. DOI: 10.2147/IJN.S241183.

Tahir, K., Nazir, S., Ahmad, A., Li, B., Ali Shah, S. A., Khan, A. U., Khan, G. M. et al. 2016a. Biodirected synthesis of palladium nanoparticles using Phoenix dactylifera leaves extract and their size dependent biomedical and catalytic applications. *RSC Advances* 6, no. 89: 85903–85916. DOI: 10.1039/C6RA11409A.

Tahir, K., Nazir, S., Li, B., Ahmad, A., Nasir, T., Khan, A. U., Shah, S. A. A. et al. 2016b. Sapium sebiferum leaf extract mediated synthesis of palladium nanoparticles and in vitro investigation of their bacterial and photocatalytic activities. *Journal of Photochemistry and Photobiology B: Biology* 164: 164–173. DOI: 10.1016/j.jphotobiol.2016.09.030.

Tajon, C. A., Seo, D., Asmussen, J., Shah, N., Jun, Y.-W., & Craik, C. S. 2014. Sensitive and selective plasmon ruler nanosensors for monitoring the apoptotic drug response in leukemia. *ACS Nano* 8, no. 9: 9199–9208. DOI: 10.1021/nn502959q.

Teixeira, M. C., Sanchez-Lopez, E., Espina, M., Calpena, A. C., Silva, A. M., Veiga, F. J., Garcia, M. L. et al. 2018. Chapter 9: Advances in antibiotic nanotherapy: Overcoming antimicrobial resistance. In *Emerging Nanotechnologies in Immunology*, edited by R. Shegokar & E. B. Souto, 233–259. Boston: Elsevier. https://doi.org/10.1016/B978-0-323-40016-9.00009-9.

Turner, R. J. 2017. Metal-based antimicrobial strategies. *Microbial Biotechnology* 10, no. 5: 1062–1065. DOI: 10.1111/1751-7915.12785.

Van Norman, G. A. 2016. Drugs, devices, and the FDA: Part 1. An overview of approval processes for drugs. *JACC: Basic to Translational Science* 1, no. 3: 170–179. DOI: 10.1016/j.jacbts.2016.03.002.

Vergheese, M., & Vishal, S. K. 2018. Green synthesis of magnesium oxide nanoparticles using Trigonella foenum-graecum leaf extract and its antibacterial activity. *Journal of Pharmacognosy and Phytochemistry* 7: 1193–1200.

Vijayaraghavan, K., & Ashokkumar, T. 2017. Plant-mediated biosynthesis of metallic nanoparticles: A review of literature, factors affecting synthesis, characterization techniques and applications. *Journal of Environmental Chemical Engineering* 5, no. 5: 4866–4883. DOI: 10.1016/j.jece.2017.09.026.

Wang, L., Hu, C., & Shao, L. 2017. The antimicrobial activity of nanoparticles: Present situation and prospects for the future. *International Journal of Nanomedicine*, 12: 1227–1249. DOI: 10.2147/IJN.S121956.

Wang, L.-S., Gupta, A., & Rotello, V. M. 2016a. Nanomaterials for the treatment of bacterial biofilms. *ACS Infectious Diseases* 2, no. 1: 3–4. DOI: 10.1021/acsinfecdis.5b00116.

Wang, Y., Wen, G., Ye, L., Liang, A., & Jiang, Z. 2016b. Label-free SERS study of galvanic replacement reaction on silver nanorod surface and its application to detect trace mercury ion. *Scientific Reports* 6, no. 1: 19650. DOI: 10.1038/srep19650.

Yadi, M., Mostafavi, E., Saleh, B., Davaran, S., Aliyeva, I., Khalilov, R., Nikzamir, M. et al. 2018. Current developments in green synthesis of metallic nanoparticles using plant extracts: A review. *Artificial Cells, Nanomedicine, and Biotechnology* 46, no. sup3: S336–S343. DOI: 10.1080/21691401.2018.1492931.

Yu, M., Nagalingam, G., Ellis, S., Martinez, E., Sintchenko, V., Spain, M., Rutledge, P. J. et al. 2016. Nontoxic metal–cyclam complexes, a new class of compounds with potency against drug-resistant mycobacterium tuberculosis. *Journal of Medicinal Chemistry* 59, no. 12: 5917–5921. DOI: 10.1021/acs.jmedchem.6b00432.

Zhang, Y., Shareena Dasari, T. P., Deng, H., & Yu, H. 2015. Antimicrobial activity of gold nanoparticles and ionic gold. *Journal of Environmental Science and Health, Part C* 33, no. 3: 286–327. DOI: 10.1080/10590501.2015.1055161.

Zimbone, M., Buccheri, M. A., Cacciato, G., Sanz, R., Rappazzo, G., Boninelli, S., Reitano, R. et al. 2015. Photocatalytical and antibacterial activity of TiO2 nanoparticles obtained by laser ablation in water. *Applied Catalysis B: Environmental* 165: 487–494. DOI: 10.1016/j.apcatb.2014.10.031.

8 Recent Advances in the Utilization of Bioengineered Plant-Based Nanoparticles

A Sustainable Nanobiotechnology for the Management of Extensively Drug-Resistant Tuberculosis

Charles Oluwaseun Adetunji,
Olugbenga Samuel Michael,
Muhammad Akram, Kadiri Oseni,
Olerimi Samson E, Osikemekha Anthony Anani,
Wilson Nwankwo, Hina Anwar,
Juliana Bunmi Adetunji, and
Akinola Samson Olayinka

CONTENTS

8.1 Introduction 150
8.2 Transmission of TB Disease, Pathogenesis and Its Types 151
8.3 Conventional Chemotherapy of TB, Its Limitations and Emergence of Drug Resistance 151
8.4 Nanotechnology and Their Role in the Management of Tuberculosis Diseases 152
8.5 Nanotechnology in Treatment of Tuberculosis 152
8.6 Management of *Mycobacterium tuberculosis* Using Phytonanotechnology 153
8.7 Management of Drug-Resistant Tuberculosis 155
8.8 Management of Drug-Resistant Tuberculosis with Synthetic Drugs 157

8.9 Management of Multidrug-Resistant Tuberculosis with Plant Extracts....... 159
8.10 Conclusion and Recommendations.. 160
References.. 160

8.1 INTRODUCTION

The world's population is increasing every day. The utilization of modern drugs to curb the incidences of some deadly diseases has improved tremendously of late. Contrary to this, some infectious microorganisms have developed highly resistant vigour to these modern drugs. However, there is always a way out in every condition. The use of nanotechnology has been proven to be more effective in the therapeutic management of highly resistant diseases such as tuberculosis (TB) caused by the strain *Mycobacterium tuberculosis* (Ogbole and Ajaiyeoba, 2010; Gupta et al., 2017). The application of nanobiotechnology in medicine for the treatment and prevention of mild, severe or chronic diseases and sickness is called nanomedicine. Nanophytomedicine, which is the application of plant extracts in the therapeutics of diseases, has been in the forefront of traditional medicine and promising to be a sustainable health tool for life longevity against several ill-health. This is a novel therapeutic area in medicine with wide-ranging regulation and ethical and safety approach in solving health care issues.

Recent studies from epidemiological data showed that one-third of the world's population is infested with *M. tuberculosis*, leading to about 8 million cases yearly and 2 million deaths globally (Gautam et al., 2007). In the year 2007, about 8.8 ×10 million new-fangled cases were documented and 1.6 ×10 million demise recorded (WHO, 2007). It has also been documented that in 2015 about 1.4 ×10 million demise of humans occurred as a result of tuberculosis infection, which basically affected the developed and developing world's populace (South Africa, Pakistan, Nigeria, China, Indonesia and India) with 60% incidence rate as compared to the global rate (World Health Organization 2016). The strain *M. tuberculosis* was accounted as one of the deadliest respiratory diseases amongst others and the main cause of mortality in human race (Kumar and Clark, 2002).

However, the efficacy of *M. tuberculosis* in terms of resistance towards assorted drugs (rifampicin and isoniazid) in the management of TB needs an alternative approach (Sacchettini et al., 2008). Botanical approach has been the major focus in the therapeutic management of TB in recent times. Ogbole and Ajaiyeoba (2010) and Gupta et al. (2017) have paid special attention to ethnobotany. However, drug extracts from plants have faced various incidence of general acceptance because of the modes of preparation and administration as well as the lack of characterization, pharmacological assessment and proximate analysis of contents.

The effective therapeutic management of TB using plants extracts have been established by many researchers – Ajaiyeoba et al. (2003, 2004 and 2006), Ogbole and Ajaiyeoba (2010), Bheemanagouda et al. (2016), Gupta et al. (2017) – by utilizing phytonanotechnology. Plants extract synthesis using nanotechnology has been proven to be effective, eco-friendly and simple to administer and also permits control of precise shape and size of the particle, reduction and capping when compared with its synthetic chemical counterparts (Bheemanagouda et al., 2016).

Several biofabrication techniques have been used in the production of decoctions for the treatment of *M. tuberculosis* using nanoparticles with plant extracts from roots, leaves, stem and seeds (Gladwin et al., 1998; Bastian et al., 2000; Pandey et al., 2003a, b, 2006; Sharma et al., 2004b; Chan and Iseman, 2008; Griffiths et al., 2010). This technical know-how (technology) is very cost-effective and does not require specialized skills to be utilized and have high capacity and stability (Bheemanagouda et al., 2016).

Currently, several modified engineered nanoparticles are being applied for effective treatment of TB because of their unique characteristics (bioadhesive and bioabsorption) with plant extracts/decoctions. Examples of such are biogenic silver particles, polythene glycol, rifampicin and chitosan nanoparticles (Grange and Zumla, 2002; CDC, 2006; WHO, 2011).

In view of all these, this chapter tends to look at the application of next-generation nanobiofabricated particles from plant origin for the management of *Mycobacterium tuberculosis* with special focus and highlights on the synthesis of nanoparticles from plant, the characterization of nanoparticles, the management of *Mycobacterium tuberculosis* by using phytonanotechnology and the modes of action of plant-based nanoparticles against *Mycobacterium tuberculosis.*

8.2 TRANSMISSION OF TB DISEASE, PATHOGENESIS AND ITS TYPES

Basically, TB is an airborne disease transmitted by MTB from diseased persons to normal persons. MTB are present in the form of non-spore-forming and non-motile rod-shaped bacteria; basically this is anaerobic which is highly resistant to alcohol and acid. During speaking, coughing and sneezing, the infection is transmitted by diseased persons from droplet; pulmonary TB accounts for 80% of the total cases (Pethe et al., 2004). The pathogenic strain of TB is drug-susceptible: MDR and XDR strain. The drug-susceptible TB strain is treated by first-line drugs, specifically ethambutol, isoniazid, pyrazinamide and rifampicin. The MDR-TB strain which is resistant to the INH and RIF is difficult to be managed. Further resistance in XDR strain represents a major challenge in the treatment of TB. INH and RIF are the first-line drugs which can also be used as second-line drugs; for example, fluoroquinolones can be used as an injectable drug.

8.3 CONVENTIONAL CHEMOTHERAPY OF TB, ITS LIMITATIONS AND EMERGENCE OF DRUG RESISTANCE

The first-line ATDs are the first choice in the drug-susceptible TB therapy due to their high efficacy and least side effects. The drug-susceptible TB therapy consists of intensive and continuance phase: the first phase consists of the first-line ATDs for the passé of four months while the latter phase lasts for four to six months (Pandey et al., 2003b). The treatment is performed under the scheme of directly observed therapy known as DOTS in India.

The ill-planned monotherapy causes the development of a resistant strain of bacteria against the first-line ATDs (Pandey et al., 2003b; Sharma et al., 2004a). The idea of cure as a factor or determinant for early termination of the therapy creates a desirable condition for drug resistance.

The combination therapy causes the problems of overdosing or underdosing related to monotherapy (Pandey et al., 2003a). The combination of first-line ATDs can be used as the fixed dose for treatment (Sharma et al., 2004a). The quality of drugs in FDC product affects the treatment of TB disease. The misuse of such drugs in FDC products leads to the MDR strain of bacteria which is required for second-line ATDs.

8.4 NANOTECHNOLOGY AND THEIR ROLE IN THE MANAGEMENT OF TUBERCULOSIS DISEASES

Tuberculosis is a major chronic infectious disease, and it affected over one-third of the people on this planet. *Mycobacterium tuberculosis* is a major cause of mortality and they are responsible for single pathogen (Dye et al., 2005). The developed countries are mostly affected through this infection (Zumla et al., 2013). The prevalence of the TB still exists despite many decades of various treatment plans and the presence of active drugs for this disease. The patient dying due to TB infection is alarming in this decade and its effective and preventive measures should be taken by concerned bodies and authorities around the world (Meena, 2010). The resistant TB infection has been a trial for the control of TB across the world (Sarkar and Suresh, 2011). The administration of the drug through nanocarrier systems can be used to deliver drugs through the parenteral, oral, nasal and pulmonary route. Besides, pulmonary drug delivery (PDD) of these ATD-overloaded nanocarriers helps in depositing site-specific drugs at high concentrations within the diseased lungs thereby reducing the overall dose of a drug to patients which can reduce the systemic side effects (Armstrong and Hart, 1971; O'hara and Hickey, 2000; Suarez et al., 2001; Vyas et al., 2004; Pandey and Khuller, 2005; Sung et al., 2009).

8.5 NANOTECHNOLOGY IN TREATMENT OF TUBERCULOSIS

The small size of nanomaterials of 10–100 nm and large surface area contribute to its unique biological, physical and chemical properties (WHO, 2010; Jain and Pradeep, 2005). Silver nanoparticles, for instance, have been reported to receive substantial attention in pharmacological and medical research because of their action against microbial activity against drug-resistant and drug-sensitive bacteria and bacteriostatic effect. Bactericidal effect is preferred to bacteriostatic effect because it provides quicker relief from infection and displays little incidence of drug resistance. Likewise, there have been reported action of silver nanoparticles against hepatitis B and the HIV-1, though there are no reports of this nanoparticle against drug resistance and *Mycobacterium* other than tuberculosis so far (Elechiguerra et al., 2005; Lu et al., 2008).

Chemotherapy treatment of TB is quite a tedious process due to multidrug regimens, which need administration over a period of time and the low compliance of patient to this drug regimens is a major reason for its unsuccessful application (Prabakaran et al., 2004). The micro-encapsulation of substances of medicinal importance in polymers which are biodegradable in the delivery of drugs is an emerging technology. Typical examples of delivery or carrier systems developed for anti-TB drugs delivery include the microspheres and liposomes. This system has been shown to exert effective chemotherapeutic efficacy in animal models. Undetected counts of bacteria were detected in the spleens and lungs of mice infected with *M. tuberculosis* for 21 days. After a series of experiments with nanoparticles made from lectin-functionalized poly(lactide-*co*-glycolide), Schmidt and Bodmeier (1999) reported their potential as antitubercular drug carriers which are effective against TB. In a study by Johnson et al. (2005), the effectiveness of an anti-tuberculosis nanoparticle-based encapsulated drug was tested against the *M. tuberculosis* infection in guinea pigs for 10 days with non-encapsulated drugs as a control. A significant reduction in bacteria count was recorded in both cases. Their study suggested the possibility of tuberculosis treatment using a nano-based drug delivery system (Johnson et al., 2005). The US firm Biosante developed a nanotechnology-based vaccine adjuvant for TB in 2002 (Maclurcan, 2005).

Kumar et al. (2012) developed a novel drug: poly(lactic-*co*-glycolic acid) (PLGA)-based nanoformulation of levofloxacin. This was aimed to achieve a sustained release in plasma and combat against drug-resistant TB. The plasma level of PLGA nanoformulation of levofloxacin was about four to six days when compared with 24 hours clearance for a free levofloxacin. There was no significant adverse effect on the mouse treated with this formulation.

8.6 MANAGEMENT OF *MYCOBACTERIUM TUBERCULOSIS* USING PHYTONANOTECHNOLOGY

It has been affirmed that numerous plants possess several numerous biological active constituents with many medical benefits (Adetunji et al., 2011a–d; Adetunji and Olaleye, 2011). In view of this, several scientists have established the benefits of these active biological components for the synthesis of nanodrugs for the management of *Mycobacterium tuberculosis*. Bheemanagouda et al. (2016) tested and evaluated the synthesis of ZnO nanoparticles with *Limonia acidissima* plant extracts for the treatment of *Mycobacterium tuberculosis* growth in humans. Various standard techniques were used to characterize and standardize the ZnO nanoparticles and the *Limonia acidissima* plant extracts. The results from their study revealed that the ultraviolet absorbance of 374 nm indicates the formation of ZnO of various sizes (12–53 nm). The nanoparticles were able to regulate the growth and development of the strain *Mycobacterium tuberculosis* at a concentration of 12.5 lg/ml. The authors concluded by opining that ZnO nanoparticles with *Limonia acidissima* plant extracts are a promising tool for the management of *Mycobacterium tuberculosis* and it is a cheap, eco-friendly and a green technology that can also be utilized as an innovative medicine in the management of different strains of respiratory tract infections.

Lawal et al. (2014) in a pilot study did a review on the knowledge of various plants used in the treatment of tuberculosis in South Africa. The authors recounted the current status of the threat of medicinal plants in South Africa by deforestations. They suggested that conservation of these plants should be put into practice for the sustainable future usage in the management of certain diseases like TB. A pilot study on the ethnobotanical importance of plants for the therapeutics of TB was carried out in some areas in eastern South Africa (Gquamashe, Sheshegu, Ncera and Hala) where 100 informants were questioned on the herbs used in the treatment of TB. The results from the pilot study showed that 30 plants were documented belonging to 21 families that are commonly utilized by the herbalists. The most commonly used ones are *Artemisia afra*, *Haemanthus albiflos* and *Clausena anisata*. Findings from their study showed that most of the herbalists are aged and their legacy needs to be sustained. In conclusion, the authors recommend that further works should be carried on the highlighted therapeutic herbs by assaying their anti-tuberculosis properties in order to authenticate their ethnopharmacology importance in the management of tuberculosis diseases and other respiratory infections.

Kerry et al. (2018) did a review of the treatment of tuberculosis utilizing nanotechnology. The authors recounted the health impact of *Mycobacterium tuberculosis*. They stated the mode of action of the bacterium: it attacks the host macrophage and distorts the immune system, thus preventing the creation of phagolysome, lysosome transport protein and blocking the receptor-dependent TNF apoptosis of the host monocytes. They also recounted the utilization of conventional drugs and their low biodistribution and bioavailability in the management of high infection as well as in the prevention of human mortality. More highlights on the utilization of phytochemicals extracts with nanotechnology such as terpenoids, tannins, steroids, saponins, phenols and alkaloids sourced from aquatic plants (mangroves) and terrestrial plants in the management of TB were reported via various mechanisms of molecular evaluations. The understanding of proteomics and genomics of the pathogenic microorganism can aid in overcoming many gaps of research in developing an appropriate treatment for TB.

Sarvamangala et al. (2015) tested and evaluated the treatment efficiency of using nanoconjugates for the management of TB. The authors stated that the increase of multi- medications antagonism by TB has paved way for the development of several drugs using nanotechnology. They stressed the need for a combination of natural nanotechnology (green/phytonanotechnology) with conventional nanomedicine as conjugate drugs in the management of TB. The standard method was used to assay *M. tuberculosis* strain H37RV and the conjugated drug. The results from the study indicated that there was an increase in the set-up efficiency from 85% to 95% as well as the formation of the conjugated drugs from 264 to 356 nm. The activity of persistence as shown by the relative units of light was 93% for the inhibition of isoniazid. Thus, for cost-effective techniques of the production of nanodrugs, nanoconjugates of isoniazid are proven potent for the management of TB, while the natural nanotechnology could be used to decrease the dosage, which is also a potent feature for suitable diagnosis of TB as well as in the management of possible outcome of MDR tuberculosis. Therefore, the nanoconjugate which is eco-friendly can be used to replace the ecotoxic and biohazards of the conservative nanoparticles.

The incidence of herbal medicine in the management of certain chronic diseases is on the increase in the field of phytomedicine using nanoparticles and nanodevices. Chakraborty et al. (2016) did a review of the role of herbal medicine in the treatment of chronic diseases like cancer, HIV and TB. The authors stated the different problems and obstacles faced when utilizing conventional drugs such as undefined regions of action, poor alimentary absorption and insolubility in aqueous medium, *in vivo* constancy and low bioavailability. The integration of traditional medicine and nanotechnology can enrich the possibility of natural drugs from plant extracts in the management of chronic diseases. They opined that the manufacturing of nanoparticles can be more effective with the adoption of new techniques like metallic nanoparticles, magnetic nanoparticles and polymer nanoparticles. The choice is based on the characterization of the toxicity of the chemical and physical constituents of the nanoparticles. The prospective effects of nanotechnology on herbal medicine were highlighted.

8.7 MANAGEMENT OF DRUG-RESISTANT TUBERCULOSIS

Since the human civilization, plants have been the major suppliers for humans. They are major food sources, medicines and several livelihood requirements. Plant phytochemicals have been reported by several scientists to have effectiveness against TB. Phytochemicals such as phenols, steroids, tannins, phlorotannins, resins and several gums are some typical examples. Conventional drugs are limited in their usage due to the intra-macrophage characteristics of *M. tuberculosis* (Yah and Simate, 2015). Conventional drugs lack efficient biodistribution or intracellular penetration capability. Nanodrugs are able to circumvent this limitation and be effective in *M. tuberculosis*. There has been report of the effectiveness of nanodrugs in delivery to targeted cell (Singh et al., 2015).

Due to its size, its particle shape and its biological and chemical properties, nanoparticle has been extensively applied in research and invention in the biomedical sciences. Their effectiveness to act as an antibacterial agent is also a result of their potent ligand conjugates and high surface-to-volume ratio, which improves their ability to penetrate cell membranes and act against microbial cells (Yah and Simate, 2015). In studies by Rawashed and Haik (2009), Li et al. (2014) and Yah and Simate (2015), silver, gold and copper nanoparticles were reported to be effective against *Mycobacterium* spp. In spite of the various limitations faced, nanotechnology has provided a viable alternative for the treatment of infectious and deadly disease with less or no adverse effect on its host. The nanoparticle system has a low permissible level of toxicity and side effect, and it can also be degraded: a major advantage of this type of drug system. Some examples of nano-based drug delivery system against TB include niosomes, liposomes, liquid crystal, polymeric nanoparticles, solid lipid nanoparticles, microemulsion, metal-based nanoparticles and carbon-based nanoparticles (Nikalje, 2015; Costa et al., 2016; da Silva et al., 2016; Nasiruddin et al., 2017).

Drug-resistant tuberculosis is a major health concern around the globe (WHO, 2018a). The WHO recommends two standard drug types for the treatment of drug-resistant tuberculosis (i.e., a short- and a long-treatment regimen). The major difference between them is the duration of treatment and the drug combination. Treatment duration of 18–20 months is suggested for the long-treatment regimen while 9–11 months is suggested for the short-treatment regimen. However, in region where drug resistance index to fluoroquinolones or other second-line agents is high, the feasibility of short-term regimen is less likely (Lange et al., 2016).

The culture status conversion at 6 months of treatment is a distinct verified marker which shows the completion of a treatment outcome (Kurbatova et al., 2015). Investigation on the use of several biomarkers in the prediction of the duration of treatment, including transcriptomic approaches, is ongoing (Thompson et al., 2017).

Cure for TB has been defined by WHO guidelines as follows: when there are three culture or more whose results are negative, and which were completed in a 30-day duration apart after an intensive treatment phase. This definition however has some limitation which includes patient inability to produce sputum after months of treatment which are presumed to be effective (Lange et al., 2018). Likewise, failure rate is not properly assessed due to improper evaluation of relapse after treatment due to poor or inadequate follow-up (Lange et al., 2018).

A definition presumed to be more accurate in which patients are checked on after treatment inclusive of a 6-month culture status (Günther et al., 2016; Lange et al., 2018). Due to the challenges of extended follow-up, the engagement of tuberculosis programme is overbearing.

Due to pill burden and frequency of drug toxicities, supervised therapy is most desirable. In a situation where malabsorption is suspected, therapeutic drug monitoring should be given consideration. For the optimization of a successful outcome, nutritional, socio-economic and psychosocial supports are of importance. Patient needs to be treated with some sense of dignity and compassion. The major goal is to have a system which is patient-centred, where the needs of every patient are paramount. This will result in substantial benefits to tuberculosis patients globally. For HIV patients with tuberculosis ailments, there should be urgent treatment with multidrug resistance when resistance to rifampicin is confirmed (WHO, 2014). There should be initiation of antiretroviral therapy as soon as there is tolerance to tuberculosis treatment and within eight weeks, irrespective of baseline counts of CD4 with mortality rate as high as 91–100% in situation where there is absence of antiretroviral therapy (Havlir et al., 2011). Nonetheless, HIV patients with multidrug-resistant tuberculosis are susceptible to advanced tuberculosis disease. Drug toxicities are quite challenging (WHO, 2014).

In recent times, repurposed and new drugs have shown marked improvement in the management of HIV patients with multidrug-resistant tuberculosis (Esmail et al., 2018). A typical example of such drug is bedaquiline. Bedaquiline should however not be used with efavirenz (Pandie et al., 2016). Careful monitoring nonetheless is required when bedaquiline is used with lopinavir–ritonavir (Svensson et al., 2014). Dolutegravir has no known interaction with either bedaquiline or antiretroviral therapy (WHO, 2013; Mallikaarjun et al., 2016). Recent information on drug interactions between anti-tuberculosis and antiretroviral therapy is available (USD, 2019).

Of the 25,000–32,000 children affected annually with multidrug-resistant tuberculosis, only about 3–4% receive treatment for this illness. Fatality cases in the region of 21% have been reported. Bacteriological confirmation is absent in the clinical diagnosis of paediatric tuberculosis. Though pregnancy does not interfere with the treatment of multidrug-resistant tuberculosis, it however poses possible threats to foetus and mother (Schnippel et al., 2016; Rohilla et al., 2016). The foetus and mother are usually excluded from drug trials, resulting in the dearth of information on the safety of these drugs on pregnant mothers (Gupta et al., 2016). Despite these limitations, bedaquiline and fluoroquinolones are drugs given consideration, though drugs such as aminoglycosides, ethionamide and linezolid should be avoided due to its complication resulting from usage. There should be reinforcement of maternal treatment immediately after childbirth. Injectable agents should be administered if possible, while BCG vaccines should be administered likewise. Mothers whose sputum smear are positive to TB should wear a surgical mask in an area well ventilated and should not be prevented from being with her infant or family members. Tuberculosis medications are present in minimal concentration in breast milk and there are no adverse effects on exposure to it. Surgery is recommended by WHO when there are experienced thoracic surgeons, good postoperative care and when there is limitation of the multidrug-resistant tuberculosis disease (WHO, 2019). Improved success treatment was reported in a partial lung resection surgery (Harris et al., 2016). Tuberculosis preventive therapy (TPT) delivery could be employed in averting multidrug-resistant tuberculosis within the context of household contact tracing as endorsed by WHO (WHO, 2018b). Before the initiation of TPT, there should be exclusion of tuberculosis in contacts. Tuberculosis preventive therapy was reported to reduce multidrug-resistant tuberculosis risk by 90% (Mark et al., 2017).

8.8 MANAGEMENT OF DRUG-RESISTANT TUBERCULOSIS WITH SYNTHETIC DRUGS

The loading of a second-line anti-tuberculosis drug, ethionamide (ETH), with a synthesized carboxylic acid functionalized thermally hydrocarbonized porous silicon nanoparticles (UnTHCPSi NP) is highly effective. This study was conducted by Vale et al. (2017) who showed an increased activity of conjugated ethionamide nanoparticles when compared to free ethonamide. This reduces the frequency of administration of ethonamide (Vale et al., 2017).

Another effective metal-based nanoparticle which has been seen to be effective against multidrug resistance tuberculosis is titanium dioxide (TiO_2) nanoparticles, as described by Ramalingam (2019). The synthesis results from sol–gel method (Ramalingam, 2019).

Encapsulation of alginated modified-PLGA nanoparticles with two antitubercular drugs, amikacin and moxifloxacin, for the treatment of multidrug-resistant tuberculosis was shown to have quick internalization into the macrophages and is often seen in perinuclear region. The antimicrobial activity of the entrapped nanoparticles revealed increased inhibition of viable mycobacterium entrapped in the macrophages *in vitro* (Abdelghany et al., 2019).

Silver nanoparticles remain one of the most frequently used nanoparticles and have gained popularity in medicine and pharmacological field. They have been found to exhibit bactericidal effect on microorganisms. In a study conducted by Agarwal et al., silver nanoparticles were evaluated against about 26 strains of different standard strains, including drug-sensitive (DS), multidrug-resistant, extensive drug-resistant (XDR) and MOTT strains through BACTEC 460TB radiometric analysis. The results showed the inhibition of all isolates used for this test. Thus, silver nanoparticles from the plant extract of cucmium were found to have a potent anti-mycobacterial effect on multidrug-resistant tuberculosis (Agarwal et al., 2013).

Individually, econazole and moxifloxacin are potent drugs for the treatment of multidrug-resistant tuberculosis and latent phase of *Mycobacterium tuberculosis*. In drug-resistant tuberculosis, multiple drugs are administered concurrently, poly(DL-lactide-*co*-glycolide) (PLG) nanoparticles were prepared by solvent evaporation technique, as described by Ahmad et al. (2007), was encapsulated with the econazole and moxifloxacin. It was discovered that the drugs stayed about four to six days compared to 12–24 hours if administered individually in an infected rat. It was proved to be more efficacious compared to individual drugs which is a useful drug for multidrug-resistant tuberculosis (Ahmad et al. 2008).

Millions of people are affected by tuberculosis every year and it is the second leading cause of death from an infectious disease after acquired immune deficiency syndrome (Lin and Flynn, 2010). The appearance of drug-resistant strains of *Mycobacterium tuberculosis* is substantially challenging the goal of elimination of TB in the 21st century. It is caused by *Mycobacterium tuberculosis*, which normally affects the lungs. Mainly people recover from major TB infection without additional evidence of the disease. However, the infection may remain dormant for years, and in some cases it can be reactivated (Lin and Flynn, 2010). Men are more infected with the disease than women (Lin and Flynn, 2010).

The following drugs have been used previously to manage TB:

1. Rifampicin: This works by inhibiting the synthesis of bacterial RNA polymerase, thus inhibiting the nucleic acid synthesis.
2. Isoniazid: It inhibits the synthesis of mycolic acid (an essential cell wall component), thus interfering with cell wall synthesis in mycobacterium.
3. Ethambutol: It inhibits bacterium arabinogalactan synthesis and finally cell wall synthesis gets inhibited.
4. Pyrazinamide: It depletes and inhibits the membrane energy.
5. Aminoglycosides: It inhibits protein synthesis irreversibly.
6. Amino salicylic acid: It inhibits the incorporation of PABA into folic acid and iron metabolism.
7. Cycloserine: It inhibits the incorporation of D-alanine into peptidoglycan (Singh et al., 2016). However, they are not without serious adverse effects. Newer drugs for TB treatment have been developed to combat the resistance in TB and its associated complications. It is extremely important to give proper management and treatment to MDR-TB and XDR-TB in order to limit the spread, complications and resistance. The drugs include nitroimidazole

(TBA-354), oxazolidinone AZD5847, bedaquiline TMC207), delamanid OPC67683), moxifloxacin, benzothiazone (BTZ-043), capuramucin (SQ-641) and imidazopyridine (Q201) (Mohan et al., 2013).

Treatment of resistant strains of MDR-TB presents a challenge globally because it involves use of more than three drugs for a long period, which results in non-adherence to and development of even more resistant strains (Sotgiu et al., 2009). While new drugs like bedaquiline and delamanid have recently been discovered and approved for use in MDR-TB (Cox and Laessig, 2014), they are expensive and mostly unavailable in poor-resource countries.

8.9 MANAGEMENT OF MULTIDRUG-RESISTANT TUBERCULOSIS WITH PLANT EXTRACTS

Bunalema et al. (2017) demonstrated that *Zanthoxylum leprieurii* possesses anti-mycobacterial activity against a rifampicin- and isoniazid-resistant strains of *M. tuberculosis*. Both crude extracts from *Z. leprieurii* showed some inhibitory activity with the methanol extract being more active against all the *M. tuberculosis* strains. Hence, *Z. leprieurii* could potentially lead to the discovery of new anti-TB agents. Two compounds (acridone alkaloids) with antitubercular activity against *M. tuberculosis*-resistant strains were isolated from *Z. leprieurii*.

In addition, Camacho-Corona et al. (2008) revealed that some plants such as *Citrus aurantifolia*, *Citrus sinensis* and *Olea europaea* possess antitubercular activity against isoniazid-resistant *M. tuberculosis*. Therefore, these plants could be an important source of compounds with anti-mycobacterial activity against multidrug-resistant *M. tuberculosis*. Dey et al. (2014) reported that quinonoid have both antitubercular and antibacterial activity against drug-resistant mycobacterium. Furthermore, Dey et al. (2015) revealed that extracts as well as the pure constituents of pomegranate exhibited greater activity against *M. tuberculosis*, suggesting anti-tubercular efficacy of pomegranate against drug-resistant TB. Interestingly, garlic (*Allium sativum*) has been shown to have antitubercular properties *in vitro* and *in vivo* against multidrug-resistant TB (Dini et al., 2011).

Approximately 10.4 million people have been diagnosed with tuberculosis worldwide, and an estimated 1.79 million people have died from the disease in 2015 (WHO, 2016). *Zanthoxylum schinifolium* Siebold & Zucc. (Rutaceae) is a medicinal plant that is usually found in Northeast Asia, including Korea, China and Japan (Wang et al., 2011). Bioassay-guided fractionation of leaves extracts of *Z. schinifolium* was performed and collinin was found to possess an exceptional and explicit antitubercular activity against both the drug-susceptible and the drug-resistant strains of *M. tuberculosis* with low cytotoxicity. The growth of *M. tuberculosis* was considerably repressed by collinin without any cytotoxicity. Therefore, the decrease seen in the number of live *M. tuberculosis* was actually associated with the activity of collinin and not the number of cells present (Kim et al., 2018).

Furthermore, Lakshmanan et al. (2011) isolated an active anti-TB molecule, ethyl *p*-methoxycinnamate (EPMC), from *Kaempferia galangal* plant. The ethyl *p*-methoxycinnamate isolated was found to inhibit the growth of drug-susceptible as well as MDR clinical isolates of *M. tuberculosis*. EPMC is one of the major constituents present in the essential oil of *K. galanga* rhizomes, which has been previously reported to have larvicidal (Choochote et al., 1999), nematicidal (Choi et al., 2006), antifungal (Jantan et al., 2003) and anticarcinogenic activities (Xue et al., 2002) as well as inhibitory activity on monoamine oxidase enzyme (Noro et al., 1983). Essentially, along with the ability of EPMC to inhibit the growth of MDR strains of *M. tuberculosis*, it is a possible lead molecule worthy of further investigation. The plant *K. galanga* indeed was revealed by Lakshmanan et al. (2011) to possess an active molecule which could inhibit *M. tuberculosis*.

8.10 CONCLUSION AND RECOMMENDATIONS

This chapter has shown a comprehensive report on the application of nanodrugs obtained for natural products mainly from active ingredients derived from plants for the management of extensively drug-resistant tuberculosis. Various synthetic drugs with all their numerous adverse effects which are commonly used for the treatments of extensively drug-resistant tuberculosis were also highlighted. The modes of action of these biologically synthesized nanodrugs were also discussed. The different types of nanobioengineered nanodrugs alongside their merits and demerits were also highlighted. There is a need for a multidisciplinary approach between scientists from different backgrounds to search for a novel technique that could be used for the synthesis of more effective drugs when compared to synthetic drugs commonly used for the management of extensively drug-resistant tuberculosis. Moreover, there is a need to increase consumer awareness and benefits of natural products derived from plant extracts for the management of numerous diseases.

REFERENCES

Abdelghany S., Parumasivam T., Pang A., Roediger B., Tang P., Jahn K., Britton W.J., Chan H.K. 2019. Alginate modified-PLGA nanoparticles entrapping amikacin and moxifloxacin as a novel host-directed therapy for multidrug-resistant tuberculosis. *Journal of Drug Delivery Science and Technology* 52: 642–651.

Adetunji, C.O. and Olaleye, O.O. 2011. Phytochemical screening and antimicrobial activity of the plant extracts of *Vitellaria paradoxa* against selected microbes. *Journal of Research in Biosciences* 7(1): 64–69.

Adetunji, C.O., Arowora K.A, Afolayan S.S., Olaleye O.O., and Olatilewa M.O. 2011a. Evaluation of antibacterial activity of leaf extract of *Chromolaena odorata*. *Science Focus* 16(1): 1–6.

Adetunji, C.O., Kolawole, O.M., Afolayan, S.S., Olaleye, O.O., Umanah, J.T., and Anjorin E. 2011b. Preliminary phytochemical and antibacterial properties of Pseudocedrela *kotschyi*: A potential medicinal plant. *Journal of Research in Bioscience. African Journal of Bioscience* 4(1): 47–50.

Adetunji, C.O., Olaleye, O.O., Adetunji, J.B., Oyebanji, A.O. Olaleye, O.O., and Olatilewa, M.O. 2011c. Studies on the antimicrobial properties and phytochemical screening of methanolic extracts of *Bambusa vulgaris* leaf. *International Journal of Biochemistry* 3(1): 21-26.

Adetunji, C.O., Olaleye, O.O., Umanah, J.T., Sanu, F.T., Nwaehujor, I.U. 2011d. *In vitro* antibacterial properties and preliminary phytochemical of *Kigelia Africana*. *Journal of Research in Physical Sciences* 7(1): 8–11.

Adetunji, C.O, Olatunji, O.M, Ogunkunle, A.T.J, Adetunji, J.B., and Ogundare, M.O. 2014. Antimicrobial activity of ethanolic extract of *Helianthus annuus* stem. *Sikkim Manipal University Medical Journal* 1(1): 79–88.

Agarwal, P., Mehta, A., Kachhwaha, S., and Kothari, S.L. 2013. Green synthesis of silver nanoparticles and their activity against *Mycobacterium tuberculosis*. *Advanced Science, Engineering and Medicine* 5(7): 709–714.

Ahmad Z, Sharma S, Khuller GK. 2007. Chemotherapeutic evaluation of alginate nanoparticle-encapsulated azole antifungal and antitubercular drugs against murine tuberculosis Nanomedicine. Nanotechnology, Biology and Medicine 3 (3): 239–243.

Ahmad Z, Pandey R, Sharma S, Khuller GK. 2008. Novel chemotherapy for tuberculosis: chemotherapeutic potential of econazole-and moxifloxacin-loaded PLG nanoparticles *International journal of antimicrobial agents* 31 (2): 142–146.

Ajaiyeoba, E.O., Oladepo, O., Fawole, O.I., Bolaji, O.M., Akinboye, D.O., Ogundahunsi, O.A.T., Falade, C.O. et al. 2003. Cultural categorization of febrile illnesses in correlation with herbal remedies used for treatment in Southwestern Nigeria. *Journal of Ethnopharmacology* 85: 179–185.

Ajaiyeoba, E., Ashidi, J., Abiodun, O., Okpako, L., Ogbole, O., Akinboye, D., Falade, C. et al. 2004. Anti-malarial ethnobotany: In vitro antiplasmodial activity of seven plants identified in the Nigerian middle belt. *Pharmaceutical Biology* 42(8): 588–591.

Ajaiyeoba, E.O., Ogbole, O.O., and Ogundipe, O.O. 2006. Ethnobotanical survey of plants used in the traditional management of viral infections in Ogun State of Nigeria. *European Journal of Scientific Research* 13(1): 64–73.

Armstrong, J., and Hart, P.D.A. 1971. Response of cultured macrophages to *Mycobacterium tuberculosis*, with observations on fusion of lysosomes with phagosomes. *The Journal of Experimental Medicine* 134(3): 713–740.

Bastian, I., Rigouts, L., Van Deun, A., and Portaels, F. 2000. Directly observed treatment, short-course strategy and multidrug-resistant tuberculosis: Are any modifications required? *Bulletin of the World Health Organization* 78(2): 238–251.

Bheemanagouda, N., Patil, Tarikere C., Taranath. 2016. Limonia acidissima L. leaf mediated synthesis of zinc oxide nanoparticles: a potent tool against Mycobacterium tuberculosis. *Int. J. Mycobacteriol* 5: 197–204.

Bunalema L, Fotso GW, Waako P, Tabuti J, Yeboah SO. 2017. Potential of Zanthoxylum leprieurii as a source of active compounds against drug resistant Mycobacterium tuberculosis. *BMC Complement Altern Med.* 17(1): 89.

Camacho-Corona Mdel R, Ramírez-Cabrera MA, Santiago OG, Garza-González E, Palacios Ide P, Luna-Herrera J. 2008. Activity against drug resistant-tuberculosis strains of plants used in Mexican traditional medicine to treat tuberculosis and other respiratory diseases. *Phytother Res.* 22(1): 82–5.

CDC (Centre for Disease Control). Emergence of mycobacterium tuberculosis with extensive resistance to second-line drug-Worldwide, 2000–2004. *Morbidity and Mortality Weekly Report* 55(11): 301–305.

Chakraborty, K., Arun Shivakumar, and Sundaram Ramachandran. 2016. Nanotechnology in herbal medicines: A review. *International Journal of Herbal Medicine* 4(3): 21–27.

Chan E.D., and Iseman M.D. 2008. Multidrug-resistant and extensively drug-resistant tuberculosis: A review. *Current Opinion in Infectious Diseases* 21(6): 587–595.

Choi IH, Park JY, Shin SC, Park K-II. 2006. Nematicidal activity of medicinal plant extracts and two cinnamates isolated from Kaempferia galanga L. (Proh Hom) against the pine wood nematode, Bursaphelenchus xylophilus. *Nematology* 8(3): 359–365.

Choochote W, Kanjanapothi D, Panthong A, Taesotikul T, Jitpakdi A, Chaithong U, Pitasawat B. 1999. Larvicidal, adulticidal and repellent effects of Kaempferia galangal. *Southeast Asian J Trop Med Public Health* 30(3): 470–476.

Costa, A., Pinheiro, M., Magalhães, J., Ribeiro, R., Seabra, V., Reis, S., and Sarmento, B. 2016. The formulation of nanomedicines for treating tuberculosis. *Advanced Drug Delivery Reviews* 102: 102–115.

Cox E, Laessig K. 2014. FDA approval of bedaquiline--the benefit-risk balance for drug-resistant tuberculosis. *N Engl J Med* 371(8): 689–691.

da Silva, P.B., de Freitas, E.S., Bernegossi, J., Gonçalez, M.L., Sato, M.R., Leite, C.Q.F., Pavan, F.R., and Chorilli, M. 2016. Nanotechnology-based drug delivery systems for treatment of tuberculosis: A review. *Journal of Biomedical Nanotechnology* 12(2): 241–260.

Dey D, Ray R, Hazra B. 2014. Antitubercular and antibacterial activity of quinonoid natural products against multi-drug resistant clinical isolates. *Phytother Res.* 28(7): 1014–1021.

Dey D, Ray R, Hazra B. 2015. Antimicrobial activity of pomegranate fruit constituents against drug-resistant Mycobacterium tuberculosis and β-lactamase producing Klebsiella pneumoniae. *Pharm Biol.* 53(10): 1474–1480.

Dini C, Fabbri A, Geraci A. 2011. The potential role of garlic (Allium sativum) against the multi-drug resistant tuberculosis pandemic: a review. *Ann Ist Super Sanita.* 47(4): 465–73.

Dye, C., Watt, C.J., Bleed, D.M., Hosseini, S.M., and Raviglione, M.C. 2005. Evolution of tuberculosis control and prospects for reducing tuberculosis incidence, prevalence, and deaths globally. *JAMA* 293(22): 2767–2775.

Elechiguerra, J.L., Burt, J.L., Morones, J.R., Camacho-Bragado, A., Gao, X., Lara, H. H., and Yacaman, M.J. 2005. Interaction of silver nanoparticles with HIV-1. *Journal of Nanobiotechnology* 3(1): 6.

Esmail, A., Sabur, N.F., Okpechi, I., and Dheda, K. 2018. Management of drug-resistant tuberculosis in special sub-populations including those with HIV co-infection, pregnancy, diabetes, organ-specific dysfunction, and in the critically ill. *Journal of Thoracic Disease* 10(5): 3102.

Gautam, R.A. Saklani, A., and Jachak, S.M. 2007. Indian medicinal plants as a source of antimycobacterial agent. *Journal of Ethnopharmacology* 1109: 200–234.

Gladwin, M.T., Plorde, J.J., and Martin, T.R. 1998. Clinical application of the *Mycobacterium tuberculosis* direct test: Case report, literature review, and proposed clinical algorithm. *Chest* 114(1): 317–323.

Grange, J.M., and Zumla, A. 2002. The global emergency of tuberculosis: What is the cause? *The Journal of the Royal Society for the Promotion of Health*122(2): 78–81.

Griffiths, G., Nystrom, B., Sable, S.B., and Khuller, G.K. 2010. Nanobead-based interventions for the treatment and prevention of tuberculosis. *Nature Reviews Microbiology* 8(11): 827–834.

Gupta, A., Mathad, J.S., Abdel-Rahman, S.M., Albano, J.D., Botgros, R., Brown, V., Browning, R.S. et al. 2016. Toward earlier inclusion of pregnant and postpartum women in tuberculosis drug trials: consensus statements from an international expert panel. *Clinical Infectious Diseases* 62(6): 761–769.

Gupta, V.K., Madhan Kumarb, M., Deepa Bishta, and Anupam Kaushika. 2017. Plants in our combating strategies against Mycobacterium tuberculosis: Progress made and obstacles met. *Pharmaceutical Biology* 55(1): 1536–1544. DOI: 10.1080/13880209.2017.1309440.

Harris, R.C., Khan, M.S., Martin, L.J., Allen, V., Moore, D.A., Fielding, K., and Grandjean, L. 2016. The effect of surgery on the outcome of treatment for multidrug-resistant tuberculosis: A systematic review and meta-analysis. *BMC Infectious Diseases* 16(1): 262.

Havlir, D.V., Kendall, M.A., Ive, P., Kumwenda, J., Swindells, S., Qasba, S.S., Luetkemeyer, A.F. et al. 2011. Timing of antiretroviral therapy for HIV-1 infection and tuberculosis. *The New England Journal of Medicine* 365(16): 1482–1491.

Jain, P., and Pradeep, T. 2005. Potential of silver nanoparticle-coated polyurethane foam as an antibacterial water filter. *Biotechnology and Bioengineering* 90(1): 59–63.

Jantan IB, Mohd Salleh Mohd Yassin, Chen Bee Chin, Lau Lee Chen and Ng Lee Sim. 2003. Antifungal Activity of the Essential Oils of Nine Zingiberaceae Species. *Pharmaceutical Biology* 41:5: 392–397.

Johnson, C.M., Pandey, R., Sharma, S., Khuller, G. K., Basaraba, R. J., Orme, I. M., and Lenaerts, A. J. 2005. Oral therapy using nanoparticle-encapsulated antituberculosis drugs in guinea pigs infected with Mycobacterium tuberculosis. *Antimicrobial Agents and Chemotherapy* 49(10): 4335–4338.

Kerry, R.G., Sushanto Gouda, Bikram Sil, Gitishree Das, Han-Seung Shin, Gajanan Ghodake, and Jayanta Kumar Patra. 2018. Cure of tuberculosis using nanotechnology: An overview. *Journal of Microbiology* 56(5): 287–299. DOI: 10.1007/s12275-018-7414-y.

Kim S, Seo H, Mahmud HA, Islam MI, Lee BE, Cho ML, Song HY. 2018. In vitro activity of collinin isolated from the leaves of Zanthoxylum schinifolium against multidrug- and extensively drug-resistant Mycobacterium tuberculosis. *Phytomedicine* 46: 104–110.

Kumar, P.J., and Clark, M. 2002. 5th Ed. Respiratory diseases. *Clinical Medicine*: 134–156.

Kumar, G., Sharma, S., Shafiq, N., Khuller, G. K., & Malhotra, S. 2012. Optimization, in vitro–in vivo Evaluation, and Short-term Tolerability of Novel Levofloxacin-loaded PLGA Nanoparticle Formulation. *J Pharm Sci.* 101: 2165–2176.

Kurbatova, E.V., Cegielski, J.P., Lienhardt, C., Akksilp, R., Bayona, J., Becerra, M.C., and Demikhova, O.V. 2015. Sputum culture conversion as a prognostic marker for end-of-treatment outcome in patients with multidrug-resistant tuberculosis: A secondary analysis of data from two observational cohort studies. *The Lancet Respiratory Medicine* 3(3): 201–209.

Lange, C., Duarte, R., Fréchet-Jachym, M., Guenther, G., Guglielmetti, L., Olaru, I.D., and van Leth, F. 2016. Limited benefit of the new shorter multidrug-resistant tuberculosis regimen in Europe. *American Journal of Respiratory and Critical Care Medicine* 194(8): 1029–1031.

Lange, C., van Leth, F., Mitnick, C.D., Dheda, K., and Günther, G. 2018. Time to revise WHO-recommended definitions of MDR-TB treatment outcomes. *The Lancet Respiratory Medicine* 6(4): 246–248.

Lawal I.O., Grierson, D.S., and Afolayan, A.J. 2014. Phytotherapeutic information on plants used for the treatment of tuberculosis in Eastern Cape Province, South Africa. *Evidence-Based Complementary and Alternative Medicine* 2014: 735423, 11 pages. DOI: 10.1155/2014/735423.

Lakshmanan D, Werngren J, Jose L, Suja K.P., Nair M.S., Varma R. L., Mundayoor S., Hoffner S., Kumar R.A. 2011. Ethyl p-methoxycinnamate isolated from a traditional anti-tuberculosis medicinal herb inhibits drug resistant strains of Mycobacterium tuberculosis in vitro. *Fitoterapia* 82: 757–761.

Li, X., Robinson, S.M., Gupta, A., Saha, K., Jiang, Z., Moyano, D.F., and Rotello, V.M. 2014. Functional gold nanoparticles as potent antimicrobial agents against multi-drug-resistant bacteria. *ACS Nano* 8(10): 10682–10686.

Lin P.L. Flynn J.L. 2010. Understanding Latent Tuberculosis: A Moving Target. *J Immunol* 185 (1): 15–22.

Lu, L., Sun, R.W., Chen, R., Hui, C.K., Ho, C.M., Luk, J.M., Lau, G.K., and Che, C.M. 2008. Silver nanoparticles inhibit hepatitis B virus replication. *Antiviral Therapy* 13(2): 253.

Maclurcan, D.C. 2005. Nanotechnology and developing countries part 1: What possibilities. *AZojono-Journal of Nanotechnology Online*, 1 (a0103): 1-10.

Mallikaarjun, S., Wells, C., Petersen, C., Paccaly, A., Shoaf, S.E., Patil, S., and Geiter, L. 2016. Delamanid co-administered with antiretroviral drugs or antituberculosis drugs shows no clinically relevant drug-drug interactions in healthy subjects. *Antimicrobial Agents and Chemotherapy* 60(10): 5976–5985.

Marks, S.M., Mase, S.R., and Morris, S.B. 2017. Systematic review, meta-analysis, and cost-effectiveness of treatment of latent tuberculosis to reduce progression to multidrug-resistant tuberculosis. *Clinical Infectious Diseases* 64(12): 1670–1677.

Meena, L.S. 2010. Survival mechanisms of pathogenic Mycobacterium tuberculosis H37Rv. *The FEBS Journal* 277(11): 2416–2427.

Mohan K., Rawall, S., Pawar, UM, Sadani, M., Nagad, P., Nene, P., Nene A. 2013. Drug resistance patterns in 111 cases of drug-resistant tuberculosis spine. *Eur Spine J.* 22(Suppl 4): 647–652.

Nasiruddin, M., Neyaz, M., and Das, S. 2017. Nanotechnology-based approach in tuberculosis treatment. *Tuberculosis Research and Treatment.* vol. 2017, Article ID 4920209, 12 pages, 2017. https://doi.org/10.1155/2017/4920209

Nikalje, A.P. 2015. Nanotechnology and its applications in medicine. *Medicinal Chemistry* 5(2): 081–089.

Noro T, Miyase T, Kuroyanagi M, Ueno A, Fukushima S. 1983. Monoamine oxidase inhibitor from the rhizomes of Kaempferia galanga L. *Chem Pharm Bull (Tokyo).* 31(8):2708–11.

Ogbole, O., and Ajaiyeoba, E. 2010. Traditional management of tuberculosis in Ogun State of Nigeria: The practice and ethnobotanical survey. *African Journal of Traditional Complementary and Alternative Medicines* 7(1): 79–84.

O'hara, P., and Hickey, A.J. 2000. Respirable PLGA microspheres containing rifampicin for the treatment of tuberculosis: Manufacture and characterization. *Pharmaceutical Research* 17(8): 955–961.

Pandey R., Sharma, A., Zahoor, A., Sharma, S., Khuller, G.K., and Prasad, B. 2003a. Poly (DL-lactide-co-glycolide) nanoparticle-based inhalable sustained drug delivery system for experimental tuberculosis. *Journal of Antimicrobial Chemotherapy* 52(6): 981–986.

Pandey R., Zahoor, A., Sharma, S., and Khuller, G.K. 2003b. Nanoparticle encapsulated anti-tubercular drugs as a potential oral drug delivery system against murine tuberculosis. *Tuberculosis* 83(6): 373–378.

Pandey, R., and Khuller, G.J. 2005. Antitubercular inhaled therapy: Opportunities, progress and challenges. *Journal of Antimicrobial Chemotheraphy* 55(4): 430–435.

Pandey R., Sharma, S., and Khuller, G.K. 2006. Chemotherapeutic efficacy of nanoparticle encapsulated antitubercular drugs. *Drug Delivery* 13(4): 287–294.

Pandie, M., Wiesner, L., McIlleron, H., Hughes, J., Siwendu, S., Conradie, F., and Maartens, G. 2016. Drug–drug interactions between bedaquiline and the antiretrovirals lopinavir/ritonavir and nevirapine in HIV-infected patients with drug-resistant TB. *Journal of Antimicrobial Chemotherapy* 71(4): 1037–1040.

Pethe, K., Swenson, D.L., Alonso, S., Anderson, J., Wang, C., and Russell, D.G. 2004. Isolation of mycobacterium tuberculosis mutants defective in the arrest of phagosome maturation. *Proceedings of the National Academy of Sciences of the United States of America* 101(37): 13642–13647.

Prabakaran, D., Singh, P., Jaganathan, K.S., and Vyas, S.P. 2004. Osmotically regulated asymmetric capsular systems for simultaneous sustained delivery of anti-tubercular drugs. *Journal of Controlled Release* 95(2): 239–248.

Ramalingam, S. 2019. Synthesis of nanosized titanium dioxide by sol-gel method. *International Journal of Innivative Technology and Exploring Engineering* 9: 732–735.

Ramanlingam, V., Sundaramahalingam, S., and Rajaram, R. 2019. Size-dependent antimycobacterial activity of titanium oxide nanoparticles against mycobacterium tuberculosis. *Journal of Materials Chemistry B*: 7; 4338–4346.

Rawashdeh, R., and Haik, Y. 2009. Antibacterial mechanisms of metallic nanoparticles: A review. *Dynamic Biochemistry, Process Biotechnology and Molecular Biology* 3(2); 12–20.

Rohilla, M., Joshi, B., Jain, V., Kalra, J., and Prasad, G.R.V. 2016. Multidrug-resistant tuberculosis during pregnancy: Two case reports and review of the literature. *Case Reports in Obstetrics and Gynecology,* vol. 2016, Article ID 1536281, 4 pages, 2016. https://doi.org/10.1155/2016/1536281.

Sacchettini, J.C., Rubin, E.J., and Freundich, J.S. 2008. Drugs versus bugs: In pursuit of the persistent predator mycobacterium tuberculosis. *Nature Reviews Microbiology* 6: 41–52.

Sarkar S, Suresh MR. (2011). An overview of tuberculosis chemotherapy - a literature review. *J Pharm Pharm Sci.* 14(2):148–61. doi: 10.18433/j33591.

Sarvamangala D, Nagasejitha, P., Seenivasan, S.P., Srinivas, L., and Murthy, U.S.N. 2015. Preparation and evaluation of isoniazide nano-conjugates for improving the therapeutic efficiency. *International Journal of Pharmaceutical Sciences and Research* 6(2): 739–745.

Schmidt, C., and Bodmeier, R. 1999. Incorporation of polymeric nanoparticles into solid dosage forms. *Journal of Controlled Release* 57(2): 115–125.

Schnippel, K., Ndjeka, N., Conradie, F., Berhanu, R., Claasen, Z., Banoo, S., and Firnhaber, C. 2016. A call to action: Addressing the reproductive health needs of women with drug-resistant tuberculosis. *South African Medical Journal* 106(4): 333–334.

Sharma A., Pandey, R., Sharma, S., and Khuller, G.K. 2004a. Chemotherapeutic efficacy of poly (dl-lactide-co-glycolide) nanoparticle encapsulated antitubercular drugs at sub-therapeutic dose against experimental tuberculosis. *International Journal of Antimicrobial Agents* 24(6): 599–604.

Sharma, A., Sharma, S., and Khuller, G.J. 2004b. Lectin-functionalized poly (lactide-co-glycolide) nanoparticles as oral/aerosolized antitubercular drug carriers for treatment of tuberculosis. *Journal of Antimicrobial Chemotherapy* 54(4): 761–766.

Singh, A., Gopinath, K., Sharma, P., Bisht, D., Sharma, P., Singh, N., and Singh, S. 2015. Comparative proteomic analysis of sequential isolates of Mycobacterium tuberculosis from a patient with pulmonary tuberculosis turning from drug sensitive to multidrug resistant. *The Indian Journal of Medical Research* 141(1): 27.

Singh, R., Nawale, L., Arkile, M., Wadhwani, S., Shedbalkar, U., Chopade, S., Sarkar, D., and Chopade, B.A. 2016. Phytogenic silver, gold, and bimetallic nanoparticles as novel antitubercular agents. *International Journal of Nanomedicine* 11: 1889–1897.

Sotgiu G., Ferrara G., Matteelli A., Richardson M. D., Centis R., Ruesch-Gerdes S., Toungoussova O., Zellweger J.P., Spanevello A., Cirillo D., Lange C., Migliori, G. B. 2009. Epidemiology and clinical management of XDR-TB: a systematic review by TBNET. *European Respiratory Journal* 33: 871–881.

Suarez, S., O'Hara, P., Kazantseva. M., Newcomer, C.E., Hopfer, R., McMurray, D.N., and Hickey, A.J. 2001. Respirable PLGA microspheres containing rifampicin for the treatment of tuberculosis: Screening in an infectious disease model. *Pharmaceutical Research* 18(9): 1315–1319.

Sung, J.C., Padilla, D.J., Garcia-Contreras, L., VerBerkmoes, J.L., Durbin, D., Peloquin, C.A., Elbert, K.J., Hickey, A.J., and Edwards, D.A. 2009. Formulation and pharmacokinetics of self-assembled rifampicin nanoparticle systems for pulmonary delivery. *Pharmaceutical Research* 26(8): 1847–1855.

Svensson, E.M., Dooley, K.E., and Karlsson, M.O. 2014. Impact of lopinavir-ritonavir or nevirapine on bedaquiline exposures and potential implications for patients with tuberculosis-HIV coinfection. *Antimicrobial Agents and Chemotherapy* 58(11): 6406–6412.

Thompson, E.G., Du, Y., Malherbe, S.T., Shankar, S., Braun, J., Valvo, J., Ronacher, K. et al. 2017. Host blood RNA signatures predict the outcome of tuberculosis treatment. *Tuberculosis* 107: 48–58.

US Department of Health and Human Services. 2019. AIDS info—drugs. https://aidsinfo.nih.gov/drugs (accessed Aug 29, 2019).

Vale, N., Correia, A., Silva, S., Figueiredo, P., Mäkilä, E., Salonen, J., Hirvonen, J., Pedrosa, J., Santos, H.A., and Fraga, A. 2017. Preparation and biological evaluation of ethionamide-mesoporous silicon nanoparticles against Mycobacterium tuberculosis. *Bioorganic & Medicinal Chemistry Letters* 27: 403–405.

Vyas, S., Kannan, M., Jain, S., Mishra, V., and Singh, P. 2004. Design of liposomal aerosols for improved delivery of rifampicin to alveolar macrophages. *International Journal of Pharmaceutics* 269(1): 37–49.

Wang CF, Yang K, Zhang HM, Cao J, Fang R, Liu ZL, Du SS, Wang YY, Deng ZW, Zhou L. 2011. Components and insecticidal activity against the maize weevils of Zanthoxylum schinifolium fruits and leaves. *Molecules.* 16(4): 3077–3088.

WHO (World Health Organization). 2007. Global Tuberculosis Control: Surveillance Planning, Financing. WHO Report, Geneva, Switzerland (WHO/HTM/TB/2007.376).

WHO. 2010. Tuberculosis Report, Global Tuberculosis Control. Available: www.who.int/tb/publications/global_report/ 2010/en/index.html (accessed November 2010).

WHO. 2011. Partners call for increased commitment to tackle MDR-TB. http://www.who.int/mediacentre/news/releases/2011/TBday 20110322/en/index.html.

WHO. 2013. *Consolidated Guidelines on the Use of Antiretroviral Drugs for Treating and Preventing HIV Infection: Recommendations for a Public Health Approach.* Geneva, Switzerland: WHO.

WHO. 2014. *Companion Handbook to the WHO Guidelines for the Programmatic Management of Drug-Resistant Tuberculosis.* Geneva, Switzerland: WHO.

WHO. 2016. *Global Tuberculosis Report.* Geneva, Switzerland: WHO.

WHO. 2018a. *Global Tuberculosis Report.* Geneva, Switzerland: WHO.

WHO. 2018b. *Latent Tuberculosis Infection. Updated and Consolidated Guidelines for Programmatic Management.* Geneva, Switzerland: WHO.

WHO. 2019. *Consolidated Guidelines on Drug-Resistant Tuberculosis Treatment.* Geneva, Switzerland: WHO.

Xue Y, Chen H. 2002. Study on the anti-carcinogenic effects of three compounds in Kaempferia galanga L. *Wei Sheng Yan Jiu.* 31(4): 247–248.

Yah, C.S., and Simate, G.S. 2015. Nanoparticles as potential new generation broad spectrum antimicrobial agents. *DARU Journal of Pharmaceutical Sciences* 23(1): 43.

Zumla A, George A, Sharma V, Herbert N, Baroness Masham of Ilton. 2013. WHO's 2013 global report on tuberculosis: successes, threats, and opportunities. *Lancet* 382(9907): 1765–1767.

9 Green Synthesis of Nanoparticles and Their Antimicrobial Efficacy against Drug-Resistant *Staphylococcus aureus*

Nonhlanhla Tlotleng, Jiya M. John, Dumisile W. Nyembe and Wells Utembe

CONTENTS

9.1 Introduction ... 168
9.2 Metal Nanomaterials: Efficacy against *S. aureus* and Mechanism of Action ... 169
9.3 Biological Synthesis of Metal Nanoparticles ... 169
  9.3.1 Metal Oxide Nanoparticles: Efficacy against *S. aureus* and Mechanism of Action ... 170
  9.3.2 Biological Synthesis of Metal Oxide Nanomaterials ... 170
9.4 Carbon-Based Nanoparticles: Efficacy against *S. aureus*, Mechanism of Action and Synthesis ... 171
  9.4.1 Carbon-Based Nanomaterials ... 171
    9.4.1.1 Fullerenes ... 171
    9.4.1.2 Carbon Nanotubes ... 172
  9.4.2 Green Synthesis of Carbon Nanotubes ... 172
9.5 Organic Carbon-Based Nanomaterials ... 173
  9.5.1 Liposomes: Efficacy against *S. aureus* and Mechanism of Action ... 173
  9.5.2 Synthesis of Liposomes ... 173
  9.5.3 Dendrimers: Efficacy against *S. aureus* and Mechanism of Action ... 173
  9.5.4 Green Synthesis of Dendrimers ... 174
9.6 Summary and Conclusion ... 174
References ... 175

9.1 INTRODUCTION

Microorganisms, such as bacteria, viruses and fungi, develop resistance to antimicrobial agents as they find adaptive ways to fight and prevail against various treatments. According to the World Health Organization (WHO), there is an increase in bacterial resistance against conventional drugs because of the intensive use and misuse of antibiotics (WHO, 2014). Among antimicrobial-resistant microorganisms of concern is *Staphylococcus aureus*, which is a cause of many serious hospital-acquired and community-acquired infections that are a great burden in terms of morbidity, mortality and healthcare costs (Schito, 2006). Among hospital-acquired *S. aureus* infections in children's hospital in the United States, 61% were attributed to *S. aureus* infection (Hultén et al., 2010), causing lower respiratory tract infections, surgical site infections, as well as nosocomial bacteraemia, pneumonia and cardiovascular infections. Furthermore, *S. aureus* is a leading cause of community-acquired food poisoning in the world (Kadariya et al., 2014).

S. aureus strains have developed resistance against widely used antibiotics such as β-lactams, β-lactamase-resistant penicillin, methicillin as well as vancomycin (Gardete and Tomasz, 2014). Therefore, there is an urgent need to address the lack of effective treatments for *S. aureus*, especially multidrug-resistant *S. aureus* (MDRSA). Among the few solutions against MDRSA is the use of nanomaterials (NMs) such as metal nanoparticles (NPs) metal oxide, and carbon-based NMs. The small size, large surface-to-volume ratio and the ability of NMs to target different bacterial structures enhance their biological activity and interaction with microbial cells, while intracellularly, NMs act by reducing the activity of enzymes and disturbing the vital cell functions leading to cell death (Zheng et al., 2017).

Widespread use of NMs in antimicrobial formulations requires large-scale synthesis of NMs (Erjaee et al., 2017). While chemical synthesis methods have extensively been used in nanotechnology, green synthesis methods are desirable because of their simplicity, cost-effectiveness, environment-friendliness and the ability to rapidly synthesize stable NMs of defined sizes and morphologies (Logeswari et al., 2015; Shankar et al., 2016). For these reasons, biosynthetic methods have emerged as alternatives to chemical synthetic methods.

This chapter explores the use of various types of NMs synthesized by green methods to combat the drug-resistant pathogen *S. aureus*. In particular, the chapter discusses the application of metallic NMs, metal oxides NMs as well as carbon-based NMs (both inorganic and organic) on treatment of *S. aureus* infections, in that order, focusing on the efficacy of the NMs against *S. aureus*, mechanisms of actions (MOAs), biocompatibility (toxicity) and green methods of synthesis of the NMs. In the chapter, efficacy of the NMs against *S. aureus* is not only limited to antimicrobial activity of the NMs, but also the (potential) use of the NMs as drug delivery systems to enhance the efficacy of conventional antibiotics against *S aureus*. Biodurability is one of the most important factors that may limit the application of some NMs in drug delivery (Utembe et al., 2015). In this regard, NMs that dissolve slowly, such as AuNMs, would be expected to bioaccumulate *in vivo* and cause chronic adverse effects (Oberdörster, 2000). On the other hand, nanoparticles that dissolve fast, such as Ag and Cu, may exert acute toxic effects arising from the released ions (Beer et al., 2012). Therefore, in addition to antimicrobial activity, it will be important to assess the effects of biodurability and bioaccumulation of the NMs.

9.2 METAL NANOMATERIALS: EFFICACY AGAINST *S. AUREUS* AND MECHANISM OF ACTION

Metal NMs exert their antimicrobial effect by damaging cell membranes and by causing oxidative stress and injury to proteins and DNA, primarily through the release of metal ions (Brandelli et al., 2017). The metal NMs can selectively affect bacterial cells due to the larger surface area to volume ratio, which permits rapid uptake and intracellular distribution of the NMs (Ayala-Núñez et al., 2009). In addition to their antimicrobial activity, metal NPs are utilized as delivery platforms for antimicrobial agents as they increase the concentration of antibiotics at the site of bacterium–antibiotic interaction (Jiang et al., 2018).

Silver nanoparticles (AgNPs) with sizes ranging from 5 to 32 nm have been shown to increase the antibacterial activities of conventional antibiotics, including penicillin and erythromycin (Shahverdi et al., 2007). In the study, the zone of inhibition against *S. aureus* was shown to increase from 10 mm to 14 mm in the presence of erythromycin-AgNPs synthesized with the bacterial culture supernatant of *Klebsiella pneumonia* (Shahverdi et al., 2007). Antibacterial activity has also been demonstrated for other metal NMs, including copper (Cu) (Esteban-Cubillo et al., 2006) and iron (Fe) (Naseem and Farrukh, 2015). An inhibitory effect could be achieved for CuNPs against *S. aureus* (99.6%) at a concentration of 3,200 μg/ml.

9.3 BIOLOGICAL SYNTHESIS OF METAL NANOPARTICLES

Prokaryotic bacteria have been the most extensively researched bacteria in the biological synthesis of metal NPs (Thakkar et al., 2010). Several authors have demonstrated the synthesis of AgNPs using the filamentous cyanobacterium *Plectonema boryanum* UTEX 485 (Lengke et al., 2007; Thakkar et al., 2010). Fungi and marine algae have also been used in the synthesis of AgNPs with controlled size and morphology. For example, Mukherjee et al. (2008) demonstrated the biosynthesis of nanocrystalline AgNPs using the non-pathogenic fungi species *Trichoderma asperellum*. Similarly, AgNPs were synthesized using the fungus *Verticillium* isolated from the *Taxus* plant, where exposure of the fungal cells to aqueous Ag^+ ions resulted in the intracellular reduction of the metal ions and formation of intracellular 25-nm AgNPs (Mukherjee et al., 2001). The metal NMs synthesized using microbiological methods have been shown to possess antimicrobial similar to those prepared by other methods (Shahverdi et al., 2007).

Plant extracts have also been utilized as reducing, stabilizing and coating agent to produce stable NMs (Shankar et al., 2016). The source of the plant extract has been shown to play a role in determining the morphological characteristics of the NMs due to the differences in concentrations and varieties of phytochemicals found in various plant extracts (Mittal et al., 2013). AgNPs have been synthesized using plant extracts from *Ocimum tenuiflorum*, *Solanum tricobatum*, *Syzygium cumini*, *Centella asiatica* and *Citrus sinensis* (Logeswari et al., 2015) as well as *Chamaemelum nobile* (Erjaee et al., 2017). The antibacterial activity of the synthesized AgNPs was assessed in both studies against *S. aureus* where the minimum inhibitory concentration (MIC) defined as the lowest concentration where growth of the bacteria was inhibited was reported as 31 μg/ml (Erjaee et al., 2017). In addition, the lowest concentration (MBC) that was shown to kill the bacteria in the study was 62.5 μg/ml. Similarly, Logeswar et al. (2015) reported a zone of inhibition of 30 mm in *S. aureus* cultures treated with 50, 75 and 100 μl of AgNPs synthesized using natural plant extract.

9.3.1 Metal Oxide Nanoparticles: Efficacy against *S. aureus* and Mechanism of Action

There is an emerging interest in the synthesis of metal oxide nanoparticles (MO-NPs) in the biomedical field as disinfectants, antimicrobial agents and drug delivery carriers due to their unique optical, electronic and physiochemical properties. A number of MO-NPs have been shown to have antibacterial activity against *S. aureus*. For instance, antimicrobial activity against *S. aureus* has been demonstrated with ZnO (Azam et al., 2012), iron(III) oxide (Fe_2O_3; Tran et al., 2010) and copper oxide (CuO) NPs (Jadhav et al., 2011). In the study conducted by Tran et al. (2010), the antibacterial activity of Fe_2O_3 against *S. aureus* was shown to be concentration dependent. The results showed that the ratio of live/dead bacteria was significantly lower with the highest dose, at time points of 4, 12 and 24 hours when incubated with 30 μg/ml, 300 μg/ml and 3 mg/ml of NPs.

MO-NPs have also been conjugated with conventional antibiotics to enhance the efficacy of conventional antibiotics. In that regard, TiO_2-NPs enhanced the antimicrobial activity of beta lactams, cephalosporins, aminoglycosides, glycopeptides, macrolids and lincosamides, as well as tetracycline against multidrug-resistant *S. aureus* (Roy et al., 2010). Similarly, Fe_2O_3 were functionalized with vancomycin (Lai and Chen 2013) and methicillin (Geilich et al., 2017) to penetrate into bacteria cells and biofilm masses. In the latter case, a formulation of 40 μg/ml Fe_2O_3-NPs and 20 μg/ml methicillin was selectively toxic towards methicillin-resistant biofilm cells, but not towards mammalian cells. The exact MOA of MO-NPs is still not clearly understood, owing to the different parameters investigated in different studies. However, the formation of ROS, intracellular release of metal ions, disruption of cell wall membrane, particle internalization as well as damage to DNA and proteins are some of the proposed MOA for MO-NPs (Stankic et al., 2016).

9.3.2 Biological Synthesis of Metal Oxide Nanomaterials

Bacteria are widely used in the preparation of MO-NPs since they have the ability to reduce metal ions. Fungi-mediated biosynthesis offers an efficient way to produce large quantities of NPs compared to bacteria; however, plant-mediated synthesis is preferred as it is simpler and cost-effective. A variety of plant extracts have been used to synthesize MO-NPs such as ZnO, CuO, Fe_3O_4, TiO_2, NiO and CeO_2 NPs were their antimicrobial activity were assessed (Sharmila et al., 2018; Singh et al., 2018; Teow et al., 2018).

Vijayakumar et al. (2016) demonstrated the antibacterial activity of green-synthesized ZnO-NPs against *S. aureus*. The ZnO-NPs, with an average size of 47.27 nm and flower-like structures, were synthesized using the leaf extract of *Laurus nobilis* (*Ln*-ZnO-NPs). The zone of inhibition (ZOI) of *Ln*-ZnO-NPs against *S. aureus* was 11.4, 12.6 and 14.2 mm at 25, 50 and 75 μg/ml; whereas the ZOI of *Ln*-ZnO-NPs against *P. aeruginosa* was 9.8, 10.2 and 11.3 mm at the same concentrations. Results indicate that ZnO-NPs have greater antibacterial activity against *S. aureus* than *P. aeruginosa*.

Maqbool et al. (2016) showed the efficacy of green-synthesized cerium dioxide (CeO_2) NPs against *S. aureus*. The 24-nm CeO_2-NPs synthesized using *Olea europaea* leaf extract showed antibacterial activity against both gram-negative and gram-positive (*S. aureus*) bacteria. The addition of CeO_2-NPs (20 μg/5 μl) resulted in moderate antibacterial activity at a ZOI of 9, 10 and 8 mm, respectively. Green-synthesized nickel oxide (NiO) NPs also showed potential biomedical application (Ezhilarasi et al., 2016). The spherical-shaped NiO-NPs with a size of 9.69 nm were synthesized using *Moringa oleifera* plant extract. In this study, NiO-NPs were found to be more effective against gram-positive *S. aureus* (ZOI 15 mm) than gram-negative bacteria.

9.4 CARBON-BASED NANOPARTICLES: EFFICACY AGAINST *S. AUREUS*, MECHANISM OF ACTION AND SYNTHESIS

Carbon-based NMs (CBNMs) have found applications towards the improvement of physico-chemical and therapeutic effectiveness of drugs. In this section, we shall present the applications of both inorganic NMs (such as fullerenes and carbon nanotubes – CNTs – and organic NMs such as liposomes and dendrimers) in the fight against drug-resistant *S. aureus*.

9.4.1 Carbon-Based Nanomaterials

9.4.1.1 Fullerenes

Fullerenes, also known as C_{60}, are closed-cage carbon allotrope soccer ball-shaped NMs of different shapes, sizes and charges, due to derivatization with various functional groups. Fullerenes have been shown to possess antimicrobial activity against many species of bacteria, including *S. aureus*. In a study by Kumar and Menon (2009), fullerenes derivatized with s-triazine analogues were shown to possess antimicrobial activity against *S. aureus* at an MIC of 6.5 μg/ml, comparable to the antimicrobial drug, ciprofloxacin. Generally, the fullerenes antimicrobial activity observed has been attributed to cell membrane disruption as well as inhibition of energy metabolism and respiration following internalization by bacteria (Dizaj et al., 2015).

The antimicrobial activity of fullerenes can be enhanced by the use of light because of their ability to absorb visible light in their extended π-conjugation system, resulting in the generation of singlet oxygen as well as hydroxyl and superoxide radicals (Tegos et al., 2005; Mroz et al., 2007). These ROS can cleave DNA of bacteria, enabling fullerenes to act as photosensitizers that can have applications in antimicrobial PDT (Bosi et al., 2003). As an example, Mizuno et al. (2011) utilized cationic-substituted fullerene derivatives for the *in vitro* antimicrobial PDT against *S. aureus* by increasing their effectiveness through irradiation with white light. The bioaccumulation of fullerenes has been reported to be lower than that of CNTs (Wang et al., 2014), and therefore may have higher prospects in treatment of *S. aureus* than CNTs (discussed in the next section).

9.4.1.2 Carbon Nanotubes

Carbon nanotubes, including single-walled CNTs (SWCNTs) and multiwalled CNTs (MWCNTs), are allotropes of carbon arranged in layers of cylindrical shapes, with outer diameters ranging from 3 to 30 nm (Saifuddin et al., 2012). SWCNTs possess a single layer of graphene, while MWCNTs possess multiple layers. Both SWCNTs and MWCNTs have been shown to possess antimicrobial activity. For example, Liu et al. (2009) demonstrated the antimicrobial activity of individually dispersed pristine SWCNTs. Similarly, SWCNTs were shown to inhibit the growth of *S. aureus* at concentrations of 5, 10 and 20 μg/ml after three days of incubation (Al-Shaeri, 2019). Arias and Yang (2009) reported the inactivation of *S. aureus* in suspensions of SWCNTs and MWCNTs, while Zardini et al. (2012) demonstrated enhanced antibacterial activity of MWCNTs through functionalization with amino acids.

CNTs do not only possess antimicrobial activity but can also be functionalized with conventional antibiotics for enhanced efficacy. For example, ciprofloxacin-functionalized SWNCTs exhibited a 16-fold efficacy against *S. aureus* (Assali et al., 2017). While membrane damage appears to be the dominant MOA, Vecitis et al. (2010) proposed a three-step MOA involving initial contact, cell membrane damage and electronic structure-dependent oxidation of bacterial lipids, proteins and DNA. Applications of CNTs in treatment of *S. aureus* infections may however be limited by the tendency to bioaccumulate in living things (Yang et al., 2008; Qin et al., 2017)

9.4.2 Green Synthesis of Carbon Nanotubes

Carbon nanotubes are usually prepared using energy-intensive and environment-unfriendly methods such as laser vaporization, arc discharge, pyrolysis and chemical vapour deposition (CVD). However, in a recent study, Tripathi et al. (2017) utilized a plant-derived catalyst to synthesize MWCNTs. The notable results obtained were that the green catalyst synthesized CNT yield was much higher than that of metal catalyst grown CNTs. Moreover, the temperature used for the green catalyst grown CNTs was much less, i.e. 575°C, whilst that of metal catalyst grown CNTs range 700–1200°C (Tripathi et al., 2017).

9.5 ORGANIC CARBON-BASED NANOMATERIALS

9.5.1 Liposomes: Efficacy against *S. aureus* and Mechanism of Action

Liposomes are spherically shaped microscopic phospholipid vesicles that form spontaneously when phospholipids are added to an aqueous solution (Eloy et al., 2014). Due to the amphiphilic nature of the phospholipids, a hydrophilic phosphate head group is formed on one side and a hydrophobic acyl hydrocarbon chains on the other, resulting in the formation of polar shells. The unique structure of liposomes allows encapsulation of drugs with different lipophilicities, where hydrophobic molecules are entrapped into the lipid membrane, while hydrophilic molecules can be enclosed in the aqueous core (Akbarzadeh et al., 2013).

Liposomes have been used in the delivery of many antimicrobial agents, including those utilized against *S. aureus*. Pumerantz et al. (2011) showed that treatment of MRSA-infected macrophages with 200 and 800 g/ml concentrations of conventional liposomal vancomycin formulation resulted in a significant reduction in intracellular viability of MRSA, while treatment with PEGylated liposomal vancomycin had no effect on intracellular survival of MRSA after 24 hours.

Liposomes were also shown to improve the therapeutic efficacy of oleic acid (OA) in eliminating MRSA skin infections (Huang et al., 2011), where the minimal bactericidal concentration of OA was 0.8 μg/ml, which was about 12 times more effective than that of free OA, which had an MBC value of 10 μg/ml. Similarly, encapsulation with liposomes improved the efficacy of chloramphenicol against MRSA follicular targeting bacteria (Hsu et al., 2017). In that regard, an inhibition zone of about twofold was achieved by liposomes as compared to the free control (0.5 mg/ml chloramphenicol). These studies show that as drug delivery systems, liposomes can improve antimicrobial activity. Compared to synthetic-based liposomes, liposomes made of natural polymers offer more versatile and effective drug delivery systems because of their biodegradability *in vivo.*

9.5.2 Synthesis of Liposomes

Numerous techniques exist for laboratory-scale and large-scale preparation of liposomes, often involving lipid hydration, the use of mechanical procedures and subsequent replacement of organic solvents by an aqueous media (Alavi et al., 2017). However, there appears to be no literature on green methods available for the synthesis of liposomes, which should involve the replacement of organic solvents, elimination of high-energy processes and the use of near-neutral pH.

9.5.3 Dendrimers: Efficacy against *S. aureus* and Mechanism of Action

Dendrimers are among a few macromolecules that are used to improve the efficacy of drugs by increasing solubility, circulation times, permeability across barriers and metabolic rates in circulation (Lee et al., 2005). They possess a unimolecular micellar structure with a hydrophilic layer at the core and hydrophilic layer at the periphery, which can enhance the solubility of poorly soluble drugs by forming covalent and non-covalent complexes with hydrophobic groups (Malik et al., 2000).

Furthermore, the highly branched and globular dendritic structure can function as a multivalent biocide or a nanoscale platform with a large surface area to volume ratio that provides high reactivity and conjugation to a variety of a microorganism's receptors (Gillies and Frechet, 2005; Rastegar et al., 2017). Among dendrimers, poly(amidoamine) (PAMAM) dendrimers have been widely studied as drug delivery systems and as antimicrobial agents, where electrostatic interaction between cationic dendrimers and anionic bacteria cell surfaces results in the disruption of the lipid bilayer, cell lysis and death (Rastegar et al., 2017).

In this regard, a number of PAMAM dendrimers have been shown to possess antimicrobial activity against *S. aureus*. Amino-terminated Generation 6 PAMAM (PAMAM-G6) was shown to be effective against *S. aureus* at the concentration of 25 μg/ml, with little cytotoxicity to host cells (Rastegar et al., 2017). PAMAM-G7 also showed antimicrobial activity against *S. aureus* with an MIC at 4 μg/ml (Gholami et al., 2017). Antimicrobial activity against *S. aureus* has also been demonstrated in other types of dendrimers, including G1 poly(propyl ether imine) (PETIM), at MICs of 52.1, 41.7 and 20.8 mg/ml for three types of PETIM compounds (Suleman et al., 2015), poly(propylene imine) (PPI) dendrimers, with MIC and MBC values of between 5 μg/mL and 50 μg/ml, respectively (Ahmadi et al., 2015), peptide-based dendrimers, with MICs ranging from 0.93 t0 12.9 μM (Lind et al., 2015), organometallic dendrimers, with MICs ranging from 3.9 to 18.0 μM (Abd-El-Aziz et al., 2015), and lysine dendrimers, with MICs ranging from 64 to 144 μM (Janiszewska and Urbanczyk-Lipkowska, 2006).

9.5.4 Green Synthesis of Dendrimers

Dendrimers can be synthesized either through convergent routes, which involve prior construction of dendrimer segments followed by integration into a single structure, or through divergent routes, which involve sequential addition of generations of monomeric arms to a central core (Abbasi et al., 2014). While these methods result in monodisperse dendrimers, they are often tedious and time-consuming (Carlmark et al., 2009). For these reasons, there is a particular interest in green synthesis approaches for dendrimers, especially the use of the so-called click reactions, which are "modular, wide in scope, give very high yields, and generate only inoffensive by-products" (Kolb et al., 2001). It is important to design dendrimers that circulate in blood long enough to accumulate at target sites, but at the same time undergo elimination from the body at a reasonable rate to avoid bioaccumulation (Gillies and Frechet, 2005). Accumulation and subsequent cytotoxicity result primarily from dendrimer non-degradability under physiological conditions as well as poor elimination rates which depend on the nature of dendrimer generation size, molecular weights and surface charge (Ryan et al., 2013; Kaminskas et al, 2011).

9.6 SUMMARY AND CONCLUSION

Nanomaterials, including metal, metal oxide and carbon-based NMs (such as fullerenes, CNTs, liposomes and dendrimers) are among the most promising solutions to combat drug-resistant pathogens, including *S. aureus*. These NMs either possess microbiocidal activity in themselves or enhance the efficacy of conventional *S. aureus* antibiotics not reachable by standard drug formulations. However, the widespread use of NMs against drug-resistant *S. aureus* will require more careful assessment of their biocompatibility and their bioaccumulation in human cells following long-term administration. While green synthesis methods are widely available for metal and metal oxide NPS, much fewer studies have been conducted on the green synthesis of fullerenes, liposomes and dendrimers, and therefore there is need to develop methods for green synthesis of these NMs.

Wide-scale application of NMs in the treatment of *S. aureus* will result in large amounts of nanomaterial-based pharmaceutical and healthcare waste that will require special care. Nano waste from these facilities will have to be handled as hazardous waste requiring, at the minimum, double-bagging and enclosure in rigid impermeable containers. Depending on their physico-chemical characteristics, the nanowaste may be amenable to incineration, biocomposting or chemical digestion. In conclusion, NMs offer one of the most promising approaches in the war against drug-resistant *S. aureus*, firstly as antimicrobial agents and secondly as drug delivery systems for conventional antibiotics. Green synthesis is becoming a preferred method for the manufacturing of NPs of any size and shape. However, further research is needed to optimize yields for large-scale synthesis of NMs using green methods.

REFERENCES

Abbasi, Elham, Sedigheh Fekri Aval, Abolfazl Akbarzadeh, Morteza Milani, Hamid Tayefi Nasrabadi, Sang Woo Joo, Younes Hanifehpour, Kazem Nejati-Koshki, and Roghiyeh Pashaei-Asl. 2014. "Dendrimers: Synthesis, applications, and properties." *Nanoscale Research Letters* 9(1): 247. DOI: 10.1186/1556-276X-9-247.

Abd-El-Aziz, Alaa S, Christian Agatemor, Nola Etkin, David P Overy, Martin Lanteigne, Katherine McQuillan, and Russell G Kerr. 2015. "Antimicrobial organometallic dendrimers with tunable activity against multidrug-resistant bacteria." *Biomacromolecules* 16(11): 3694–3703. DOI: 10.1021/acs.biomac.5b01207.

Ahmadi Jebelli, Mohammad, Enayatollah Kalantar, Afshin Maleki, Hassan Izanloo, Fardin Gharibi, Hiua Daraei, Bagher Hayati, Ehsan Ghasemi, and Ali Azari. 2015. "Antimicrobial activities of the polypropylene imine dendrimer against bacteria isolated from rural water resources." *Jundishapur Journal of Natural Pharmaceutical Products* 10(3): 1–10. DOI: 10.17795/jjnpp-20621.

Akbarzadeh, Abolfazl, Rogaie Rezaei-Sadabady, Soodabeh Davaran, Sang Woo Joo, Nosratollah Zarghami, Younes Hanifehpour, Mohammad Samiei, Mohammad Kouhi, and Kazem Nejati-Koshki. 2013. "Liposome: Classification, preparation, and applications." *Nanoscale Research Letters* 8(1): 102. DOI: 10.1186/1556-276X-8-102.

Alavi, Mehran, Naser Karimi, and Mohsen Safaei. 2017. "Application of various types of liposomes in drug delivery systems." *Advanced Pharmaceutical Bulletin* 7(1): 3. DOI: 10.15171/apb.2017.002.

Al-Shaeri, Majed Ahmed. 2019. "Antimicrobial activity of single-walled carbon nanotubes (SWCNTs) against marine bacteria staphylococcus aureus." *Acta Scientific Pharmaceutical Sciences* 3(12): 43–46. DOI: 10.31080/ASPS.2019.03.0442.

Arias, L Renea, and Liju Yang. 2009. "Inactivation of bacterial pathogens by carbon nanotubes in suspensions." *Langmuir* 25(5): 3003–3012. DOI: 10.1021/la802769m.

Assali, Mohyeddin, Abdel Naser Zaid, Farah Abdallah, Motasem Almasri, and Rasha Khayyat. 2017. "Single-walled carbon nanotubes-ciprofloxacin nanoantibiotic: Strategy to improve ciprofloxacin antibacterial activity." *International Journal of Nanomedicine* 12: 6647–6659. DOI: 10.2147/IJN.S140625.

Ayala-Núñez, Nilda Vanesa, Humberto H Lara Villegas, Liliana del Carmen Ixtepan Turrent, and Cristina Rodríguez Padilla. 2009. "Silver nanoparticles toxicity and bactericidal effect against methicillin-resistant Staphylococcus aureus: Nanoscale does matter." *Nanobiotechnology* 5(1–4): 2–9. DOI: 10.1007/s12030-009-9029-1.

Azam Ameer, Ahmed S Arham, Oves Mohammed, Khan, Mohammad S Khan, Habib S Sami, and Memic Adnan, 2012. "Antimicrobial activity of metal oxide nanoparticles against Gram-positive and Gram-negative bacteria: a comparative study". *International journal of nanomedicine 7*: 6003–6009. Doi: 10.2147/IJN.S35347.

Beer, Christiane, Rasmus Foldbjerg, Yuya Hayashi, Duncan S. Sutherland, and Herman Autrup. 2012. "Toxicity of silver nanoparticles: Nanoparticle or silver ion?" *Toxicology Letters* 208(3): 286–292. DOI: 10.1016/j.toxlet.2011.11.002.

Bosi, Susanna, Tatiana Da Ros, Giampiero Spalluto, and Maurizio Prato. 2003. "Fullerene derivatives: An attractive tool for biological applications." *European Journal of Medicinal Chemistry* 38(11–12): 913–923. DOI: 10.1016/j.ejmech.2003.09.005.

Brandelli, Adriano, Ana Carolina Ritter, and Flávio Fonseca Veras. 2017. "Antimicrobial activities of metal nanoparticles." In *Metal Nanoparticles in Pharma*, 337–363. Cham: Springer. DOI: 10.1007/978-3-319-63790-7_15.

Carlmark, Anna, Craig Hawker, Anders Hult, and Michael Malkoch. 2009. "New methodologies in the construction of dendritic materials." *Chemical Society Reviews* 38 (2):352–362. Doi.org/10.1039/B711745K.

Eloy, Josimar Oliveira, Marina Claro de Souza, Raquel Petrilli, Juliana Palma Abriata Barcellos, Robert J Lee, and Juliana Maldonado Marchetti. 2014. "Liposomes as carriers of hydrophilic small molecule drugs: Strategies to enhance encapsulation and delivery." *Colloids and surfaces B: Biointerfaces* 123: 345–363. DOI: 10.1016/j.colsurfb.2014.09.029.

Erjaee, Hoda, Hamid Rajaian, and Saeed Nazifi. 2017. "Synthesis and characterization of novel silver nanoparticles using Chamaemelum nobile extract for antibacterial application." *Advances in Natural Sciences: Nanoscience and Nanotechnology* 8(2): 025004. DOI: 10.1088/2043-6254/aa690b.

Esteban-Cubillo Antonio, Carlos Pecharromán, Eduardo Aguilar, Julio Santarén, and José S Moya. 2006. "Antibacterial activity of copper monodispersed nanoparticles into sepiolite." *Journal of materials science* 41 (16):5208–5212. https://doi.org/10.1007/s10853-006-0432-x.

Ezhilarasi A Angel, Vijaya J Judith, Kaviyarasu K, Maaza, M, Ayeshamariam A, and Kennedy L John, 2016. "Green synthesis of NiO nanoparticles using Moringa oleifera extract and their biomedical applications: Cytotoxicity effect of nanoparticles against HT-29 cancer cells". *Journal of Photochemistry and Photobiology B: Biology 164*: 352–360. doi.org/10.1016/j.jphotobiol.2016.10.003.

Gardete Susana, and Alexander Tomasz. 2014. "Mechanisms of vancomycin resistance in Staphylococcus aureus." *The Journal of clinical investigation* 124 (7):2836–2840. https://doi.org/10.1172/JCI68834.

Geilich, Benjamin M, Ilia Gelfat, Srinivas Sridhar, Anne L van de Ven, and Thomas J Webster. 2017. "Superparamagnetic iron oxide-encapsulating polymersome nanocarriers for biofilm eradication." *Biomaterials* 119: 78–85. DOI: 10.1016/j.biomaterials.2016.12.011.

Gholami, Mitra, Rashin Mohammadi, Mohsen Arzanlou, Fakhraddin Akbari Dourbash, Ebrahim Kouhsari, Gharib Majidi, Seyed Mohsen Mohseni, and Shahram Nazari. 2017. "In vitro antibacterial activity of poly (amidoamine)-G7 dendrimer." *BMC Infectious Diseases* 17(1): 395. DOI: 10.1186/s12879-017-2513-7.

Gillies, Elizabeth R., and Jean M.J. Frechet. 2005. "Dendrimers and dendritic polymers in drug delivery." *Drug Discovery Today* 10(1): 35–43. DOI: 10.1016/S1359-6446(04)03276-3.

Hsu, Ching-Yun, Shih-Chun Yang, Calvin T Sung, Yi-Han Weng, and Jia-You Fang. 2017. "Anti-MRSA malleable liposomes carrying chloramphenicol for ameliorating hair follicle targeting." *International Journal of Nanomedicine* 12: 8227. DOI: 10.2147/IJN.S147226.

Huang, Chun-Ming, Chao-Hsuan Chen, Dissaya Pornpattananangkul, Li Zhang, Michael Chan, Ming-Fa Hsieh, and Liangfang Zhang. 2011. "Eradication of drug resistant Staphylococcus aureus by liposomal oleic acids." *Biomaterials* 32(1): 214–221. DOI: 10.1016/j.biomaterials.2010.08.076.

Hultén, Kristina, Kaplan Sheldon, Lamberth Linda. 2010. "Hospital-acquired *Staphylococcus aureus* Infections at Texas Children's Hospital, 2001–2007." *Infection Control and Hospital Epidemiology* 31(2): 183–190. DOI: 10.1086/649793.

Jadhav, Sunita, Suresh Gaikwad, Madhav Nimse, and Anjali Rajbhoj. 2011. "Copper oxide nanoparticles: Synthesis, characterization and their antibacterial activity." *Journal of Cluster Science* 22(2): 121–129. DOI: https://doi.org/10.1007/s10876-011-0349-7.

Janiszewska, Jolanta, and Zofia Urbanczyk-Lipkowska. 2006. "Synthesis, antimicrobial activity and structural studies of low molecular mass lysine dendrimers." *Acta Biochimica Polonica* 53(1): 77. http://www.actabp.pl/pdf/1_2006/77.pdf.

Jiang, Lai, Jia Lin, Clifford C Taggart, José A Bengoechea, and Christopher J Scott. 2018. "Nanodelivery strategies for the treatment of multidrug-resistant bacterial infections." *Journal of Interdisciplinary Nanomedicine* 3(3): 111–121. DOI: 10.1002/jin2.48.

Kadariya, Jhalka, Smith Tara, and Thapaliya Dipendra. 2014. *Staphylococcus aureus* Staphylococcal food borne disease: An ongoing challenge in public health." *Biomed Research International* 2014: 1-10. DOI: 10.1155%2F2014%2F827965.

Kaminskas, Lisa M, Ben J Boyd, and Christopher JH Porter. 2011. "Dendrimer pharmacokinetics: The effect of size, structure and surface characteristics on ADME properties." *Nanomedicine* 6(6): 1063–1084. DOI: https://doi.org/10.2217/nnm.11.67.

Kolb, Hartmuth C, MG Finn, and K Barry Sharpless. 2001. "Click chemistry: Diverse chemical function from a few good reactions." *Angewandte Chemie International Edition* 40(11): 2004–2021. DOI: 10.1002/1521-3773(20010601)40:11<2004::aid-anie2004>3.3.co;2-x.

Kumar Anish, and Menon Shobhana Karuveettil. 2009. Fullerene derivatized s-triazine analogues as antimicrobial agents. *European journal of medicinal chemistry 44*(5): 2178–2183. https://doi.org/10.1016/j.ejmech.2008.10.036.

Lai, Bo-Hung, and Dong-Hwang Chen. 2013. "Vancomycin-modified LaB6@ SiO_2/Fe_3O_4 composite nanoparticles for near-infrared photothermal ablation of bacteria." *Acta Biomaterialia* 9(7): 7573–7579. DOI: 10.1016/j.actbio.2013.03.023.

Lee, Cameron C, John A MacKay, Jean MJ Fréchet, and Francis C Szoka. 2005. "Designing dendrimers for biological applications." *Nature Biotechnology* 23(12): 1517. DOI: 10.1038/nbt1171.

Lengke, Maggy F, Michael E Fleet, and Gordon Southam. 2007. "Biosynthesis of silver nanoparticles by filamentous cyanobacteria from a silver (I) nitrate complex." *Langmuir* 23(5): 2694–2699. DOI: 10.1021/la0613124.

Lind, Tania K, Piotr Polcyn, Paulina Zielinska, Marité Cárdenas, and Zofia Urbanczyk-Lipkowska. 2015. "On the antimicrobial activity of various peptide-based dendrimers of similar architecture." *Molecules* 20(1): 738–753. DOI: 10.3390/molecules20010738.

Liu, Shaobin, Li Wei, Lin Hao, Ning Fang, Matthew Wook Chang, Rong Xu, Yanhui Yang, and Yuan Chen. 2009. "Sharper and faster 'nano darts' kill more bacteria: A study of antibacterial activity of individually dispersed pristine single-walled carbon nanotube." *ACS Nano* 3(12): 3891–3902. DOI: 10.1021/nn901252r.

Logeswari, Peter, Sivagnanam Silambarasan, and Jayanthi Abraham. 2015. "Synthesis of silver nanoparticles using plants extract and analysis of their antimicrobial property." *Journal of Saudi Chemical Society* 19(3): 311–317. DOI: 10.1016/j.jscs.2012.04.007.

Malik, Noeen, R Wiwattanapatapee, R Klopsch, K Lorenz, H Frey, JW Weener, EW Meijer, W Paulus, and R Duncan. 2000. "Dendrimers: Relationship between structure and biocompatibility in vitro, and preliminary studies on the biodistribution of 125I-labelled polyamidoamine dendrimers in vivo." *Journal of Controlled Release* 65(1–2): 133–148. DOI: 10.1016/s0168-3659(99)00246-1.

Maqbool, Qaisar, Mudassar Nazar, Sania Naz, Talib Hussain, Nyla Jabeen, Rizwan Kausar, Sadaf Anwaar, Fazal Abbas, and Tariq Jan. 2016. "Antimicrobial potential of green synthesized CeO_2 nanoparticles from Olea europaea leaf extract." *International Journal of Nanomedicine* 11: 5015. DOI: 10.2147/IJN.S113508.

Mittal, Amit Kumar, Yusuf Chisti, and Uttam Chand Banerjee. 2013. "Synthesis of metallic nanoparticles using plant extracts." *Biotechnology Advances* 31(2): 346–356. DOI: 10.1016/j.biotechadv.2013.01.003.

Mizuno, Kazue, Timur Zhiyentayev, Liyi Huang, Sarwat Khalil, Faria Nasim, George P Tegos, Hariprasad Gali, Ashlee Jahnke, Tim Wharton, and Michael R Hamblin. 2011. "Antimicrobial photodynamic therapy with functionalized fullerenes: Quantitative structure-activity relationships." *Journal of Nanomedicine & Nanotechnology* 2(2): 1. DOI: 10.4172/2157-7439.1000109.

Mroz, Pawel, George P Tegos, Hariprasad Gali, Tim Wharton, Tadeusz Sarna, and Michael R Hamblin. 2007. "Photodynamic therapy with fullerenes." *Photochemical & Photobiological Sciences* 6(11): 1139–1149. DOI: 10.1039/b711141j.

Mukherjee, Priyabrata, Absar Ahmad, Deendayal Mandal, Satyajyoti Senapati, Sudhakar R Sainkar, Mohammad I Khan, Renu Parishcha, PV Ajaykumar, Mansoor Alam, and Rajiv Kumar. 2001. "Fungus-mediated synthesis of silver nanoparticles and their immobilization in the mycelial matrix: A novel biological approach to nanoparticle synthesis. *Nano Letters* 1(10): 515–519. DOI: 10.1021/nl0155274.

Mukherjee, P, M Roy, BP Mandal, GK Dey, PK Mukherjee, J Ghatak, AK Tyagi, and SP Kale. 2008. "Green synthesis of highly stabilized nanocrystalline silver particles by a non-pathogenic and agriculturally important fungus T. asperellum." *Nanotechnology* 19(7): 075103. DOI: 10.1088/0957-4484/19/7/075103.

Naseem, Tayyaba, and Muhammad Akhyar, Farrukh. 2015. "Antibacterial activity of green synthesis of iron nanoparticles using Lawsonia inermis and Gardenia jasminoides leaves extract." *Journal of Chemistry* 2015: 1-8. DOI: https://doi.org/10.1155/2015/912342.

Oberdörster, G. 2000. "Determinants of the pathogenicity of man-made vitreous fibers (MMVF)." *International Archives of Occupational and Environmental Health* 73(1): S60–S68. DOI: https://doi.org/10.1007/PL00014628.

Pumerantz, Andrew, Krishna Muppidi, Sunil Agnihotri, Carlos Guerra, Vishwanath Venketaraman, Jeffrey Wang, and Guru Betageri. 2011. "Preparation of liposomal vancomycin and intracellular killing of meticillin-resistant Staphylococcus aureus (MRSA)." *International Journal of Antimicrobial Agents* 37(2): 140–144. DOI: 10.1016/j.ijantimicag.2010.10.011.

Qin Yue, Suning Li, Gan Zhao, Xuanhao Fu, Xueping Xie, Yiyi Huang, Xiaojing Cheng, Jinbin Wei, Huagang Liu, and Zefeng Lai. 2017. "Long-term intravenous administration of carboxylated single-walled carbon nanotubes induces persistent accumulation in the lungs and pulmonary fibrosis via the nuclear factor-kappa B pathway." *International journal of nanomedicine* 12: 263–277. Doi: 10.2147/IJN.S123839.

Rastegar, Ayoob, Shahram Nazari, Ahmad Allahabadi, Farahnaz Falanji, Fakhreddin Akbari Dourbash Akbari Dourbash, Zahra Rezai, Soudabeh Alizadeh Matboo, Reza Hekmat-Shoar, Seyed Mohsen Mohseni, and Gharib Majidi. 2017. "Antibacterial activity of amino-and amido-terminated poly (amidoamine)-G6 dendrimer on isolated bacteria from clinical specimens and standard strains." *Medical Journal of the Islamic Republic of Iran* 31: 64. DOI: 10.14196/mjiri.31.64.

Roy, Aashis S, Ameena Parveen, Anil R Koppalkar, and MVN Ambika Prasad. 2010. "Effect of nano-titanium dioxide with different antibiotics against methicillin-resistant Staphylococcus aureus." *Journal of Biomaterials and Nanobiotechnology* 1(1): 37. DOI: 10.4236/jbnb.2010.11005.

Ryan, Gemma M, Lisa M Kaminskas, Brian D Kelly, David J Owen, Michelle P McIntosh, and Christopher JH Porter. 2013. "Pulmonary administration of PEGylated polylysine dendrimers: Absorption from the lung versus retention within the lung is highly size-dependent." *Molecular Pharmaceutics* 10(8): 2986–2995. DOI: 10.1021/mp400091n.

Saifuddin, N, AZ Raziah, and AR Junizah. 2012. "Carbon nanotubes: A review on structure and their interaction with proteins.' *Journal of Chemistry* 2013: 1-19. DOI: 10.1155/2013/676815.

Schito, GC. 2006. "The importance of the development of antibiotic resistance in Staphylococcus aureus." *Clinical Microbiology and Infection* 12: 3–8. DOI: 10.1111/j.1469-0691.2006.01343.x.

Shahverdi, Ahmad R, Ali Fakhimi, Hamid R Shahverdi, and Sara Minaian. 2007. "Synthesis and effect of silver nanoparticles on the antibacterial activity of different antibiotics against Staphylococcus aureus and Escherichia coli." *Nanomedicine: Nanotechnology, Biology and Medicine* 3(2): 168–171. DOI: 10.1016/j.nano.2007.02.001.

Shankar, P Dheeban Sutha Shobana, Indira Karuppusamy, Arivalagan Pugazhendhi, Vijayan Sri Ramkumar, Sundaram Arvindnarayan, and Gopalakrishnan Kumar. 2016. "A review on the biosynthesis of metallic nanoparticles (gold and silver) using bio-components of microalgae: Formation mechanism and applications." *Enzyme and Microbial Technology* 95: 28–44. DOI: 10.1016/j.enzmictec.2016.10.015.

Sharmila, G., R. Sakthi Pradeep, K. Sandiya, S. Santhiya, C. Muthukumaran, J. Jeyanthi, N. Manoj Kumar, and M. Thirumarimurugan. 2018. "Biogenic synthesis of CuO nanoparticles using Bauhinia tomentosa leaves extract: Characterization and its antibacterial application." *Journal of Molecular Structure* 1165: 288–292. DOI: 10.1016/j.molstruc.2018.04.011.

Singh, Jagpreet, Tanushree Dutta, Ki-Hyun Kim, Mohit Rawat, Pallabi Samddar, and Pawan Kumar. 2018. "'Green' synthesis of metals and their oxide nanoparticles: Applications for environmental remediation." *Journal of Nanobiotechnology* 16(1): 84. DOI: 10.1186/s12951-018-0408-4.

Solmaz Maleki Dizaj, Afsaneh Mennati, Samira Jafari, Khadejeh Khezri, and Khosro Adibkia, 2015. Antimicrobial activity of carbon-based nanoparticles. *Advanced pharmaceutical bulletin* 5(1): 19–23. doi: 10.5681/apb.2015.003.

Stankic, Slavica, Sneha Suman, Francia Haque, and Jasmina Vidic. 2016. "Pure and multi metal oxide nanoparticles: Synthesis, antibacterial and cytotoxic properties." *Journal of Nanobiotechnology* 14(1): 73. DOI: 10.1186/s12951-016-0225-6.

Suleman, Nadia, Rahul S Kalhapure, Chunderika Mocktar, Sanjeev Rambharose, Moganavelli Singh, and Thirumala Govender. 2015. "Silver salts of carboxylic acid terminated generation 1 poly (propyl ether imine) (PETIM) dendron and dendrimers as antimicrobial agents against S. aureus and MRSA." *RSC Advances* 5(44): 34967–34978. DOI: 10.1039/C5RA03179F.

Tegos, George P., Tatiana N. Demidova, Dennisse Arcila-Lopez, Haeryeon Lee, Tim Wharton, Hariprasad Gali, and Michael R. Hamblin. 2005. "Cationic fullerenes are effective and selective antimicrobial photosensitizers." *Chemistry & Biology* 12(10): 1127–1135. DOI: 10.1016/j.chembiol.2005.08.014.

Teow, Sin-Yeang, Magdelyn Mei-Theng Wong, Hooi-Yeen Yap, Suat-Cheng Peh, and Kamyar Shameli. 2018. "Bactericidal properties of plants-derived metal and metal oxide nanoparticles (NPs)." *Molecules* 23(6): 1366. DOI: 10.3390/molecules23061366.

Thakkar, Kaushik N, Snehit S Mhatre, and Rasesh Y Parikh. 2010. "Biological synthesis of metallic nanoparticles." *Nanomedicine: Nanotechnology, Biology and Medicine* 6(2): 257–262. DOI: 10.1016/j.nano.2009.07.002.

Tran, Nhiem, Aparna Mir, Dhriti Mallik, Arvind Sinha, Suprabha Nayar, and Thomas J Webster. 2010. "Bactericidal effect of iron oxide nanoparticles on *Staphylococcus aureus*." *International Journal of Nanomedicine* 5: 277. DOI: 10.2147/ijn.s9220.

Tripathi, Nishant, Vladimir Pavelyev, and S. S. Islam. 2017. "Synthesis of carbon nanotubes using green plant extract as catalyst: Unconventional concept and its realization." *Applied Nanoscience* 7(8): 557–566. DOI: 10.1007/s13204-017-0598-3.

Utembe Wells, Kariska Potgieter, Aleksandr Byron Stefaniak, and Mary Gulumian. 2015. "Dissolution and biodurability: Important parameters needed for risk assessment of nanomaterials." *Particle and Fibre Toxicology* 12(1): 11. DOI: 10.1186/s12989-015-0088-2.

Vecitis, Chad D, Katherine R Zodrow, Seoktae Kang, and Menachem Elimelech. 2010. "Electronic-structure-dependent bacterial cytotoxicity of single-walled carbon nanotubes." *ACS Nano* 4(9): 5471–5479. DOI: https://doi.org/10.1021/nn101558x.

Vijayakumar, Sekar, Baskaralingam Vaseeharan, Balasubramanian Malaikozhundan, and Malaikkarasu Shobiya. 2016. "Laurus nobilis leaf extract mediated green synthesis of ZnO nanoparticles: Characterization and biomedical applications." *Biomedicine & Pharmacotherapy* 84: 1213–1222. DOI: 10.1016/j.biopha.2016.10.038.

Wang Jiafan, Wages Mike, Yu Shuangying, Maul D Jonathan, Mayer Greg, Hope-Weeks Louisa, and Cobb P George. 2014. Bioaccumulation of fullerene (C60) and corresponding catalase elevation in *Lumbriculus variegatus. Environmental toxicology and chemistry 33*(5): 1135–1141. Doi.org/10.1002/etc.2540.

WHO (World Health Organization). 2014. *Antimicrobial Resistance: Global Report on Surveillance*: World Health Organization. https://www.who.int/drugresistance/documents/surveillancereport/en/.

Yang Seng-Tao, Wang, Xiang, Jia, Guang Gu, Yiqun Wang, Tiancheng Nie, Haiyu Ge, Cuicui Wang, Haifang, and Liu Yuanfrang. 2008. "Long-term accumulation and low toxicity of single-walled carbon nanotubes in intravenously exposed mice". *Toxicology letters 181*(3): 182–189. Doi.org/10.1016/j.toxlet.2008.07.020.

Zardini, Hadi Zare, Ahmad Amiri, Mehdi Shanbedi, Morteza Maghrebi, and Majid Baniadam. 2012. "Enhanced antibacterial activity of amino acids-functionalized multi walled carbon nanotubes by a simple method." *Colloids and Surfaces B: Biointerfaces* 92: 196–202. DOI: https://doi.org/10.1016/j.colsurfb.2011.11.045.

Zheng, Kaiyuan, Magdiel I Setyawati, David Tai Leong, and Jianping Xie. 2017. "Antimicrobial gold nanoclusters." *ACS Nano* 11(7): 6904–6910. DOI: 10.1021/acsnano.7b02035.

10 Green Metal-Based Nanoparticles Synthesized Using Medicinal Plants and Plant Phytochemicals against Multidrug-Resistant *Staphylococcus aureus*

Abeer Ahmed Qaed Ahmed, Lin Xiao,
Tracey Jill Morton McKay and Guang Yang

CONTENTS

10.1 Introduction ... 182
10.2 *S. aureus* and Emerging Multiple Resistance to Drugs ... 185
10.3 *S. aureus* and Resistance to Antibiotics ... 186
10.3.1 Penicillin-Resistant *S. aureus* (PRSA) ... 187
10.3.2 Methicillin-Resistant *S. aureus* (MRSA) ... 188
10.3.3 Vancomycin-Resistant *S. aureus* and Vancomycin Intermediate Resistant *S. aureus* ... 188
10.4 Potential of Medicinal Plants as Antimicrobial Therapeutics ... 189
10.5 Typical Approaches for Green Synthesis of Metal-Based NPs Using Plants ... 191
10.6 Green Metal-Based Nanoparticles (G-NPs) Using Medicinal Plants against Multidrug-Resistant *S. aureus* ... 192
10.6.1 Green Silver Nanoparticles (GAgNPs) Synthesized Using Medicinal Plants ... 193
10.6.2 Green Gold Nanoparticles (GAuNPs) Synthesized Using Medicinal Plants ... 209

10.6.3 Other Green Metal-Based Nanoparticles Synthesized Using Medicinal Plants ... 212
10.7 Mechanism of Antimicrobial Activity of Green Metallic Nanoparticles ... 220
10.8 Potential Medicinal Plants for Green Synthesis of NPs against Multidrug Resistance *S. aureus* ... 220
10.9 Plant Phytochemicals Responsible for NPs Formation and Stabilization ... 224
10.10 Conclusion ... 227
Acknowledgements ... 227
References ... 227

10.1 INTRODUCTION

In recent years, much research attention has been paid to nanotechnology. Nanotechnology involves the creation of nanosized particles that hold specific functions using biological, chemical, mechanical and physical methods. The synthesis of noble metal nanoparticles (NPs) such as gold, silver, palladium and platinum, as well as inorganic oxides such as titanium and zinc oxide has attracted increasing interest in fields such as biomedical science, environmental science, food science, physiochemical science and agricultural science (Singh et al., 2016; Salouti and Derakhshan, 2019; Karthik et al., 2020).

Nanoparticles are increasingly used in a variety of applications such as therapeutics for acute and chronic diseases, diagnostics, biosensors, cancer detection and treatment, drug delivery, cosmetics, antimicrobial agents, textiles, medical devices and instruments, electronics, catalysis, food, food packaging industry, food security, food quality enhancement, water treatment and paints (Karthik et al., 2020). This is due to the unique properties of NPs, such as their magnetic, electronic, chemical, mechanical and optical properties. In addition, NPs have desirable properties such as their availability in a variety of shapes, sizes and ratio of surface area to volume. Therefore, the preparation of a specific NP with specific characteristics and desirable activity is of great value. Numerous methods have been adopted to synthesize NPs. Although chemical, mechanical and physical methods are currently mainstream methods for NPs preparation, these methods are highly toxic and negatively impact on human and environmental health (Kratošová et al., 2019). This is due to the use of hazardous materials during the production and synthesis of NPs. For example, hazardous chemicals are used as capping and reducing agents for NPs synthesis, leaving toxic residues on the NPs surface (Fahimirad et al., 2019). Moreover, these methods are expensive to upscale due to high energy inputs but low material conversion rates. Hence, many of these methods are unpopular.

Biological methods to synthesize NPs (Fig. 10.1) have advantages over chemical, mechanical and physical ones, and natural sources such as plants, bacteria, fungi and algae are usually used for the synthesis of metal-based NPs (Fig. 10.2).

Biowastes, bioproducts and biomolecules are also used for the synthesis of NPs acting as reducing and stabilizing agents (Kratošová et al., 2019). Using plants and microorganisms allow for an eco-friendly method for the synthesis of metal-based NPs, as they do not require the use of hazardous chemicals. Thus, they do not burden

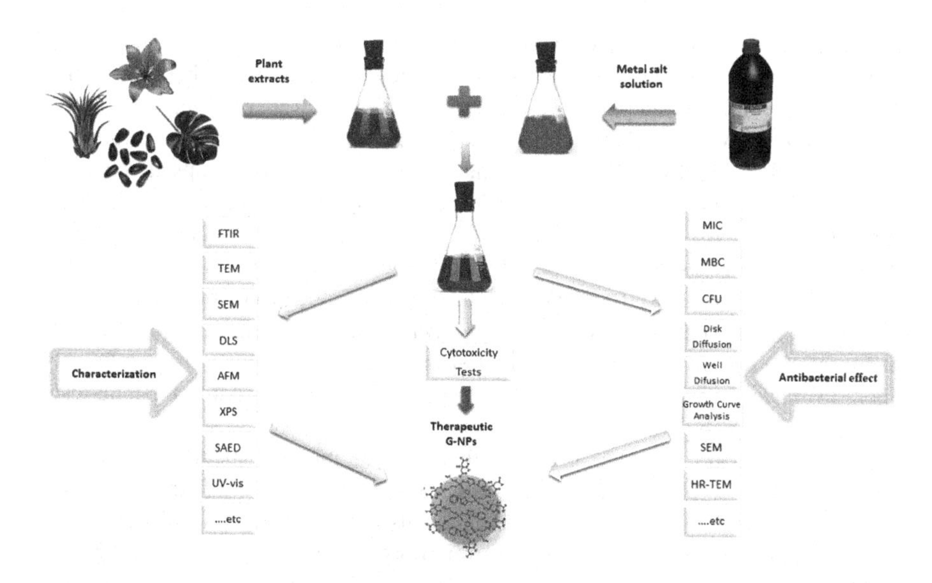

FIGURE 10.1 Typical various approaches for the synthesis and evaluation of green metal-based NPs.

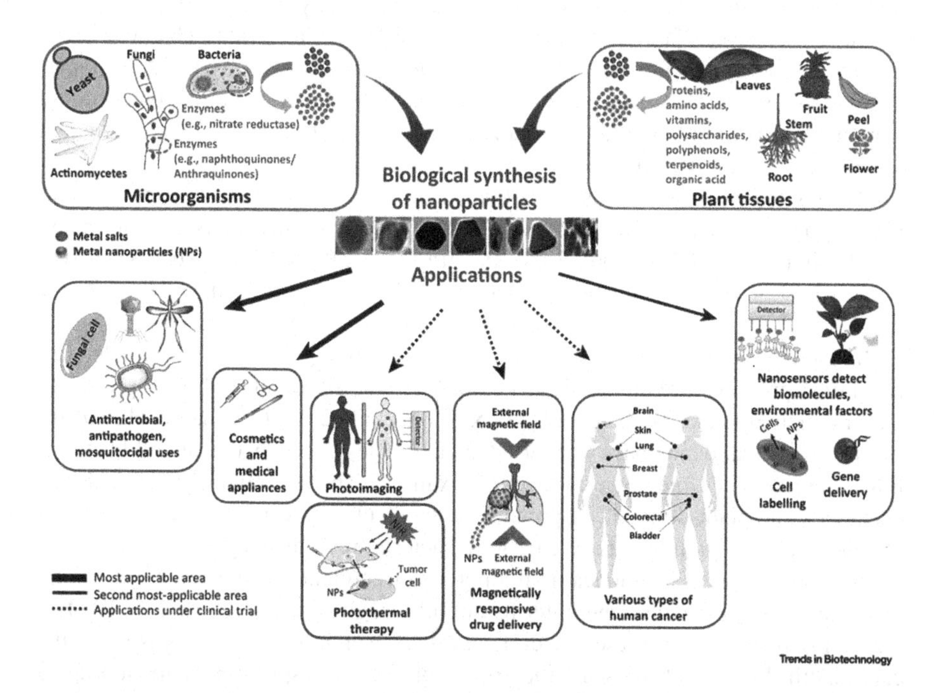

FIGURE 10.2 Biological (green) synthesis and the applications of metal nanoparticles in biomedical and environmental fields. (Reproduced with permission from Singh et al., 2016.)

the biophysical environment with toxic waste and effluent, which is the case for chemical, mechanical and physical methods. They are also cheap, because unlike chemical, mechanical and physical methods, they have low energy inputs.

It is well known that plant extracts and microbial cultures contain metabolites that have different biological activities. These biologically active groups attach to the synthesized NPs surfaces, creating an additional capping layer (Zhang et al., 2015; Yugandhar et al., 2017; Sinsinwar et al., 2018). This is likely to enhance the efficacy of biologically synthesized metal-based NPs. However, not all biological sources are a good natural source for NPs synthesis. This is because the production process must ensure conditions such as (1) safety; (2) the availability and continuity of the natural source; (3) cost-effectiveness; (4) least effort and energy inputs; (5) be relatively quick; (6) have the least number of steps/procedures; (7) be suitable for long-term production and (8) biocompatibility.

Of all biological resources, plants are the most suitable biological source to synthesize metal-based NPs (Fig. 10.2) in part because a great variety of large quantities of plants are widely available in nature. Plants also do not require special culturing or maintenance techniques compared to microorganisms. Fungi, algae and bacteria (depending on each strain and species) need distinct maintenance procedures and specific culturing techniques. In addition, highly aseptic conditions are required to avoid contamination with undesirable compounds. Culture conditions such as temperature, pH, humidity and gases must be rigorously maintained, according to the precise needs of each microorganism. This is usually both energy and human resource intensive. Plant-based metal NPs have many favourable characteristics, making them popular in terms of medical applicability, scalability and biocompatibility (Salouti and Derakhshan, 2019). Non-toxic edible and medicinal plants could play major part in this regard, especially as they are cost-effective, energy-efficient and not human resource intensive.

Various plants, whole or as parts (fruits, peels, roots, flower, stem bark, latex, leaves, gum, stems, rhizomes and seeds) can be used to synthesize NPs (Salouti and Derakhshan, 2019; Iqbal et al., 2020; Pereira et al., 2020). Their extracts are rich in a variety of phytochemicals that act as stabilizing and reducing agents. These two effects are mainly due to secondary metabolites (polysaccharides, flavonoids, heterocyclic compounds, alkaloids, terpenoids and polyphenols) and biomolecules (vitamins, proteins, organic acids and amino acids). These have already proven to have medicinal values (Zhang et al., 2015; Singh et al., 2016; Chauhan et al., 2019).

For millennia, plants have been used as medicines to combat viral and bacterial infections, diabetes and some cancers for example. The potential of plants' phytochemicals and plant-derived compounds as therapeutic material is well documented (Anand et al., 2019). In an era of emerging bacterial resistance against a wide range of antibiotics, it is vital to focus on finding new treatments. *Staphylococcus aureus* is one of the most troublesome in terms of multidrug-resistant bacteria (MDR). The increasing resistance of these bacteria against conventional antibiotics is a significant health threat, and the scientific community has a responsibility to step up to the challenge. Thus, this chapter focuses on the potential role of metal-based NPs biosynthesized using medicinal plants to combat MDR *S. aureus*.

10.2 *S. AUREUS* AND EMERGING MULTIPLE RESISTANCE TO DRUGS

MDR bacteria are resistant to more than one antimicrobial agent, usually to the most commonly available antimicrobial agents, although there are some exceptions. This includes microorganisms resistant to at least one antimicrobial agent in three or more categories of antimicrobial agents. The increase of MDR bacteria and its associated infections is a global threat and needs to be urgently addressed by the scientific community. The MDR bacteria in question include: methicillin-resistant *S. aureus* (MRSA), vancomycin-resistant *S. aureus* (VRSA) and vancomycin intermediate resistant *S. aureus* (VISA), multidrug-resistant *Streptococcus pneumoniae* (MDRSP), carbapenem-resistant *Enterobacteriaceae* (CRE), multidrug-resistant *Acinetobacter baumannii*, vancomycin-resistant *Enterococci* (VRE) and extended spectrum beta-lactamases producing gram-negative bacilli (ESBLs) (Okwu et al. 2019). MDR bacteria are of great concern as they are of a family of bacteria that have been associated with death and morbidity for centuries (Nii-Trebi, 2017). MDR bacteria are the cause of numerous infectious diseases, presenting a global challenge in terms of negatively affecting not just public health but also economic stability. Furthermore, new and emerging or unrecognized infectious diseases will undoubtedly come to the fore, as well demonstrated by the emergence of COVID-19.

Despite the use of many different antibiotics to treat bacterial infections, bacteria are exhibiting increasing resistance to them. This is of global concern as there may come a time when no antibacterial agent is effective against specific bacterial infections. For instance, *S. aureus* is an aerobic gram-positive bacterium, which lives commensal as normal flora in skin, pharynx, perineum and anterior nares, could also lead to bacteremia, infections of skin and soft tissue, hospital-acquired infections, osteomyelitis, joint and bone infections and endocarditis (Sakr et al., 2018; Turner et al., 2019; Troeman et al., 2019). Although *S. aureus* is commensal, a large number of *S. aureus* infections could emerge from the patient's own flora, with a link between *S. aureus* residing normally in common carriages sites leading to *S. aureus* infections (Von Eiff et al., 2001; Wertheim et al., 2005; de Jong et al., 2019). *S. aureus* can evolve into different strains such as MRSA, VISA and VRSA. This occurs due to antibiotic resistance associated with having genes that encode antibiotic-inactivating enzymes, obtaining pre-existing resistance factors though genetic exchange pathways between bacterial cells and producing low-affinity antibiotic-binding proteins that reduces the binding affinities of antibiotics and stepwise mutations in the genes, among others. These mechanisms are described in Sections 3.1–3.3.

These MRD bacteria cause numerous infectious diseases and are a global threat. For example, Methicillin-resistant *S. aureus* causes serious MRSA infections. In addition, other strains of community-acquired methicillin-resistant *S. aureus* (CA-MRSA) are involved in serious infections of hospital-onset methicillin-resistant *S. aureus.* MRSA is becoming increasingly resistant to first- and second-line antibiotics, making MRSA infections hard to treat. A major challenge is the unique ability of *S. aureus* to rapidly develop resistance to antibiotics by developing different resistance mechanisms, including the first discovered penicillin to daptomycin and linezolid (Kaur and Chate, 2015). In the case of MRSA, the resistance developed when methicillin-sensitive *S. aureus* (MSSA) strains obtained *Staphylococcal* Cassette Chromosome *mec* (SCCmec). This carries the *mec*A gene that encodes the low-affinity penicillin-binding protein (PBP2' or PBP2a), causing resistance to all of the β-lactam antibiotics, including methicillin (McGuinness et al., 2017; Okwu et al., 2019). The treatment of MRSA infections now uses vancomycin as a last resort, making it the drug of choice to treat severe MRSA infections. It is a glycopeptide antibiotic, inhibiting the biosynthesis of bacterial cell wall (McGuinness et al., 2017).

A major concern is that several *S. aureus* strains show resistance to vancomycin with many reports of clinical isolates exhibiting a decreased susceptibility to vancomycin. *S. aureus* strains exhibiting complete resistance to vancomycin were first reported in the United States in 2002. These isolates are known as VISA. In VISA, the molecular basis of resistance is polygenic along with the involvement of stepwise mutations in the genes that encode the molecules predominantly involved in the biosynthesis of the cell envelop. The resistance of VRSA also lies with the *vanA* gene and operon that exists on a plasmid. VISA burdens are comparatively high, and the resistance of molecular mechanisms less well known. VISA is associated with poor clinical outcomes, such as the failure of vancomycin treatments and persistent infections (McGuinness et al., 2017). Although VRSA strains now make treating VRSA infections challenging, the total number of human VRSA infections is fortunately limited. Currently, isolated VRSA show susceptibility to some antibiotics accompanied by different treatment regimens (Cong et al., 2020). In this regard, linezolid and daptomycin are now the most commonly used antibiotics to treat VRSA infections. Concerningly, several other bacterial strains exhibit resistance to daptomycin and linezolid (Kaur and Chate, 2015). Thus, as *S. aureus* strains can rapidly acquire antibiotic resistance, it is highly likely that VRSA will eventually develop resistance to these antibiotics as well. To deal with such an inevitability, extensive pharmaceutical research is required to develop novel therapeutics that can be deployed as soon as possible.

10.3 *S. AUREUS* AND RESISTANCE TO ANTIBIOTICS

The discovery of penicillin by the Scottish physician-scientist Sir Alexander Fleming saved millions of lives and opened up a huge research field aimed at finding antibiotics to combat many infectious diseases (Fig. 10.3) (Tan and Tatsumura, 2015).

Many infectious diseases, including *S. aureus*, became curable due to antibiotics. However, the treatment of *S. aureus* infections is complex as *S. aureus* can acquire resistance to antibiotics due to mobile genetic elements (MGEs), the primary way that genetic information can be exchanged by horizontal gene transfer between bacterial cells. MGEs facilitate horizontal genetic exchange, enabling *S. aureus* to become antibiotic resistant (Partridge et al., 2018). Through this, bacterial cells overcome antibiotics by obtaining pre-existing resistance factors from the bacterial gene pool of bacteria. Mobile genetic elements can move between or within DNA molecules via transposons, gene cassettes/integrons and insertion sequences. Some, such as integrative conjugative elements and plasmids, can be transferred between bacterial cells. *S. aureus* has MGEs that explain their ability to develop resistance to antibiotics over the years (Fig. 10.3).

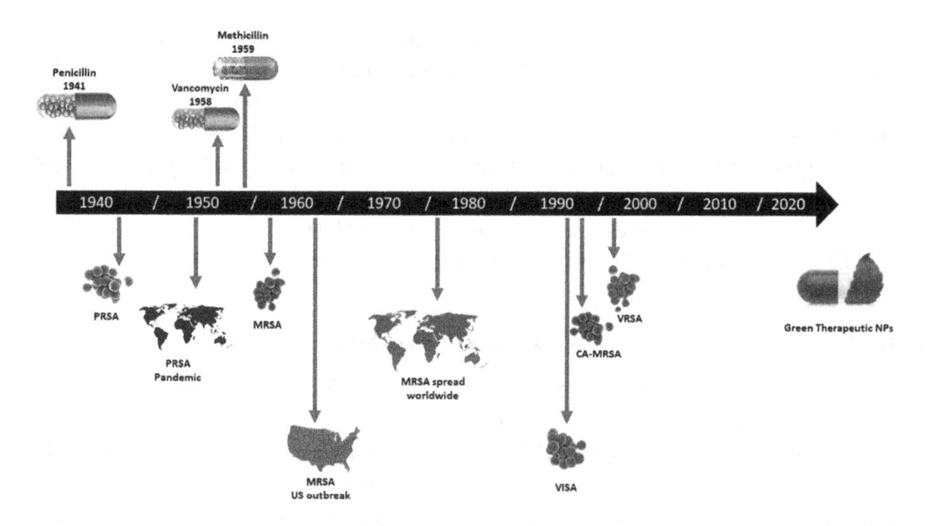

FIGURE 10.3 Historical review of MDR *S. aureus* strains and their emerging multiple resistance to drugs.

10.3.1 Penicillin-Resistant *S. aureus* (PRSA)

Penicillin was discovered in 1928 and the first penicillin-resistant bacteria appeared in 1947, only one year after the introduction of penicillin into clinics, and only two years after penicillin became widely available. Patients began presenting with four *Staphylococci* strains exhibiting resistance to penicillin (Rammelkamp and Maxon, 1942; Fleming, 1980). In the early 1940s, penicillin-resistant *S. aureus* (PRSA) was reported (Fig. 10.3), followed by widespread PRSA infections in both community and hospital settings during the 1950s and early 1960s (Rammelkamp and Maxon, 1942; Plough, 1945, Rountree and Freeman, 1955).

During the 1950s, large infections of PRSA, known as phage type 80/81 (pandemic *S. aureus* phagetype 80/81 strain) appeared and caused serious illnesses worldwide (Fig. 10.3). *S. aureus* with a phage-lysis pattern (phage type 80) was first isolated from neonatal infections in Australia in 1953. The isolates were resistant to benzylpenicillin and are considered responsible for 81% of all major neonatal infection outbreaks in Australia between the years 1954 and 1957. In 1954, Canada identified similar isolates (phage type 81) (Rountree and Beard, 1958). The Australian and Canadian isolates were thought to belong to the same clone as they were lysed via both phages named phage type 80/81. This type became pandemic in the 1950s, causing skin lesions, pneumonia and sepsis (Fig. 10.3). During the pandemic, many physicians noted these isolates were highly virulent and remarkably transmissible. They suspected these isolates had a unique virulence factor. Later, it was suggested that the ability to produce leucotoxin (resembling the Panton Valentine leukocidin) was the cause of the increased virulence of the 80/81 strain (Donahue and Baldwin, 1966).

Subsequently, the resistance of PRSA to penicillin was primarily attributed to the *blaZ* gene that encodes β-lactamase. The β-lactam ring of penicillin is hydrolyzed by β-lactamase, which inactivates the penicillin (Olsen et al., 2006). The low similarity between plasmid- and chromosomal-encoded *blaZ* is an indication of the vertical (clonal) spread of the *blaZ* gene in *Staphylococcus*, and that there is an exchange of *blaZ* gene between strains. In the late 1950s, a semi-synthetic beta-lactam antibiotic known as methicillin (and its derivatives) was introduced to treat PRSA infections (Jevons, 1961). It was the leading cure of PRSA infections throughout the 1960s (Fig. 10.3).

10.3.2 Methicillin-Resistant *S. aureus* (MRSA)

Methicillin was developed in 1959 by the pharmaceutical industry as a response to penicillin resistance and reached the market in 1960. Despite the efficacy of methicillin to treat PRSA infections, MRSA was reported in the United Kingdom only one year later, in 1961 (Fig. 10.3). Thus, *S. aureus* speedily develops antibiotic resistance (Jevons, 1961; Enright et al., 2002). Worldwide, the challenges of MRSA have increased, accounting for a huge proportion of hospital-associated *S. aureus* infections, with significant morbidity and mortality (Klevens et al., 2006; Jarvis et al., 2007; Klein et al., 2007, Dantes et al., 2013). Furthermore, MRSA is no longer confined to hospitals. Community or community-associated MRSA (CA-MRSA) can spread quickly among healthy individuals (Chambers and Deleo 2009). Outbreaks of CA-MRSA infections have been reported worldwide and are widespread in the United States (Chambers and Deleo 2009). CA-MRSA is now a global problem (Herold et al., 1998; Centers for Disease Control Prevention, 1999; Chambers and Deleo, 2009).

MRSA resistance is rooted in the production of a novel low-affinity penicillin-binding protein (PBP2a or PBP2'), which reduces the binding affinities of β-lactam antibiotics (Hartman and Tomasz, 1984; Reynolds and Brown, 1985; Utsui and Yokota, 1985). The *mecA* gene has been reported responsible for encoding the PBP2a (Jevons, 1961; Katayama et al., 2000). MRSA can continue with cell wall synthesis regardless of the presence of high concentrations of β-lactam antibiotics (Matthews and Tomasz, 1990). The *mecA* gene that encodes PBP2a is located on MRSA chromosome. The sequence of *mecA* gene was determined in Japan in 1987, when it was cloned from a MRSA strain (*Staphylococcus aureus*, TK784) (Song et al., 1987). The *mecA* gene was found to be distributed among coagulase-negative *Staphylococcus strains* as well as *S. aureus* (Hürlimann-Dalel et al., 1992; Suzuki et al., 1992). Therefore, the methicillin resistance determinant (*mec*) was hypothesized as to being able to transfer freely among *Staphylococcal* species. This *mecA* might have entered *S. aureus* only once or twice (Barry et al., 1993).

10.3.3 Vancomycin-Resistant *S. aureus* and Vancomycin Intermediate Resistant *S. aureus*

In 1958, vancomycin was approved for use in humans and by the late 1980s, it became the treatment of choice for MRSA infections (Sorrell et al., 1982; Levine, 2006; D'Agata et al., 2009). At the same time, resistance to vancomycin was found in

Enterococci, raising serious concerns about the effectiveness of vancomycin to treat MRSA (Murray, 2000). Soon after, *S. aureus* strains with reduced susceptibility to teicoplanin (a structural vancomycin relative) was reported on the European continent (Kaatz et al., 1990, Manquat et al., 1992). In 2002, the first isolate of VRSA was discovered in the United States (Centers for Disease Control and Prevention, 2002; Chang et al., 2003). Since then, a total of 14 VRSA isolates were found in the United States (Walters et al., 2015). Resistance to vancomycin in *S. aureus* is attributed to the *vanA* operon that encodes on the transposon Tn1546, a primary part of VRE conjugative plasmid (Arthur et al., 1993). During discrete conjugation process, *S. aureus* can obtain *Enterococcal* plasmids. Therefore, *S. aureus* resistance to vancomycin could be acquired by conserving the original *Enterococcal* plasmid or by the transposition process of Tn1546 obtained from VRE plasmid into a *Staphylococcal* native plasmid (Perichon and Courvalin, 2009; Zhu et al., 2010). In 1997, Japan reported the first VISA clinical isolate exhibiting reduced susceptibility against vancomycin (Fig. 10.3) (Hiramatsu et al., 1997). Some retrospective reports indicate the possibility of *S. aureus* resistant vancomycin present in the United States in 1987 (Jackson and Hicks, 1987). VISA is mostly associated with persistent infections, hospitalization and prolonged treatment with vancomycin or even treatment failure.

Some of the fundamental phenotypical characteristics of VISA include increased thickness of the cell wall as a result of differentially regulated stimulatory pathways and cell wall biosynthesis (Daum et al., 1992; Moreira et al., 1997; Hanaki et al., 1998; Boyle-Vavra et al., 2001). This reduces peptidoglycan cross-linking, affecting cell wall turnover enzymes by decreasing their autolytic activity (Vaudaux et al., 2001; Boyle-Vavra et al., 2003; Howden et al., 2006,). There is also a change in the protein profile of the surface, compromising the *agr* function and changing growth characteristics (Pfeltz et al., 2000; Sakoulas et al., 2002; McCallum et al., 2006). Several methods have been used to explore the VISA phenotype molecular genetic basis. Multiple genes, as well as mutations thereof, appear to be responsible for vancomycin intermediate phenotype (Pfeltz et al., 2000; Boyle-Vavra et al., 2001; Muthaiyan et al., 2004; Scherl et al., 2006; Mwangi et al., 2007). Multiple studies show that VISA could have originated from a multistep process. Most probably numerous pathways are involved with intermediatory vancomycin resistance (Howden et al., 2010). A complete reconstitution of VISA from vancomycin-susceptible *S. aureus* (strain N315ΔIP) concluded that all six mutated genes are needed to develop the VISA phenotype (Katayama et al., 2016). These genes are involved in maintaining cell physiology. Therefore, VISA phenotype is achieved via multiple genetic incidents leading to extreme alteration in cell physiology.

10.4 POTENTIAL OF MEDICINAL PLANTS AS ANTIMICROBIAL THERAPEUTICS

While there are many antibiotics, bacteria constantly evolve, building resistance to many. As plants have been used in the fight against diseases for centuries, the scientific community is showing increasing interest in using plant extracts for their antimicrobial potential. Reports confirm the antimicrobial activity of plants' extracts, offering potential alternatives to conventional antibiotics, specifically for the treatment of MDR infections (Kali, 2015; Subramani et al., 2017). The emergence of MDR pathogenic bacteria is a major public health threat, especially for the developing world. This is made worse by few effective MDR antibacterial drugs being available. In some cases, there are no available antimicrobial agents to treat pathogenic MDR bacterial infections (Boucher et al., 2009; Giamarellou, 2010). Thus, there is an urgent need to find alternative antibacterial agents, although the rapid emergence of resistance to new antibiotics makes for grim reading. As a result, even new families of antibacterial drugs display only a short life expectancy (Coates et al., 2002; Marasini et al., 2015). It is estimated that by the year 2050, the death rate associated with antimicrobial resistance will increase to 10 million per annum (de Kraker et al., 2016; O'Neill, 2014). The potential catastrophe of a post-antibiotic era should not be underestimated, and efforts need to be urgently put into securing effective and alternative solutions.

The spread of MDR bacteria is, in part, related to the misuse of antibiotics. The World Health Organization (WHO) recognizes the problem and has called for new approaches such as a shift to medicinal plants (World Health Organization, 2002). Medicinal plants have long contributed positively to human health and welfare. They are valuable natural sources of new drugs and phytomedicine (Mahady, 2005). In addition, plants represent an abundant source of potentially novel pharmaceutical substances (Okwu et al., 2019). Many medicinal plants are important resources of natural antibacterial compounds, and some show promising treatment potential against pathogenic bacteria (Iwu et al., 1999). Plants have medicinal and antimicrobial characteristics, attributed to their phytochemicals that synthesize during secondary metabolism (Medina et al., 2005; Romero et al., 2005). Plants are abundant with secondary metabolites that include flavonoids, tannins, phenolic compounds and alkaloids. All recognized as having *in vitro* antimicrobial properties (Duraipandiyan et al., 2006, Djeussi et al., 2013). Several phytotherapy manuals report diverse medicinal plants to treat infectious diseases such as cutaneous infections, urinary tract infections, respiratory diseases and gastrointestinal disorders (Manandhar et al., 2019). Plant-derived antimicrobials are a rich source of medicinal phytochemicals and considered safe to use due to their natural origin compared to synthetic compounds. Many plant-derived components (extracts, hydrosols and essential oils) are used as skin care products, flavouring agents, and anti-infectious products (Degirmenci and Erkurt, 2020).

With regard to finding antimicrobial agents, effort has been put into determining the antimicrobial activity of medicinal plant extracts, particularly the compounds of such plant extracts that could be separated to detect its microbial activity alone. These compounds include, but are not limited to, flavonoids, alkaloids, tannins and phenolic compounds. Thus, plants and their phytochemicals have great potential for use in various industries such as the pharmaceutical industry. Using medicinal plants in therapeutic practices is considered safer than synthetic drugs. In addition, medicinal plants are used and highly accepted by people not only for therapeutic purposes, but also for cosmetics, health products and food flavouring. Medicinal plants have a long and geographical widespread acceptance. Moreover, plant-based materials are relatively low in cost (Ke et al., 2012).

The plant kingdom has variety of species rich in phytochemicals displaying antimicrobial activity towards MDR bacteria such as MRSA, VISA and VRSA. For example, plants belonging to families of Asteraceae, Apocynaceae, Lamiaceae, Polygonaceae, Solanaceae, Urticaceae, Piperaceae, Ranunculaceae, Poaceae, Myrtaceae, Rosaceae, Lauraceae, Combretaceae, Theaceae, Rutaceae, Fagaceae, Fabaceae and Loranthaceae were reported to have the ability to inhibit MDR *S. aureus* (Askarinia et al., 2019).

Plants in Asteraceae family were reported to have phytochemicals such as sesquiterpene lactones, saponins, flavonoids, alkaloids and glycosides that showed antibacterial activity (Constabel et al., 1988; Da Costa et al., 2005). *Xanthium strumarium* (Asteraceae family) extract exhibited remarkable antibacterial effects on MSSA and MRSA, attributed to the presence of flavonoids, phenolic acids, tannins and terpenoids (Rad et al., 2013). The antibacterial activity of *Saussurea lappa* root (*Asteraceae* family) ethanolic extracts against MDR bacteria, including MRSA, were evaluated. The results exhibited bacteriostatic effects against MDR bacteria at lower concentrations and bactericidal effects at higher concentrations (Hasson et al., 2013).

Extracts from a variety of plants such as *Tabernaemontana alternifolia* (Roxb) (Apocynaceae family), *Salvia officinalis* (Sage) (Lamiaceae family), *Rumex nervosus* (Polygonaceae family), *Withania somnifera* (Solanaceae family), *Hydrastis canadensis* L. (Ranunculaceae family), lemongrass (Poaceae family), *Rosa canina* L. (Rosaceae family), *Persea americana* (Lauraceae family) and *Camellia sinensis* (Theaceae family) have been reported to have antibacterial activities against MDR bacteria (MRSA, VISA and VRSA) (Askarinia et al., 2019). Hence, screening of bioactive molecules in medicinal plants could lead to the discovery of new therapeutic active agents such as antibacterial pharmaceuticals.

10.5 TYPICAL APPROACHES FOR GREEN SYNTHESIS OF METAL-BASED NPS USING PLANTS

The preparation of plant extract is the first step in green synthesis of NPs (Fig. 10.1). The whole plant or parts of the plants should be collected from the designated area and thoroughly washed (several times) with tap water to remove all visible dirt, soil and dead tissue. Plant materials must then be washed with sterile distilled water. Further shade or sunlight drying for up to 20 days is required (Abbasi et al., 2020). Alternatively, several hours of oven drying at high temperatures are also required (Suresh et al., 2018). The dry plant materials should be powdered using a blender. Store plant powder in a dry airtight container until needed. Plant extract is prepared as required. For instance, if aqueous extract is needed, dried plant powder (around 1–10 g) can be boiled with a suitable volume of deionized distilled water (20–100 ml) at 70°C for 10 minutes. The obtained infusion is then filtered with Whatman filter paper until all insoluble materials are removed. The filtrate can be stored at 4°C until required (Ahmad et al., 2016; Iqbal et al., 2020). Other plant extracts such as ethyl acetate, chloroform, methanol and hexane extracts are prepared according to the literature (Manubolu et al., 2013; Elango and Roopan, 2015).

Next, metal NPs are synthesized by mixing the aqueous metal salt solution (e.g. $HAuCl_4$, $AgNO_3$ or $CuSO_4$), which was prepared previously in Erlenmeyer flask, with plant extract. The reaction mixture is further heated in an incubator, hot plate or water bath with continuous stirring. The amount of metal salt and plant extract should be calculated as needed. The reaction mixture should further incubate to reduce the metal salts. Colour must be observed and evaluated as it is the first qualitative confirmation of NP formation. For instance, when the reaction mixture appears brownish red, faint yellow, green, dark brown or pale yellow appears, this is initial confirmation of the formation of AuNPs, AgNPs, CuNPs, PdNPs or ZnO-NPs (Khan et al., 2016a; Salouti and Derakhshan, 2019a).

Thereafter centrifugation of the fully reduced mixture is necessary. The nanopellets are collected after discarding the supernatant. Obtained nanopellets must be resuspended in deionized water. For washing any excessive substances absorbed on the NPs, centrifugation should be repeated at least two to three times. The resultant precipitate should be repeatedly washed with distilled water, followed by ethanol washing to eradicate any impurities (Salouti and Derakhshan, 2019). Nanopellets could be washed with other solvents such as distilled water and acetone (Alagesan and Venugopal, 2018). The precipitate must be dried at 60°C in an oven for 12 hours to obtain the metal NPs powder.

Finally, synthesized NPs need to be characterized according to their shape, size, dispersity and surface area, using several qualitative and quantitative techniques such as UV-vis spectroscopy, TEM, EDS, XRD and FTIR (Fig. 10.1). The factors affecting the production of NPs, their properties and quantities must be evaluated. These factors include plant metabolites (concentration and type), culturing factors (e.g. pH and reaction temperature) and metal salt concentration, among others (Salouti and Derakhshan, 2019).

10.6 GREEN METAL-BASED NANOPARTICLES (G-NPS) USING MEDICINAL PLANTS AGAINST MULTIDRUG-RESISTANT *S. AUREUS*

Three important parameters must be considered when adopting metallic NPs biosynthesis using plant extracts: a stabilizing or capping agent, a reducing agent and a metal salt to control the size of NPs and prevent their aggregation (Abou El-Nour et al., 2010). In the case of the biosynthesis of G-NPs, plant phytochemicals are both stabilizing and reducing agents (Agarwal et al., 2017). The extract of the whole plant or plant parts such as leaves, seeds and flowers are mixed with a metal salt solution such as $HAuCl_4$, $AgNO_3$ and $ZnNO_3$ to obtain the green NPs. Plants are the most natural source for the synthesis of NPs on account of the large-scale production achievable, specifically in terms of the production of stable NPs that vary in size and shape. Plant parts produce phytochemicals with desirable characteristics, which can supplement the desirable properties of the synthesized NPs. Plant extracts are natural, cheap and eco-friendly, and do not require the use of intermediate base groups. Many plants have been extensively investigated for their role in NPs biosynthesis, and the potential of G-NPs in variety of applications has also been explored. The potential role of G-NPs in combating MDR such as *S. aureus* is an important area of study. MRSA is one of the bacteria tested for susceptibility using medicinal plants (Mahady, 2005). Therefore, it is important to explore the role of NPs biosynthesized using medicinal plants as potential antibacterial agents to treat MRSA infections (Tables 10.1–10.3).

10.6.1 Green Silver Nanoparticles (GAgNPs) Synthesized Using Medicinal Plants

Ansari and Alzohairy (2018) reported the green synthesis of GAgNPs using seed extract of *Phoenix dactylifera* ranging in size from 14 to 30 nm. The green-synthesized GAgNPs were characterized using UV-vis spectroscopy, DLS, HR-TEM and SEM (Table 10.1). The bactericidal potential of the GAgNPs against MRSA was tested by agar diffusion methods, MIC/MBC and electron microscopy. TEM images of the GAgNPs revealed that they were spherical in shape. The MBC and MIC of GAgNPs for MRSA were 17.33 ± 1.89 and 10.67 ± 0.94 µg/ml, respectively. The MRSA antibacterial activities increase with concentration levels of GAgNPs. The inhibition zone was highest when the concentrations of GAgNPs (500 µg/ml) was greater (24 mm). The smallest zone of inhibition (11 mm) was at the lowest concentrations (7.8 µg/ml). The SEM images of the GAgNPs-treated MRSA cells showed wrinkled, damaged cell walls (Fig. 10.4), an indication of disorganization and disruption of the cell membrane. HR-TEM analysis showed massive injuries and entire disintegration of the cell membrane and cell wall (Fig. 10.5). In the cytoplasm were large translucent zones, attributed to either complete or localized separation of the membrane from the cell wall. This study suggests green-synthesized GAgNPs could be an effective prevention and treatment option for the infections associated with MRSA as well as other MDR pathogens.

The green synthesis of GAgNPs was previously done using *Cleome viscosa* plant extract which were further characterized (Table 10.1) (Lakshmanan et al., 2018). The presence of biosynthesized GAgNPs (410–430 nm) was determined by UV-vis spectroscopy. FTIR spectrum tested the existence of various functional groups that act as capping agents for the GAgNPs. The GAgNPs morphology was analysed by SEM and the existence of silver was determined using elemental analysis. TEM revealed that the GAgNPs were well-dispersed and predominantly spherical in shape. Some GAgNPs exhibited irregular shapes between 5 and 30 nm. The antibacterial activity of GAgNPs was tested using the well diffusion method on Mueller–Hinton agar plates against *S. aureus, Bacillus subtilis, Klebsiella pneumoniae* and *Escherichia coli*. GAgNPs showed good antibacterial activity towards both gram-positive and gram-negative bacteria. The maximum inhibition zone for *S. aureus* was 17 ± 0.8 mm. The antibacterial activity of GAgNPs is concentration dependent. High levels of GAgNPs performed well in terms of inhibition activity against bacterial growth. Additionally, biosynthesized GAgNPs using *Cleome viscosa* plant extract proved effective against the growth of cancer cells *in vitro*, a possible indicator of an anticancer effect.

TABLE 10.1
Green Synthesis of GAgNPs from Different Medicinal Plants Extracts against *S. aureus* Strains

Plant species	Part used	Shape	Characterization	Size (nm)	References
Phoenix dactylifera	Seed	Spherical	UV-vis, SEM, HR-TEM, DLS	14–30	Ansari and Alzohairy (2018)
Drosera indica,	Fresh plant tissue	Quasi-spherical	TEM, SEM, DLS, FTIR, XPS	5–10	Banasiuk et al. (2020)
Cleome viscosa L.	Fruit	Spherical/irregular	UV-vis, FTIR, XRD, FESEM-EDAX, TEM	20–50	Lakshmanan et al. (2018)
Thymus vulgaris L.	Leaves	Spherical	SEM and AFM, XRD, FTIR, DLS	75	Manukumar et al. (2020)
Phyllanthus pinnatus	Stem	Cubical	UV-vis, SEM, XRD, FTIR	<100	Balachandar et al. (2019)
Coleus forskohlii	Root	Spherical	UV-vis, XRD, FTIR, TEM	30–40/35–55	Naraginti and Sivakumar (2014)
Hibiscus cannabinus	Stem	Spherical	XRD, TEM and FTIR, EDX, SPR	10	Bindhu et al. (2014)
Solanum lycopersicums	Fruit	Spherical	UV-vis, TEM, EDS, XRD, FTIR	14–33	Bindhu and Umadevi (2014b)
Rivea hypocrateriformis	Aerial	Spherical	UV-vis, XRD, FTIR, FESEM/TEM, TGA, EDAX	10–50	Godipurge et al. (2016)
Drosera binate	Fresh plant tissue	Quasi-spherical	TEM, SEM, DLS, FTIR, XPS	5–10	Banasiuk et al. (2020)
Gloriosa superba	Leaves	Triangular/spherical	FTIR, XRD, EDX AFM, TEM	20–50	Gopinath et al. (2016)
Mammea suriga	Root bark	Square/spherical	UV-vis, SEM, EDX, XRD, FTIR	50	Poojary et al. (2016)
Gnidia glauca	Flower/leaf/stem	Monodisperse	HRTM, EDS, DLS, XRD	10–100	Shinde et al. (2018)
Anthemis atropatana	Aerial	Spherical	UV-vis, TEM, SEM, XRD, FTIR	10–8	Dehghanizade et al. (2018)

(*Continued*)

TABLE 10.1 (CONTINUED)
Green Synthesis of GAgNPs from Different Medicinal Plants Extracts against *S. aureus* Strains

Plant species	Part used	Shape	Characterization	Size (nm)	References
Rosmarinus officinalis	Leaf	Spherical	UV-vis, FTIR, SEM, TEM, XRD	10–33	Ghaedi et al. (2015)
Cocos nucifera	Shell	Spherical	UV-vis, TEM, FTIR, XRD	14.2–22.96	Sinsinwar et al. (2018)
Rumex acetosa	Leaf	Spherical	UV-vis, FTIR, XRD, SEM-EDX, TEM	5–80	Kota et al. (2017)
Eulophia herbacea (Lindl.)	Tuber	Face-centred cubic geometry	UV-vis, FTIR, SEM, EDX, XRD	11.70 ± 2.43	Pawar and Patil (2020)
Drosera spatulata	Fresh plant tissue	Quasi-spherical	TEM, SEM, DLS, FTIR, XPS	5–10	Banasiuk et al. (2020)
Alysicarpus monilifer	Leaf	Spherical	UV-vis, TEM, EDXR, SAED, XRD, FTIR	15 ± 2	Kasithevar et al. (2017)
Syzygium aromaticum L.	Essential oil	Diverse/ spherical	UV-vis DLS, TEM	31–72	Maciel et al. (2019)
Handroanthus heptaphyllus (Vell.) Mattos	Diverse	Diverse	UV-vis, DLS, zeta potential, AFM, XRD, Raman spectroscopy	15/35	Pereira et al. (2020)
Murraya koenigii	Leaves	Spheroidal	UV-vis, FTIR, XRD, SEM, TEM	5–20	Qais et al. (2019)
Dionaea muscipula	Fresh plant tissue	Quasi-spherical	TEM, SEM, DLS, FTIR, XPS	5–10	Banasiuk et al. (2020)
Panax ginseng	Leaves	Spherical	UV-vis, FE-TEM, EDX, XRD, DLS	5–15	Singh et al. (2016)
Pistacia atlantica	Seed	Spherical	UV-vis, XRD, FTIR, TEM, SEM, EDAX	10–50	Sadeghi, Rostami, et al. (2015)
Olax nana	Leaves	Divers	XRD, FTIR, SEM, TEM, DLS, EDX, SAED	26	Ovais et al. (2018)

(Continued)

TABLE 10.1 (CONTINUED)
Green Synthesis of GAgNPs from Different Medicinal Plants Extracts against *S. aureus* Strains

Plant species	Part used	Shape	Characterization	Size (nm)	References
Red ginseng	Root	Spherical	FE-TEM, EDX, elemental mapping	10–30	Singh et al. (2016)
Alpinia katsumadai	Seed	Quasi-spherical	UV-vis, FETEM, EDX, SAED, XRD, DLS, FTIR	12.6	He et al. (2017)
Ilex paraguariensis	Leaves	Diverse	TEM, AFM, DLS, zeta potential, UV-vis, FTIR	4–30/24–60	Silveira et al. (2018)
Lippia nodiflora	Aerial	Spherical	UV-vis, FTIR and XRD	30–60	Sudha et al. (2017)
Bergenia ciliate	Whole	Spherical	UV-vis, FTIR, FESEM	35	Phull et al. (2016)
Crocus sativus L.	Wast-ages	Spherical	UV-vis, HR-TEM, XRD, FTIR	12–20	Bagherzade et al. (2017)

Methicillin-resistant *S. aureus* (MRSA) has a complicated defence mechanism such that it easily avoids innate immune systems. Manukumar et al. (2020) biosynthesized thyme-loaded silver NPs (GTAgNPs) to ascertain the exact mechanism of action and the toxicity of green silver NPs to bacterial cells. GTAgNPs were characterized and the toxicity to MRSA (MRSA 090 strain) was evaluated with the agar well diffusion method (Ghanwate et al., 2016; Manukumar and Umesha, 2017). The biosynthesized GTAgNPs showed a dominant particle size of 75 nm and exhibited antimicrobial activity at 1 mg/ml, validated using membrane destabilization, alterations of surface through AFM, SEM and bioelectrochemistry. The GTAgNPs exhibited anticancer activity against MCF-7 and A549 cell lines with insignificant toxicity to PBMC. GTAgNPs exhibited excellent blood compatibility, significantly delaying MRSA coagulation, and confirmed the downregulation of virulence genes such as *Coa* and *SpA*. This study suggests that designing drugs from active molecules of thyme (e.g. GTAgNPs) can contribute significantly to MRSA treatment.

Godipurge et al. (2016) reported the synthesis of spherical-shaped 10–50 nm AgNPs (average 30 nm) in size from an aqueous solution of metal precursor ($AgNO_3$) and extract of the aerial parts of *Rivea hypocrateriformis* (that act as both stabilizing and reducing agents) under microwave irradiation. The formation of AgNPs was evident from the surface plasmon resonance peak (observed at 450 nm). XRD pattern showed that *fcc* structure, meanwhile the FTIR spectra indicated the presence of the phytochemicals adsorbed on AgNPs. The biofunctionalized GAgNPs were

TABLE 10.2
Green Synthesis of GAuNPs from Different Medicinal Plants Extracts against *S. aureus* Strains

Plant species	Part used	Shape	Characterization	Size (nm)	References
Cressa cretica	Leaf	Divers	UV-vis, FTIR, SEM, XRD, EDX	15–22	Balasubramanian et al. (2019)
Acer pentapomicum	Leaves	Spherical	UV-vis, SEM, EDX, XRD, FTIR	19–24	Khan et al. (2018)
Alternanthera bettzickiana	Leaf	Spherical	UV-vis, XRD, FTIR, SEM, EDX, zeta potential, TEM	80–120	Nagalingam et al. (2018)
Areca catechu	Nuts	Spherical	UV-VIS, TEM, XRD and FTIR	13.7	Rajan et al. (2017)
Olax nana	Leaves	Diverse	XRD, FTIR, SEM, TEM, DLS, EDX, SAED	47	Ovais et al. (2018)
Panax ginseng	Leaves	Spherical	UV-vis, FE-TEM, EDX, XRD, DLS	10–20	Singh et al. (2016)
Acorus calamus	Rhizome	Spherical	SPR, UV–Vis, XRD, FTIR	<100	Ganesan and Gurumallesh Prabu (2019)
Gloriosa superba	Leaves	Triangular/ spherical	FTIR, XRD and EDX AFM, TEM	20–50	Gopinath et al. (2016)
Mammea suriga	Root	Square/ spherical	UV-vis, SEM, EDX, XRD, FTIR	22	Poojary et al. (2016)
Aloysia triphylla	Leaves	Spherical	UV-vis, FTIR, XRD, TEM	40–60	Lopez-Miranda et al. (2019)
Ananas comosus	Fruit	Tetrahedral	UV-vis, TEM, XRD	16	Bindhu and Umadevi (2014a)
Coleus forskohlii	Root	Triangular/ spherical	UV-vis, XRD, FTIR, TEM	25–40/15–35	Naraginti and Sivakumar (2014)
Zizyphus mauritiana	Leaves	Spherical	UV-vis, SEM, XRD, FTIR	20–40	Sadeghi (2015)
Rivea hypocrateriformis	Aerial	Spherical	UV-vis, XRD, FTIR, FESEM, TEM, TGA, EDAX	10–50	Godipurge et al. (2016)

(*Continued*)

TABLE 10.2 (CONTINUED)
Green Synthesis of GAuNPs from Different Medicinal Plants Extracts against *S. aureus* Strains

Plant species	Part used	Shape	Characterization	Size (nm)	References
Peganum harmala L	Leaf/ seed	Cubic/ spherical	UV-vis, FT-IRS, FESEM, XRD, EDX, TEM	43.44/52.04	Moustafa and Alomari (2019)
Diospyros Kaki L.	Fruit	Diverse	UV-vis, TEM, DLS	12.0 ± 5.0	Huo et al. (2019)
Solanum lycopersicums	Fruit	Spherical/ nanotri-angles	UV-vis, TEM, EDS, XRD, FTIR	5–19	Bindhu and Umadevi (2014b)
Salicornia brachiate	Areal	Spherical	XRD, TEM, FFT, FESEM	22–35	Ayaz Ahmed et al. (2014)
Solanum nigrum	Leaves	Spherical	UV-vis, TEM, XRD FTIR, zeta potential, DLS	50	Muthuvel et al. (2014)
Punica granatum	Fruit	Spherical/ triangular	HRTEM, XRD, TGA, FTIR	5–17	Lokina et al. (2014)
Red ginseng	Root	Spherical	FE-TEM, EDX, elemental mapping	10–30	Singh, Kim, Wang, et al. (2016)

characterized via their weight loss due to the thermal degradation of the phytochemicals noted in TG analysis, while EDAX analysis approved the predicted elemental composition. The antimicrobial activity of synthesized GAgNPs was tested against different human pathogens, including *S. aureus* at various concentrations (25–100 μg/ml) using agar well diffusion. Results show that phytochemicals functionalized GAgNPs presented a good antibacterial effect against *S. aureus* (mean zone of inhibitions 06, 08, 09, 11 mm for concentrations of 25, 50, 75 and 100 μg/ml, respectively). The synthesis of GAgNPs using *R. hypocrateriformis* extract demonstrated that it could compete with commercially available antibacterial agents. Moreover, these synthesized GAgNPs showed improved antimicrobial, anticancer and antioxidant activities, attributed to the combination effect of phytochemicals and Ag metals. Thus, Godipurge et al. (2016) concluded that biosynthesized GAgNPs using *R. hypocrateriformis* extract has the potential to be an effective growth inhibitor in a variety of biomedical applications.

Gloriosa superba leaf extract was used to green synthesize AgNPs (Gopinath et al., 2016). AFM and TEM analyses showed that GAgNPs had spherical and triangular morphologies, with a mean size of 20 nm. GAgNPs showed high antibacterial and antibiofilm activities towards gram-negative and gram-positive bacteria, including *S. aureus*. The obtained GAgNPs showed significant inhibition of biofilm growth

TABLE 10.3
Synthesis of Different Green NPs from Variety of Medicinal Plants Extracts against *S. aureus* Strains

Plant species	Part used	Type of NPs	Shape	Characterization	Size (nm)	References
Cassia siamea	Leaves	ZnO-NPs	Ellipsoidal/spherical	UV-vis, FTIR, SEM	<100	Chauhan et al. (2019)
Lawsonia inermis,	Leaves	I-NPs	Hexagonal	TGA, FTIR, TEM, SEM, AFM, XRD	21	Naseem and Farrukh (2015)
Gardenia jasminoides	Leaves	I-NPs	Shattered rock-like	TGA, FTIR, TEM, SEM, AFM, XRD	32	Naseem and Farrukh (2015)
Artemisia haussknechtii	Leaf	CuNPs	Spheres	UV-vis, FTIR, XRD, AFM	35.36 ± 44.4	Alavi and Karimi (2018)
Eucalyptus globulus	Leaf	NiO-NPs	Pleomorphism	UV-vis, FTIR, XRD, EDX, EM	10–20	Saleem et al. (2017)
Ocimum americanum	Whole	ZnO-NPs	Spherical	XRD, SEM, DLS, FTIR	21	Kumar et al. (2019)
Artemisia haussknechtii	Leaf	TiO_2-NPs	Spheres	UV-vis, FTIR, XRD, AFM	92.58 ± 56.98	Alavi and Karimi (2018)
Millettia pinnata	Flower	CuNPs	Spherical	UV-vis, XRD, FT-IR, SEM, TEM, SAED	23 ± 1.10	Thiruvengadam et al. (2019)
Cocos nucifera L	Whole	PbNPs	Spherical	UV-vis, XRD, TEM, EDAX	47	Elango and Roopan (2015)
Syzygium alternifolium	Stem bark	CuNPs	Spherical	UV-vis, FTIR, DLS, zeta potential, XRD, TEM	5–13	Yugandhar et al. (2017)
Cochlospermum religiosum (L.)	Leaf	ZnO-NPs	Hexagonal wurtzite	XRD, DLS, SEM, FTIR	<100	Mahendra et al. (2017)
Aloe vera	Leaf	CuNPs	Spherical	SEM-EDS, XRD, TEM, FTIR, UV-vis	20–30	Kumar et al. (2015)

(*Continued*)

TABLE 10.3 (CONTINUED)
Synthesis of Different Green NPs from Variety of Medicinal Plants Extracts against *S. aureus* Strains

Plant species	Part used	Type of NPs	Shape	Characterization	Size (nm)	References
Musa paradaisica	Peels	TiNPs	–	UV-vis, AFM, XRD	65–115	Hameed et al. (2019)
Galeopsidis herba	Whole	CuNPs	Spherical	UV-vis, FTIR, SEM, SEM-EDS, TEM	10 ± 5	Dobrucka (2017)
Azadirachta indica	Leaf	TiO_2-NPs	Spherical	FTIR, XRD, TEM, SEM	15–50/25–87	Thakur et al. (2019)
Stevia rebaudiana	Leaf	NiO-NPs	Spherical	UV-vis, XRD, FE-SEM, FTIR, SEM, TEM	20–50	Srihasam et al. (2020)
Sageretia thea (Osbeck.)	Leaves	NiO-NPs	Spherical	XRD, ATR-FTIR, EDS, SAED, HRSEM/TEM, Raman spectroscopy	~18	Khalil et al. (2018)
Phyllanthus emblica	Seeds	PdNPs	Spherical	UV-vis, FT-IR, XRD, Zeta potential, SEM, TEM	28 ± 2	Dinesh et al. (2017)
Neem (Azadirachta indica)	Leaf.	NiO-NPs	Oblong	XRD, TEM, EDAX	12	Helan et al. (2016)
Rhamnella gilgitica	Leaves	IO-NPs	Spherical	SEM, TEM, XRD, DLS, FT-IR, EDX, Raman spectroscopy	~21	Iqbal et al. (2020)
Terminalia bellirica	Fruit	I-NPs	Spherical	UV-vis, FTIR, SEM, TEM, XRD	21.32	Jegadeesan et al. (2019)
Rhamnus triquetra	Leaves	IO-NPs	Spherical	SEM, TEM, XRD, DLS, FT-IR, EDX, Raman spectroscopy	~ 21	Abbasi et al. (2020)

(Continued)

TABLE 10.3 (CONTINUED)
Synthesis of Different Green NPs from Variety of Medicinal Plants Extracts against *S. aureus* Strains

Plant species	Part used	Type of NPs	Shape	Characterization	Size (nm)	References
Papaver somniferum L.	Pods	PbONPs	Irregular	XRD, FTIR, EDX, SEM	23 ± 11	Muhammad et al. (2019)
Pisonia alba	Leaf	MgO-NPs	Spherical	UV-vis, TEM, EDX, XRD, FTIR	<100	Sharmila et al. (2019)
Moringa oleifera	Fruit	I-NPs	Irregular	UV-vis, FTIR, SEM, TEM, XRD	45	Jegadeesan et al. (2019)
Withania somnifera	Leaves	SeNPs	Spherical	FTIR, XRD, EDX, XRD, FE-SEM, TEM, UV-vis	45–90	Alagesan and Venugopal (2018)
Santalum album	Leaf	PdNPs	Spherical	UV-vis, TEM, XRD, FTIR	10–40	Sharmila, Haries, et al. (2017b)
Moringa oleifera	Leaves	I-NPs	Irregular	UV-vis, FTIR, SEM, TEM, XRD	45	Jegadeesan et al. (2019)
Papaver somniferum L.	Pods	I-NPs	Elliptical/spherical	XRD, FTIR, EDX, SEM	38 ± 13	Muhammad et al. (2019)
Filicium decipiens	Leaf	PdNPs	spherical	UV-vis, TEM, XRD, FTIR	2–22	Sharmila, Farzana Fathima, et al. (2017a)
Pisidium guvajava Aloe vera	Leaf	MgO-NPs	cubic	UV-vis, FTIR, XRD, FESEM, EDAX, XPS	50–90	Umaralikhan and Jamal Mohamed Jaffar (2016)
Costus pictus D. Don	Leave	MgO-NPs	Hexagonal	FT-IR, XRD, SEM, TEM	50	Suresh et al. (2018)

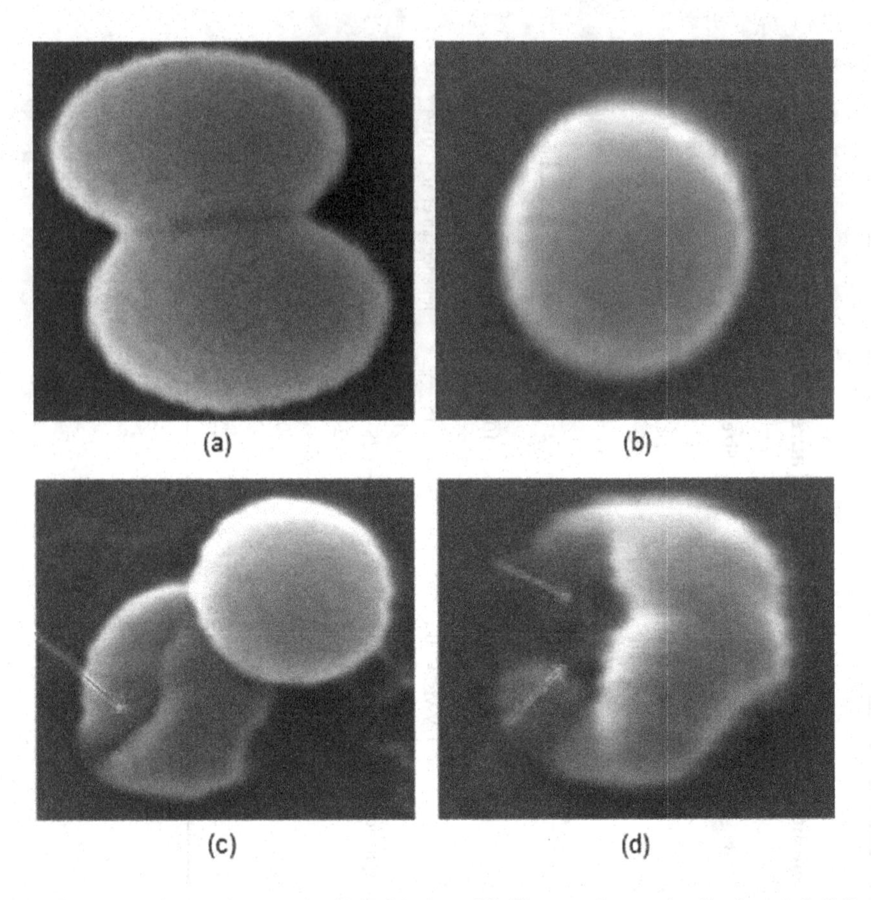

FIGURE 10.4 SEM micrograph of MRSA. (a and b) Untreated control cells. (c and d) Cells treated with 25 and 50 μg/ml of AgNPs; red arrows illustrating structural deformities and irregular cell surface. (Reproduced with permission from Ansari, M.A., and Alzohairy, M.A. 2018. "One-pot facile green synthesis of silver nanoparticles using seed extract of *Phoenix dactylifera* and their bactericidal potential against MRSA". *Evidence-Based Complementary and Alternative Medicine* 2018:1–9. doi: https://doi.org/10.1155/2018/1860280.)

thickness at 15 μm for *S. aureus*. It is well known that a mechanistic interaction is present between AgNPs and the bacterial cell membrane, causing cell death due to the electrostatic attraction of the AgNPs. The bacterial cell membrane experiences damage, leading to the formation of pits on its surface, creating structural changes and affecting cell respiration (Sondi and Salopek-Sondi, 2004). In addition, AgNPs react with thiol groups present in proteins. This results in the inhibition of protein synthesis in the bacterial cell as well as DNA replication (Feng et al., 2000). Similarly, oxygen links with Ag and interacts with the sulfhydryl groups (—S—H) located on the cell wall, resulting in the hydrogen ions being removed in the water form. This induces the sulfur atoms to create R—S—S—R bonds leading to blocking cell respiration (Kumar et al., 2004). AgNPs attack cell division and the respiratory

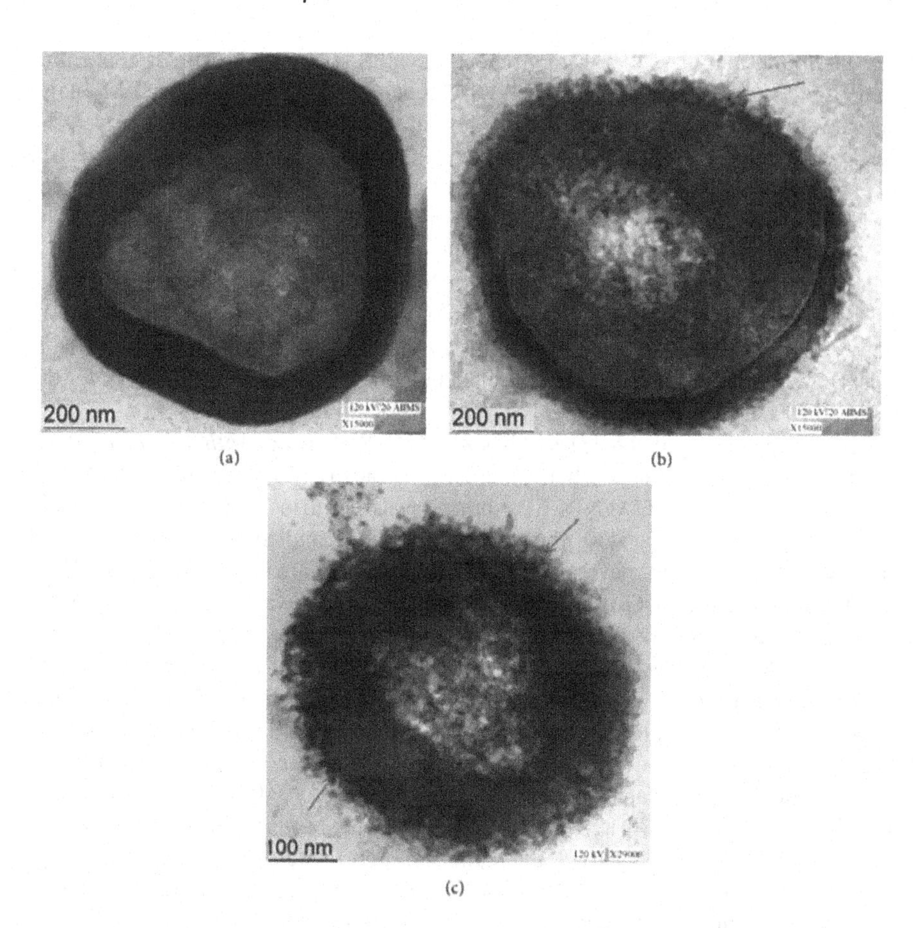

FIGURE 10.5 HR-TEM micrograph of MRSA. (a) Untreated control cell. (b and c) Treated with 25 and 50 μg/ml AgNPs. Red arrows indicate the attachment and penetration of NPs and degradation and destruction of the outermost layers of cell wall and cytoplasmic membrane. (Reproduced with permission from Ansari, M.A., and Alzohairy, M.A. 2018. "One-pot facile green synthesis of silver nanoparticles using seed extract of *Phoenix dactylifera* and their bactericidal potential against MRSA". *Evidence-Based Complementary and Alternative Medicine* 2018:1–9. doi: https://doi.org/10.1155/2018/1860280.)

chain, causing cell death (Luo et al., 2010). Moreover, it was reported that biofilms treated with AgNP affect the abundance of bacterial groups, with a longer-term potential impact to reduce biofilm function and development (Fabrega et al., 2011). Therefore, Gopinath et al. (2016) suggested that the proposed route of the biosynthesis of GAgNPs using *Gloriosa superba* leaf extract should be investigated by the pharmaceutical industry as potential drugs to treat microbial infections.

Green AgNPs from the respective precursors $AgNO_3$ using root bark extract of *Mammea suriga* were biosynthesized (Poojary et al., 2016). This study reported the effect of different reaction parameters (temperature, pH, volume of the extract and precursor concentration) on the size and morphology of the synthesized GAgNPs.

Results found that GAgNPs were efficiently synthesized with precursor concentration ($AgNO_3$) of 1 mM, at pH 10 and 80°C reaction temperature. The SEM analysis showed that GAgNPs size decreases as extract volume increases. Moreover, the XRD study revealed GAgNPs formation with an average size of 50 nm. The spectral data of FTIR analysis determined the role of different functional groups of the biomolecules in the extract involving capping as well as bioreduction of AgNPs. The *in vitro* antibacterial screening was tested using the Kirby–Bauer disc diffusion susceptibility test. The results showed that GAgNPs synthesized using root bark extract of *Mammea suriga* are a potential antibacterial agent. There was good antibacterial activity against *S. aureus* with an inhibition zone of 14 ± 1.0 mm. The conclusion is that *M. suriga* root bark extract is an excellent, non-toxic, eco-friendly source for the biosynthesis of biologically active NPs (at ideal conditions).

Dehghanizade et al. (2018) synthesized GAgNPs using *Anthemis atropatana* extract, and then evaluated their chemical characteristics, cytotoxic effects and antimicrobial activity. GAgNPs biosynthesis was confirmed by UV-vis spectrum (maximum absorption at 430 nm wavelength). TEM and SEM results confirmed that the GAgNPs have an average size of 38.89 nm and are spherical in shape. The crystalline structure of GAgNPs in optimum conditions was determined by XRD analysis, where XRD peaks pattern of face-centred cubic (*fcc*) 1 1 1, 2 0 0, 2 2 0, 3 1 1 and 2 2 2 was observed. FTIR results confirmed the green synthesis of AgNPs using *Anthemis atropatana* extract. For biological analysis, the MTT findings showed AgNPs cytotoxic effects on colon cancer cell lines (HT29) were dose dependent. GAgNPs showed a maximum cytotoxicity at concentration of 100 μg/ml: statistically significant compared to control cells ($p < 0.001$). The antibacterial activity of GAgNPs was tested against four pathogenic bacteria, including *S. aureus.* Effectiveness of GAgNPs varied, for example *S. aureus* was among bacteria belonging to the highest and lowest MIC values. Several interpretations were presented to explain the antimicrobial mechanism of GAgNPs. The GAgNPs were able to destroy the membrane of bacterial cell, leading to bacterial cell death (Syeda et al., 2016, Fouad et al., 2017). In addition, it was thought that the Ag ions released from AgNPs interact with biomolecules that contain oxygen, nitrogen and sulfur, killing the bacterial cell (Składanowski et al., 2016). Another hypothesis relates to the reaction of Ag with DNA and cellular proteins associated with the ATP synthesis (Fig. 10.6) (Magalhães et al., 2016). The study aligned to similar reports of GAgNPs using plants demonstrating antibacterial activity on pathogenic bacteria (Alsalhi et al., 2016; Salehi et al., 2016; Dehghanizade et al., 2018).

The extract of coconut shell (*Cocos nucifera*) was used to biosynthesize AgNPs and their antibacterial activity was evaluated against human pathogens *S. aureus, E. coli, L. monocytogenes* and *S. typhimurium* (Sinsinwar et al., 2018). The biosynthesized AgNPs using extract of coconut shell (CSE-AgNPs) were determined by UV-vis spectroscopy (432 nm peak absorption), TEM (size from 14.2 to 22.96 nm with spherical shape), FTIR spectroscopy revealed coconut shell extract capping around the AgNPs (with peaks at 1384, 1609 and 3418; correspond to the biomolecules) and XRD (with a peak at 32.078 as well as 2-Theta). CSE-AgNPs showed 15 mm inhibition zone against *S. aureus*, while the minimum inhibitory concentration

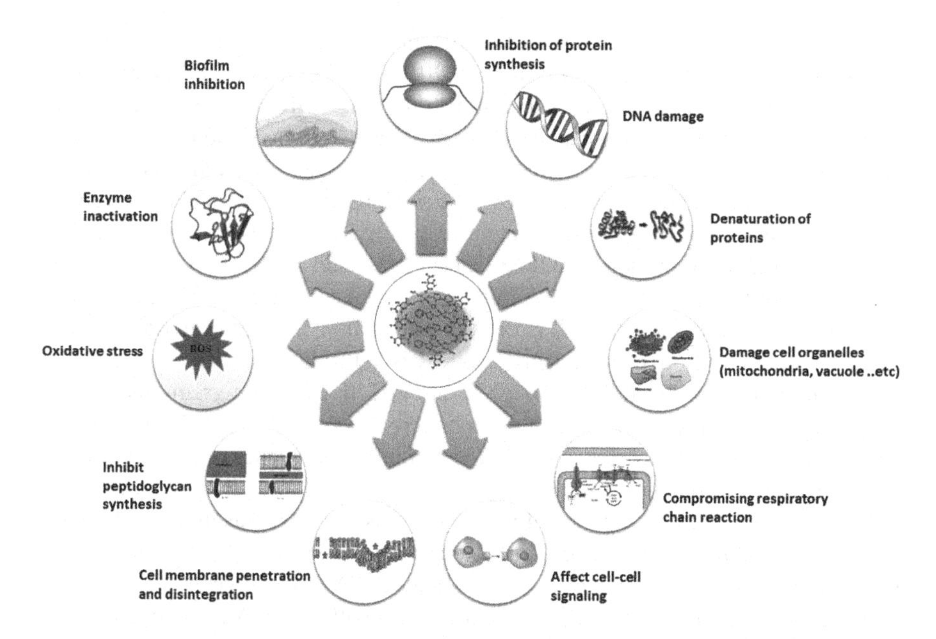

FIGURE 10.6 Possible mechanism of actions for green metal-based NPs.

(MIC) was 26 μg/ml. CSE-AgNPs ability to inhibit pathogens was compared to extract and ampicillin control using growth curve assay. The effect of CSE as well as CSE-AgNPs on human pathogens' growth pattern was evaluated on the absorbance at 600 nm. CSE-AgNPs noticeably inhibited bacterial cells' growth, even at the start of incubation. Both CSE and ampicillin (positive control) slowed the inhibition up to 10 hours, but subsequently failed to contain the growth of the tested bacteria.

Proposed mechanisms for the antimicrobial activity of CSE-AgNPs were adherence to the cell membrane, capability to penetrate the bacterial cell and damage the ribosomes, Golgi apparatus, biomolecules, vacuoles, mitochondria and interfere with thiol group, all leading to the compromising of the respiratory chain reaction, cell division and eventually cell death (Fig. 10.6) (Rajkuberan et al., 2015; Dakal et al., 2016). SEM results showed that the degradation of the cell wall might be due to the antibacterial activity of CSE-AgNPs. CSE-AgNPs exhibit a high level of antibacterial effect against *S. aureus*. The cell morphology was further analysed under SEM. *S. aureus* are usually in cocci shape. For CSE-AgNPs-treated cells, damage to the cell wall was observed morphologically. These results indicated the tendency of CSE-AgNPs to damage the *S. aureus* cell wall. The toxicity of both CSE and CSE-AgNPs were analysed in PBMC cell line. No toxicity was reported between 0.078 and 2.5 mg/ml concentrations. Sinsinwar et al. (2018) concluded that as chemically synthesized AgNPs are environmentally hazardous, toxic and expensive, green-synthesized CSE-AgNPs are possibly an alternative safe antibacterial agent for human therapeutic applications.

Silver NPs were green synthesized using leaf extract of *Rumex acetosa* (sorrel) and subsequently evaluated for their *in vitro* antibacterial activity against 16 bacterial clinical isolates, their antioxidant potential and cytotoxicity towards human osteosarcoma (HOS) cell lines (Kota et al., 2017). AgNPs surface plasmon spectra for the brownish black colour appeared about 448 nm. FTIR analysis confirmed the appearance of reactive O—H and N—H groups that reduce Ag(I) ions to Ag(0) that can further react with the extract contents to $AgCl/Ag_2C_2O_4$. SEM and TEM analyses results showed the resultant particles were predominantly spherical in shape with a size between 5 and 80 nm. An abundance in size from 15 to 20 nm was noted. GAgNPs exhibited remarkable antioxidant activity, and HOS cell lines were significantly inhibited ($p < 0.05$) at GAgNPs extract concentration of 25%, exhibiting a marginal revival at concentrations of 50% and 100%. The antibacterial effect of GAgNPs was tested on 16 pathogenic clinical isolates and the measured zones of inhibition obtained indicated that GAgNPs are effective on both gram-positive and gram-negative bacteria. Zones of inhibition were between 12 to 18.5 mm. Among gram-positive bacteria, *S. aureus* was better inhibited with zones of inhibition ranging from 18 ± 0 to 18.5 ± 0.5 mm.

Pawar and Patil (2020) reported the biosynthesis of AgNPs using *Eulophia herbacea* Lindl. (Orchidaceae) tuber aqueous extract and evaluated the antimicrobial and catalytic activities. The extract reduced aqueous Ag ions to generate stable and bioactive NPs. When 2% w/v extract was incubated with 1 mM $AgNO_3$ for 5 hours, a maximum reduction of $AgNO_3$ was obtained. The biosynthesized GAgNPs showed a surface plasma resonance of 447 nm, while the zeta potential was −15 mV. SEM analysis exhibited non-agglomerated GAgNP with an average size of 11.70 ± 2.43 nm. An energy-dispersive X-ray analysis proved the existence of elemental silver. The catalytic activity of the obtained GAgNPs was done via methylene blue and Congo red reduction. Results exhibited that the GAgNPs using *Eulophia herbacea* Lindl have catalytical and biological activities. XRD analysis proved the crystallinity of GAgNPs, namely that they have a face-centred cubic geometry. The obtained GAgNPs exhibited remarkable antifungal and antibacterial activities against *S. aureus* and other common human pathogens. These activities were comparable with standard antibiotics. The antibacterial activity of the biosynthesized GAgNPs was tested by the agar well diffusion method. The biosynthesized GAgNPs exhibited antibacterial activity against both gram-positive and gram-negative bacteria. GAgNPs showed stronger antibacterial activity compared to the corresponding $AgNO_3$ solution and the *Eulophia herbacea* tuber extract. Interestingly, the *Eulophia herbacea* tuber extract (2% w/v) alone had no antibacterial effects. The MIC of the GAgNPs towards pathogenic bacteria (including *S. aureus*) was also determined. For MIC test of the GAgNPs, the MIC value was 10 μg/ml for *S. aureus*. The effect of the combination of GAgNPs with standard antibiotic (streptomycin and chloramphenicol) bacterial strains was determined via well diffusion method. The GAgNPs were effective bactericide agents. Pawar and Patil (2020) observed that gram-positive bacteria had larger inhibition zones compared to gram-negative bacteria, due to the differences in the cell wall composition.

Banasiuk et al. (2020) explored the antioxidative properties of *Drosera indica*, *Drosera binata*, *Drosera spatulata* and *Dionaea muscipula* to produce uniform as well as biologically active AgNPs. Using polyvinylpyrrolidone assist the synthesis of quasi-spherical NPs characterized through stability and high uniformity. TEM and scanning SEM images revealed differences in GAgNPs morphology depending on the plant extract type. Moreover, different sizes of NPs were observed depending on the plant extract type. FTIR indicates that GAgNPs bind to plant extract compounds due to the interactions between the C=O and the —OH groups of the suitable polyphenolic compounds present in plant extracts. To evaluate the antioxidative activity of aqueous extracts of *D. spatulate*, *D. muscipula*, *D. indica* and *D. binate*, their capacity to scavenge DPPH free radicals was tested. Both *D. indica* and *D. muscipula* have similar high radical scavenging characteristics (IC_{50} = 0.857± 0.023 mg FW/ml) in spite of their different secondary metabolite composition. The DPPH test showed that *D. binata* extract has the lowest radical scavenging effect among the four tested extracts. The radical scavenging characteristics of plants extracts were strongly associated with the antimicrobial properties of GAgNPs acquired using the studied plant extracts. The MIC values varied depending on the microorganism and were between 5.3 mg/ml (for *P. aeruginosa* and *D. dadantii*) and 170 mg/ml or more (for *C. albicans* and *S. aureus*). The highest antimicrobial effect occurred by GAgNPs synthesized using the microwave method along with *D. muscipula* and *D. indica* extracts. The ranges of MBC were between 5.3 mg/ml and 42.5 mg/ml. The study suggests that the total antioxidative potential of the plant extract, and the method of Ag ions reduction, was critical for antimicrobial activity. The structure of bacterial cell is an important factor in terms of antimicrobial effectivity. Thus, gram-negative bacteria are more susceptible to AgNPs than gram-positive bacteria due to their cell membrane and wall structure (Dakal et al., 2016). Negatively charged cell membranes and walls are the major sites for attaching positively charged AgNPs. The nanostructures Ag aggregate on the bacterial cell surface, disrupting its functions. In addition, AgNPs can penetrate cell membranes and enter the cytoplasm. From the surface, highly reactive Ag+ ions release intracellularly and Ag+ ions affect the cell. The membrane penetration rate makes gram-negative bacteria more susceptible to AgNPs.

Murraya koenigii aqueous leaves extract was used for the biosynthesis of AgNPs and named MK-AgNPs (Qais et al., 2019). Furthermore, their antibacterial activity was tested against multiple ESβL-producing enteric bacteria and MRSA. MK-AgNPs were found to be predominantly spheroidal with size range of 5–20 nm. About 60.86% Ag content was found in MK-AgNPs. The evaluation of antibacterial effects was conducted by using the disc diffusion method. MK-AgNPs inhibited the tested pathogens' growth with varying inhibition zones. MICs of obtained MK-AgNPs against MRSA, as well as methicillin-sensitive *S. aureus* (MSSA) strains, were found to be 32 μg/ml, while for ESβL-producing *E. coli* it ranged between 32 and 64 μg/ml. *E. coli* ECS (control strain) was more sensitive with 16 μg/ml MIC value. MBCs values were in accordance with their respective MICs. Growth kinetics analysis showed that all tested *S. aureus* strains were inhibited (~90%) in the presence of MK-AgNPs with 32 μg/ml concentration. *E. coli* (ECS) (sensitive strain) exhibited the least resistance to MK-AgNPs with inhibition percent of 31 and 81.8 at 8 and 16 g/ml, respectively. Concentration of above 16 g/ml inhibited ECS by more than 90%. Marslin et al. (2018) reported encouraging findings on *in vitro* efficacy of biosynthesized MK-AgNPs suggesting further *in vivo* studies to test the therapeutic efficacy with respect to MDR bacteria.

Silver NPs were obtained through green synthesis using aqueous extracts of petiole and leaflets (vegetative) as well as flower and flower bud (reproductive) parts of *Handroanthus heptaphyllus* (Vell.) Mattos (Pereira et al., 2020). Pereira et al. (2020) reported some specific properties of the plant that may lead to modulating physico-chemical characteristics of GAgNPs, such as diameter and zeta potential, and even the possibility of altering the antimicrobial potential. Thus, if the aim is to biosynthesize GAgNPs with lower colloidal stability, higher hydrodynamic diameters and antibacterial activity at lower concentrations, the aqueous extracts of vegetative plant parts are better than the reproductive plant parts. The vegetative plant parts resulted in cubic and pyramidal forms of GAgNPs. For higher GAgNPs production with greater stability and small hydrodynamic diameters, reproductive plant parts are best. To produce GAgNPs with lower stability, larger hydrodynamic diameters and higher antibacterial activity, vegetative plant parts must be used. The physico-chemical characteristics, synthesis reaction progress and antibacterial activity against *E. coli* and *S. aureus* of GAgNPs were evaluated. Leaves produced better antibacterial activity against *S. aureus* than other plant parts and did so at lower concentrations than other plant parts. These differences could be attributed to the types and concentrations of the metabolites present in the leaves (Geisler et al., 2013). Secondary metabolites which are responsible for GAgNPs (vegetative) stabilization give less stability to the GAgNPs, making the silver easily released from the AgNPs resulting in stronger antibacterial activity when aqueous extract of the leaves was used for the green synthesis. Thus, Pereira et al. (2020) concluded that there are peculiarities even within one plant with respect to green synthesis of GAgNPs, and this must be taken into account as it modulates physico-chemical and biological activities.

Kasithevar et al. (2017) reported a rapid and simple green synthesis of AgNPs using leaf aqueous extract of *Alysicarpus monilifer.* The antibacterial activity against MRSA and coagulase-negative *Staphylococci* (CoNS) isolates from HIV patients was evaluated using the modified Kirby–Bauer disc diffusion method. MRSA and CoNS were found to be highly sensitive to the antimicrobial action of GAgNPs with an average inhibition zone of 19 and 17.5 mm, respectively, at GAgNPs concentration of 100 μg/ml. Meanwhile, for 50 μg/ml concentration of GAgNPs, the zone of inhibition was 14 and 16 mm, for both MRSA and CoNS, respectively. Bactericidal effects of the obtained GAgNPs against MRSA and CoNS, as well as the bacterial growth kinetics were also evaluated. Stable, mostly spherical, well-defined GAgNPs with a mean size of 15 ± 2 nm were formed within an hour. GAgNPs showed significant dose-dependent antibacterial action patterns against MRSA and CoNS isolates. With higher concentrations of GAgNPs, almost total reduction of bacterial colonies was observed. Many reports show that GAgNPs inhibit bacterial growth in a dose-dependent pattern (Suganya et al., 2015b; Różalska et al., 2016). Moreover, the obtained GAgNPs showed a remarkable antibacterial efficacy in comparison to antibiotics amoxicillin and erythromycin. When bacteria were treated with GAgNPs, their cell membrane permeability was compromised leading to cellular leakage, inhibiting replication ability and cell growth. GAgNPs were also found to be biocompatible with normal Vero cell line. Thus, Kasithevar et al. (2017) concluded that GAgNPs has proven to be potent antibacterial agents compared to conventional antibiotics (Table 10.1).

10.6.2 Green Gold Nanoparticles (GAuNPs) Synthesized Using Medicinal Plants

A green method was reported to synthesize gold AuNPs through bioreduction reactions of aqueous corresponding Au precursors using extracts of the aerial parts of *Rivea hypocrateriformis* acting as stabilizing and reducing agents, under microwave irradiation (Godipurge et al., 2016). *R. hypocrateriformis* was selected due to its good antimicrobial, anticancer, antidiabetic and analgesic properties (Shivalingappa et al., 1999). The formation of GAuNPs is indicated from the surface plasmon resonance peak noticed at λ_{max} = ~550 nm for GAuNPs. XRD patterns showed an *fcc* structure, while the FTIR spectra showed the existence of phytochemicals GAuNPs. The spherical-shaped GAuNPs averaged in size at 30 nm (10–50 nm) analysed by FE-SEM/TEM images. GAuNPs exhibited improved antimicrobial, anticancer and antioxidant activities. The antimicrobial activity of the synthesized GAuNPs was tested against different human pathogens, including *S. aureus*. Antibacterial activity of biosynthesized GAuNPs against *S. aureus* showed inhibition zones of 06, 07, 08 and 10 mm for different GAuNPs concentrations at 25, 50, 75 and 100 μg/ml. Godipurge et al. (2016) concluded that green synthesis of metal NPs using the extract of *R. hypocrateriformis* could compete with commercial antimicrobial agents (Std. ciprofloxacin). Therefore, green-synthesized NPs have great potential as growth inhibitors in a variety of biomedical applications.

Diospyros Kaki L.f. (Persimmon) is a common fruit abundant in flavonoids, polyphenols and sugars. Huo et al. (2019) biosynthesized GAuNPs using *Diospyros Kaki* L.f. fruit extract. The results of UV-vis spectra, TEM and DLS showed that the shape and size of GAuNPs are highly dependent on the $HAuCl_4 \cdot 4H_2O$ concentration and reaction temperature. When using concentrations of 1.0 and 1.5 mM of $HAuCl_4 \cdot 4H_2O$, hexagonal and triangular GAuNPs were obtained, whereas using 2.0 mM of $HAuCl_4 \cdot 4H_2O$ promoted nanoaggregates formation. Despite the high temperatures (50, 60 and 70°C) promoting the isotropic growth of GAuNPs, there was no significant impact on GAuNPs size. The XRD analysis showed that GAuNPs exhibited *fcc* structure. FTIR shows that different functional groups (mainly hydroxyl group) are involved in GAuNPs fabrication. The green-synthesized GAuNPs showed good catalytic and antioxidant activities. AuNPs are usually biologically inert and have not been considered as antibacterial agents, but surface modification can enhance their antibacterial ability (Zhang et al., 2015). Medicinal plants contain components such as polyphenols and flavonoids which could make them capping agents that can be adsorbed on the synthesized GAuNPs surface. This could possibly give the GAuNPs an inhibition ability. Synthesized GAuNPs using extract of *Diospyros Kaki* L.f. fruit showed no inhibition activity against *S. aureus* and no impact on the microorganism's proliferation when GAuNPs concentrations were increased. The GAuNPs synthesized using medicinal plants *Citrus maxiam* and *Euphorbia hirta* showed good antimicrobial activity (Yuan et al., 2017). Therefore, Huo et al. (2019) posits that the various effects of GAuNPs coated with medicinal plant extracts may not only be related to the characteristics of plant's extract, but is affected by the shape, size and concentration of NPs.

Moustafa and Alomari (2019) biosynthesized polydispersed anisotropic and monodispersed isotropic spherical GAuNPs from seed and leaf extracts of *Peganum harmala* L. (Ph. L). The antimicrobial potential of GAuNPs against human pathogens (*S. aureus*) were assessed. Seed- and leaf-derived GAuNPs had localized surface plasmon resonances about 578 and 530 nm, respectively. TEM, EDX, FE-SEM and XRD showed face-centred cubic spherical polydispersed anisotropic GAuNPs of mean size 52.04 nm and monodispersed isotropic GAuNPs with mean size 43.44 nm from seed and leaf extracts, respectively. FTIR showed alcohols and polyphenols are responsible for GAuNP reduction, capping and protection. Anisotropic GAuNPs exhibited no antibacterial effect, while isotropic GAuNPs exhibited good inhibition of *S. aureus* and *E. coli*. The catalytic and biological activities of AuNPs rely on their shape. Regarding agar medium, the anisotropic GAuNPs exhibited no antibacterial effect, in agreement with previous studies (Shankar et al., 2004). Isotopic GAuNPs exhibited maximum inhibition against *S. aureus* (30 mm) at a concentration of 200 µg/ml, although results were dose dependent. Results of liquid culture indicate that anisotropic GAuNPs inhibited *S. aureus* growth by 31.5% at a concentration of 100 µg/ml. Isotropic GAuNPs were much more efficient in inhibiting *S. aureus* (at 91.8%) than anisotropic GAuNPs. Isotropic GAuNPs can inhibit bacteria in agar and can completely inhibit gram-positive bacteria in liquid cultures. Isotropic GAuNPs are more efficient antibacterial agents than anisotropic GAuNPs. The higher inhibition ability of isotropic GAuNPs indicates its antibacterial effect is shape dependent.

Fruit juice of *Euphoria longana* Lam. (Longan) juice was found to proficiently produce GAuNPs (Khan et al., 2016b). The antibacterial effect of GAuNPs was carried out against gram-positive and gram-negative bacteria through agar well diffusion and the determination of MIC values. GAuNPs exhibited significant antibacterial activity against *S. aureus* with a clear zone of inhibition at 18 ± 00.6 mm and MIC value of 50 µg. This may be due to the uniform dispersion of GAuNPs, surface capping with active phytochemicals and their small size (25 nm). Such findings are in line with other studies (Khan et al., 2016a; Patra and Baek, 2016; Abdel-Raouf et al., 2017). The spherical shape and uniform dispersion of GAuNPs results in a large surface area exhibiting significant antibacterial effects (Khan et al., 2016). It is suggested that the anionic biomacromolecular moieties and DNAs interact with the cationic noble metallic ions leading to damage to and modification of the structure of cell membranes and walls, disrupting metabolic processes and ultimately causing cell death (Feng et al., 2000). NPs are able to enter the bacterial cell membrane due to their small size, subsequently causing cell death (Abdel-Raouf et al., 2017). GAuNPs showed significant photocatalytic degradation (about 76% of methylene blue in 55 minutes), indicating the efficient photocatalytic characteristic of the biosynthesized GAuNPs. The effective photocatalytic and antibacterial activities of the green-synthesized GAuNPs is due to their spherical morphology, small size and uniform dispersion (Khan et al., 2016b).

Lopez-Miranda et al. (2019) green-synthesized GAuNPs using *Aloysia triphylla* aqueous extracts and evaluated their catalytic and antibacterial properties. The antibacterial activity of GAuNPs was evaluated using *S. aureus* and *E. coli*. The catalytic activity was tested by the degradation of Congo red and methylene blue. UV-vis analysis revealed an increase in GAuNPs concentrations by increasing the volume extract, extract concentration and the concentration of precursor salt. GAuNPs crystalline nature was corroborated by XRD and TEM analysis showed NPs with mostly spherical in shape and sized from 40 nm to 60 nm. The Lopez-Miranda et al. (2019) study obtained GAuNPs that were narrow in size distribution and homogeneous in morphology, uncommon when using green synthesis methods. The analysis of antibacterial activity was performed using the disc diffusion method. Results exhibited an inhibition zone of 11.3 mm for S. *aureus*, showing the bactericidal potential of GAuNPs. In addition, the bactericidal ability of the GAuNPs stemmed from their ability to damage cell walls and hinder critical functions such as permeability and respiration. They also release Au ions that result in bacterial cell damage (Suganya et al., 2015a; Rajan et al., 2017). The degradation periods for Congo red and methylene blue were 11 and 5 minutes, respectively, a very short period compared to previous studies. These results have a great significance for different catalytic applications. Therefore, using the extracts of *A. triphylla* to green synthesize AuNPs with novel desirable characteristics indicates that GAuNPs could offer antibacterial and catalytic agents with desirable properties.

Zizyphus mauritiana extracts and chloroauric acid ($HAuCl_4$) were used for the green synthesis of AuNPs, with sizes ranging between 20 and 40 nm (Sadeghi, 2015). The growth of GAuNPs ceases after 25 minutes of reaction time. In the UV-vis spectrum, the GAuNPs peaked at 535 nm. XRD proved the crystalline nature of the GAuNPs of 27 nm size. The size and morphology of the synthesized GAuNPs were evaluated by TEM. The NPs were spherical in shape and in the range of 20–40 nm (average size 32 nm). The GAuNPs were monodispersed, with a few NPs of different sizes. The FTIR results indicated that the *Zizyphus mauritiana* extracts containing OH functional group play a role in capping NPs during green synthesis. Antibacterial activities of GAuNPs were evaluated against the growth of *S. aureus* using SEM. The SEM results showed most *S. aureus* cells were damaged and gone when treated with GAuNPs. Therefore, the green-synthesized AuNPs using *Zizyphus mauritiana* extracts can be considered a good candidate for medicinal and biological applications.

Islam et al. (2019) reported a green extracellular biosynthetic method of highly stable AuNPs. When the aqueous gold ions were mixed with the leaf extract of *Salix alba* L., they were bioreduced and gold nanoparticles (Au-WAs) were formed. Their stability was tested against different volumes of sodium chloride and pH along with elevated temperature as well as enzyme inhibition, antifungal, antibacterial, muscle relaxant, anti-nociceptive and sedative activities. UV-vis spectra of the Au-WAs showed surface plasmon resonance at a wavelength of 540 nm. Meanwhile, the SEM and AFM Au-WAs analyses showed the NPs size of 63 nm and 50–80 nm, respectively. FTIR spectra affirmed amide, amines and aromatic groups were involved in the reduction and capping of the gold NPs. Au-WAs exhibited remarkable stability in a variety of salt volumes and various pH solutions. Meanwhile, Au-WAs were found to be relatively unstable at higher temperatures. Au-WAs exhibited good antibacterial activity, antifungal activity, as well as significant antinociceptive ($P < 0.001$) and muscle relaxant properties ($P < 0.05$). The antibacterial activity was evaluated by using the well diffusion method against several bacterial and fungal strains. Au-WAs showed good antibacterial activity against *S. aureus* (10 ± 0.58 mm), and significant antifungal activity against *A. niger* and *A. solani*, although it was less effective against *A. flavus*. AuNPs were reported to possess a well-developed surface chemistry, exhibited high chemical stability and have a large surface-to-volume ratio. This could enhance the number of drug/phytochemical molecules that can be adsorbed, by means of electrostatic attraction between the NPs and the amine groups of drugs/phytochemicals, on their surface. The AuNPs enclosed by several drug moieties could behave as a one single group against microorganisms and as a result of the microbial activity increased (Burygin et al., 2009). Islam et al. (2019) concluded that *S. alba* leaf extract is a good bio-reductant in the green synthesis of AuNPs and could have great potential for a variety of pharmaceutical and biomedical applications.

One-pot biosynthesis of green AuNPs was performed using *Catharanthus roseus* (CR) and *Carica papaya* (CP) leaf extracts as well as the combination of the both extracts (CPCRM) (Muthukumar et al., 2016). The green-synthesized AuNPs were investigated for their morphology, crystallinity, size and activity on bacterial and cancer cell lines. Electron microscopy revealed that AuNPs were mostly spherical, although hexagonal- and triangle-shaped GAuNPs were also found. Microscopies and infrared studies showed that flavonoids, alkaloids and proteins were found along with AuNPs, which contributed to their non-agglomeration, stabilization and biological properties, including anticancer and antibacterial activities. The AuNPs prepared using CPCRM extract showed improved activity, attributed to the synergistic effects of the two plant extracts used. The antimicrobial effect of GAuNPs prepared using CP, CR and CPCRM extracts was tested by disc diffusion as well as serial dilution method (MIC) against several pathogenic bacteria, including *S. aureus*. The MIC values of the obtained AuNPs-CP, AuNPs-CR and AuNPs-CPCRM against *S. aureus* were 125, 250 and 62.5 μg/ml. All the prepared NPs, AuNPs-CPCRM were observed to have antibacterial activity against all tested bacterial strains (MIC values of 15.625 μg/ml for *E. coli*, 62.5 μg/ml for *P. vulgaris*, 125 μg/ml for *B. subtilis* and 62.5 μg/ml for *S. aureus*). Table 10.2 reviews GAuNPs synthesized using medicinal plants against MDR bacteria.

10.6.3 Other Green Metal-Based Nanoparticles Synthesized Using Medicinal Plants

The synthesis of noble metal (gold, silver, palladium and platinum) NPs and inorganic oxides (such as titanium and zinc oxide) NPs have attracted great attention in the field of therapeutics, disease treatments, disease diagnosis, biosensors, cancer detection, cancer treatment, drug delivery, cosmetics, antimicrobial agents, antifungal agents, medical devices and medical instruments (Singh et al., 2016; Salouti and Derakhshan 2019; Karthik et al. 2020). In addition to the gold and silver NPs reviewed in this chapter, copper, zinc oxide, iron, titanium dioxide, nickel oxide, and lead have also been used (Table 10.3).

In this regard, Chauhan et al. (2019) evaluated the antimicrobial activity of green-synthesized zinc oxide NPs against pathogenic bacteria. The leaf extract of *Cassia siamea* (Kassod) was used for green synthesis of zinc oxide NPs. Furthermore, these GZnO-NPs were evaluated of their antimicrobial activity. FTIR results reveal the presence of several functional groups. SEM results indicated that the synthesized GZnO-NPs ranged in size, but all were under 100 nm, while the morphology of the GZnO-NPs was slightly ellipsoidal-spherical. At room temperature, the synthesized GZnO-NPs were stable with no agglomeration. The synthesized GZnO-NPs show significant antimicrobial activity against tested pathogens. Antimicrobial activity of GZnO-NPs against wound-associated bacteria demonstrated that biosynthesized GZnO-NPs have significant antibacterial activity. The culture tubes containing GZnO-NPs ranging between 0.1 and 0.5 mg/ml exhibited bacterial growth, while no growth was observed in cultures that contained GZnO-NPs ranging in size between 0.7 and 0.9 mg/ml. Hence, it is concluded that MIC of GZnO-NPs is effective at 0.7 mg/ml of concentrations. The formation of stable GZnO-NPs was attributed to the existence of stabilizing and capping phytochemicals such as terpenoids, flavonoids and phenols in the plant extract. Results showed that GZnO-NPs have potential antibacterial activity which could be better therapeutic agent against wound-associated bacteria providing new directions towards effective and safer therapies. Zinc oxide NPs were also green synthesized using aqueous extract of *Ocimum americanum* (Kumar et al., 2019). Green synthesis of ZnO-NPs is of great importance due to their non-toxic nature and applicability for a wide range of therapeutic applications. Kumar et al. (2019) had green-synthesized spherical-shaped ZnO-NPs with average size of 21 nm. The GZnO-NPs exhibited a strong absorption peak occurred at 316 nm showing a clear characteristic property of ZnO-NPs. Biophysical characterization of GZnO-NPs was performed using FTIR in comparison with *O. americanum* extract to reveal the potential functional groups involved in GZnO-NPs formation. XRD approved the crystalline nature that was in accordance with JCPDS-ID for ZnO-NPs. SEM determined the morphology of GZnO-NPs which was found to be spherical in shape; meanwhile, DLS determined GZnO-NPs size to be an average of 21 nm. The zeta potential was 12.6 mV with positive polarity (Clogston and Patri, 2011; Mahendra et al., 2017). The aqueous extract of *O. americanum* and GZnO-NPs were further analysed by disc diffusion method for their antimicrobial potential against fungal and bacterial pathogens. The results exhibited significant inhibition towards tested pathogens. The inhibition zone for GZnO-NPs against *S. aureus* was observed as 29 mm. Interestingly, the aqueous plant extract and ZnO alone did not show any antimicrobial effect. The possible mechanism of action for GZnO-NPs was suggested to be due to damaging the cell membrane; DNA damage associated with the reactive oxygen species (ROS); ATP production disruption or DNA replication inhibition (Reddy et al., 2007; Rasmussen et al., 2010; Kaviyarasu et al., 2017). Mahendra et al. (2017) green-synthesized zinc oxide NPs using aqueous leaf extract of *Cochlospermum religiosum*. Obtained GZnO-NPs were hexagonal wurtzite-shaped NPs below 100 nm in size (confirmed through XRD analysis). The GZnO-NPs exhibited an absorption peak happened at 305 nm. The GZnO-NPs were high in purity with a mean size of ~76 nm (confirmed through DLS analysis), which was in line with XRD findings. SEM results proved the same results with agglomeration of smaller NPs. FTIR explored the composition of *C. religiosum* aqueous leaf extract and GZnO-NPs. The plant extract as well as GZnO-NPs showed antibacterial activity against pathogenic bacterial activity, including *S. aureus*. Zones of inhibitions against *S. aureus* for aqueous leaf extract, ZnO and GZnO-NPs were 15.30 ± 0.17, 00.00 ± 0.00 and 22.20 ± 0.11 mm, respectively. Results showed that GZnO-NPs exhibited higher antibacterial activity compared to aqueous leaf extract, and even ZnO that did not show any antibacterial activity. It was suggested that the inhibition effect of GZnO-NPs is due to the rupturing of the outer and inner bacterial walls, leading to leakage and disorganization of the cell membrane. MIC values of GZnO-NPs and *C. religiosum* extract were found to be from 4.8 to 625 μg/ml against tested pathogens. The best MIC of the GZnO-NPs was observed against *S. aureus* at 4.8 mg μg/ml. All these reports concluded that the green synthesis of ZnO-NPs have great potential in variety of applications due to possessing various biological properties.

Alavi and Karimi (2018) reported the green synthesis of titanium dioxide (TiO_2) and copper (Cu) NPs using leaf aqueous extract of *Artemisia haussknechtii* in terms of antibacterial effects against MDR bacterial species. Three different concentrations of 0.001, 0.01 as well as 0.1 M of $CuSO_4$ and TiO $(OH)_2$ were evaluated for testing the optimum green NPs synthesis. The average size of the obtained GTiO_2-NPs and GCuNPs were 92.58 ± 56.98 and 35.36 ± 44.4 nm, respectively. For NPs morphology, GTiO_2-NPs and GCuNPs showed a spherical shape. For antibacterial assay, disc diffusion assay, MIC, MBC, bacterial morphology and growth of four MDR (*S. aureus* ATCC 43300, *S. epidermidis* ATCC 12258, *S. marcescens* ATTC13880 and *E. coli* ATCC 25922) were analysed. Results showed that leaf extract of *A. haussknechtii* with different groups of phytochemicals could green synthesize CuNPs and TiO_2-NPs. Moreover, GCuNPs had a greater antibacterial effect than GTiO_2-NPs. Results show antibacterial activity depends on NPs concentrations. GCuNPs had higher antibacterial activity in 0.1 as well as 0.01 M concentrations. The inhibition zones for GCuNPs against *S. aureus* were 4 ± 1 and 8 ± 1.73 mm at concentrations of 0.1 and 0.01 M, respectively. There was no antibacterial activity at 0.001 M concentration. TiO_2-NPs did not have a significant effect on *S. aureus* ATCC 43300. With regard to NPs concentrations, the bacterial growth curve continuously decreased. At high concentrations of GCuNPs, bacterial growth was delayed and inhibited at a concentration of 60 μg/ml GCuNPs. Therefore, at low concentrations of GCuNPs, there was bacteriostatic effect, but a bactericidal effect at high concentrations of GCuNPs.

Copper nanoparticles CuNPs have astonishing properties and their importance in pharmaceutical industry has attracted much scholarly interest. Thiruvengadam et al. (2019) green-synthesized CuNPs with the *Millettia pinnata* flower aqueous extract, and their characteristics were further studied. Reduction of copper acetate to CuNPs was conducted, and then evaluated by UV-vis spectrophotometer. The maximum absorption occurred at 384 nm which confirmed the surface plasmon resonance of the NPs. FTIR results of the GCuNPs confirmed the involvement of organic moieties of the *M. pinnata* flower extract in the green synthesis. The synthesized GCuNPs were highly durable, spherical in shape, with an average size of 23 ± 1.10 nm. The GCuNPs showed a strong anti-inflammatory effect by using membrane stabilization and albumin denaturation. The GCuNPs showed high inhibition on DPPH nitric oxide and radical scavenging effects. The antibacterial activity of GCuNPs was tested against gram-negative and gram-positive bacteria. The antibacterial effect of GCuNPs and plant extract was analysed by the disc diffusion method. Results showed that the maximum inhibition was observed on gram-positive bacteria (*S. aureus* and *B. subtilis*) compared to gram-negative bacteria (*E. coli* and *P. aeruginosa*). The bacterial cell membrane interacts with the GCuNPs decreasing the trans-membrane electrochemical potential, which affects membrane integrity (Alzahrani, 2016; Thiruvengadam et al., 2019).

Yugandhar et al. (2017) synthesized GCuNPs by using the stem bark of *Syzygium alternifolium* and characterized it. GCuNPs changed colour upon synthesis (peak at 285 nm analysed by UV-vis spectroscopy). FTIR showed that primary amines and phenols were mainly involved in the stabilization and capping of GCuNPs. DLS as well as zeta potential analysis showed a narrow size of NPs with great stability. XRD analysis showed the crystallographic nature of GCuNPs with an average size of 17.2 nm. TEM revealed the GCuNPs sizes range from 5 to 13 nm, were spherical shape, polydispersed and non-agglomerated in condition. Antimicrobial tests of GCuNPs showed the highest inhibitory activity against *E. coli* followed by *S. aureus*. Similarly, Alzahrani (2016) synthesized CuNPs and found significant antibacterial effect against *E. coli* compared to *S. aureus*. In summary, among all other metal NPs, CuNPs attract much attention and are used in numerous pharmaceutical and biomedical applications. This could be attributed to their superior strength, good thermal, optical and electrical properties, as well as their use as catalysts and sensors. The lower cost of CuNPs compared to noble metals (Ag and Au) make it cost-effective, sustainable and innovative.

Titanium NPs (TiO-NPs) were synthesized by Hameed et al. (2019) due to their desirable characteristics such as being non-toxic, bioactivate and eco-friendly. Banana peel extract was used with titanium dioxide to prepare GTiO-NPs. The obtained GTiO-NPs were of average diameter of 88.45 nm and volume of 31.5 nm (confirmed by AFM and XRD). The antimicrobial activity of GTiO-NPs was tested against several pathogenic bacteria. The GTiO-NPs exhibited antimicrobial activity against pathogenic bacteria while banana peel extract did not. All the tested concentrations (25%, 50% and 100%) showed antibacterial activity against *S. aureus* with inhibition zones of 6, 27 and 16 mm. The most effective concentration of GTiO-NPs against *S. aureus* was found to be at 50% concentration. The possible mechanisms might involve the interaction between GTiO-NPs with the bacterial cell biomolecules, with the negatively charged bacterial cell attracting the positively charged GTiO-NPs. This electromagnetic attraction caused oxidation and rapid bacterial death. Another explanation is that the positive ions in GTiO-NPs form a bond with the thiol group (—SH). This thiol group is protein part critical to the membrane of the bacterial cell. The bond would destruct the bacterial cell membranes by increasing permeability of membrane and causing unorganized mass transport through cell membranes (Zhang and Chen, 2009; Hameed et al., 2019). Green synthesis of TiO-NPs using *Azadirachta indica* leaf extract was performed and their antibacterial activity was evaluated (Thakur et al., 2019). FTIR analysis showed the presence of flavonoids, terpenoids and proteins responsible for the formation and stabilization of TiO-NPs. The XRD analysis showed the crystalline nature of TiO-NPs. TEM revealed the spherical shape of GTiO-NPs with sizes ranging from 15 to 50 nm, while SEM analysis revealed that GTiO-NPs were spherical in shape and size ranged from 25 to 87 nm. The antibacterial effect of synthesized GTiO-NPs and TiO_2 compound was evaluated against several pathogenic bacteria. Results show that GTiO-NPs inhibited the growth of all the tested bacteria. The antibacterial effect was stronger for GTiO-NPs versus TiO_2 compound. The lowest MIC value (10.42 μg/ml) of GTiO-NPs was against *S. typhi* and *E. coli*. Meanwhile, the lowest MBC value (83.3 μg/ml) was with respect to *K. pneumoniae*. In general, synthesized GTiO-NPs showed broad spectrum antimicrobial effects against a wide range of pathogenic bacteria (Alavi and Karimi, 2018; Thakur et al., 2019; Hameed et al., 2019). With respect to MDR bacteria, GTiO-NPs are novel and promising antibacterial agents that could be used against MDR bacteria such as MRSA.

Nickel oxide nanoparticles (NiO-NPs) were green synthesized using neem leaves (Helan et al., 2016). The synthesized GNiO-NPs were evaluated for morphological, structural, magnetic, optical properties. XRD analysis showed the polycrystalline nature of samples with hexagonal crystal phase. TEM reveals the synthesized GNiO-NPs were oblong in shape and sized 12 nm. Furthermore, the GNiO-NPs antibacterial activity was evaluated against *S. aureus* (MTCC 1430) and *E. coli* (MTCC 739) by the agar diffusion method. GNiO-NPs exhibited significant antibacterial activity, albeit concentration dependent. The activity of GNiO-NPs and its annealing temperatures were found to be moderate. In addition, GNiO-NPs can penetrate the bacterial cell wall, alter the inner cellular components as well as the cell membrane, causing cell death. Similarly, Khalil et al. (2018) synthesized GNiO-NPs (~18 nm) using aqueous leave extracts of *Sageretia thea* (Osbeck.) and tested their biological activities. Antibacterial effects were evaluated against several pathogenic bacteria, including *S. aureus* ATCC. Their corresponding MICs were also analysed. Most tested bacterial strains inhibited antibacterial effects across concentration ranges between 1,000 and 31.25 μg/ml. Saleem et al. (2017) green-synthesized NiO-NPs using leaf extract of *Eucalyptus globulus* and tested their bactericidal activity against several pathogenic bacteria including MRSA. The GNiO-NPs were pleomorphic in size ranging from 10 nm to 20 nm. Using XRD analysis, the average size of GNiO-NPs was shown to be 19 nm. The antibacterial effect of GNiO-NPs against tested pathogenic bacteria varied considerably. GNiO-NPs against MRSA (15 and 14 mm) and MSSA (15 and 13 mm) showed good antibacterial activity. The MIC and MBC values of GNiO-NPs against MRSA and MSSA were 0.8 and 1.6 mg/ml, respectively. In general, similar studies concluded the effectiveness of GNiO-NPs as promising antimicrobial agents against MDR bacteria (Behera et al., 2019; Srihasam et al., 2020).

Regarding iron NPs with potential biomedical properties, Iqbal et al. (2020) green-synthesized IO-NPs using leaf extract of *Rhamnella gilgitica* as both stabilizing and reducing agents. GIO-NPs were confirmed through the presence of peak at 341 nm. TEM analysis revealed the average size of the GIO-NPs to be ~21 nm and spherical shaped. GIO-NPs inhibited various pathogenic bacteria, including *S. aureus.* This strong antibacterial effect might be attributable to the active biomolecules attached to the GIO-NPs surface. ROS production is the most important mechanism in terms of the antimicrobial potential of GIO-NPs (Fig. 10.6). Damaging the cell membranes due to GIO-NPs adsorption on the cell surface is also possible. Jegadeesan et al. (2019) also reported the green synthesis of GIO-NPs using three aqueous plant extracts of *Moringa oleifera* fruit (MOF), *Terminalia bellirica* (TB) and *Moringa oleifera* leaves (MOL). TEM and SEM analyses showed single spherical NPs for T-GIO-NPs (21.32 nm), and irregular shaped MOF-GIO-NPs and MOL-GIO-NPs (particle size of 45 nm). The antimicrobial effect of GIO-NPs was higher than the extract alone, with T-GIO-NPs (inhibition zone; 15 ± 2 mm) and MOL-GIO-NPs (inhibition zone; 15 ± 1 mm) found to be potent against *S. aureus.* Abbasi et al. (2020) reported the green synthesis of GIO-NPs using leaves extract of *Rhamnus triquetra.* Results showed significant inhibition effects of GIO-NPs against various pathogenic bacteria at concentrations ranging from 37.5 to 1,200/ml, where GIO-NPs were the most effective against *S. aureus* (MIC: 37.5 μg/ml). In addition, results determined that antibacterial activity increased with the increase in GIO-NPs concentrations.

Similarly, Muhammad et al. (2019) demonstrated mediated green synthesis of iron oxide (Fe_2O_3) and lead oxide (PbO) NPs using *Papaver somniferum* L. extract. XRD analysis confirmed the crystalline nature of NPs. FTIR results confirmed NPs capping by plant phytochemicals. The results of SEM indicated the size of GFe_2O_3-NPs and GPbO-NPs to be 38 ± 13 nm and 23 ± 11 nm, respectively. Tested bacterial strains were susceptible towards both GFe_2O_3-NPs and GPbO-NPs. However, GPbO-NPs exhibited potency against all the tested pathogenic bacteria, with a superior effect compared to GFe_2O_3-NPs. This was attributed to its smaller size (confirmed using SEM analysis). It is well known that smaller sized NPs are better able to penetrate bacterial cell membranes. Upon penetration, they dissociate into their corresponding ions, generating oxidative stress and efficiently killing the bacteria. Overall, both GFe_2O_3-NPs and GPbO-NPs play a vital role in multiple biological processes. Lead nanoparticles (PbNPs) were also green synthesized using methanolic extract of *Cocos nucifera* L. extract (Elango and Roopan, 2015). Phytochemicals present in methanolic extract responsible for reducing and capping PbNPs were determined by GC-MS. Antimicrobial activity for GPbNPs against pathogenic bacteria such as *S. aureus* were evaluated. Results showed that the GPbNPs are spherical in shape with the particle distribution at 47 ± 2 nm. The inhibition zone against *S. aureus* occurred at a concentration of 50 μl compared to other tested bacteria. This confirmed the antimicrobial activity of synthesized GPbNPs as well as their superior antibacterial ability against *S. aureus.*

Palladium nanoparticles (PdNPs) were green synthesized using *Phyllanthus emblica* (*P. emblica*) seeds extracts (Dinesh et al. (2017). Results showed that

GPdNPs were of an average particle size of 28 ± 2 nm and spherical in shape. The synthesized GPdNPs and plant extracts were subjected to antibacterial analysis against various disease-causing bacteria (*S. aureus*, *B. subtilis*, *P. aeruginosa* and *P. mirabilis*) by agar well diffusion assay. Streptomycin (Standard antibiotic) was used at the concentration of 50 μg/ml. The FTIR results for Pd $(OAc)_2$, *P. emblica* seed extract as well as GPdNPs confirmed the formation of NPs. *P. emblica* seed extract demonstrated maximum inhibition against *B. subtilis* at 8.9± 1.46 mm followed by *S. aureus* with inhibition zone of 8.2 ± 1.10 mm (at concentrations of 100 mg/ml). Meanwhile, for GPdNPs, the best inhibition zone against *S. aureus* was 9.6 ± 1.10 mm and 8.5 ± 1.21 mm for concentrations of 100 and 75 mg/ml, respectively. The streptomycin (positive control) exhibited a maximum inhibition zone at 11.6 ± 1.05 mm at 75 μg/ml.

Likewise, Sharmila et al. (2017a) reported green synthesis of PdNPs using leaf extract from *Filicium decipiens*. The GPdNPs synthesis was validated by UV-vis spectrophotometer. Morphology and size were determined as spherical GPdNPs with sizes ranging from 2 to 22 nm, confirmed using TEM analysis. EDS analysis proved the presence of palladium in the green-synthesized NPs. XRD patterns confirmed the crystalline nature of GPdNPs. The proteins and phytochemicals were identified in FTIR by their functional groups and showed amide and amine groups present in *Filicium decipiens* which might play a role in the bioreduction reaction for GPdNPs synthesis. The antibacterial effect of the synthesized GPdNPs was tested by the well diffusion method. GPdNPs showed good antibacterial activity against both gram-positive (*S. aureus and B. subtilis*) and gram-negative (*E. coli and P. aeruginosa*) bacterial strains. The metal NPs mechanism of antibacterial activity seems to be due to the interaction and the internalization of NPs with cell membranes, disrupting the transportation process or/and interactions with enzymes and DNA, as well as inhibiting the respiratory system (Fig. 10.6) (Sharmila et al., 2017a). Leaf extract of *Santalum album* was used to green synthesize GPdNPs (Sharmila et al. 2017b). Spherical-shaped GPdNPs were observed in TEM analysis. The XRD pattern analysis revealed the face-centred cubic crystalline nature of GPdNPs. Antibacterial activity results show that *S. album* derived PdNPs have better bactericidal action against gram-negative bacteria than gram-positive bacteria. This green synthesis approach using *S. album* leaf extract for PdNPs synthesis is efficient and economical, enabling large-scale production. The synthesized PdNPs are an effective catalyst and antibacterial agent. The antibacterial effect of GPdNPs against *P. aeruginosa* (30 mm) and *E. coli* (31 mm) was greater than against *S. aureus* (18 mm) and *B. subtilis* (12 mm). Thus, the antibacterial effect of the synthesized GPdNPs is also more effective against gram-negative bacteria than gram-positive bacteria. This observation was true for several green-synthesized palladium NPs (Sharmila et al., 2017a; Sharmila et al., 2017b). The higher antibacterial activity of GPdNPs against gram-negative bacteria might be attributed to their ease of penetration of gram-negative bacterial cell membranes, whereas gram-positive bacteria have a thick outer layer of peptidoglycan. Antibacterial effects of PdNPs are an active research area with respect to different biological applications in the nanotechnology field. The antibacterial properties of PdNPs and GPdNPs are useful in various pharmaceutical formulations, biomedical engineering and medical fields.

Umaralikhan and Jamal Mohamed Jaffar (2016) reported a simple yet effective approach for the green synthesis of MgO nanoparticles (MgO-NPs) using aqueous leaf extracts of both *Pisidium guvajava* (*P. guvajava*) and *Aloe vera*. Further antimicrobial effects of GMgO-NPs were studied extensively for *E. coli* and *S. aureus*. UV absorbance appeared at 221 nm, which confirmed that $MgNO_3$ was reduced as MgO. FTIR showed that both plant extracts acted as reducing and capping agents. The XRD analysis confirmed that the final products were pure in the *fcc* crystal phase. EDAX with FESEM showed the synthesized GMgO-NPs to be cubic in shape, while XPS analysis showed that the GMgO-NPs were composed of MgO and plant extracts. Antibacterial activity results showed the inhibition zones of GMgO-NPs prepared from *P. guvajava* extract against *S. aureus* were 11, 12 and 16 mm at different concentrations of 1, 3, 5 mg/ml, respectively. Meanwhile, for GMgO-NPs prepared from *A. vera* extract against *S. aureus*, the inhibition zones were at 12, 13 and 15 mm for the same tested concentrations. Zones of inhibitions were found to increase as the concentration of the GMgO-NPs increased. The mechanism behind the antibacterial effect was determined to be due to ROS and lipid peroxidation (Umaralikhan and Jaffar, 2016). Similarly, Suresh et al. (2018) reported green synthesis and characterization of hexagonal-shaped GMgO-NPs using *Costus pictus* D. Don (insulin plant) leaf extract and further evaluated their antimicrobial and anticancer activity. The formation of GMgO-NPs was verified by different characterization techniques. The presence of metal oxides and plant's biomolecules were confirmed by FTIR analysis. The XRD revealed the synthesis of pure cubic GMgO-NPs crystalline NPs. The surface morphology of GMgO-NPs observed by SEM revealed hexagonal-shaped GMgO-NPs crystallites. The average size of GMgO-NPs was approximately 50 nm by TEM. The GMgO-NPs exhibited good antimicrobial activity with a maximum inhibition rate for GMgO-NPs at 200 μg in terms of anticancer activity. The results for antibacterial activity of GMgO-NPs against *S. aureus* showed the inhibition zone to be 32.5 ± 0.707 mm. Thus, biosynthesized GMgO-NPs show strong anticancer and antimicrobial activities. This type of synthesis is simple, cost-effective, eco-friendly, non-toxic alongside many desirable biological properties, making it beneficial for many pharmaceutical applications.

Tran and Webster (2011) provided the first evidence of strong growth inhibition of *S. aureus* when treated with selenium nanoparticles (SeNPs) after 3, 4 and 5 hours at different concentrations of 7.8, 15.5, and 31 μg/ml. The percentage of live bacteria decreased when SeNPs was present. Therefore, SeNPs have great potential to treat and prevent *S. aureus* infections. Therefore, SeNPs should be further investigated for novel and effective solutions against *S. aureus* associated infections and MRSA. In this regard, Alagesan and Venugopal (2018) green-synthesized SeNPs by mixing leaf extract of *Withania somnifera* and selenious acid (H_2SeO_3) solution. Screening analysis unveiled vast quantities of phytochemicals in leaf aqueous extract of *W. somnifera*. The green-synthesized SeNPs by FTIR analysis indicated the presence of bioactive molecules' functional groups. The UV analysis confirmed the formation of SeNPs peaking at 310 nm. XRD and TEM analyses showed the amorphous nature of GSeNPs. GSeNPs were found to be spherical in shape and range in size from 45 to 90 nm (analysed by FE-SEM). The maximum inhibition zone was observed in *S. aureus* (19.66 mm) treated with GSeNPs. These findings provide a clear indication that GSeNPs could serve as an effective antibacterial agent against MSSA and MRSA infections (Tran and Webster 2011; Alagesan and Venugopal, 2018).

10.7 MECHANISM OF ANTIMICROBIAL ACTIVITY OF GREEN METALLIC NANOPARTICLES

As discussed earlier, metallic NPs are the focus of several biomedical applications, including antimicrobial agents because of their size-and-shape-subordinate tuneable characteristics. Metallic NPs, for example, gold, silver, titanium, copper, zinc, selenium and iron demonstrate strong activity against MDR bacteria (Marslin et al., 2018; Alavi and Karimi, 2018; Srihasam et al., 2020). Significantly, green-synthesized NPs are suitable for different antimicrobial applications due to their strong stability and biocompatibility. NPs' mechanisms of action responsible for their antimicrobial effect flow from the release of metal ions, oxidative stress as well as non-oxidative stress that happens simultaneously (Zaidi et al., 2017). These green NPs can target the bacterial cells and disrupt critical functions of the cell membranes, causing severe impacts on the respiration and permeability of membranes. In addition, G-NPs interact with intracellular components (proteins and nucleic acids), which inhibit many biological processes such as gene transfer and cell division (Dakal et al., 2016; Slavin et al., 2017). The mechanism of action of G-NPs and antibiotics appear to be similar in terms of interfering with the synthesis of protein, RNA, DNA and membrane disruption (Kohanski et al., 2010; Dakal et al., 2016). However, these metallic NPs and G-NPs antimicrobial activities hail from multiple mechanisms, invaluable in terms of limiting the ability bacteria developing resistance to them (Slavin et al., 2017). To develop resistance against these nanoparticles, the bacterial cells must obtain multiple gene mutations concurrently, which is highly unlikely. In addition, synthesizing metallic NPs using green methods allow bioactive compounds, proteins and polysaccharides to bind with nanoparticles. This enhances NPs antimicrobial activity against MDR bacteria. Figure 10.6 summarizes the possible mechanisms of action of G-NPs against bacteria.

10.8 POTENTIAL MEDICINAL PLANTS FOR GREEN SYNTHESIS OF NPS AGAINST MULTIDRUG RESISTANCE *S. AUREUS*

Staphylococcus aureus strains are pathogenic bacteria that have developed resistance to multiple drugs and so cause severe infections, in both hospitals and the wider community. Worse is that this bacterium can continually adapt, becoming even more resistant over time. Thus, attention has turned to finding novel treatment solutions to overcome this bacterium and its associated diseases, specifically using medicinal plants. Many reports have explored medicinal plants and/or their phytochemicals to cure *S. aureus* and its associated diseases. The antimicrobial potential of medicinal plants and phytochemicals against *S. aureus* (MRSA, VISA, VRSA and MSSA) have been addressed in several scientific studies. Previously tested plants showed antibacterial activity against both gram-positive and gram-negative bacteria, including MDR *S. aureus*.

In this regard, Talib and Mahasneh (2010) reported that the MIC value of *Rosa damascene* receptacles butanol and aqueous extracts were found to be 500 μg/ml, while MIC for *Inula viscosa* flowers butanol extract was 250 μg/ml. MRSA was found to be inhibited using butanol, *Rosa damascena* receptacles aqueous extracts as well as *Inula viscosa* flowers butanol extract (MIC values equals 500, 500 and 250 μg/ml), respectively. Meanwhile, MRSA was sensitive to *Rosa damascena* receptacles' ethanol extract with 95% inhibition, *Verbascum sinaiticum* flowers ethanol extract with 70% inhibition and *Inula viscosa* flowers ethanol extract with 92% inhibition. These results were attributed to phytochemicals present in plant extracts such as terpenoids and flavonoids, as well as alkaloids found in *Narcissus tazetta* aerial parts and *Ononis hirta* aerial parts. Great attention has been paid to the phytochemicals of plants, especially those linked to antimicrobial activity such as flavonoids, triterpenes, alkaloids, sesquiterpene lactones, diterpenes and naphthoquinones. Phytochemicals were isolated and tested for their antimicrobial activities (Ríos and Recio, 2005). For instance, flavonoids are commonly found in plant parts such as seeds, fruits, flowers, vegetables and stems. Many investigations have been undertaken to determine the flavonoids antibacterial mechanisms. In this regard, quercetin activity was found to be partly responsible for DNA gyrase inhibition. Epigallocatechin gallate and sophoraflavone hampered the metabolic functions of the cell membrane functions (Cushnie and Lamb, 2005). Meanwhile, flavonostilbenes exhibited unique antibacterial and antibiofilm effects against *S. epidermidis* with MIC of 3.1–12.5 μg/ml (Wan et al., 2015).

Rad et al. (2013) reported that *Xanthium strumarium* extract (at the 300 μl concentration) had the highest effect against MSSA (25 mm) and MRSA (20 mm) strains. Moreover, there was a correlation between the plant extract's concentration and bacterial growth inhibition. The antibacterial activity of *X. strumarium* extract could possibly be attributed to the flavonoids, phenolic acids, terpenoids and tannins present. The ethanolic extract of *Saussurea lappa* root was tested against MDR bacteria, including MRSA (Hasson et al., 2013). For MRSA, the bacteriostatic effect occurred at 2,000 μg/ml concentration, while the bactericidal effect occurred at 6,000 μg/ml concentration. Manubolu et al. (2013) investigated and identified the antibacterial *S. aureus* components from ethyl acetate, chloroform, methanol hexane and aqueous extracts of *Senecio tenuifolius* Burm. F. (*S. tenuifolius*) and their antibacterial effect was tested against *S. aureus* (ATCC 25923), MRSA and MSSA. Methanol extracts were found to significantly decrease the growth of *S. aureus* (ATCC 25923), MRSA and MSSA with the maximum inhibition zone at 16.23, 14.06 and 15.23 mm and MIC of 426.16, 683.22 and 512.12 μg/ml, respectively. To investigate the active component, the methanol extract was purified using column chromatography. Four fractions (St1, St2, St3 and St4) were obtained. St3 fraction was the most effective fraction against *S. aureus* (ATCC 25923), MRSA and MSSA, with the maximum inhibition zone at 15.09, 13.25 and 14.12 mm and the maximum MIC of 88.16, 128.11 and 116.12 μg/

ml, respectively. GC-MS analysis revealed that St3 fraction contains hydroquinone, 3-[methyl-6,7-dihydro benzofuran-4(5*H*)-one], 1,2-benzenedicarboxylic acid, methyl ester as well as three unknown compounds. The study concluded that medicinal plants (*S. tenuifolius*) have the potential to treat skin infections and combat MRSA and its associated diseases (Manubolu et al., 2013). Similarly, compounds with antibacterial activity were isolated from the roots of *Atractylodes japonica* (*A. japonica*) and further characterized. Four compounds were isolated and identified as **(1)**: atractylenolide III, **(2)**: atractylenolide I, **(3)**: diacetylatractylodiol [(6E,12E)-tetradeca-6,12-diene-8,10 -diyne-1,3-diol diacetate, TDEYA, and **(4)**: (6E,12E)-tetradecadiene-8,10-diyne1,3-di ol (TDEA). The compound number **(4)** showed antibacterial activity against all tested MRSA isolates with MIC values from 4 to 32 μg/ml. Similar findings were noted with other solvents fractions. However, the chloroform fraction exhibited the highest antibacterial activity against MRSA, which could be attributed to its bioactive components (Jeong et al., 2010). Results reveal the antibacterial effect of medicinal plants (*A. japonica*) and their extracts against MRSA.

In the search for alternative solutions against MRSA, the stem bark and leaf of *Tabernaemontana alternifolia* (Roxb) (an indigenous Indian medicine used to treat skin infections) were tested to determine their antibacterial activity against MRSA. *T. alternifolia* stem bark aqueous extracts showed antibacterial effects against MRSA and VRSA. The MIC values ranged from 600 to 800 μg/ml for MRSA. The phytochemical profiling showed that saponins, alkaloids, coumarins, flavonoids and steroids were present. Moreover, *T. alternifolia* extract did not show any cytotoxic activity towards Vero cells, making the extract a good candidate to treat MRSA infections (Marathe et al., 2013). This is also true for *Tabernaemontana stapfiana* (Britten) where different solvents of root and stem extracts showed good antibacterial activity. The phytochemical profiling showed that tannins, alkaloids, coumarins, flavonoids and saponins were present, all of which are associated with antimicrobial effects. The methanolic extract exhibited good antibacterial effects against the tested bacterial strains, including MRSA with MIC ranging between 15.6 and 500 mg/ml. MBC ranged between 31.25 and 500 mg/ml (Ruttoh et al., 2009).

The antibacterial effect of several solvents, as well as aqueous extracts of oregano, neem, bryophyllum, tulsi, aloe vera, rosemary, lemongrass and thyme were evaluated on 10 MDR clinical isolates (Dahiya and Purkayastha, 2012). Methanol and ethanol extracts showed significant inhibitory effects against most tested bacteria. *S. aureus* were the most inhibited bacteria in 24 of the extracts (60%). The MIC values of tulsi, rosemary, oregano, and aloe vera extracts were found to be in the range of 1.56–6.25 mg/ml when tested against MRSA. Phytochemical profiling showed the presence of saponins and tannins in all tested plants. Bioautography agar overlay analysis and TLC of ethanol extracts of tulsi, neem and aloe vera demonstrated that tannins and flavonoids are the main active compounds against MRSA (Dahiya and Purkayastha, 2012). Ursolic and oleanolic acids both were isolated from the leaves of *Salvia officinalis* (Sage) and these acids exhibited antibacterial effects against VRE, MRSA and *Streptococcus pneumoniae*. The antimicrobial effect of ursolic acid on VRE, MRSA and *S. pneumoniae* were double that of oleanolic acid (determined by calculating values from MIC) (Horiuchi et al., 2007).

Premna resinosa (Hochst.) Schauer is a medicinal plant used in the treatment of respiratory illnesses. *P. resinosa* has strong antibacterial effects against tuberculous with MIC value of <6.25 μg/ml in the fraction of ethyl acetate. Dichloromethane fraction, however, exhibited the maximum antibacterial MIC of 31.25 μg/ml towards MRSA. Meanwhile, ethyl acetate fraction showed the best inhibition zone of 22.3 ± 0.3 towards *S. aureus*. The antibacterial activity was associated with the detected anthraquinones, alkaloids, terpenoids, flavonoids and phenols in plant extracts. The study showed that *P. resinosa* is a high-potential source for novel antibacterial, antituberculous and antifungal drugs, with a possible strong effect against MRSA, *C. albicans*, *S. aureus* and *Mycobacterium tuberculosis* (MTB), which are well known as public health challenge (Njeru et al., 2015).

Hydrastis canadensis L. (golden seal) is a medicinal plant traditionally used in skin infections treatment. Cech et al. (2012) reported the activity of *H. canadensis* leaf extracts against MRSA. *H. canadensis* extract exhibited a higher antimicrobial effect than the alkaloid berberine alone as MICs values were found to be 75 and 150 μg/ml, respectively. LC-MS analysis detected alkaloids and flavonoids (efflux-pump inhibitory phytochemical) in the extract, which may explain the improved efficacy of *H. canadensis* in comparison to the berberine alone. The extract of *H. canadensis* has an anti-quorum sensing effect on MRSA, which can be attributed to the cell signalling reduction of the *Agr*CA two-component regulatory system (TCS). This extract was also found to inhibit the MRSA toxin production and damage keratinocytes. The antimicrobial activity of *Rumex nervosus* (aerial part) was evaluated against *S. aureus*, MRSA, *E. faecalis*, *E. coli*, *P. aeruginosa* and *Candida albicans*. The results showed that medicinal plants' polar extracts have significant antimicrobial effects. Among gram-positive bacteria, *R. nervosus* extract showed a dose-dependent growth inhibition effect against *S. aureus* and MRSA (Al-Asmari et al., 2015). In contrast, the antibacterial effect of non-polar extracts exhibited higher antimicrobial effect against MRSA compared to polar extracts (Ahmad et al., 2014). Lemongrass chloroform and hexane extracts were ineffective on the studied bacteria (MRSA, *S. aureus*, *K. pneumoniae*, *E. coli* and *P. mirabilis*). The reason for this minimal antibacterial effect could be attributed to the low concentration of phytochemicals in these extracts (Dahiya and Purkayastha, 2012). The medicinal plant of *Eleucine indica* is traditionally used to treat diseases of the kidneys and liver. Its therapeutic effect is often linked to their antioxidant characteristics. Hexane extract in particular showed a remarkable antibacterial effect against MRSA (Al-Zubairi et al., 2011). Similarly, Aliahmadi et al. (2014) reported that the highest antibacterial activities were found in the hexane extract of *Bromus inermis* Leyss Inflorescences. Hexane extract had a significant impact on MRSA with 8 μg/ml MIC value.

Rhodomyrtus tomentosa leaf ethanolic extract showed significant antibacterial effect against both MRSA and *S. aureus* ATCC 29213. Its MIC values range between 31.25 and 62.5 μg/ml, and the MBC was 250 μg/ml. The impact of rhodomyrtone on the expression of MRSA's cellular protein has been studied using proteomic approaches in order to provide insights into the antibacterial mechanisms involved. Proteome analyses show that the subinhibitory concentration of rhodomyrtone (0.174 μg/ml) influences the expression of multiple main functional classes of MRSA whole cell proteins. Transmission electron micrographs revealed rhodomyrtone impacts in the treated MRSA, notably ultrastructural and morphological alterations. Important biological processes in cell division and cell wall biosynthesis were found to be interrupted. Significant changes included cellular disintegration, cell wall alterations, formation of abnormal septum and cell lysis. Abnormal shape and size of *Staphylococcal* cells were observed in the treated MRSA cells (Sianglum et al., 2011).

The ability of MRSA to form biofilms is one of the major attributes in its pathogenicity. In this regard, de Araujo et al. (2015) evaluated the antibiofilm, antibacterial and cytotoxic impacts of *Terminalia fagifolia* stem bark ethanol extract (EtE) as well as the three extract fractions (HaF, AqF and WSF). It was noticed that antibacterial effect MICs values of EtE and the fractions were in the range of 25–200 μg/ml; meanwhile, the MBCs values ranged from 200 to 400 μg/ml. The antibiofilm activity of both the EtE and the HaF, AqF and WSF fractions exhibited remarkable biofilm formation inhibition, in over 80% of the tested strains. Microscopic images from the AFM show changes in morphology in addition to significant size alterations of *S. aureus* ATCC 29213 surface, caused by AqF.

Quercus infectoria G. Olivier nutgall is well known in traditional Thai medicine as an efficient drug for skin and wound infections. Chusri and Voravuthikunchai (2011) reported the effect of different *Q. infectoria* fractions and its purified compounds against MRSA and *S. aureus*, which caused hypersensitivity to low and high osmotic pressure. The synergistic effect of *Q. infectoria* extract with β-lactam antibiotics have shown that *Q. infectoria* can interfere with the *Staphylococcal* enzymes, including β-lactamase and autolysins (Chusri and Voravuthikunchai, 2009). Results also show that *Q. infectoria* extract as well as tannic acid influences the biofilm formation, bacterial cell surface hydrophobicity and the cell wall, which might impact on the anti-formation activity of biofilm (Chusri et al., 2012).

In addition to the above-mentioned studies, many reports evaluated the antimicrobial potential of medicinal plants as well as its purified phytochemicals against MRSA such as *Rosa canina L. (rose red), Cinnamomum iners, Camellia sinensis, Juglans regia, Psoralea corylifolia, Abrus schimperi, Atuna racemose, Tectona grandis,* and *Plectranthus amboinicus* (Lour.) among others (Shiota et al., 2000; Neamatallah et al., 2005; Buenz et al., 2007; Mustaffa et al., 2011; Rahman et al., 2011; de Oliveira et al., 2013; Farooqui et al., 2015; Cui et al,. 2015).

10.9 PLANT PHYTOCHEMICALS RESPONSIBLE FOR NPS FORMATION AND STABILIZATION

Several phytochemicals in plant extracts such as polysaccharides, enzymes, vitamins, amino acids, proteins as well as organic acids (Table 10.4) were found to have the ability to reduce metal ions. The interactions between plant extracts and NPs occur chemically or physically. The chemical interactions usually responsible for reactive oxygen species production interfere with ion membrane transport activity, peroxidation of lipid and oxidative damage. Upon entering the cell, NPs act like metal ions and react with carboxyl and sulfhydryl groups affecting

TABLE 10.4
Plant Phytochemicals Involved in the Formation and Stabilization of G-NPs

Plant species	Type of NPs	Identified phytochemicals	Reference
Catharanthus roseus (CR) and *Carica papaya (CP)*	AuNPs	Flavonoids, alkaloids and proteins	Muthukumar et al. (2016)
Cassia siamea (Kassod)	ZnO-NPs	Terpenoids, flavonoids and phenols	Chauhan et al. (2019)
Azadirachta indica	TiO-NPs	Flavonoids, terpenoids and proteins	Thakur et al. (2019)
Macrotyloma uniflorum	AgNPs	Phenols	Vidhu et al. (2011)
Rumex hymenosepalus	AgNPs	Polyphenols (e.g. stilbenes and catechins)	Rodríguez-León et al. (2013)
Mirabilis jalapa	AgNPa	Alkaloids and saponins	Asha et al. (2017)
Scutellaria barbata	AuNPs	Steroids, diterpenoids, polysaccharides, flavonoids and alkaloids	Wang et al. (2009)
Punica granutum	AuNPs	Phenolic hydroxyls and aromatic rings	Ganeshkumar et al. (2013)
Calotropis procera	ZnO-NPs	Carboxylic acids, ketones, amines, aldehydes and hydroxyl groups	Gawade et al. (2017)
Rosmarinus officinalis	AgNPs	Polyphenols	Ghaedi et al. (2015)
Stevia rebaudiana	AuNPs	Biomolecules with primary carbonyl and amine groups	Sadeghi, Mohammadzadeh, et al. (2015)
Salix alba	AuNPs	Proteins and metabolites having functional groups of amines, alcohols, ketones, aldehydes, carboxylic acids (salicin)	Islam et al. (2019)
Galaxaura elongata	AuNPs	Glutamic acid, hexadecanoic acid, oleic acid, 11-eicosenoic acid, stearic acid, gallic acid, epigallocatechin, catechin and epicatechin gallate	Abdel-Raouf et al. (2017)
Stevia rebaudiana	AgNPs	Ketones	Yilmaz et al. (2011)
Rosa rugosa	AgNPs	Carboxylate and amine groups	Dubey et al. (2010)

protein activity. The difference in NPs characteristics such as size and shape relies on the existence of phytochemicals such as polysaccharides, phenolic amides, piperine and other reducing components responsible for the formation of metal NPs. Various phytochemicals helped metal NPs formation from their corresponding ionic compounds. Phytochemicals involved in the reduction reaction of metal ions in order to synthesize NPs are not limited to proteins, sugars, phenolic acids, organic compounds, pigments, polyphenols, alkaloids, terpenoids and flavonoids (Makarov et al., 2014; Rajeshkumar, 2016; Mohammadlou et al., 2016; Abdelghany et al., 2017). These phytochemicals also have the ability to treat several diseases (Dubey et al., 2010). For instance, phenolic acids with hydroxyl cinnamic or hydroxyl benzoic structures have been reported to possess carbonyl and hydroxyl groups that can bind to metals. It was reported previously that phenolic compounds in *Macrotyloma uniflorum* seed extract could be a key factor in the formation of AgNPs (Vidhu et al., 2011). Moreover, *R. hymenosepalus* root extract was reported to be rich in polyphenols (e.g. stilbenes and catechins) and plays a role in stabilizing and reducing agents for the formation of G-AgNPs. The main mechanisms proposed was the abstraction of hydrogen due to the OH groups present in the polyphenol molecules (Rodríguez-León et al., 2013). Phytochemical screening of *Mirabilis jalapa* aqueous leaf extracts showed the presence of phenols, tannins, alkaloids and flavonoids. In contrast, there were no glycosides, saponins or steroids present. Alkaloids and saponins played role in AgNPa synthesized using leaf extract, while steroids, flavonoid, glycosides, phenol and tannins were not present (Asha et al., 2017).

The bioreduction involved in reducing the ions of zinc oxide to form ZnO-NPs were confirmed with the use of phytochemicals such as terpenoids, polysaccharides, alkaloids, phosphorous compounds, secondary sulfonamides, polyphenolic compounds, amino acids, vitamins, amino acids, diketone and vinyl (Agarwal et al., 2017). Steroids, diterpenoids, polysaccharides, flavonoids and alkaloids from *Scutellaria barbata* were found to be involved in AuNPs bioreduction (Wang et al., 2009). Ketones were reported to play an important role in the synthesis of AgNPs using plant extracts (Yilmaz et al., 2011). *Punica granutum* fruit peel extracts were subjected to FTIR analysis to show the presence of phenolic hydroxyls (O—H band) as well as an aromatic ring that acted as a functional group in capping AuNPs (Ganeshkumar et al., 2013). The amine groups from *Hibiscus rosa sinensis* leaf extract bonded with AuNPs. Moreover, carboxylate ion groups from the same plant extract (*Hibiscus rosa sinensis*) were found to bind with AuNPs and AgNPs (Philip, 2010). AuNPs were functionalized with stabilizing functional groups from *Stevia rebaudiana* extracts such as biomolecules with primary carbonyl group or amine group ($-NH_2$) (Sadeghi et al., 2015).

FTIR results of *Sargassum muticum* showed soluble elements present in the aqueous extract (such as the C–O from C–O–SO_3 group, C—C groups from aromatic rings and carbonyl [–C=O] group) acted as a capping agent preventing NPs aggregation in the solution. This played a major role in the extracellular synthesis as well (Mahdavi et al., 2013). Likewise, FTIR results of *Calotropis procera* aqueous leaf extract showed that carboxylic acids, ketones, amines, aldehydes and hydroxyl groups were responsible for the biochemical reaction (Gawade et al., 2017). Polyphenols from plants lose electrons easily, resulting in zinc–ellagate complex formation. Zinc ions ligated with aromatic hydroxyl groups found in *Nephelium lappaceum* peel extracts formed zinc–ellagate complex at pH 5–7 to synthesize ZnO-NPs (Yuvakkumar et al., 2014). In summary, various plant phytochemicals such as phenolic acids, alkaloids terpenoids and polyphenols can bioreduce metal ions to form NPs (Table 10.4). They are both capping and stabilizing agents. Additionally, they have desirable properties such as antimicrobial, antioxidant and anticancer activities. These characteristics make them highly suitable for pharmaceutical applications.

10.10 CONCLUSION

The emergence of antibiotic resistance among the bacterial population has called attention to the need to develop and design new antimicrobial drugs. There is an urgent need to find novel, effective antibacterial agents to treat MDR bacteria such as MRSA, VISA and VRSA. This could be feasible when seeking novel systems such as combining two antibacterial agents or combining medicinal plant phytochemicals with metal nanoparticles such that their synergistic effects work simultaneously to create antibacterial drugs with many desirable characteristics. This chapter presents different medicinal plants and their phytochemicals which could inhibit MDR *S. aureus*, making them a good natural source of new effective drugs to combat MRSA, VISA and VRSA. Metal NPs have various applications in variety of fields, including disease treatment and diagnostics. Moreover, metal NPs are synthesized using different physical, chemical, biological and mechanical methods. However, biological or green methods are cost-effective, eco-friendly, non-toxic, use less energy, time-efficient, easy to adopt and suitable for large-scale production. Medicinal plants are a biological source that could be used to green synthesize NPs (G-NPs), for both reducing and stabilizing agents for the green synthesis of controlled shape and size NPs. G-NPs have many desirable characteristics such as stability, biocompatibility and low levels of toxicity, as well as capping with bioactive phytochemicals that could enhance their medicinal properties. Several studies have focused on the green syntheses of metal-based NPs by using plant extracts and their phytochemicals such as the ones discussed in this chapter. However, there is still an abundance of possible medicinal plant species that are yet unexplored. Moreover, this chapter explored the previously tested medicinal plants against multidrug-resistant *S. aureus*, which were already proven to have antibacterial activity yet not been used for green synthesis of G-NPs in order to be tested against multidrug-resistant *S. aureus*. Many of these could be excellent candidates for green-synthesized metal-based NPs and so enable the discovery of novel and effective antibacterial agents against MDR bacteria.

ACKNOWLEDGEMENTS

This work was supported by the National Natural Science Foundation of China (Grant Nos. 51803067, 51973076 and 21774039).

REFERENCES

Abbasi, B. A., Iqbal, J., Zahra, S. A., Shahbaz, A., Kanwal, S., Rabbani, A., and Mahmood, T. 2020. "Bioinspired synthesis and activity characterization of iron oxide nanoparticles made using *Rhamnus Triquetra* leaf extract." *Materials Research Express* 6(12): 1250–1257. DOI: 10.1088/2053-1591/ab664d.

Abdelghany, A. M., Abdelrazek, E. M., Badr, S. I., Abdel-Aziz, M. S., and Morsi, M. A. 2017. "Effect of Gamma-irradiation on biosynthesized gold nanoparticles using *Chenopodium murale* leaf extract." *Journal of Saudi Chemical Society* 21(5): 528–537. DOI: 10.1016/j.jscs.2015.10.002.

Abdel-Raouf, N., Al-Enazi, N. M., and Ibraheem, I. B. M. 2017. "Green biosynthesis of gold nanoparticles using *Galaxaura elongata* and characterization of their antibacterial activity." *Arabian Journal of Chemistry* 10: S3029–S3039. DOI: 10.1016/j.arabjc.2013.11.044.

Abou El-Nour, K. M. M., Eftaiha, A., Al-Warthan, A., and Ammar, R. A. A. 2010. "Synthesis and applications of silver nanoparticles." *Arabian Journal of Chemistry* 3(3): 135–140. DOI: 10.1016/j.arabjc.2010.04.008.

Agarwal, H., Venkat Kumar, S., and Rajeshkumar, S. 2017. "A review on green synthesis of zinc oxide nanoparticles: An eco-friendly approach." *Resource-Efficient Technologies* 3(4): 406–413. DOI: 10.1016/j.reffit.2017.03.002.

Ahmad, R., Baharum, S. N., Bunawan, H., Lee, M., Mohd Noor, N., Rohani, E. R., Ilias, N., and Zin, N. M. 2014. "Volatile profiling of aromatic traditional medicinal plant, *Polygonum minus* in different tissues and its biological activities." *Molecules* 19(11): 19220–19242.

Ahmad, T., Irfan, M., and Bhattacharjee, S. 2016. "Parametric study on gold nanoparticle synthesis using aqueous *Elaise Guineensis* (Oil palm) leaf extract: Effect of precursor concentration." *Procedia Engineering* 148: 1396–1401. DOI: 10.1016/j.proeng.2016.06.558.

Ahmed, K. B. A., Subramanian, S., Sivasubramanian, A., Veerappan, G., and Veerappan, A. 2014. "Preparation of gold nanoparticles using Salicornia brachiata plant extract and evaluation of catalytic and antibacterial activity". *Spectrochimica Acta Part A: Molecular and Biomolecular Spectroscopy* 130: 54–58.

Alagesan, V., and Venugopal, S. 2018. "Green synthesis of selenium nanoparticle using leaves extract of *Withania somnifera* and its biological applications and photocatalytic Activities." *BioNanoScience* 9(1): 105–116. DOI: 10.1007/s12668-018-0566-8.

Al-Asmari, A. R. K., Siddiqui, Y. M., Athar, M. T., Al-Buraidi, A., Al-Eid, A., and Horaib, G. B. 2015. "Antimicrobial activity of aqueous and organic extracts of a Saudi medicinal plant: *Rumex nervosus*." *Journal of Pharmacy & Bioallied Sciences* 7(4): 300–303.

Alavi, M., and Karimi, N. 2018. "Characterization, antibacterial, total antioxidant, scavenging, reducing power and ion chelating activities of green synthesized silver, copper and titanium dioxide nanoparticles using *Artemisia haussknechtii* leaf extract". *Artificial Cells, Nanomedicine, and Biotechnology* 46 (8): 2066–2081. DOI: 10.1080/21691401.2017.1408121.

Aliahmadi, A., Mirzajani, F., Ghassempour, A., and Sonboli, A. 2014. "Bioassay guided fractionation of an anti-methicillin-resistant *Staphylococcus aureus* flavonoid from *Bromus inermis* Leyss Inflorescences." *Jundishapur Journal of Microbiology* 7(12): 1–4. DOI: 10.5812/jjm.12739.

Alsalhi, M. S., Devanesan, S., Alfuraydi, A. A., Vishnubalaji, R., Munusamy, M. A., Murugan, K., Nicoletti, M., and Benelli, G. 2016. "Green synthesis of silver nanoparticles using *Pimpinella anisum* seeds: Antimicrobial activity and cytotoxicity on human neonatal skin stromal cells and colon cancer cells." *International Journal of Nanomedicine* 11: 4439–4449. DOI: 10.2147/IJN.S113193.

Alzahrani, E. 2016. "Synthesis of copper nanoparticles with various sizes and shapes: Application as a superior non-enzymatic sensor and antibacterial agent." *International Journal of Electrochemical Science* 11(6): 4712–4723. DOI: 10.20964/2016.06.83.

Al-Zubairi, A. S., Abdul, A. B., Abdelwahab, S. I., Peng, C. Y., Mohan, S., and Elhassan, M. M. 2011. "*Eleucine indica* possesses antioxidant, antibacterial and cytotoxic properties." *Evidence-Based Complementary and Alternative Medicine* 2011: 1–6. DOI: 10.1093/ecam/nep091.

Anand, U., Jacobo-Herrera, N., Altemimi, A., and Lakhssassi, N. 2019. "A comprehensive review on medicinal plants as antimicrobial therapeutics: Potential avenues of biocompatible drug discovery." *Metabolites* 9(11): 1–13. DOI: 10.3390/metabo9110258.

Ansari, M. A., and Alzohairy, M. A. 2018. "One-pot facile green synthesis of silver nanoparticles using seed extract of *Phoenix dactylifera* and their bactericidal potential against MRSA." *Evidence-Based Complementary and Alternative Medicine* 2018: 1–9. DOI: 10.1155/2018/1860280.

Arthur, M., Molinas, C., Depardieu, F., and Courvalin, P. 1993. "Characterization of Tn1546, a Tn3-related transposon conferring glycopeptide resistance by synthesis of depsipeptide peptidoglycan precursors in *Enterococcus faecium* BM4147." *Journal of Bacteriology* 175(1): 117–127. DOI: 10.1128/jb.175.1.117-127.1993.

Asha, S., Thirunavukkarasu, P., and Shanmugam, R. 2017. "Green synthesis of silver nanoparticles using *Mirabilis jalapa* aqueous extract and their antibacterial activity against respective microorganisms." *Research Journal of Pharmacy and Technology* 10(3): 811–817. DOI: 10.5958/0974-360X.2017.00153.6.

Askarinia, M., Ganji, A., Jadidi-Niaragh, F., Hasanzadeh, S., Mohammadi, B., Ghalamfarsa, F., Ghalamfarsa, G., and Mahmoudi, H. 2019. "A review on medicinal plant extracts and their active ingredients against methicillin-resistant and methicillin-sensitive *Staphylococcus aureus*." *Journal of Herbmed Pharmacology* 8(3): 173–184. DOI: 10.15171/jhp.2019.27.

Balachandar, R., Gurumoorthy, P., Karmegam, N. et al. 2019. "Plant-mediated synthesis, characterization and bactericidal potential of emerging silver nanoparticles using stem extract of phyllanthus pinnatus: A recent advance in phytonanotechnology". *J Clust Sci.* 30: 1481–1488. https://doi.org/10.1007/s10876-019-01591-y

Banasiuk, R., Krychowiak, M., Swigon, D., Tomaszewicz, W., Michalak, A., Chylewska, A., Ziabka, M. et al. 2020. "Carnivorous plants used for green synthesis of silver nanoparticles with broad-spectrum antimicrobial activity." *Arabian Journal of Chemistry* 13(1): 1415–1428. DOI: 10.1016/j.arabjc.2017.11.013.

Barry, K., Kornblum, J., D. Arbeit, R., Eisner, W., JMaslow, o. N., McGeer, A., Low, D. E., and Novick, R. P. 1993. "Evidence for a clonal origin of methicillin resistance in *Staphylococcus aureus*." *Science* 259(5092): 227–230. DOI: 10.1126/science.8093647.

Bagherzade, G., Tavakoli, M. M., and Namaei, M. H. 2017. "Green synthesis of silver nanoparticles using aqueous extract of saffron (Crocus sativus L.) wastages and its antibacterial activity against six bacteria". *Asian Pacific Journal of Tropical Biomedicine* 7(3): 227–233.

Balasubramanian, S., Kala, S. M. J., Pushparaj, T. L., and Kumar, P. 2019. "Biofabrication of gold nanoparticles using cressa cretica leaf extract and evaluation of catalytic and antibacterial efficacy". *Nano Biomedicine and Engineering* 11(1): 58–66.

Behera, N., Arakha, M., Priyadarshinee, M., Pattanayak, B. S., Soren, S., Jha, S., and Mallick, B. C. 2019. "Oxidative stress generated at nickel oxide nanoparticle interface results in bacterial membrane damage leading to cell death." *RSC Advances* 9(43): 24888–24894. DOI: 10.1039/c9ra02082a.

Bindhu, M. R., and Umadevi, M. 2014a. "Antibacterial activities of green synthesized gold nanoparticles". *Materials Letters* 120: 122–125.

Bindhu, M. R., and Umadevi, M. 2014b. "Silver and gold nanoparticles for sensor and antibacterial applications". *Spectrochimica Acta Part A: Molecular and Biomolecular Spectroscopy* 128: 37–45.

Boucher, H. W., Talbot, G. H., Bradley, J. S., Edwards, J. E., Gilbert, D., Rice, L. B., Scheld, M., Spellberg, B., and Bartlett, J. 2009. "Bad bugs, no drugs: No ESKAPE! An update from the infectious diseases society of America." *Clinical Infectious Diseases* 48(1): 1–12. DOI: 10.1086/595011.

Boyle-Vavra, S., Carey, R. B., and Daum, R. S. 2001. "Development of Vancomycin and lysostaphin resistance in a methicillin-resistant *Staphylococcus aureus* isolate." *Journal of Antimicrobial Chemotherapy* 48(5): 617–625. DOI: 10.1093/jac/48.5.617.

Boyle-Vavra, S., Challapalli, M., and Daum, R. S. 2003. "Resistance to autolysis in Vancomycin-selected *Staphylococcus aureus* isolates precedes Vancomycin-intermediate resistance." *Antimicrobial Agents and Chemotherapy* 47(6): 2036–2039. DOI: 10.1128/aac.47.6.2036-2039.2003.

Buenz, E. J., Bauer, B. A., Schnepple, D. J., Wahner-Roedler, D. L., Vandell, A. G., and Howe, C. L. 2007. "A randomized phase I study of *Atuna racemosa*: A potential new anti-MRSA natural product extract." *Journal of Ethnopharmacology* 114(3): 371–376. DOI: 10.1016/j.jep.2007.08.027.

Burygin GL, Khlebtsov BN, Shantrokha AN, Dykman LA, Bogatyrev VA, Khlebtsov NG. 2009. "On the enhanced antibacterial activity of antibiotics mixed with gold nanoparticles". *Nanoscale Res Lett.* 4(8): 794–801. doi:10.1007/s11671-009-9316-8

Cech, N. B., Junio, H. A., Ackermann, L. W., Kavanaugh, J. S., and Horswill, A. R. 2012. "Quorum quenching and antimicrobial activity of goldenseal (*Hydrastis canadensis*) against methicillin-resistant *Staphylococcus aureus* (MRSA)." *Planta Medica* 78(14): 1556–1561. DOI: 10.1055/s-0032-1315042.

Centers for Disease Control Prevention. 1999. "Four pediatric deaths from community-acquired methicillin-resistant *Staphylococcus aureus*—Minnesota and North Dakota, 1997–1999." *MMWR. Morbidity and Mortality Weekly Report* 48: 707.

Centers for Disease Control and Prevention. 2002. "*Staphylococcus aureus* resistant to vancomycin—United States, 2002." *MMWR Morbidity and Mortality Weekly Report* 51(26): 565–567.

Chambers, H. F., and Deleo, F. R. 2009. "Waves of resistance: *Staphylococcus aureus* in the antibiotic era." *Nature Reviews Microbiology* 7(9): 629–641. DOI: 10.1038/nrmicro2200.

Chang, S., Sievert, D. M., Hageman, J. C., Boulton, M. L., Tenover, F. C., Downes, F. P., Shah, S. et al. 2003. "Infection with vancomycin-resistant *Staphylococcus aureus* containing the *van*A resistance gene." *New England Journal of Medicine* 348(14): 1342–1347. DOI: 10.1056/NEJMoa025025.

Chauhan, P. S., Shrivastava, V., and Tomar, R. S. 2019. "Biosynthesis of zinc oxide nanoparticles using *Cassia siamea* leaves extracts and their efficacy evaluation as potential antimicrobial agent." *Journal of Pharmacognosy and Phytochemistry* 8(3): 162–166. DOI: 10.13140/RG.2.2.13678.87367.

Chusri, S., and Voravuthikunchai, S. P. 2009. "Detailed studies on *Quercus infectoria* Olivier (nutgalls) as an alternative treatment for methicillin-resistant *Staphylococcus aureus* infections." *Journal of Applied Microbiology* 106(1): 89–96. DOI: 10.1111/j.1365-2672.2008.03979.x.

Chusri, S., and Voravuthikunchai, S. P. 2011. "Damage of *Staphylococcal* cytoplasmic membrane by *Quercus infectoria* G. Olivier and its components." *Letters in Applied Microbiology* 52(6): 565–572. DOI: 10.1111/j.1472-765X.2011.03041.x.

Chusri, S., Phatthalung, P. N., and Voravuthikunchai, S. P. 2012. "Anti-biofilm activity of *Quercus infectoria* G. Olivier against methicillin-resistant *Staphylococcus aureus.*" *Letters in Applied Microbiology* 54(6): 511–517. DOI: 10.1111/j.1472-765X.2012.03236.x.

Clogston, J. D., and Patri, A. K. 2011. "Zeta potential measurement." In *Characterization of Nanoparticles Intended for Drug Delivery*, 63–70. Cham: Springer.

Coates, A., Hu, Y., Bax, R., and Page, C. 2002. "The future challenges facing the development of new antimicrobial drugs." *Nature Reviews Drug Discovery* 1(11): 895–910. DOI: 10.1038/nrd940.

Cong, Y., Yang, S., and Rao, X. 2020. "Vancomycin resistant *Staphylococcus aureus* infections: A review of case updating and clinical features." *Journal of Advanced Research* 21: 169–176. DOI: 10.1016/j.jare.2019.10.005.

Constabel, C. P., Balza, F., and Towers, G. N. 1988. "Dithiacyclohexadienes and thiophenes of *Rudbeckia hirta*." *Phytochemistry* 27(11): 3533–3535. DOI: 10.1016/0031-9422(88)80762-3.

Cui, Y., Taniguchi, S., Kuroda, T., and Hatano, T. 2015. "Constituents of *Psoralea corylifolia* fruits and their effects on methicillin-resistant *Staphylococcus aureus*." *Molecules* 20(7): 12500–12511. DOI: 10.3390/molecules200712500.

Cushnie, T. P., and Lamb, A. J. 2005. "Antimicrobial activity of flavonoids." *International Journal of Antimicrobial Agents* 26(5): 343–356. DOI: 10.1016/j.ijantimicag.2005.09.002.

Da Costa, F. B., Terfloth, L., and Gasteiger, J. 2005. "Sesquiterpene lactone-based classification of three *Asteraceae* tribes: A study based on self-organizing neural networks applied to chemosystematics." *Phytochemistry* 66(3): 345–353. DOI: 10.1016/j.phytochem.2004.12.006.

D'Agata, E. M., Webb, G. F., Horn, M. A., Moellering, R. C., and Ruan, S. 2009. "Modeling the invasion of community-acquired methicillin-resistant *Staphylococcus aureus* into hospitals." *Clinical Infectious Diseases* 48(3): 274–284. DOI: 10.1086/595844.

Dahiya, P., and Purkayastha, S. 2012. "Phytochemical screening and antimicrobial activity of some medicinal plants against multi-drug resistant bacteria from clinical isolates." *Indian Journal of Pharmaceutical Sciences* 74(5): 443–450. DOI: 10.4103/0250-474X.108420.

Dakal, T. C., Kumar, A., Majumdar, R. S., and Yadav, V. 2016. "Mechanistic basis of antimicrobial actions of silver nanoparticles." *Frontiers in Microbiology* 7: 1831. DOI: 10.3389/fmicb.2016.01831.

Dantes, R., Mu, Y., Belflower, R., Aragon, D., Dumyati, G., Harrison, L. H., Lessa, F. C. et al. 2013. "National burden of invasive methicillin-resistant *Staphylococcus aureus* infections, United States, 2011." *JAMA Internal Medicine* 173(21): 1970–1978. DOI: 10.1001/jamainternmed.2013.10423.

Daum, R. S., Gupta, S., Sabbagh, R., and Milewski, W. M. 1992. "Characterization of *Staphylococcus aureus* isolates with decreased susceptibility to Vancomycin and Teicoplanin: Isolation and purification of a constitutively produced protein associated with decreased susceptibility." *The Journal of Infectious Diseases* 166(5): 1066–1072. DOI: 10.1093/infdis/166.5.1066.

de Araujo, A. R., Quelemes, P. V., Perfeito, M. L. G., de Lima, L. I., Sá, M. C., Nunes, P. H. M., Joanitti, G. A., Eaton, P., dos Santos Soares, M. J., and de Almeida, J. R. d. S. 2015. "Antibacterial, antibiofilm and cytotoxic activities of *Terminalia fagifolia* Mart. extract and fractions." *Annals of Clinical Microbiology and Antimicrobials* 14(25): 1–10. DOI: 10.1186/s12941-015-0084-2.

Degirmenci, H., and Erkurt, H. 2020. "Relationship between volatile components, antimicrobial and antioxidant properties of the essential oil, hydrosol and extracts of *Citrus aurantium* L. flowers." *Journal of Infection and Public Health* 13(1): 58–67. DOI: 10.1016/j.jiph.2019.06.017.

Dehghanizade, S., Arasteh, J., and Mirzaie, A. 2018. "Green synthesis of silver nanoparticles using *Anthemis atropatana* extract: characterization and in vitro biological activities." *Artificial Cells, Nanomedicine, and Biotechnology* 46(1): 160–168. DOI: 10.1080/21691401.2017.1304402.

de Jong, N. W. M., van Kessel, K. P. M., and van Strijp, J. A. G. 2019. "Immune evasion by *Staphylococcus aureus*." *Microbiology Spectrum* 7(2): GPP3-0061-2019. DOI: 10.1128/microbiolspec.GPP3-0061-2019.

de Kraker, M. E., Stewardson, A. J., and Harbarth, S. 2016. "Will 10 million people die a year due to antimicrobial resistance by 2050?" *PLoS Medicine* 13(11): 1–6. DOI: 10.1371/journal.pmed.1002184.

de Oliveira, F. F. M., Torres, A. F., Gonçalves, T. B., Santiago, G. M. P., de Carvalho, C. B. M., Aguiar, M. B., Camara, L. M. C., Rabenhorst, S. H., Martins, A. M. C., and Valença Junior, J. T. 2013. "Efficacy of *Plectranthus amboinicus* (Lour.) Spreng in a murine model of methicillin-resistant *Staphylococcus aureus* skin abscesses." *Evidence-Based Complementary and Alternative Medicine* 2013: 1–9. DOI: 10.1155/2013/291592.

Dinesh, M., Roopan, S. M., Selvaraj, C. I., and Arunachalam, P. 2017. "*Phyllanthus emblica* seed extract mediated synthesis of PdNPs against antibacterial, heamolytic and cytotoxic studies." *Journal of Photochemistry and Photobiology B: Biology* 167: 64–71. DOI: 10.1016/j.jphotobiol.2016.12.012.

Djeussi, D. E., Noumedem, J. A., Seukep, J. A., Fankam, A. G., Voukeng, I. K., Tankeo, S. B., Nkuete, A. H., and Kuete, V. 2013. "Antibacterial activities of selected edible plants extracts against multidrug-resistant Gram-negative bacteria." *BMC Complementary and Alternative Medicine* 13(164): 1–8. DOI: https://doi.org/10.1186/1472-6882-13-164.

Dobrucka, R. 2018. "Antioxidant and catalytic activity of biosynthesized CuO nanoparticles using extract of Galeopsidis herba". *Journal of Inorganic and Organometallic Polymers and Materials* 28(3): 812–819.

Donahue, J. A., and Baldwin, J. N. 1966. "Hemolysin and leukocidin production by 80/81 strains of *Staphylococcus aureus*." *The Journal of Infectious Diseases*: 116(3):324–328.

Dubey, S. P., Lahtinen, M., and Sillanpää, M. 2010. "Green synthesis and characterizations of silver and gold nanoparticles using leaf extract of *Rosa rugosa*." *Colloids and Surfaces A: Physicochemical and Engineering Aspects* 364(1–3): 34–41. DOI: 10.1016/j.colsurfa.2010.04.023.

Duraipandiyan, V., Ayyanar, M., and Ignacimuthu, S. 2006. "Antimicrobial activity of some ethnomedicinal plants used by Paliyar tribe from Tamil Nadu, India." *BMC Complementary and Alternative Medicine* 6(35): 1–7. DOI: 10.1186/1472-6882-6-35.

Elango, G., and Roopan, S. M. 2015. "Green synthesis, spectroscopic investigation and photocatalytic activity of lead nanoparticles." *Spectrochimica Acta Part A: Molecular and Biomolecular Spectroscopy* 139: 367–373. DOI: 10.1016/j.saa.2014.12.066.

Enright, M. C., Robinson, D. A., Randle, G., Feil, E. J., Grundmann, H., and Spratt, B. G. 2002. "The evolutionary history of methicillin-resistant *Staphylococcus aureus* (MRSA)." *Proceedings of the National Academy of Sciences of the United States of America* 99(11): 7687–7692. DOI: 10.1073/pnas.122108599.

Fabrega, J., Zhang, R., Renshaw, J. C., Liu, W.-T., and Lead, J. R. 2011. "Impact of silver nanoparticles on natural marine biofilm bacteria." *Chemosphere* 85(6): 961–966. DOI: 10.1016/j.chemosphere.2011.06.066.

Fahimirad, S., Ajalloueian, F., and Ghorbanpour, M. 2019. "Synthesis and therapeutic potential of silver nanomaterials derived from plant extracts." *Ecotoxicology and Environmental Safety* 168: 260–278. DOI: 10.1016/j.ecoenv.2018.10.017.

Farooqui, A., Khan, A., Borghetto, I., Kazmi, S. U., Rubino, S., and Paglietti, B. 2015. "Synergistic antimicrobial activity of *Camellia sinensis* and *Juglans regia* against multidrug-resistant bacteria." *PloS one* 10(2): 1–14. DOI: 10.1371/journal.pone.0118431.

Feng, Q. L., Wu, J., Chen, G. Q., Cui, F. Z., Kim, T. N., and Kim, J. O. 2000. "A mechanistic study of the antibacterial effect of silver ions on Escherichia coli and *Staphylococcus aureus*." *Journal of Biomedical Materials Research* 52(4): 662–668. DOI: 10.1002/1097-4636(20001215)52:4<662::aid-jbm10>3.0.co;2-3.

Fleming, A. 1980. "On the antibacterial action of cultures of a Penicillium, with special reference to their use in the isolation of *B. influenzae*." *Reviews of Infectious Diseases* 2(1): 129–139.

Fouad, H., Hongjie, L., Yanmei, D., Baoting, Y., El-Shakh, A., Abbas, G., and Jianchu, M. 2017. "Synthesis and characterization of silver nanoparticles using *Bacillus amyloliquefaciens* and *Bacillus subtilis* to control filarial vector *Culex pipiens* pallens and its antimicrobial activity." *Artificial Cells, Nanomedicine, and Biotechnology* 45(7): 1369–1378. DOI: 10.1080/21691401.2016.1241793.

Ganeshkumar, M., Sathishkumar, M., Ponrasu, T., Dinesh, M. G., and Suguna, L. 2013. "Spontaneous ultra fast synthesis of gold nanoparticles using *Punica granatum* for cancer targeted drug delivery." *Colloids and Surfaces B: Biointerfaces* 106: 208–216. DOI: 10.1016/j.colsurfb.2013.01.035.

Ganesan, R. M., and Prabu, H. G. 2019. "Synthesis of gold nanoparticles using herbal Acorus calamus rhizome extract and coating on cotton fabric for antibacterial and UV blocking applications". *Arabian Journal of Chemistry* 12(8): 2166–2174.

Gawade, V. V., Gavade, N. L., Shinde, H. M., Babar, S. B., Kadam, A. N., and Garadkar, K. M. 2017. "Green synthesis of ZnO nanoparticles by using *Calotropis procera* leaves for the photodegradation of methyl orange." *Journal of Materials Science: Materials in Electronics* 28(18): 14033–14039. DOI: 10.1007/s10854-017-7254-2.

Geisler, K., Hughes, R. K., Sainsbury, F., Lomonossoff, G. P., Rejzek, M., Fairhurst, S., Olsen, C.-E. et al. 2013. "Biochemical analysis of a multifunctional cytochrome P450 (CYP51) enzyme required for synthesis of antimicrobial triterpenes in plants." *Proceedings of the National Academy of Sciences* 110(35): E3360–E3367. DOI: 10.1073/pnas.1309157110.

Ghanwate, N., Thakare, P., Bhise, P., and Gawande, S. 2016. "Colorimetric method for rapid detection of Oxacillin resistance in *Staphylococcus aureus* and its comparison with PCR for *mecA* gene." *Scientific Reports* 6(1): 1–5. DOI: 10.1038/srep23013.

Ghaedi, M., Yousefinejad, M., Safarpoor, M., Khafri, H. Z., and Purkait, M. K. 2015. "Rosmarinus officinalis leaf extract mediated green synthesis of silver nanoparticles and investigation of its antimicrobial properties". *Journal of Industrial and Engineering Chemistry* 31: 167–172.

Giamarellou, H. 2010. "Multidrug-resistant Gram-negative bacteria: how to treat and for how long." *International Journal of Antimicrobial Agents* 36(Suppl 2): S50–S54. DOI: 10.1016/j.ijantimicag.2010.11.014.

Godipurge, S. S., Yallappa, S., Biradar, N. J., Biradar, J. S., Dhananjaya, B. L., Hegde, G., Jagadish, K., and Hegde, G. 2016. "A facile and green strategy for the synthesis of Au, Ag and Au-Ag alloy nanoparticles using aerial parts of *R. hypocrateriformis* extract and their biological evaluation." *Enzyme and Microbial Technology* 95: 174–184. DOI: 10.1016/j.enzmictec.2016.08.006.

Gopinath, K., Kumaraguru, S., Bhakyaraj, K., Mohan, S., Venkatesh, K. S., Esakkirajan, M., Kaleeswarran, P. et al. 2016. "Green synthesis of silver, gold and silver/gold bimetallic nanoparticles using the *Gloriosa superba* leaf extract and their antibacterial and antibiofilm activities." *Microbial Pathogenesis* 101: 1–11. DOI: 10.1016/j.micpath.2016.10.011.

Hameed, R. S., Fayyad, R. J., Nuaman, R. S., Hamdan, N. T., and Maliki, S. A. J. 2019. "Synthesis and characterization of a novel titanium nanoparticals using banana peel extract and investigate its antibacterial and insecticidal activity." *Journal of Pure and Applied Microbiology* 13(4): 2241–2249. DOI: 10.22207/jpam.13.4.38.

Hanaki, H., Kuwahara-Arai, K., Boyle-Vavra, S., Daum, R. S., Labischinski, H., and Hiramatsu, K. 1998. "Activated cell-wall synthesis is associated with Vancomycin resistance in methicillin-resistant *Staphylococcus aureus* clinical strains Mu3 and Mu50." *Journal of Antimicrobial Chemotherapy* 42(2): 199–209. DOI: 10.1093/jac/42.2.199.

Hartman, B. J., and Tomasz, A. L. 1984. "Low-affinity penicillin-binding protein associated with beta-lactam resistance in *Staphylococcus aureus*." *Journal of Bacteriology* 158(2): 513–516. DOI: 10.1128/JB.158.2.513-516.1984.

Hasson, S. S. A., Al-Balushi, M. S., Al-Busaidi, J., Othman, M. S., Said, E. A., Habal, O., Sallam, T. A., Aljabri, A. A., and AhmedIdris, M. 2013. "Evaluation of anti–resistant activity of Auklandia (*Saussurea lappa*) root against some human pathogens." *Asian Pacific Journal of Tropical Biomedicine* 3(7): 557–562. DOI: 10.1016/S2221-1691(13)60113-6.

He, Y., Wei, F., Ma, Z., Zhang, H., Yang, Q., Yao, B., ... and Zhang, Q. 2017. "Green synthesis of silver nanoparticles using seed extract of Alpinia katsumadai, and their antioxidant, cytotoxicity, and antibacterial activities". *RSC Advances* 7(63): 39842–39851.

Helan, V., Prince, J. J., Al-Dhabi, N. A., Arasu, M. V., Ayeshamariam, A., Madhumitha, G., Roopan, S. M., and Jayachandran, M. 2016. "Neem leaves mediated preparation of NiO nanoparticles and its magnetization, coercivity and antibacterial analysis." *Results in Physics* 6: 712–718. DOI: 10.1016/j.rinp.2016.10.005.

Herold, B. C., Immergluck, L. C., Maranan, M. C., Lauderdale, D. S., Gaskin, R. E., Boyle-Vavra, S., Leitch, C. D., and Daum, R. S. 1998. "Community-acquired methicillin-resistant Staphylococcus aureus in children with no identified predisposing risk." *JAMA* 279(8): 593–598. DOI: 10.1001/jama.279.8.593.

Hiramatsu, K., Hanaki, H., Ino, T., Yabuta, K., Oguri, T., and Tenover, F. C. 1997. "Methicillin-resistant *Staphylococcus aureus* clinical strain with reduced vancomycin susceptibility." *Journal of Antimicrobial Chemotherapy* 40(1): 135–136. DOI: 10.1093/jac/40.1.135.

Horiuchi, K., Shiota, S., Hatano, T., Yoshida, T., Kuroda, T., and Tsuchiya, T. 2007. "Antimicrobial activity of oleanolic acid from *Salvia officinalis* and related compounds on Vancomycin-resistant enterococci (VRE)." *Biological and Pharmaceutical Bulletin* 30(6): 1147–1149. DOI: 10.1248/bpb.30.1147.

Howden, B. P., Johnson, P. D., Ward, P. B., Stinear, T. P., and Davies, J. K. 2006. "Isolates with low-level Vancomycin resistance associated with persistent Methicillin-resistant *Staphylococcus aureus* bacteremia." *Antimicrobial Agents and Chemotherapy* 50(9): 3039–3047. DOI: 10.1128/aac.00422-06.

Howden, B. P., Davies, J. K., Johnson, P. D., Stinear, T. P., and Grayson, M. L. 2010. "Reduced Vancomycin susceptibility in *Staphylococcus aureus*, including Vancomycin-intermediate and heterogeneous Vancomycin-intermediate strains: resistance mechanisms, laboratory detection, and clinical implications." *Clinical Microbiology Reviews* 23(1): 99–139. DOI: 10.1128/cmr.00042-09.

Huo, C., Khoshnamvand, M., Liu, P., Liu, C., and Yuan, C.-G. 2019. "Rapid mediated biosynthesis and quantification of AuNPs using persimmon (*Diospyros Kaki* L.f) fruit extract." *Journal of Molecular Structure* 1178: 366–374. DOI: 10.1016/j.molstruc.2018.10.044.

Hürlimann-Dalel, R. L., Ryffel, C., Kayser, F. H., and Berger-Bächi, B. 1992. "Survey of the Methicillin resistance-associated genes *mec*A, mecR1-*mec*I, and *fem*A-*fem*B in clinical isolates of Methicillin-resistant *Staphylococcus aureus*." *Antimicrobial Agents and Chemotherapy* 36(12): 2617–2621. DOI: 10.1128/aac.36.12.2617.

Iqbal, J., Abbasi, B. A., Ahmad, R., Shahbaz, A., Zahra, S. A., Kanwal, S., Munir, A., Rabbani, A., and Mahmood, T. 2020. "Biogenic synthesis of green and cost effective iron nanoparticles and evaluation of their potential biomedical properties." *Journal of Molecular Structure* 1199: 1–13. DOI: 10.1016/j.molstruc.2019.126979.

Islam, N. U., Jalil, K., Shahid, M., Rauf, A., Muhammad, N., Khan, A., Shah, M. R., and Khan, M. A. 2019. "Green synthesis and biological activities of gold nanoparticles functionalized with *Salix alba*." *Arabian Journal of Chemistry* 12(8): 2914–2925. DOI: 10.1016/j.arabjc.2015.06.025.

Iwu, M. W., Duncan, A. R., and Okunji, C. O. 1999. "New antimicrobials of plant origin in perspectives on new crops and new uses." In *Lant Breeding Reviews*, edited by J. Janick. Alexandria, VA: ASHS Press.

Jackson, M. A., and Hicks, R. A. 1987. "Vancomycin failure in *Staphylococcal endocarditis*." *The Pediatric Infectious Disease Journal* 6(8): 750–752. DOI: 10.1097/00006454-198708000-00011.

Jarvis, W. R., Schlosser, J., Chinn, R. Y., Tweeten, S., and Jackson, M. 2007. "National prevalence of Methicillin-resistant *Staphylococcus aureus* in inpatients at US health care facilities, 2006." *American Journal of Infection Control* 35(10): 631–637. DOI: 10.1016/j.ajic.2007.10.009.

Jegadeesan, G. B., Srimathi, K., Santosh Srinivas, N., Manishkanna, S., and Vignesh, D. 2019. "Green synthesis of iron oxide nanoparticles using *Terminalia bellirica* and *Moringa oleifera* fruit and leaf extracts: Antioxidant, antibacterial and thermoacoustic properties." *Biocatalysis and Agricultural Biotechnology* 21: 1–11. DOI: 10.1016/j.bcab.2019.101354.

Jeong, S. I., Kim, S. Y., Kim, S. J., Hwang, B. S., Kwon, T. H., Yu, K. Y., Hang, S. H., Suzuki, K., and Kim, K. J. 2010. "Antibacterial activity of phytochemicals isolated from *Atractylodes japonica* against Methicillin-resistant *Staphylococcus aureus*." *Molecules* 15(10): 7395–7402. DOI: 10.3390/molecules15107395.

Jevons, M. P. 1961. "'Celbenin'-resistant Staphylococci." *British Medical Journal* 1(5219): 124–125.

Kaatz, G. W., Seo, S. M., Dorman, N. J., and Lerner, S. A. 1990. "Emergence of teicoplanin resistance during therapy of *Staphylococcus aureus* endocarditis." *The Journal of Infectious Diseases* 162(1): 103–108. DOI: 10.1093/infdis/162.1.103.

Kali, A. 2015. "Antibiotics and bioactive natural products in treatment of Methicillin resistant Staphylococcus aureus: A brief review." *Pharmacognosy Reviews* 9(17): 29–34. DOI: 10.4103/0973-7847.156329.

Karthik, L., Vishnu Kirthi, A., Shivendu, R., and Srinivasan, V. M. 2020. *Biological Synthesis of Nanoparticles and Their Applications*. 1st ed. Boca Raton, FL: CRC Press.

Kasithevar, M., Saravanan, M., Prakash, P., Kumar, H., Ovais, M., Barabadi, H., and Shinwari, Z. K. 2017. "Green synthesis of silver nanoparticles using *Alysicarpus monilifer* leaf extract and its antibacterial activity against MRSA and CoNS isolates in HIV patients." *Journal of Interdisciplinary Nanomedicine* 2(2): 131–141. DOI: 10.1002/jin2.26.

Katayama, Y., Ito, T., and Hiramatsu, K. 2000. "A new class of genetic element, *Staphylococcus* cassette chromosome mec, encodes Methicillin resistance in *Staphylococcus aureus*." *Antimicrobial Agents and Chemotherapy* 44(6): 1549–1555. DOI: 10.1128/aac.44.6.1549-1555.2000.

Katayama, Y., Sekine, M., Hishinuma, T., Aiba, Y., and Hiramatsu, K. 2016. "Complete reconstitution of the vancomycin-intermediate *Staphylococcus aureus* phenotype of strain Mu50 in vancomycin-susceptible *S. aureus*." *Antimicrobial Agents and Chemotherapy* 60(6): 3730–3742. DOI: 10.1128/aac.00420-16.

Kaur, D. C., and Chate, S. S. 2015. "Study of antibiotic resistance pattern in Methicillin resistant Staphylococcus aureus with special reference to newer antibiotic." *Journal of Global Infectious Diseases* 7(2): 78–84. DOI: 10.4103/0974-777x.157245.

Kaviyarasu, K., Geetha, N., Kanimozhi, K., Magdalane, C. M., Sivaranjani, S., Ayeshamariam, A., Kennedy, J., and Maaza, M. 2017. "In vitro cytotoxicity effect and antibacterial performance of human lung epithelial cells A549 activity of zinc oxide doped TiO_2 nanocrystals: investigation of bio-medical application by chemical method." *Materials Science and Engineering: C* 74: 325–333. DOI: 10.1016/j.msec.2016.12.024.

Ke, F., Yadav, P. K., and Ju, L. Z. 2012. "Herbal medicine in the treatment of ulcerative colitis." *Saudi Journal of Gastroenterology* 18(1): 3–10. DOI: 10.4103/1319-3767.91726.

Khalil, A. T., Ovais, M., Ullah, I., Ali, M., Shinwari, Z. K., Hassan, D., and Maaza, M. 2018. "*Sageretia thea* (Osbeck.) modulated biosynthesis of NiO nanoparticles and their in vitro pharmacognostic, antioxidant and cytotoxic potential." *Artificial Cells, Nanomedicine, and Biotechnology* 46(4): 838–852. DOI: 10.1080/21691401.2017.1345928.

Khan, A. U., Yuan, Q., Wei, Y., Khan, G. M., Khan, Z. U. H., Khan, S., Ali, F., Tahir, K., Ahmad, A., and Khan, F. U. 2016a. "Photocatalytic and antibacterial response of biosynthesized gold nanoparticles." *Journal of Photochemistry and Photobiology B: Biology* 162: 273–277. DOI: 10.1016/j.jphotobiol.2016.06.055.

Khan, A. U., Yuan, Q., Wei, Y., Khan, S. U., Tahir, K., Khan, Z. U. H., Ahmad, A., Ali, F., Ali, S., and Nazir, S. 2016b. "Longan fruit juice mediated synthesis of uniformly dispersed spherical AuNPs: cytotoxicity against human breast cancer cell line MCF-7, antioxidant and fluorescent properties." *RSC Advances* 6(28): 23775–23782. DOI: 10.1039/c5ra27100b.

Klein, E., Smith, D. L., and Laxminarayan, R. 2007. "Hospitalizations and deaths caused by Methicillin-resistant *Staphylococcus aureus*, United States, 1999–2005." *Emerging Infectious Diseases* 13(12): 1840–1846. DOI: 10.3201/eid1312.070629.

Klevens, R. M., Edwards, J. R., Tenover, F. C., McDonald, L. C., Horan, T., Gaynes, R., and System, N. N. I. S. 2006. "Changes in the epidemiology of methicillin-resistant *Staphylococcus aureus* in intensive care units in US hospitals, 1992–2003." *Clinical Infectious Diseases* 42(3): 389–391. DOI: 10.1086/499367.

Khan, S., Bakht, J., and Syed, F. 2018. "Green synthesis of gold nanoparticles using Acer pentapomicum leaves extract its characterization, antibacterial, antifungal and antioxidant bioassay". *Dig. J. Nanomater. Biostruct* 2(13): 579–589.

Kohanski, M. A., Dwyer, D. J., and Collins, J. J. 2010. "How antibiotics kill bacteria: From targets to networks." *Nature Reviews Microbiology* 8(6): 423–435. DOI: 10.1038/nrmicro2333.

Kota, S., Dumpala, P., Anantha, R. K., Verma, M. K., and Kandepu, S. 2017. "Evaluation of therapeutic potential of the silver/silver chloride nanoparticles synthesized with the aqueous leaf extract of *Rumex acetosa*." *Scientific Reports* 7(1): 1–11. DOI: 10.1038/s41598-017-11853-2.

Kratošová, G., Holišová, V., Konvičková, Z., Ingle, A. P., Gaikwad, S., Škrlová, K., Prokop, A., Rai, M., and Plachá, D. 2019. "From biotechnology principles to functional and low-cost metallic bionanocatalysts." *Biotechnology Advances* 37(1): 154–176. DOI: 10.1016/j.biotechadv.2018.11.012.

Kumar, H. K. N., Mohana, N. C., Nuthan, B. R., Ramesha, K. P., Rakshith, D., Geetha, N., and Satish, S. 2019. "Phyto-mediated synthesis of zinc oxide nanoparticles using aqueous plant extract of *Ocimum americanum* and evaluation of its bioactivity." *SN Applied Sciences* 1(651): 1–9. DOI: 10.1007/s42452-019-0671-5.

Kumar, V. S., Nagaraja, B. M., Shashikala, V., Padmasri, A. H., Madhavendra, S. S., Raju, B. D., and Rao, K. S. R. 2004. "Highly efficient Ag/C catalyst prepared by electro-chemical deposition method in controlling microorganisms in water." *Journal of Molecular Catalysis A: Chemical* 223(1): 313–319. DOI: 10.1016/j.molcata.2003.09.047.

Kumar, P. V., Shameem, U., Kollu, P., Kalyani, R. L., and Pammi, S. V. N. 2015. "Green synthesis of copper oxide nanoparticles using Aloe vera leaf extract and its antibacterial activity against fish bacterial pathogens". *BioNanoScience* 5(3): 135–139.

Lakshmanan, G., Sathiyaseelan, A., Kalaichelvan, P. T., and Murugesan, K. 2018. "Plant-mediated synthesis of silver nanoparticles using fruit extract of *Cleome viscosa* L.: Assessment of their antibacterial and anticancer activity." *Karbala International Journal of Modern Science* 4(1): 61–68. DOI: 10.1016/j.kijoms.2017.10.007.

Levine, D. P. 2006. "Vancomycin: A history." *Clinical Infectious Diseases* 42(Suppl 1): S5–S12. DOI: 10.1086/491709.

Lokina, S., Suresh, R., Giribabu, K., Stephen, A., Sundaram, R. L., and Narayanan, V. 2014. "Spectroscopic investigations, antimicrobial, and cytotoxic activity of green synthesized gold nanoparticles". *Spectrochimica Acta Part A: Molecular and Biomolecular Spectroscopy* 129: 484–490.

Lopez-Miranda, J. L., Esparza, R., Rosas, G., Perez, R., and Estevez-Gonzalez, M. 2019. "Catalytic and antibacterial properties of gold nanoparticles synthesized by a green approach for bioremediation applications." *3 Biotech* 9(135): 1–9. DOI: 10.1007/s13205-019-1666-z.

Luo, J., Chan, W. B., Wang, L., and Zhong, C. J. 2010. "Probing interfacial interactions of bacteria on metal nanoparticles and substrates with different surface properties." *International Journal of Antimicrobial Agents* 36(6): 549–556. DOI: 10.1016/j.ijantimicag.2010.08.015.

Maciel, M. V. D. O. B., da Rosa Almeida, A., Machado, M. H., De Melo, A. P. Z., Da Rosa, C. G., De Freitas, D. Z., ... and Barreto, P. L. M. 2019. "Syzygium aromaticum L.(Clove) Essential Oil as a Reducing Agent for the Green Synthesis of Silver Nanoparticles". *Open Journal of Applied Sciences* 9(2): 45–54.

Magalhães, A.-P.-R., Moreira, F.-C.-L., Alves, D.-R.-S., Estrela, C.-R.-A., Estrela, C., Carrião, M.-S., Bakuzis, A.-F., and Lopes, L.-G. 2016. "Silver nanoparticles in resin luting cements: Antibacterial and physiochemical properties". *Journal of Clinical and Experimental Dentistry* 8(4): e415–e422. DOI: 10.4317/jced.52983.

Mahady, G. B. 2005. "Medicinal plants for the prevention and treatment of bacterial infections." *Current Pharmaceutical Design* 11(19): 2405–2427. DOI: 10.2174/1381612054367481.

Mahdavi, M., Namvar, F., Ahmad, M. B., and Mohamad, R. 2013. "Green biosynthesis and characterization of magnetic iron oxide (Fe_3O4) nanoparticles using seaweed (*Sargassum muticum*) aqueous extract." *Molecules* 18(5): 5954–5964. DOI: 10.3390/molecules18055954.

Mahendra, C., Murali, M., Manasa, G., Ponnamma, P., Abhilash, M. R., Lakshmeesha, T. R., Satish, A., Amruthesh, K. N., and Sudarshana, M. S. 2017. "Antibacterial and antimitotic potential of bio-fabricated zinc oxide nanoparticles of *Cochlospermum religiosum* (L.)." *Microbial Pathogenesis* 110: 620–629. DOI: 10.1016/j.micpath.2017.07.051.

Makarov, V. V., Love, A. J., Sinitsyna, O. V., Makarova, S. S., Yaminsky, I. V., Taliansky, M. E., and Kalinina, N. O. 2014. "'Green' nanotechnologies: Synthesis of metal nanoparticles using plants." *Acta Naturae* 6(1): 35–44.

Manandhar, S., Luitel, S., and Dahal, R. K. 2019. "In vitro antimicrobial activity of some medicinal plants against human pathogenic bacteria." *Journal of Tropical Medicine* 2019: 1–5. DOI: 10.1155/2019/1895340.

Manquat, G., Croize, J., Stahl, J. P., Meyran, M., Hirtz, P., and Micoud, M. 1992. "Failure of teicoplanin treatment associated with an increase in MIC during therapy of *Staphylococcus aureus* septicaemia." *The Journal of Antimicrobial Chemotherapy* 29(6): 731–732. DOI: 10.1093/jac/29.6.731.

Manubolu, M., Goodla, L., Ravilla, S., and Obulum, V. R. 2013. "Activity-guided isolation and identification of anti-Staphylococcal components from *Senecio tenuifolius* Burm. F. leaf extracts." *Asian Pacific Journal of Tropical Biomedicine* 3(3): 191–195. DOI: 10.1016/S2221-1691(13)60048-9.

Manukumar, H. M., and Umesha, S. 2017. "MALDI-TOF-MS based identification and molecular characterization of food associated methicillin-resistant *Staphylococcus aureus.*" *Scientific Reports* 7(1): 1–16. DOI: 10.1038/s41598-017-11597-z.

Manukumar, H. M., Yashwanth, B., Umesha, S., and Venkateswara Rao, J. 2020. "Biocidal mechanism of green synthesized thyme loaded silver nanoparticles (GTAgNPs) against immune evading tricky methicillin-resistant Staphylococcus aureus 090 (MRSA090) at a homeostatic environment." *Arabian Journal of Chemistry* 13(1): 1179–1197. DOI: 10.1016/j.arabjc.2017.09.017.

Marasini, B. P., Baral, P., Aryal, P., Ghimire, K. R., Neupane, S., Dahal, N., Singh, A., Ghimire, L., and Shrestha, K. 2015. "Evaluation of antibacterial activity of some traditionally used medicinal plants against human pathogenic bacteria." *BioMed Research International* 2015: 1–6. DOI: 10.1155/2015/265425.

Marathe, N. P., Rasane, M. H., Kumar, H., Patwardhan, A. A., Shouche, Y. S., and Diwanay, S. S. 2013. "In vitro antibacterial activity of *Tabernaemontana alternifolia* (Roxb) stem bark aqueous extracts against clinical isolates of Methicillin resistant Staphylococcus aureus." *Annals of Clinical Microbiology and Antimicrobials* 12(26): 1–8. DOI: 10.1186/1476-0711-12-26.

Marslin, G., Siram, K., Maqbool, Q., Selvakesavan, R. K., Kruszka, D., Kachlicki, P., and Franklin, G. 2018. "Secondary metabolites in the green synthesis of metallic nanoparticles." *Materials* 11(940): 1–25. DOI: 10.3390/ma11060940.

Matthews, P. E., and Tomasz, A. L. 1990. "Insertional inactivation of the *mec* gene in a transposon mutant of a Methicillin-resistant clinical isolate of *Staphylococcus aureus.*" *Antimicrobial Agents and Chemotherapy.* 34(9): 1777–1779. DOI: 10.1128/aac.34.9.1777.

McCallum, N., Karauzum, H., Getzmann, R., Bischoff, M., Majcherczyk, P., Berger-Bachi, B., and Landmann, R. 2006. "In vivo survival of teicoplanin-resistant *Staphylococcus aureus* and fitness cost of teicoplanin resistance." *Antimicrobial Agents and Chemotherapy* 50(7): 2352–2360. DOI: 10.1128/aac.00073-06.

McGuinness, W. A., Malachowa, N., and DeLeo, F. R. 2017. "Vancomycin Resistance in *Staphylococcus aureus.*" *Yale Journal of Biology and Medicine* 90(2): 269–281.

Medina, A. L., Lucero, M. E., Holguin, F. O., Estell, R. E., Posakony, J. J., Simon, J., and O'Connell, M. A. 2005. "Composition and antimicrobial activity of *Anemopsis californica* leaf oil." *Journal of Agricultural and Food Chemistry* 53(22): 8694–8698. DOI: 10.1021/jf0511244.

Mohammadlou, M., Maghsoudi, H., and Jafarizadeh-Malmiri, H. 2016. "A review on green silver nanoparticles based on plants: Synthesis, potential applications and eco-friendly approach." *International Food Research Journal* 23(2): 446–463.

Moreira, B., Boyle-Vavra, S., deJonge, B. L., and Daum, R. S. 1997. "Increased production of penicillin-binding protein 2, increased detection of other penicillin-binding proteins, and decreased coagulase activity associated with glycopeptide resistance in *Staphylococcus aureus.*" *Antimicrobial Agents and Chemotherapy* 41(8): 1788–1793. DOI: 10.1128/AAC.41.8.1788.

Moustafa, N. E., and Alomari, A. A. 2019. "Green synthesis and bactericidal activities of isotropic and anisotropic spherical gold nanoparticles produced using *Peganum harmala* L leaf and seed extracts." *Biotechnology and Applied Biochemistry* 66(4): 664–672. DOI: 10.1002/bab.1782.

Muhammad, W., Khan, M. A., Nazir, M., Siddiquah, A., Mushtaq, S., Hashmi, S. S., and Abbasi, B. H. 2019. "*Papaver somniferum* L. mediated novel bioinspired lead oxide (PbO) and iron oxide ($Fe2O_3$) nanoparticles: In-vitro biological applications, biocompatibility and their potential towards HepG2 cell line." *Materials Science and Engineering: C* 103: 1–11. DOI: 10.1016/j.msec.2019.109740.

Murray, B. E. 2000. "Vancomycin-resistant *Enterococcal* infections." *The New England Journal of Medicine* 342(10): 710–721. DOI: 10.1056/nejm200003093421007.

Mustaffa, F., Indurkar, J., Ismail, S., Shah, M., and Mansor, S. M. 2011. "An antimicrobial compound isolated from *Cinnamomum iners* leaves with activity against Methicillin-resistant *Staphylococcus aureus.*" *Molecules* 16(4): 3037–3047. DOI: 10.3390/molecules16043037.

Muthaiyan, A., Jayaswal, R. K., and Wilkinson, B. J. 2004. "Intact mutS in laboratory-derived and clinical glycopeptide-intermediate *Staphylococcus aureus* strains." *Antimicrobial Agents and Chemotherapy* 48(2): 623–625. DOI: 10.1128/aac.48.2.623-625.2004.

Muthukumar, T., Sudha, k., Sambandam, B., Aravinthan, A., Sastry, T. P., and Kim, J. H. 2016. "Green synthesis of gold nanoparticles and their enhanced synergistic antitumor activity using HepG2 and MCF7 cells and its antibacterial effects." *Process Biochemistry* 51(3): 384–391. DOI: 10.1016/j.procbio.2015.12.017.

Muthuvel, A., Adavallan, K., Balamurugan, K., and Krishnakumar, N. 2014. "Biosynthesis of gold nanoparticles using Solanum nigrum leaf extract and screening their free radical scavenging and antibacterial properties". *Biomedicine & Preventive Nutrition* 4(2): 325–332.

Mwangi, M. M., Wu, S. W., Zhou, Y., Sieradzki, K., de Lencastre, H., Richardson, P., Bruce, D. et al. 2007. "Tracking the in vivo evolution of multidrug resistance in *Staphylococcus aureus* by whole-genome sequencing." *Proceedings of the National Academy of Sciences of the United States of America* 104(22): 9451–9456. DOI: 10.1073/pnas.0609839104.

Nagalingam, M., Kalpana, V. N., and Panneerselvam, A. 2018. "Biosynthesis, characterization, and evaluation of bioactivities of leaf extract-mediated biocompatible gold nanoparticles from Alternanthera bettzickiana". *Biotechnology Reports* 19: e00268.

Naraginti S, Sivakumar A. 2014. "Eco-friendly synthesis of silver and gold nanoparticles with enhanced bactericidal activity and study of silver catalyzed reduction of 4-nitrophenol". *Spectrochim Acta A Mol Biomol Spectrosc.* 128:357–362. doi: 10.1016/j.saa.2014.02.083.

Naseem, T., and Farrukh, M. A. 2015. "Antibacterial activity of green synthesis of iron nanoparticles usinglawsonia inermisandgardenia jasminoidesleaves extract". *Journal of Chemistry* 2015: 1–7. doi:10.1155/2015/912342

Neamatallah, A., Yan, L., Dewar, S. J., and Austin, B. 2005. "An extract from teak (*Tectona grandis*) bark inhibited *Listeria monocytogenes* and Methicillin resistant Staphylococcus aureus." *Letters in Applied Microbiology* 41(1): 94–96. DOI: 10.1111/j.1472-765X.2005.01680.x.

Nii-Trebi, N. I. 2017. "Emerging and neglected infectious diseases: Insights, advances, and challenges." *BioMed Research International* 2017: 1–15. DOI: 10.1155/2017/5245021.

Njeru, S. N., Obonyo, M. A., Nyambati, S. O., and Ngari, S. M. 2015. "Antimicrobial and cytotoxicity properties of the crude extracts and fractions of *Premna resinosa* (Hochst.) Schauer (Compositae): Kenyan traditional medicinal plant." *BMC Complementary and Alternative Medicine* 15(295): 1–9. DOI: 10.1186/s12906-015-0811-4.

Okwu, M. U., Olley, M., Akpoka, A. O., and Izevbuwa, O. E. 2019. "Methicillin-resistant *Staphylococcus aureus* (MRSA) and anti-MRSA activities of extracts of some medicinal plants: A brief review." *AIMS Microbiology* 5(2): 117–137. DOI: 10.3934/microbiol.2019.2.117.

Olsen, J. E., Christensen, H., and Aarestrup, F. M. 2006. "Diversity and evolution of *blaZ* from *Staphylococcus aureus* and coagulase-negative *Staphylococci*." *Journal of Antimicrobial Chemotherapy* 57(3): 450–460. DOI: 10.1093/jac/dki492.

O'Neill, J. 2014. *Review on Antimicrobial Resistance Antimicrobial Resistance: Tackling a Crisis for the Health and Wealth of Nations.* London: Review on Antimicrobial Resistance.

Ovais M, Ayaz M, Khalil AT, Shah SA, Jan MS, Raza A, Shahid M, Shinwari ZK. 2018. "HPLC-DAD finger printing, antioxidant, cholinesterase, and α-glucosidase inhibitory potentials of a novel plant Olax nana". *BMC Complement Altern Med.* 18(1): 1. doi: 10.1186/s12906-017-2057-9.

Partridge, S. R., Kwong, S. M., Firth, N., and Jensen, S. O. 2018. "Mobile genetic elements associated with antimicrobial resistance." *Clinical Microbiology Reviews* 31(4): 1–61. DOI: 10.1128/CMR.00088-17.

Patra, J. K., and Baek, K.-H. 2016. "Comparative study of proteasome inhibitory, synergistic antibacterial, synergistic anticandidal, and antioxidant activities of gold nanoparticles biosynthesized using fruit waste materials". *International Journal of Nanomedicine* 11: 4691–4705. DOI: 10.2147/IJN.S108920.

Pawar, J. S., and Patil, R. H. 2020. "Green synthesis of silver nanoparticles using *Eulophia herbacea* (Lindl.) tuber extract and evaluation of its biological and catalytic activity." *SN Applied Sciences* 2(52): 1–12. DOI: 10.1007/s42452-019-1846-9.

Pereira, T. M., Polez, V. L. P., Sousa, M. H., and Silva, L. P. 2020. "Modulating physical, chemical, and biological properties of silver nanoparticles obtained by green synthesis using different parts of the tree *Handroanthus heptaphyllus* (Vell.) Mattos." *Colloid and Interface Science Communications* 34(100224): 1–8. DOI: 10.1016/j.colcom.2019.100224.

Perichon, B., and Courvalin, P. 2009. "*Van*A-type vancomycin-resistant *Staphylococcus aureus*." *Antimicrobial Agents and Chemotherapy* 53(11): 4580–4587. DOI: 10.1128/aac.00346-09.

Pfeltz, R. F., Singh, V. K., Schmidt, J. L., Batten, M. A., Baranyk, C. S., Nadakavukaren, M. J., Jayaswal, R. K., and Wilkinson, B. J. 2000. "Characterization of passage-selected Vancomycin-Resistant *Staphylococcus aureus* strains of diverse parental backgrounds." *Antimicrobial Agents and Chemotherapy* 44(2): 294–303. DOI: 10.1128/aac.44.2.294-303.2000.

Philip, D. 2010. "Green synthesis of gold and silver nanoparticles using *Hibiscus rosa* sinensis." *Physica E: Low-Dimensional Systems and Nanostructures* 42(5): 1417–1424. DOI: 10.1016/j.physe.2009.11.081.

Phull, A. R., Abbas, Q., Ali, A., Raza, H., Zia, M., and Haq, I. U. 2016. "Antioxidant, cytotoxic and antimicrobial activities of green synthesized silver nanoparticles from crude extract of Bergenia ciliata". *Future Journal of Pharmaceutical Sciences* 2(1): 31–36.

Plough, H. H. 1945. "Penicillin resistance of *Staphylococcus aureus* and its clinical implications." *American Journal of Clinical Pathology* 15(10): 446–451.

Poojary, M. M., Passamonti, P., and Adhikari, A. V. 2016. "Green synthesis of silver and gold nanoparticles using root bark extract of *Mammea suriga*: Characterization, process optimization, and their antibacterial activity." *BioNanoScience* 6(2): 110–120. DOI: 10.1007/s12668-016-0199-8.

Qais, F. A., Shafiq, A., Khan, H. M., Husain, F. M., Khan, R. A., Alenazi, B., Alsalme, A., and Ahmad, I. 2019. "Antibacterial effect of silver nanoparticles synthesized using *Murraya koenigii* (L.) against multidrug-resistant pathogens." *Bioinorganic Chemistry and Applications* 2019(4649506): 1–11. DOI: https://doi.org/10.1155/2019/4649506.

Rad, J. S., Alfatemi, S. M. H., Rad, M. S., and Iriti, M. 2013. "In-vitro antioxidant and antibacterial activities of *Xanthium strumarium* L. extracts on Methicillin-susceptible and Methicillin-resistant *Staphylococcus aureus*." *Ancient Science of Life* 33(2): 109–113. DOI: 10.4103/0257-7941.139050.

Rahman, A. A., Samoylenko, V., Jain, S. K., Tekwani, B. L., Khan, S. I., Jacob, M. R., Midiwo, J. O., Hester, J. P., Walker, L. A., and Muhammad, I. 2011. "Antiparasitic and antimicrobial isoflavanquinones from *Abrus schimperi*." *Natural Product Communications* 6(11): 1645–1650. DOI: 10.1177/1934578X1100601120.

Rajan, A., Rajan, A. R., and Philip, D. 2017. "*Elettaria cardamomum* seed mediated rapid synthesis of gold nanoparticles and its biological activities." *OpenNano* 2: 1–8. DOI: 10.1016/j.onano.2016.11.002.

Rajeshkumar, S. 2016. "Synthesis of silver nanoparticles using fresh bark of *Pongamia pinnata* and characterization of its antibacterial activity against gram positive and gram negative pathogens." *Resource-Efficient Technologies* 2(1): 30–35. DOI: 10.1016/j.reffit.2016.06.003.

Rajkuberan, C., Sudha, K., Sathishkumar, G., and Sivaramakrishnan, S. 2015. "Antibacterial and cytotoxic potential of silver nanoparticles synthesized using latex of *Calotropis gigantea* L." *Spectrochimica Acta Part A: Molecular and Biomolecular Spectroscopy* 136: 924–930. DOI: 10.1016/j.saa.2014.09.115.

Rammelkamp, C. H., and Maxon, T. 1942. "Resistance of *Staphylococcus aureus* to the action of Penicillin." *Experimental Biology and Medicine* 51(3): 386–389. DOI: 10.3181/00379727-51-13986.

Rasmussen, J. W., Martinez, E., Louka, P., and Wingett, D. G. 2010. "Zinc oxide nanoparticles for selective destruction of tumor cells and potential for drug delivery applications." *Expert Opinion on Drug Delivery* 7(9): 1063–1077.

Reddy, K. M., Feris, K., Bell, J., Wingett, D. G., Hanley, C., and Punnoose, A. 2007. "Selective toxicity of zinc oxide nanoparticles to prokaryotic and eukaryotic systems." *Applied Physics Letters* 90(213902): 1–8.

Reynolds, P. E., and Brown, D. F. 1985. "Penicillin-binding proteins of β-lactam-resistant strains of *Staphylococcus aureus*: Effect of growth conditions." *FEBS Letters* 192(1): 28–32.

Ríos, J. L., and Recio, M. C. 2005. "Medicinal plants and antimicrobial activity." *Journal of Ethnopharmacology* 100(1): 80–84. DOI: 10.1016/j.jep.2005.04.025.

Rodríguez-León, E., Iñiguez-Palomares, R., Navarro, R. E., Herrera-Urbina, R., Tánori, J., Iñiguez-Palomares, C., and Maldonado, A. 2013. "Synthesis of silver nanoparticles using reducing agents obtained from natural sources (*Rumex hymenosepalus* extracts)." *Nanoscale Research Letters* 8(318): 1–9. DOI: 10.1186/1556-276X-8-318.

Romero, C. D., Chopin, S. F., Buck, G., Martinez, E., Garcia, M., and Bixby, L. 2005. "Antibacterial properties of common herbal remedies of the southwest." *Journal of Ethnopharmacology* 99(2): 253–257. DOI: 10.1016/j.jep.2005.02.028.

Rountree, P. M., and Beard, M. A. 1958. "Further observations on infection with phage type 80 *Staphylococci* in Australia." *Medical Journal of Australia* 2(24): 789–795.

Rountree, P. M., and Freeman, B. M. 1955. "Infections caused by a particular phage type of *Staphylococcus aureus*." *Medical Journal of Australia* 2(5): 157–161.

Różalska, S., Soliwoda, K., and Długoński, J. 2016. "Synthesis of silver nanoparticles from *Metarhizium robertsii* waste biomass extract after nonylphenol degradation, and their antimicrobial and catalytic potential." *RSC Advances* 6(26): 21475–21485. DOI: 10.1039/C5RA24335A.

Ruttoh, E., Tarus, P., Bii, C., Machocho, A., Karimi, L., and Okemo, P. 2009. "Antibacterial activity of *Tabernaemontana stapfiana* Britten (*Apocynaceae*) extracts." *African Journal of Traditional, Complementary and Alternative Medicines* 6(2): 186–194 DOI: 10.4314/ajtcam.v6i2.57090.

Sadeghi, B. 2015. "*Zizyphus mauritiana* extract-mediated green and rapid synthesis of gold nanoparticles and its antibacterial activity." *Journal of Nanostructure in Chemistry* 5(3): 265–273. DOI: 10.1007/s40097-015-0157-y.

Sadeghi, B., Mohammadzadeh, M., and Babakhani, B. 2015. "Green synthesis of gold nanoparticles using *Stevia rebaudiana* leaf extracts: Characterization and their stability." *Journal of Photochemistry and Photobiology B: Biology* 148: 101–106. DOI: 10.1016/j.jphotobiol.2015.03.025.

Sadeghi B, Rostami A, Momeni SS. 2015. "Facile green synthesis of silver nanoparticles using seed aqueous extract of Pistacia atlantica and its antibacterial activity". *Spectrochim Acta A Mol Biomol Spectrosc.* 134:326–32. doi: 10.1016/j.saa.2014.05.078.

Sakoulas, G., Eliopoulos, G. M., Moellering, R. C., Wennersten, C., Venkataraman, L., Novick, R. P., and Gold, H. S. 2002. "Impact of accessory gene regulator (*agr*) dysfunction on vancomycin pharmacodynamics among Canadian community and health-care associated methicillin-resistant *Staphylococcus aureus*." *Antimicrobial Agents and Chemotherapy* 46(5): 1492–1502. DOI: 10.1128/aac.46.5.1492-1502.2002.

Sakr, A., Bregeon, F., Mege, J.-L., Rolain, J.-M., and Blin, O. 2018. "*Staphylococcus aureus* nasal colonization: an update on mechanisms, epidemiology, risk factors, and subsequent infections." *Frontiers Microbiology* 9: 1–15. DOI: 10.3389/fmicb.2018.02419.

Saleem, S., Ahmed, B., Khan, M. S., Al-Shaeri, M., and Musarrat, J. 2017. "Inhibition of growth and biofilm formation of clinical bacterial isolates by NiO nanoparticles synthesized from *Eucalyptus globulus* plants." *Microbial Pathogenesis* 111: 375–387. DOI: 10.1016/j.micpath.2017.09.019.

Salehi, S., Shandiz, S. A., Ghanbar, F., Darvish, M. R., Ardestani, M. S., Mirzaie, A., and Jafari, M. 2016. "Phytosynthesis of silver nanoparticles using *Artemisia marschalliana* Sprengel aerial part extract and assessment of their antioxidant, anticancer, and antibacterial properties." *International Journal of Nanomedicine* 11: 1835–1846. DOI: 10.2147/ijn.s99882.

Salouti, M., and Derakhshan, F. K. 2019. "Phytosynthesis of nanoscale materials." In *Advances in Phytonanotechnology from Synthesis to Application*, edited by Mansour Ghorbanpour and Shabir Hussain Wani, 45–121. Elsevier Academic Press. Elsevier Science Publishing Co Inc. London, United Kingdom.

Scherl, A., Francois, P., Charbonnier, Y., Deshusses, J. M., Koessler, T., Huyghe, A., Bento, M. et al. 2006. "Exploring glycopeptide-resistance in *Staphylococcus aureus*: A combined proteomics and transcriptomics approach for the identification of resistance-related markers." *BMC Genomics* 7(296): 1–16. DOI: 10.1186/1471-2164-7-296.

Shankar, S. S., Rai, A., Ankamwar, B., Singh, A., Ahmad, A., and Sastry, M. 2004. "Biological synthesis of triangular gold nanoprisms." *Nature Materials* 3(7): 482–488. DOI: 10.1038/nmat1152.

Sharmila, G., Farzana Fathima, M., Haries, S., Geetha, S., Manoj Kumar, N., and Muthukumaran, C. 2017a. "Green synthesis, characterization and antibacterial efficacy of palladium nanoparticles synthesized using *Filicium decipiens* leaf extract." *Journal of Molecular Structure* 1138: 35–40. DOI: 10.1016/j.molstruc.2017.02.097.

Sharmila, G., Haries, S., Farzana Fathima, M., Geetha, S., Manoj Kumar, N., and Muthukumaran, C. 2017b. "Enhanced catalytic and antibacterial activities of phytosynthesized palladium nanoparticles using *Santalum album* leaf extract." *Powder Technology* 320: 22–26. DOI: 10.1016/j.powtec.2017.07.026.

Sharmila, G., Muthukumaran, C., Sangeetha, E., Saraswathi, H., Soundarya, S., and Kumar, N. M. 2019. "Green fabrication, characterization of Pisonia alba leaf extract derived MgO nanoparticles and its biological applications". *Nano-Structures & Nano-Objects* 20: 100380.

Shiota, S., Shimizu, M., Mizusima, T., Ito, H., Hatano, T., Yoshida, T., and Tsuchiya, T. 2000. "Restoration of effectiveness of beta-lactams on methicillin-resistant *Staphylococcus aureus* by tellimagrandin I from rose red." *FEMS Microbiology Letters* 185(2): 135–138. DOI: 10.1111/j.1574-6968.2000.tb09051.x.

Shivalingappa, H., Biradar, J., and Rudresh, K. 1999. "Antiimplantation activity of alcoholic extract of *Rivea hypocrateriformis*." *Indian Journal of Pharmaceutical Sciences* 61(5): 309–310.

Sianglum, W., Srimanote, P., Wonglumsom, W., Kittiniyom, K., and Voravuthikunchai, S. P. 2011. "Proteome analyses of cellular proteins in methicillin-resistant *Staphylococcus aureus* treated with rhodomyrtone, a novel antibiotic candidate." *PLoS One* 6(2): 1–10. DOI: 10.1371/journal.pone.0016628.

Silveira, A. P., Bonatto, C. C., Lopes, C. A. P., Rivera, L. M. R., and Silva, L. P. 2018. "Physicochemical characteristics and antibacterial effects of silver nanoparticles produced using the aqueous extract of Ilex paraguariensis". *Materials Chemistry and Physics* 216: 476–484.

Singh, P., Kim, Y. J., Zhang, D., and Yang, D. C. 2016. "Biological synthesis of nanoparticles from plants and microorganisms." *Trends in Biotechnology* 34(7): 588–599. DOI: 10.1016/j.tibtech.2016.02.006.

Sinsinwar, S., Sarkar, M. K., Suriya, K. R., Nithyanand, P., and Vadivel, V. 2018. "Use of agricultural waste (coconut shell) for the synthesis of silver nanoparticles and evaluation of their antibacterial activity against selected human pathogens." *Microbial Pathogenesis* 124: 30–37. DOI: 10.1016/j.micpath.2018.08.025.

Składanowski, M., Golinska, P., Rudnicka, K., Dahm, H., and Rai, M. 2016. "Evaluation of cytotoxicity, immune compatibility and antibacterial activity of biogenic silver nanoparticles." *Medical Microbiology and Immunology* 205(6): 603–613. DOI: 10.1007/s00430-016-0477-7.

Slavin, Y. N., Asnis, J., Hafeli, U. O., and Bach, H. 2017. "Metal nanoparticles: Understanding the mechanisms behind antibacterial activity." *Journal of Nanobiotechnology* 15(65): 1–20. DOI: 10.1186/s12951-017-0308-z.

Sondi, I., and Salopek-Sondi, B. 2004. "Silver nanoparticles as antimicrobial agent: A case study on *E. coli* as a model for Gram-negative bacteria." *Journal of Colloid and Interface Science* 275(1): 177–182. DOI: 10.1016/j.jcis.2004.02.012.

Song, M. D., Wachi, M., Doi, M., Ishino, F., and Matsuhashi, M. 1987. "Evolution of an inducible penicillin-target protein in methicillin-resistant *Staphylococcus aureus* by gene fusion." *FEBS Letters* 221(1): 167–171.

Sorrell, T. C., Packham, D. R., Shanker, S., Foldes, M., and Munro, R. 1982. "Vancomycin therapy for methicillin-resistant *Staphylococcus aureus*." *Annals of Internal Medicine* 97(3): 344–350. DOI: 10.7326/0003-4819-97-3-344.

Srihasam, S., Thyagarajan, K., Korivi, M., Lebaka, V. R., and Mallem, S. P. R. 2020. "Phytogenic generation of NiO nanoparticles using *Stevia* leaf extract and evaluation of their in-vitro antioxidant and antimicrobial properties." *Biomolecules* 10(89): 1–12. DOI: 10.3390/biom10010089.

Subramani, R., Narayanasamy, M., and Feussner, K. D. 2017. "Plant-derived antimicrobials to fight against multi-drug-resistant human pathogens." *3 Biotech* 7(172): 1–15. DOI: 10.1007/s13205-017-0848-9.

Sudha, A., Jeyakanthan, J., and Srinivasan, P. 2017. "Green synthesis of silver nanoparticles using Lippia nodiflora aerial extract and evaluation of their antioxidant, antibacterial and cytotoxic effects". *Resource-Efficient Technologies* 3(4): 506–515.

Suganya, K. S. U., Govindaraju, K., Kumar, V. G., Dhas, T. S., Karthick, V., Singaravelu, G., and Elanchezhiyan, M. 2015a. "Blue green alga mediated synthesis of gold nanoparticles and its antibacterial efficacy against Gram positive organisms." *Materials Science and Engineering: C* 47: 351–356. DOI: 10.1016/j.msec.2014.11.043.

Suganya, K. S. U., Govindaraju, K., Kumar, V. G., Dhas, T. S., Karthick, V., Singaravelu, G., and Elanchezhiyan, M. 2015b. "Size controlled biogenic silver nanoparticles as antibacterial agent against isolates from HIV infected patients." *Spectrochimica Acta Part A: Molecular and Biomolecular Spectroscopy* 144: 266–272. DOI: 10.1016/j.saa.2015.02.074.

Suresh, J., Pradheesh, G., Alexramani, V., Sundrarajan, M., and Hong, S. I. 2018. "Green synthesis and characterization of hexagonal shaped MgO nanoparticles using insulin plant (*Costus pictus* D. Don) leave extract and its antimicrobial as well as anticancer activity." *Advanced Powder Technology* 29(7): 1685–1694. DOI: 10.1016/j.apt.2018.04.003.

Suzuki, E., Hiramatsu, K., and Yokota, T. 1992. "Survey of methicillin-resistant clinical strains of coagulase-negative *Staphylococci* for *mecA* gene distribution." *Antimicrobial Agents and Chemotherapy* 36(2): 429–434. DOI: 10.1128/aac.36.2.429.

Syeda, B., Prasad, M. N. N., Dhananjaya, B. L., Kumar, K. M., Yallappa, S., and Satish, S. 2016. "Synthesis of silver nanoparticles by endosymbiont *Pseudomonas fluorescens* CA 417 and their bactericidal activity." *Enzyme and Microbial Technology* 95: 128–136. DOI: 10.1016/j.enzmictec.2016.10.004.

Talib, W. H., and Mahasneh, A. M. 2010. "Antimicrobial, cytotoxicity and phytochemical screening of Jordanian plants used in traditional medicine." *Molecules* 15(3): 1811–1824. DOI: 10.3390/molecules15031811.

Tan, S. Y., and Tatsumura, Y. 2015. "Alexander Fleming (1881–1955): Discoverer of penicillin." *Singapore Medical Journal* 56(7): 366–367. DOI: 10.11622/smedj.2015105.

Thakur, B. K., Kumar, A., and Kumar, D. 2019. "Green synthesis of titanium dioxide nanoparticles using *Azadirachta indica* leaf extract and evaluation of their antibacterial activity." *South African Journal of Botany* 124: 223–227. DOI: 10.1016/j.sajb.2019.05.024.

Thiruvengadam, M., Chung, I.-M., Gomathi, T., Ansari, M. A., Gopiesh Khanna, V., Babu, V., and Rajakumar, G. 2019. "Synthesis, characterization and pharmacological potential of green synthesized copper nanoparticles." *Bioprocess and Biosystems Engineering* 42(11): 1769–1777. DOI: 10.1007/s00449-019-02173-y.

Tran, P. A., and Webster, T. J. 2011. "Selenium nanoparticles inhibit *Staphylococcus aureus* growth." *International Journal of Nanomedicine* 6: 1553–1558. DOI: 10.2147/IJN.S21729.

Troeman, D. P. R., Van Hout, D., and Kluytmans, J. A. J. W. 2019. "Antimicrobial approaches in the prevention of *Staphylococcus aureus* infections: A review." *Journal of Antimicrobial Chemotherapy* 74(2): 281–294. DOI: 10.1093/jac/dky421.

Turner, N. A., Sharma-Kuinkel, B. K., Maskarinec, S. A., Eichenberger, E. M., Shah, P. P., Carugati, M., Holland, T. L., and Fowler, V. G. 2019. "Methicillin-resistant *Staphylococcus aureus*: an overview of basic and clinical research." *Nature Reviews Microbiology* 17(4): 203–218. DOI: 10.1038/s41579-018-0147-4.

Umaralikhan, L., and Jamal Mohamed Jaffar, M. 2016. "Green synthesis of MgO nanoparticles and it antibacterial activity." *Iranian Journal of Science and Technology, Transactions A: Science* 42(2): 477–485. DOI: 10.1007/s40995-016-0041-8.

Utsui, Y. U., and Yokota, T. 1985. "Role of an altered penicillin-binding protein in methicillin-and cephem-resistant *Staphylococcus aureus*." 28(3): 397–403.

Vaudaux, P., Francois, P., Berger-Bachi, B., and Lew, D. P. 2001. "In vivo emergence of subpopulations expressing Teicoplanin or Vancomycin resistance phenotypes in a glycopeptide-susceptible, Methicillin-resistant strain of *Staphylococcus aureus*." *The Journal of Antimicrobial Chemotherapy* 47(2): 163–170. DOI: 10.1093/jac/47.2.163.

Vidhu, V. K., Aromal, S. A., and Philip, D. 2011. "Green synthesis of silver nanoparticles using *Macrotyloma uniflorum*." *Spectrochimica Acta Part A: Molecular and Biomolecular Spectroscopy* 83(1): 392–397. DOI: 10.1016/j.saa.2011.08.051.

Von Eiff, C., Becker, K., Machka, K., Stammer, H., and Peters, G. 2001. "Nasal carriage as a source of *Staphylococcus aureus* bacteremia." *New England Journal of Medicine* 344(1): 11–16. DOI: 10.1056/NEJM200101043440102.

Walters, M. S., Eggers, P., Albrecht, V., Travis, T., Lonsway, D., Hovan, G., Taylor, D., Rasheed, K., Limbago, B., and Kallen, A. 2015. "Investigation and control of Vancomycin-resistant *Staphylococcus aureus*: A guide for health departments and infection control personnel." In *MMWR. Morbidity and Mortality Weekly Report*. Department of health and human services. Centers for disease control and prevention. Atlanta, Georgia, United states. Available at: http://www.cdc.gov/hai/pdfs/VRSA-Investigation-Guide-05_12_2015.pdf

Wan, C.-X., Luo, J. G., Ren, X. P., and Kong, L. Y. 2015. "Interconverting flavonostilbenes with antibacterial activity from *Sophora alopecuroides*." *Phytochemistry* 116: 290–297. DOI: 10.1016/j.phytochem.2015.02.022.

Wang, Y., Liu, Z., Yao, G., Zhu, P., Hu, X., Yang, C., and Xu, Q. 2009. "An electrochemical assay for the determination of Se (IV) in a sequential injection lab-on-valve system." *Analytica Chimica Acta* 649(1): 75–79. DOI: 10.1016/j.aca.2009.06.055.

Wertheim, H. F. L., Melles, D. C., Vos, M. C., van Leeuwen, W., van Belkum, A., Verbrugh, H. A., and Nouwen, J. L. 2005. "The role of nasal carriage in *Staphylococcus aureus* infections." *The Lancet Infectious Diseases* 5(12): 751–762. DOI: 10.1016/s1473-3099(05)70295-4.

World Health Organization. 2002. *WHO Traditional Medicine Strategy*. Geneva: World Health Organization.

Yilmaz, M., Turkdemir, H., Kilic, M. A., Bayram, E., Cicek, A., Mete, A., and Ulug, B. 2011. "Biosynthesis of silver nanoparticles using leaves of *Stevia rebaudiana*." *Materials Chemistry and Physics* 130(3): 1195–1202. DOI: 10.1016/j.matchemphys.2011.08.068.

Yuan, C. G., Huo, C., Gui, B., and Cao, W. P. 2017. "Green synthesis of gold nanoparticles using *Citrus maxima* peel extract and their catalytic/antibacterial activities." *IET Nanobiotechnology* 11(5): 523–530. DOI: 10.1049/iet-nbt.2016.0183.

Yugandhar, P., Vasavi, T., Uma Maheswari Devi, P., and Savithramma, N. 2017. "Bioinspired green synthesis of copper oxide nanoparticles from *Syzygium alternifolium* (Wt.) Walp: Characterization and evaluation of its synergistic antimicrobial and anticancer activity." *Applied Nanoscience* 7(7): 417–427. DOI: 10.1007/s13204-017-0584-9.

Yuvakkumar, R., Suresh, J., Nathanael, A. J., Sundrarajan, M., and Hong, S. I. 2014. "Novel green synthetic strategy to prepare ZnO nanocrystals using rambutan (*Nephelium lappaceum* L.) peel extract and its antibacterial applications." *Materials Science and Engineering: C* 41: 17–27. DOI: 10.1016/j.msec.2014.04.025.

Zaidi, S., Misba, L., and Khan, A. U. 2017. "Nano-therapeutics: A revolution in infection control in post antibiotic era." *Nanomedicine: Nanotechnology, Biology and Medicine* 13(7): 2281–2301. DOI: 10.1016/j.nano.2017.06.015.

Zhang, H., and Chen, G. 2009. "Potent antibacterial activities of Ag/TiO_2 nanocomposite powders synthesized by a one-pot sol–gel method." *Environmental Science and Technology* 43(8): 2905–2910. DOI: 10.1021/es803450f.

Zhang, Y., Shareena Dasari, T. P., Deng, H., and Yu, H. 2015. "Antimicrobial activity of gold nanoparticles and ionic gold." *Journal of Environmental Science and Health, Part C* 33(3): 286–327. DOI: 10.1080/10590501.2015.1055161.

Zhu, W., Murray, P. R., Huskins, W. C., Jernigan, J. A., McDonald, L. C., Clark, N. C., Anderson, K. F. et al. 2010. "Dissemination of an *Enterococcus* Incl8-Like *vanA* plasmid associated with Vancomycin-resistant *Staphylococcus aureus*." *Antimicrobial Agents and Chemotherapy* 54(10): 4314–4320. DOI: 10.1128/aac.00185-10.

Section VI

Cross-Cutting Issues

11 Polymer-Based Protein Delivery Systems for Loco-Regional Administration

*Muhammad Haji Mansor, Emmanuel Garcion, Bathabile Ramalapa, Nela Buchtova, Clement Toullec, Marique Aucamp, Jean Le Bideau, François Hindré, Admire Dube, Carmen Alvarez-Lorenzo, Moreno Galleni, Christine Jérôme, and Frank Boury**

CONTENTS

11.1 Introduction 250
11.2 Protein Therapeutics 251
11.2.1 A Brief History and Rationale 251
11.2.2 Limitations and Challenges 254
11.3 Common Forms of Polymer-Based Protein Delivery Systems....................256
11.3.1 Micro/nanoparticles 256
11.3.2 Hydrogels 258
11.3.3 Porous Scaffolds 259
11.3.4 Fibrous Scaffolds as a Polymer-Based Protein Delivery System 260
11.3.5 Fibrous Scaffolds with Surface-Bound Protein Molecules 261
11.4 Conclusions 264
Acknowledgements 264
References 265

* To whom correspondence should be adressed (email address : frank.boury@univ-angers.fr)

11.1 INTRODUCTION

The previous two decades have seen a remarkable progress in biotechnology that enables production of many proteins for use in biomedical research. To improve their therapeutic values, much attention has been dedicated to prolonging the biological activity of these proteins after administration in patients. Due to the challenges faced by proteins in crossing biological barriers and targeting disease sites, a variety of organic and inorganic biomaterials have been developed. This ranges from simple drug delivery system such as polymer system for delivery of biomacromolecules, described in 1976, to smarter systems with capabilities of stimulating therapeutic release and local biological action in response to interactions with the surrounding environment (Farokhzad and Langer, 2009; Langer, 1976). Proof-of-concept has already been obtained in the field of regenerative medicine (Lee et al., 2014; Agrawal, 2017) and in cancer therapy (Ozdemir-Kaynak et al., 2018; Kim and Kim, 2015). A common approach involves incorporating the protein molecules into an appropriate matrix that permits gradual release of the protein load. In doing so, the matrix limits the exposure of proteins from proteases and neutralizing antibodies that may be present in the immediate physiological environment, thus preventing them from undergoing rapid degradation.

There are multiple ways to execute the above-mentioned strategy. Two great examples are (i) the incorporation of proteins into liposomes and (ii) the preparation of protein-based nanoparticles. Liposomes refer to spherical vesicles consisting of one or more lipid bilayers with a characteristic protein-loadable aqueous core (Swaminathan and Ehrhardt, 2012). Examples of proteins successfully incorporated into the aqueous compartment of this carrier include insulin (Wong et al., 2018), bovine serum albumin (Liu et al., 2015) and human gamma-globulin (García-Santana et al., 2006). Several literatures provide comprehensive reviews of the advantages and drawbacks of this carrier (Ibraheem et al., 2014; Martins et al., 2007). Another type of protein nanocarriers with promising potential in various clinical applications is the protein-based nanoparticles. Well-defined particles of nanometre scale have been successfully produced from a variety of commercially accessible proteins, including gelatin (Lee et al., 2012), human serum albumin (Tarhini et al., 2020), silk fibroin (Zhao et al., 2015), zein (Pascoli et al., 2018) and legumin (Karimi et al., 2018). Due to the unique and wide-ranging physicochemical properties of the amino acids present in a particular protein, protein-based nanoparticles can be produced using numerous well-established techniques, including desolvation, coacervation and emulsification. Likewise, the hydrophilic–hydrophobic property of the protein-based nanoparticles can be tailored conveniently as a result of the wide range of hydrophilic and hydrophobic moieties featured in the amino acid building blocks, making this carrier an excellent vehicle for a wide range of hydrophilic (e.g. proteins and peptides) and hydrophobic molecules. Extensive research and development work to promote preclinical and clinical use of this carrier has been documented in a decent number of literatures over the last decade (Tarhini et al., 2017, 2018).

Despite the promising potential of liposomes and protein-based nanoparticles as mentioned above, polymers remain as an attractive material for the formulation of protein carriers. Both naturally derived and synthetic polymers have been widely used to produce protein-loaded matrices due to the high versatility of this material group. By changing the type of monomers, controlling the polymerization conditions or functionalizing the polymer chains with chemical groups of interest, the physicochemical and biological properties of the polymer matrix, including surface charge, hydrophobicity, biodegradability and biocompatibility, can be regulated. Examples of clinically successful polymer-based protein carriers are presented in later sections of this chapter.

11.2 PROTEIN THERAPEUTICS

11.2.1 A Brief History and Rationale

Proteins have the most dynamic and diverse role of any macromolecule in the body. They act as catalysts to biochemical reactions by forming receptors and channels in the membranes. Furthermore, proteins provide intracellular and extracellular scaffolding support and they transport molecules within the cells or from one organ to another (Leader et al., 2008). It has been estimated that there are approximately 25,000–40,000 different genes in the human genome and with alternative splicing of genes and post-translational modification of proteins, the number of distinct functional proteins is likely to be much higher (Venter et al., 2001; Hattori, 2005). The great number of functional proteins could pose vast challenges to modern medicine, as disease may result when any one of these proteins contains mutations or other abnormalities or is present in abnormally high or low concentration. These proteins however may also present immense possibilities in terms of development of protein therapeutics to alleviate disease. It is for this reason that they have progressively become the forerunners in biopharmaceutics. Protein therapeutics can be grouped into molecular types that include antibody-based drugs, anticoagulants, blood factors, bone morphogenetic proteins (BMPs), enzymes, fusion proteins, growth factors, hormones, interferons, interleukins and thrombolytics (Walsh, 2006; Carter, 2011). Protein therapeutics have also been classified based on their pharmacological activity as drugs that (i) replace a protein that is deficient or abnormal, (ii) supplement an existing pathway, (iii) provide a novel function or activity, (iv) interfere with a molecule or an organism, or (iv) deliver a payload such as a radionuclide, cytotoxic drug, or protein effector (Leader et al., 2008).

Proteins first emerged as a major class of pharmaceuticals in the 1980s, with a majority of them mainly developed for therapeutics and a small number for diagnostics and vaccines (Walsh, 2006). More than three decades later, a better understanding of the molecular biology and biochemistry behind these macromolecules and their role in various body functions and pathological conditions has led to the realization of enormous therapeutic applications for proteins (Muheem et al., 2014). Advances in the development of protein therapeutics have demonstrated that these molecules offer several advantages over the more conventional small-molecule drugs (Fig. 11.1).

The first reported use of protein therapeutics was in the 1920s when insulin that was purified from bovine and porcine pancreas was used as a life-saving daily

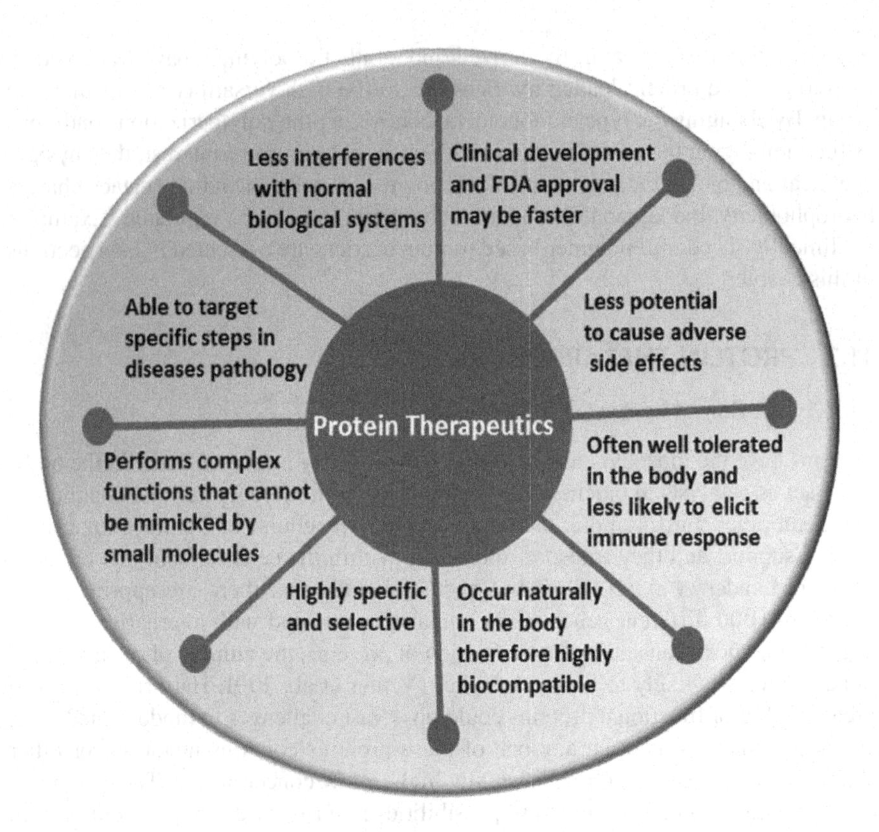

FIGURE 11.1 Advantages of protein therapeutics for clinical applications.

injection for patients suffering from type 1 diabetes mellitus (Banting et al., 1991). The low availability of animal pancreases for purification of insulin, the high cost of the purification and the immunological reaction of some patients to animal insulin hindered the widespread use of this protein (Richter et al., 2005). In 1982, insulin became the first FDA approved human protein therapeutic derived from recombinant DNA technology (//www.sciencedirect.com/topics/biochemistry-genetics-and-molecular-biology/recombinant-dna-technology) and has since become the major therapy for type 1 and type 2 diabetes mellitus (Richter et al., 2005; Göddel et al.,1979). Soon after, other recombinant human proteins were developed as therapeutics to replace the natural proteins deficient in some patients (e.g. growth hormone) or boost existing pathways (e.g. interferon-α, tissue plasminogen activator and erythropoietin) (Leader et al., 2008; Brown et al., 1997).

Recombinant production of proteins is highly favoured over purification of proteins from their native source. A small number of non-recombinant proteins purified from their native source have been reported, such as pancreatic enzymes from hog and pig pancreas (Dirksen et al., 1999) and alpha-1 proteinase inhibitor from pooled human plasma (Yamaguchi and Miyazaki, 2014). This strategy however has proven to be rather challenging and expensive. In this regard, the production of therapeutic proteins by genetic engineering using recombinant DNA technology has presented great opportunities towards overcoming the challenges faced with conventional non-recombinant proteins. In addition to availability in sufficient quantities and the reduced risk of immunological rejection, recombinant technology allows the modification of proteins or the selection of particular gene variants to improve function or specificity and enables the production of proteins that provide novel functions or activity (Banting et al., 1991). Thus, modern therapeutic proteins are largely produced by recombinant technology.

In the field of cancer treatment, it has been shown that a synergic effect with ionizing radiation could occur upon exposure to BMPs. HrBMP4, for instance, is expressed in the embryonic cortex, indicating its role in the formation of mesoderm and neurogenesis, e.g. morphological differentiation of neural stem cells (Moon et al. 2009; Vescovi et al., 2006). A recent clinical trial carried out on the brain tumour glioblastoma (GBM) is in course after resection or biopsy of the tumour, using convection enhanced delivery (CED) allowing increasing amounts of HrBMP4 solutions combined with Gd-DTPA and determining the extent of intra-tumour and interstitial drug delivery (DiMeco et al., 2016). HrBMP4 can indeed inhibit the proliferation of brain tumour stem cells, induce their morphological changes to a more differentiated phenotype and reduce their invasiveness (Liu et al., 2016; Piccirillo et al., 2006). The possibility to abolish the tumour's self-renewal potential by depleting the tumour stem cell compartment with a differentiating, non-toxic compound such as BMP4 is attractive because it could be used to render the stem cells more vulnerable to conventional post-surgery therapies. The differentiated cancer stem cells then could be better eliminated by external beam radiation or internal radiotherapy after loco-regional implantation (Bao et al., 2006). This is well illustrated by Stupp et al. (2007) differentiating strategy: GBM is a heterogeneous tumour that can be initiated and maintained by a minority of $CD133^{+}$ cancer stem-like cells that have a high tumorigenic potential and a low proliferation rate. Exposure to BMPs can force these $CD133^{+}$ tumour cells into a more differentiated phenotype characteristic of the CD133 tumour bulk, abolishing their self-renewal potential and increasing their sensitivity to radiotherapy (Stupp et al., 2007). Hence, BMP originally influences multiple signalling pathways originally involved in organogenesis and lineage-specific differentiation but also in cancer stem-like cell maintenance.

In the last decades, tissue engineering and regenerative medicine have emerged as promising strategies for bone reconstitution, with the ambition to circumvent the complications associated with traditional techniques. Bone tissue engineering aims to induce new functional bone regeneration via the synergistic combination of biomaterial scaffolds, cells and signal factor therapy. Engineered bone tissues are considered as a potential alternative to the conventional use of bone grafts, due to their limitless supply and no disease transmission. Bone scaffolds can be defined as an artificial temporary 3D matrix with micro- and nanostructures exhibiting biomimetic properties that provide a specific environment and architecture for bone growth and development (Henry et al., 2017; Lam et al., 2014). Scaffolds can be combined with different types of cells able to promote bone formation *in vivo* either by differentiating towards the osteogenic lineage or by releasing specific soluble cytokines. A challenge within scaffolds association with drugs and/or growth factors (e.g. BMPs) is that they can deliver those cytokines in the environment and exert their therapeutic/regenerative effects (e.g. proliferation, differentiation). Interestingly, clinicians have demonstrated that such polymer-based systems can be injected or implanted locally with minimal adverse effects. These two examples show that recombinant proteins of the same class may exert therapeutic effects against different diseases, depending on their ways of administration and their interactions with the biological microenvironment.

11.2.2 Limitations and Challenges

Tremendous effort has been invested in cellular engineering to optimize various hosts for protein production and there are many examples in which proteins have been used in therapy successfully. However, this kind of therapy has also presented various challenges.

The use of protein therapeutics is often limited by their instability, solubility, distribution, method of administration and side effects (Foster et al., 2010; Cleland et al., 1993). The stability of protein therapeutics is a critical issue. These molecules can suffer loss of activity in response to environmental triggers such as moisture or temperature, which can occur during storage or even when administered *in vivo* (Mitragotri et al., 2014). Nevertheless, several reports have shown that optimized processing protocols enable proteins to be encapsulated into polymer matrices or grafted to polymer scaffolds to preserve their native conformation and thus their bioactivity for several months (Wang et al., 2014; Hajavi et al. 2018). The permeability of protein therapeutics through barriers such as the skin, mucosal membranes and cellular membranes is substantially reduced due to high molecular mass, which leads to injection being the primary mode of administration (Albarran et al., 2011). As many protein molecules have their therapeutic targets inside cells, challenges arise in transporting these molecules into the target cells without them breaking down while in the bloodstream (Foster et al., 2010). The half-life of the therapeutic proteins can also be considerably reduced by proteases, protein-modifying chemicals or other clearance mechanisms in the body (Putney and Burke, 1998; Daugherty et al., 2006; Baker et al., 2010). Stromal cell-derived factor 1 (SDF-1α), for example, can be cleaved by matrix metalloproteinase-2 and metalloproteinase-9 (MMP-2/9) released during a traumatic event such as tumour resection in the case of GBM, resulting in loss of its chemotactic activity (Peng et al., 2012).

BMP-2 has been previously isolated directly from bone. However, the limited yield and potential health risks associated with its isolation from allogeneic donor bone limited its clinical application (Adelita et al., 2017; Bessho et al., 1999). The expression of the recombinant BMP-2 in mammalian cell culture such as Chinese Hamster Ovarian (CHO) cells also generates low yields of protein and the procedure is relatively expensive (Pisal et al., 2011; Israel et al., 1996; Wang et al., 1990). Eukaryotic systems such as yeast and animal cells were initially considered to produce recombinant BMP-2 to ensure adequate post-translational modifications. Indeed, BMP-2 is a naturally glycosylated protein, but it has been discovered that glycosylation is not required for its function (Ruppert et al., 1996; Long et al., 2006; Al-Hilal et al.,2013). Recently, several authors have reported the production of biologically active BMP-2 expressed in *Escherichia coli*. Although the expressed BMP-2 was insoluble and formed inclusion bodies, active BMP-2 could be successfully refolded *in vitro* using specific refolding solutions and protocols (Dalton et al., 2014; Sharapova et al., 2010; Lee et al., 2012; Sahoo et al., 2003). BMP-2 is biologically active only in a dimeric form, which is stabilized by an intermolecular disulfide bridge that connects two cysteines: Cys-114 and Cys-228 from two different BMP-2 protein molecules (Scheufler et al., 1999). The disulfide bridge that stabilizes BMP-2 dimer ensures the interaction of the BMP-2 dimer with transmembrane serine/threonine kinase receptors on osteogenic cells, which activates proliferation and differentiation of osteoblast cells (Sharapova et al., 2010; Thomasin et al., 1997).

Another challenge with protein therapy is the immune response that the body may build against the proteins. Virtually all therapeutic proteins generate some level of antibody response (Daugherty et al., 2006). There are cases where the immune response can neutralize the protein and can even cause a harmful reaction in the patient. An example of such an immune response is the activation of B cells, which produce antibodies that bind to the proteins and reduce and possibly eliminate their therapeutic effects. Such antibodies can cause complications that can be life-threatening. Thus, the immune response of therapeutic proteins is a concern for researchers, manufacturers and clinicians (Baker et al., 2010). The uses of protein therapeutics in animal trials usually do not effectively predict the response in humans. It is thus critical to evaluate the safety and efficacy of protein therapeutics and their probability to trigger antibody formation during development.

Another shortcoming associated with protein therapeutics is the high production costs. Protein therapeutics are expensive, and this may limit clinical applications as well as patient access. This high-cost issue is further aggravated for protein therapeutics where multi-gram doses are needed for a treatment course, as is the case for some antibodies (Brian, 2009).

Nano- and microsized engineered materials have received considerable attention in modern pharmaceutics due to their potential to address the challenges encountered with conventional therapeutics. These materials can address issues associated with current pharmaceuticals such as extending product life, or can add to their performance and acceptability, either by increasing efficacy or improving safety and patient compliance (Farokhzad and Langer, 2009; Schmidt and Volodkin, 2013). Targeted delivery and specific release can be achieved with these delivery systems via electrostatic interaction and pH or temperature-dependent responses to controlled stimuli *in vivo* (Honary and Zahir, 2013; Yun et al., 2014).

Current research is focused especially on developing biodegradable polymer materials that have shown significant therapeutic potential. Biodegradable polymers are natural or synthetic polymers that are able to degrade *in vivo* into biocompatible and toxicologically safe by-products that are subsequently resorbed or excreted by the body. Naturally occurring biodegradable polymers are widely explored because of their abundance in nature, biocompatibility and lower toxicity. Chitosan (Deng et al., 2017), hyaluronic acid (Giarra et al., 2018), silk fibroin (Fernández-garcía et al., 2016), cellulose (Wozniak et al., 2003) or collagen (Teixeira et al. 2010; Manavitehrani et al., 2016) have been among the most investigated natural biodegradable polymers for protein delivery applications. However, their use is challenging because of wide molecular weight distributions and batch-to-batch variability and the necessity to collaborate with companies that are able to purchase materials following clinical Good Manufacturing Practices (cGMP). On the other hand, cGMP synthetic biodegradable or bioeliminable polymers are commercially available with different and well-defined compositions, molecular weights and degradation times. Aliphatic polyesters such as poly(lactic-*co*-glycolic acid) (PLGA) and polycaprolactone (PCL) have been among the most successfully used synthetic biodegradable polymers so far (Makadia and Seigel, 2011).

11.3 COMMON FORMS OF POLYMER-BASED PROTEIN DELIVERY SYSTEMS

Common examples of polymer-based systems that have been utilized in recent years to deliver various drug molecules, including therapeutic proteins, include micro/nanoparticles, hydrogels and porous scaffolds (Fig. 11.2).

11.3.1 Micro/Nanoparticles

Micro/nanoparticles are injectable drug carriers that are usually prepared from hydrophobic polymers using straightforward processes such as solvent evaporation, phase separation and spray-drying (Sokolsky-Papkov et al., 2007). In the solvent evaporation method, an organic phase is first formed by dissolving a hydrophobic polymer and the drug molecules to be encapsulated in a water-immiscible, volatile organic solvent. This phase is then dispersed in an aqueous phase containing stabilizers such as polyvinyl alcohol under continuous mechanical agitation to form an oil-in-water (O/W) emulsion. Drug-loaded particles are formed upon evaporation of the organic solvent from the inner phase at reduced or atmospheric pressure. The particles can then be collected by filtration or centrifugation, washed to remove the stabilizing molecules adsorbed to the particle surface and lyophilized to minimize hydrolytic degradation of the particles during long-term storage. However, the single emulsion technique may not be suitable for encapsulating hydrophilic drugs such as proteins as they tend to diffuse into the external aqueous phase during the emulsification step.

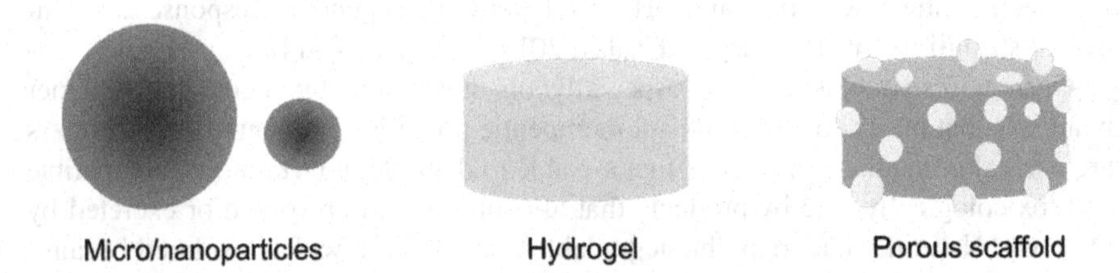

FIGURE 11.2 Common polymer-based systems for drug delivery applications.

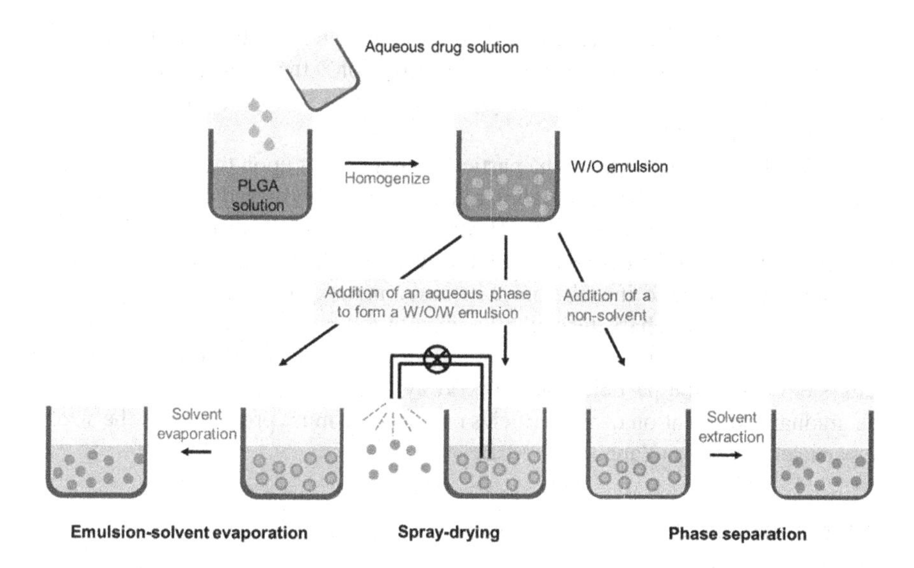

FIGURE 11.3 Examples of micro/nanoparticle preparation process.

Therefore, in a process commonly referred to as double-emulsion solvent evaporation, protein molecules are first solubilized in an aqueous solvent, and then dispersed in a polymer-containing organic phase to form a primary water-in-oil (W/O) emulsion, followed by dispersion in another aqueous solvent to form the secondary O/W emulsion (Ding and Zhu, 2018; Iqbal et al., 2015). The preparation of the primary W/O emulsion is also relevant to the phase separation method (Fig.11.3). Following this step, instead of adding an aqueous solvent, an organic solvent that is non-solvent to the dissolved polymer is gradually introduced to extract the solvent of the polymer and decrease its solubility. The phase separation of the polymer from its solution contributes to the formation of polymer-rich liquid phase (coacervate) that surrounds the inner drug-containing aqueous phase. Upon completion of the phase separation process, the coacervate solidifies to produce drug-loaded particles (Tran et al., 2012). An obvious drawback of this method is the requirement for a large volume of organic solvent. Recent work proposed the use of water-miscible organic solvents to dissolve the polymer. This replaces the need for organic solvents to induce phase separation as aqueous media can be used to extract the polymer solvent (Kohane, 2007). Finally, in the spray-drying method, the W/O emulsion is sprayed into a heated chamber that leads to a spontaneous production of drug-loaded particles. This method is more rapid and convenient and has fewer processing parameters than the other two but is limited by the adhesion of the formed particles to the inner surfaces of the drying chamber (Sokolsky-Papkov et al., 2007).

Due to their small size, micro/nanoparticles can be administered either directly to the intended site of action or into the systemic circulation to reach a desired location by passive or active targeting mechanisms (Yang and Pierstorff, 2012). Several peptide-loaded polymer-based microparticle formulations have been approved by the FDA for clinical use. The first is Lupron Depot®, which received approval in

1989 to provide sustained release of leuprolide acetate for prostate cancer treatment (Lee et al., 2016). A more recent example is Bydureon® that was approved in 2012, which releases exenatide to improve glycemic control in type 2 diabetes patients (Singh and Lillard, 2009).

In general, drug release from the particles is dependent upon the diffusion rate of the drug molecules and the degradation rate of the polymer-based matrix (Sokolsky-Papkov et al., 2007; Lee et al., 2016). However, as significant proportion of the drug load can be weakly adsorbed onto the large surface area of the micro/nanoparticles rather than incorporated into the polymer-based matrix, the drug release profile of this system is usually characterized by a huge initial burst that is followed by relatively short duration of release of the remaining drug load. Another disadvantage of this system is that the particles can move away from the targeted drug release site. The gradual translocation of the particles can become more prominent as the size of the particle decreases (Yang et al., 2012).

11.3.2 Hydrogels

Hydrogels are three-dimensional networks of cross-linked hydrophilic polymers. The cross-linking can be mediated by the physical interactions (e.g. hydrogen bonds, electrostatic interactions) between the polymer chains (Kimura et al., 2004; Ren et al., 2015) or the covalent bonds resulting from the use of chemical cross-linkers (e.g. carbodiimide, glutaraldehyde) (Lu et al., 2008; Rafat et al., 2008; Tian et al., 2016; Mirzaei et al., 2013). Most hydrogels are characterized by highly porous structure. The pore size can range from 10 μm to 500 μm and is dependent upon the degree of cross-linking in the hydrogel matrix (Chai et al., 2017; Li and Mooney, 2016). The porous structure is responsible for the deformability of hydrogels, enabling them to conform to the shape of the site to which they are applied (Hoare et al., 2008). Due to their hydrophilicity, water-soluble drug molecules can be conveniently loaded into the porous structure of a pre-formed hydrogel. However, this is not always true for high molecular weight drug molecules such as proteins, which have diffusive limitations to their partitioning into the pores of the hydrogel (Van Tomme et al., 2005). The high dependency of the drug loading process on the pore size of the hydrogel also means that the loaded drug molecules are usually released rapidly at the site of application as the release process is governed mainly by the diffusion rate of the drug molecules through the pores. In fact, the release of hydrophilic molecules from a hydrogel system typically lasts for only several hours or days, shorter than the release durations achieved with micro/nanoparticles made of hydrophobic polymers (Lee et al., 2016). To counter this, several strategies to enhance drug–hydrogel interactions have been proposed, including the introduction of charged moieties into the hydrogel to boost ionic interactions (Schneider et al., 2016) and the direct conjugation of the drug molecules to the hydrogel via covalent bond formation (Sutter et al., 2007). Another credible strategy to prolong drug release is to load the drug molecules directly into the hydrogel matrix during the hydrogel fabrication process instead of loading into the pores of a pre-formed hydrogel (Chen et al., 2004). Finally, several groups proposed the strategy of pre-encapsulating drug molecules into suitable micro/nanoparticles and co-formulating the particulate system into the hydrogel matrix to achieve sustained drug release (Gao et al., 2012; Kim et al., 2012).

As virtually any water-soluble polymer can be manipulated to produce this system, it is possible to obtain hydrogels with physicochemical and biological properties that are useful for a wide range of applications. Despite this, the number of hydrogel-based drug delivery systems approved for clinical use is still limited. An example of these is Regranex®, which consists of a carboxymethylcellulose gel that releases recombinant human platelet-derived growth factor (becaplermin) for the treatment of diabetic foot ulcers (Hoare and Kohane, 2008).

In addition to the rapid drug release issue mentioned above, hydrogels possess several drawbacks that could limit its use for applications. Their poor mechanical strengths make them susceptible to premature dissolution (Van Tomme et al., 2005), limiting the time window for acting in the microenvironment. In addition, in the absence of cell-adhesive proteins, hydrogels tend to have low capacity for cell adhesion and attachment due to their low stiffness (Sarker et al., 2014; Shen et al., 2017; Autissier et al., 2010).

11.3.3 Porous Scaffolds

Porous scaffolds refer to three-dimensional solid polymer matrices characterized by interconnected pores. Generally, they are formed by removing the solvent from a polymer solution that leads to the precipitation of the polymer molecules. Methods that have been employed to produce porous scaffolds include freeze-drying (Ochi et al., 2003), particulate leaching (Yoon and Park, 2001) and gas foaming (Tai et al., 2007). In the first method, a polymer solution is initially frozen at a sub-zero temperature inside an airtight chamber. The pressure is then gradually decreased to vaporize the frozen liquid. As more and more solvent evaporates, the polymer molecules precipitate and solidify to form a porous scaffold (Ochi et al., 2003). In the particulate leaching method, a polymer solution is first mixed with salt particles of well-defined size. The solvent is subsequently removed under vacuum, leaving behind a solid polymer matrix loaded with salt particles. The subsequent leaching of the salt particles in distilled water results in the formation of a porous scaffold (Yoon and Park, 2001). Gas foaming is another common method used to make porous scaffolds. It relies on the nucleation and growth of gas bubbles in a polymer phase. Traditionally, the gas bubbles can be formed *in situ* by adding into the polymer phase a foaming agent such as ammonium bicarbonate, which generates inert gas such as CO_2 when the pH of the system is decreased. A porous scaffold is formed upon removal of the dispersed gas bubbles from the polymer phase (Tai et al., 2007). Recently, supercritical fluids have been used as an alternative foaming agent. A supercritical fluid is any substance existing at a temperature and pressure above its critical point with an intermediate behaviour between that of a liquid and a gas. The use of supercritical fluids is useful especially in making porous scaffolds from hydrophobic polymers as it circumvents the need for organic solvents during the preparation of the polymer phase. CO_2 is widely used as a supercritical fluid due to its minimal toxicity and low cost. Initially, polymers can be dissolved or plasticized in supercritical CO_2. Upon depressurization of the system, the rapid expansion of the polymer phase as a result of the escape of CO_2 gas leads to the formation of a porous scaffold (Tai et al., 2007; Tran et al., 2015).

Similar to hydrogels, the use of porous scaffolds as a drug delivery system can be achieved by loading drug molecules into the pores of a pre-formed scaffold or incorporating them directly into the polymer phase before the scaffold fabrication process. A notable example of clinically used porous scaffold-based drug delivery systems is Infuse®, which consists of a porous collagen scaffold that can be conveniently loaded with recombinant human BMP-2 prior to administration in patients undergoing bone reconstruction procedure (McKay et al., 2007). Interestingly, the osteoinductive effect of this treatment relies on the chemotactic effect of BMP-2 that induces the infiltration of mesenchymal stem cells (MSCs) into the pores of the collagen scaffold (Liu et al., 2018). The considerable mechanical strength of the scaffold means that it can withstand the traction forces generated during cell attachment and migration, thus sustaining the cell infiltration process. After initial proliferation, the MSCs are further stimulated by BMP-2 to undergo differentiation into bone-forming osteoblasts to enable new bone formation (Luu et al., 2007). Considering its huge clinical success, Infuse® presents a working example to the idea of using a chemotactic agent and a suitable scaffold to recruit a certain cell population. During a bone reconstruction surgery, the site of bone defect can be accessed and filled with the Infuse® bone graft consisting of BMP-2-loaded porous collagen scaffolds to recruit MSC by chemotaxis. Upon infiltration into the injury site, the MSCs proliferate to increase their number before undergoing differentiation into the bone-forming osteoblasts, which secrete collagen and calcium-binding proteins to support the formation of mineralized bone tissues.

11.3.4 Fibrous Scaffolds as a Polymer-Based Protein Delivery System

Fibrous scaffolds refer to scaffolds made of fibres with diameters on the order of several micrometres down to the tens of nanometres that are stacked layer-by-layer to form a three-dimensional non-woven mesh (Fig. 11.4). Compared to micro/nanoparticles, hydrogels and porous scaffolds, the use of fibrous scaffolds as a delivery vehicle for therapeutic proteins is less common despite the multiple advantages offered by this system. This being said, the amount of research conducted to investigate the value of fibrous scaffolds in this field of application has increased steadily over the last two decades and multiple strategies for loading protein molecules into fibrous scaffolds have been proposed.

Depending on the scaffold preparation technique, protein molecules can be embedded randomly in the fibres or partitioned into a specific fibre compartment as in the case of core–shell fibres (Fig.11.4). Chew et al. incorporated β-nerve growth factor (NGF) into fibres made of poly(ε-caprolactone-ethyl ethylene phosphate) (PCLEEP) and examined the release profile. They observed that the fibrous scaffold was able to sustain NGF release over a period of 90 days. They claimed that the slow degradation of PCLEEP contributed to the sustained release profile as NGF molecules could only be released by diffusion through the hydrophobic matrix of

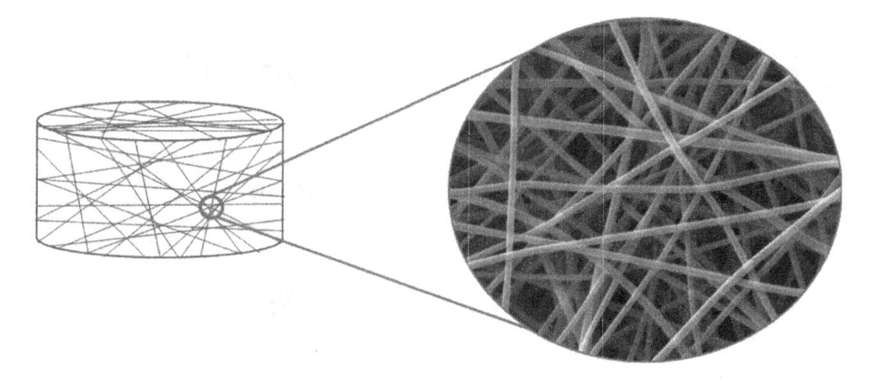

FIGURE 11.4 A simplified representation of a fibrous scaffold and its internal structure.

the fibre (Chew et al., 2005). On the other hand, Zhang et al. produced fibres with a core–shell structure as a vehicle to deliver bovine serum albumin (BSA). The outer shell was made of the hydrophobic PCL, while the core compartment dispersed with the BSA molecules was made of the hydrophilic bioeliminable poly(ethylene glycol) (PEG). They reported that the core–shell system produced lower initial burst and longer duration of BSA release than fibres made of a single blend of PCL, PEG and BSA (Zhang et al., 2006). Jiang et al. further explored the possibility of tuning the kinetics of protein release from core–shell fibres. They showed that by varying the mass ratio of PCL and PEG in the outer shell, the time to achieve complete release of BSA from the inner dextran core could be varied from one week to approximately one month. BSA release was accelerated with increasing PEG mass in the outer shell as its water solubility resulted in the formation of pores through which BSA molecules could escape from the dextran core (Jiang et al., 2006).

Protein molecules may also be encapsulated into micro/nanoparticles prior to incorporation into fibrous scaffolds. Liu et al. prepared dextran-based nanoparticles loaded with basic fibroblast growth factor (bFGF) that were subsequently embedded in poly(L-lactic acid) (PLLA) nanofibers. The duration of bFGF release provided by the nanoparticle–nanofiber composite scaffold was 10 days longer compared to what was achieved with nanofibers with directly embedded bFGF (28 vs. 18 days). In addition, the encapsulation of bFGF into the dextran-based nanoparticles was also useful in reducing bFGF structural changes during the fibre-making process (Liu et al., 2013). Qi et al. also adopted a similar approach. They incorporated BSA-loaded alginate microparticles into PLLA fibres and observed that the composite scaffold produced a longer duration of BSA release compared to the naked alginate microparticles (Qi et al., 2006).

11.3.5 Fibrous Scaffolds with Surface-Bound Protein Molecules

Alternatively, protein molecules can be loaded onto the surface of a prefabricated fibrous scaffold. This is especially useful when preparing protein-loaded fibrous scaffold using a hydrophobic polymer and there is a need to reduce the exposure of

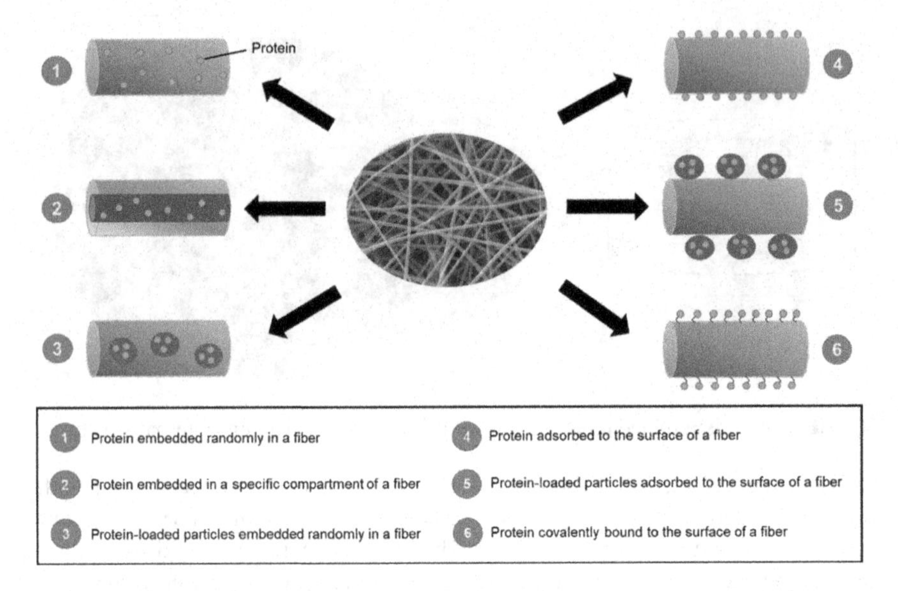

FIGURE 11.5 Different modes of protein loading into a fibrous scaffold. (Adapted from J.S. Choi, K.W. Leong, H.S. Yoo, *In vivo* wound healing of diabetic ulcers using electrospun nanofibers immobilized with human epidermal growth factor (EGF), Biomaterials. 29 (2008) 587–596 and T.G. Kim, T.G. Park, Surface functionalized electrospun biodegradable. nanofibers for immobilization of bioactive molecules, Biotechnol. Prog. 22 (2006) 1108–1113.)

the protein molecules to organic solvents that are needed to solubilize the polymer prior to the scaffold fabrication step. The nano/micro-dimension of the fibres confers a large surface area for adsorption of protein molecules (Fig.11.5). In fact, the amount of protein that can be adsorbed by a fibrous scaffold is generally four times greater than that afforded by a porous scaffold of equal volume (Woo et al., 2003).

Immobilization of protein molecules to the surface of the fibres can be mediated by non-covalent interactions, including hydrophobic interaction, van der Waals interaction, hydrogen bonding and electrostatic interaction. Heparin, a naturally occurring polysaccharide, is known to have strong binding affinity for various growth factors (e.g. VEGF, transforming growth factor-β – TGF-β, fibroblast growth factor – bFGF), morphogens (e.g. BMP-2, BMP-7, BMP-14) (Rider and Mulloy, 2017) and ECM proteins (e.g. laminin) (Utani et al. 2001) due to its ability to form non-covalent interactions with these proteins. Therefore, heparin-functionalized fibrous scaffolds can be conveniently loaded with these proteins for local delivery applications. Casper et al. prepared PEG and poly(lactic-*co*-glycolic acid) (PLGA) nanofibers functionalized with low molecular weight heparin (LMWH) that were adsorbed with bFGF. To slow down the dissociation of LMWH from the fibrous scaffold and thus prolong the duration of bFGF release, LMWH was conjugated to PEG prior to its incorporation into the nanofibers. Although the bFGF release profile was not assessed in their study, they reported that LMWH was retained in the fibrous scaffolds for at least 14 days (Casper et al., 2005). Furthermore, Patel et al. prepared heparin-functionalized

PLLA nanofibers as a delivery vehicle for bFGF and laminin. The adsorption of bFGF and laminin to the surface of PLLA nanofibers was very stable, with less than 0.1% of the total amount of immobilized protein molecules released into the surrounding solution after 20 days. The slow release of the adsorbed protein molecules could be useful in certain neuroregenerative applications as the immobilized βFGF was found to be as effective as its soluble counterpart in inducing neurite outgrowths from dorsal root ganglion tissues (Patel et al., 2007). Fibre surfaces can also be adsorbed with protein-loaded nanoparticles. Wei et al. prepared BMP-7-loaded PLGA nanoparticles that were subsequently immobilized onto PLLA nanofibers. They reported that the release kinetics of BMP-7 could be controlled by varying the degradation rate of the PLGA nanoparticles. However, as the nanoparticle surfaces were exposed to the surrounding solutions, a characteristic burst release could be observed with each formulation of BMP-7-loaded PLGA nanoparticles prepared in their study (Wei et al., 2007).

Another widely used method for functionalizing fibre surfaces with proteins is by chemical immobilization. This approach results in the formation of covalent bonds between the fibre surfaces and the protein molecules. As the covalently attached protein molecules cannot be easily desorbed from the fibres, this functionalization method is especially useful in many regenerative applications, where long-term immobilization of protein molecules in the fibrous scaffold is often necessary for the reparative actions to take place. Primary amine and carboxyl groups are the most common example of functional groups utilized in covalent conjugation of fibres and protein molecules. Many groups have prepared polymer-based nanofibers functionalized with carboxyl groups that can be activated by a combination of 1-ethyl-3-(3-dimethylaminopropyl) carbodiimide (EDC) and *N*-hydroxysuccinimide (NHS) for subsequent conjugation with primary amine groups present in protein molecules. Ye et al. prepared nanofibers from poly(acrylonitrile-*co*-maleic acid) (PANCMA) that were subsequently functionalized with lipase. However, the immobilized lipase molecules were found to have lower enzymatic activities than their soluble counterparts (Ye et al., 2006). A similar loss in activity was also reported by a group that functionalized polystyrene (PS) nanofibers with α-chymotrypsin (Jia et al., 2002). There are two possible explanations for the partial inactivation of the immobilized enzymes. First, the immobilization process may introduce covalent alterations to the active sites of the enzyme. The other is that direct conjugation of protein molecules to the fibre surfaces may cause certain parts of the immobilized molecules to be sterically inaccessible to their corresponding ligands (Yoo et al., 2009). To address the latter issue, several polymer-based linkers have been utilized to introduce a physical gap between the immobilized molecules and the fibre surfaces. To obtain these linkers, primary amine-terminated hydrophilic polymers such as PEG-diamine can be chemically conjugated to a hydrophobic polymer such as PCL and PLGA. The linker can then be mixed with an unconjugated hydrophobic polymer to prepare fibres displaying primary amine groups on their surface that can be conjugated with protein molecules. Choi et al. immobilized EGF on the surface of fibres composed of PCL and PCL-PEG-NH_2 for wound-healing applications. They showed that the EGF-functionalized fibres were able to induce differentiation of keratinocytes to a greater extent than fibres supplemented with EGF solution. The enhanced activity of the former could potentially be attributed to the fact that covalently immobilized EGF could be better retained at the wound site and thus was able to induce more durable pro-differentiation signals in the locally residing keratinocytes (Choi et al., 2008). Kim et al. also utilized a polymer-based linker to conjugate lysozyme to the surface of PLGA nanofibers. The immobilized lysozyme displayed comparable activity to its soluble counterpart (Szentivaniyi et al., 2011; Kim and Park, 2006). This is opposite to the significant loss of enzymatic activities observed with direct conjugation of enzyme molecules to the fibre surfaces as discussed above.

11.4 CONCLUSIONS

With the advent of recombinant technology, a wide variety of biocompatible therapeutic proteins can be manufactured with relative ease. These proteins would then be carefully formulated and subsequently administered in patients to address different types of diseases more effectively and selectively. At the level of formulation development, protein molecules can be physically and/or chemically conjugated to a wide array of naturally occurring, semi-synthetic and synthetic biomaterials to form different types of protein delivery systems. Depending on their architecture and the extent of protein–scaffold interactions, these delivery systems can modify the pharmacokinetic (PK) and pharmacodynamic (PD) properties of the protein molecules. The versatility of polymer-based protein delivery systems such as micro/nanoparticles, hydrogels, porous scaffolds and fibrous scaffolds means it is possible to alter the spatial distribution of the protein load within the system as well as the protein release kinetics. These can then influence the ability of the protein molecules to exert the intended effects in their immediate microenvironments, be it to kill cancer cells or to recruit stem/progenitor cells. From a pharmaceutical development perspective, the design of a protein delivery system should be commenced only after the attainment of an in-depth understanding of the PK/PD profile of the protein of interest for a given medical application. Therefore, a close communication between formulation scientists, molecular biologists, PK/PD scientists and clinicians are crucial to ensure successful development of protein delivery systems that are fit for preclinical proof-of-concept and subsequently clinical studies.

ACKNOWLEDGEMENTS

Authors would like to thank the French Ministry of Higher Education and Research for their financial support (PhD Grant) and the National Funds for Scientific Research, Belgium (FNRS) for additional financial assistance. The authors are thankful to the European financial support (EACEA) in the frame of the NanoFar program, an Erasmus Mundus Joint Doctorate (EMJD) program in nanomedicine and pharmaceutical innovation. This chapter was also supported by "La Région Pays-de-la-Loire" through "NanoFar+ international program" and Bioregate "BILBO" programs, by the "Institute National de la Santé et de la Recherche Médicale" (INSERM), by the "University of Angers" and the "University of Liège" and by the

"Cancéropôle Grand-Ouest" though the "glioblastoma" and "vectorization and radiotherapies" networks. E. Garcion is also a member of the LabEx IRON "Innovative Radiopharmaceuticals in Oncology and Neurology" as part of the French government "Investissements d'Avenir" program and head the PL-BIO 2014-2020 INCA (Institut National du Cancer) project MARENGO – "MicroRNA agonist and antagonist Nanomedicines for GliOblastoma treatment: from molecular programmation to preclinical validation". The authors acknowledge the PHC "Protea 2019" for mobilities.

REFERENCES

F. DiMeco et al., 2016. A Dose Escalation Phase I: Study of Human-Recombinant Bone Morphogenetic Protein 4 Administrated Via CED In GBM Patients. Started: August 2016, Principal investigator: Francesco DiMeco, Sponsors and Collaborators: Stemgen, ORION Clinical Services.

T. Adelita, R. Sessa, S. Won, G. Zenker, M. Porcionatto. 2017. Proteolytic processed form of CXCL12 abolishes migration and induces apoptosis in neural stem cells in vitro. *Stem Cell Res.* 22: 61–69. DOI: 10.1016/j.scr.2017.05.013.

V. Agrawal, M. Sinha. 2017. A review on carrier systems for bone morphogenetic protein-2. *J. Biomed. Mater. Res. Part B Appl. Biomater.* 105: 904–925. DOI: 10.1002/jbm.b.33599.

B. Albarran, A.S. Hoffman, P.S. Stayton. 2011. Efficient intracellular delivery of a pro-apoptotic peptide with a pH-responsive carrier. *React. Funct. Polym.* 71: 261–265.

T.A. Al-Hilal, F. Alam, Y. Byun. 2013. Oral drug delivery systems using chemical conjugates or physical complexes. *Adv. Drug Deliv. Rev.* 65: 845–864.

Autissier, C. Le Visage, C. Pouzet, F. Chaubet, D. Letourneur. 2010. Fabrication of porous polysaccharide-based scaffolds using a combined freeze-drying/cross-linking process. *Acta Biomater.* 6: 3640–3648.

M.P. Baker, H.M. Reynolds, B. Lumicisi, C.J. Bryson. 2010. Immunogenicity of protein therapeutics: The key causes, consequences and challenges. *Self. Nonself.* 1: 314–322.

F.G. Banting, C.H. Best, J.B. Collip, W.R. Campbell, A.A. Fletcher. 1991. Pancreatic extracts in the treatment of diabetes mellitus: preliminary report. 1922. *Can. Med. Assoc. J.* 145: 1281–1286.

S. Bao, Q. Wu, R. McLendon, Y. Hao, Q. Shi, A.B. Hjelmeland, M.W. Dewhirst, D.D. Bigner, J.N. Rich. 2006. Glioma stem cells promote radioresistance by preferential activation of the DNA damage response. *Nature* 444: 756–760.

K. Brian. 2009. Industrialization of MAb production technology. *MAbs* 1: 443–452.

Brown, M. Hughes, S. Tenner, P.A. Banks. 1997. Does pancreatic enzyme supplementation reduce pain in patients with chronic pancreatitis: A meta-analysis. *Am. J. Gastroenterol.* 92: 2032–2035.

P.J. Carter. 2011. Introduction to current and future protein therapeutics: A protein engineering perspective. *Exp. Cell Res.* 317: 1261–1269.

C.L. Casper, N. Yamaguchi, K.L. Kiick, J.F. Rabolt. 2005. Functionalizing electrospun fibers with biologically relevant macromolecules. *Biomacromolecules* 6: 1998–2007.

Q. Chai, Y. Jiao, X. Yu. 2017. Hydrogels for biomedical applications: Their characteristics and the mechanisms behind them. *Gels* 3: 1–15.

P.C. Chen, D.S. Kohane, Y.J. Park, R.H. Bartlett, R. Langer, V.C. Yang. 2004. Injectable microparticle-gel system for prolonged and localized lidocaine release. II. In vivo anesthetic effects. *J. Biomed. Mater. Res. Part A* 70: 459–466.

S.Y. Chew, J. Wen, E.K.F. Yim, K.W. Leong. 2005. Sustained release of proteins from electrospun biodegradable fibers. *Biomacromolecules* 6: 2017–2024.

J.S. Choi, K.W. Leong, H.S. Yoo. 2008. In vivo wound healing of diabetic ulcers using electrospun nanofibers immobilized with human epidermal growth factor (EGF). *Biomaterials*. 29: 587–596.

J.L. Cleland, M.F. Powell, S.J. Shire. 1993. The development of stable protein formulations: A close look at protein aggregation, deamidation, and oxidation. *Crit. Rev. Ther. Drug Carrier Syst.* 10: 307–377.

A.C. Dalton, W.A. Barton. 2014. Over-expression of secreted proteins from mammalian cell lines. *Protein Sci.* 23: 517–525.

A.L. Daugherty, R.J. Mrsny. 2006. Formulation and delivery issues for monoclonal antibody therapeutics. *Adv. Drug Deliv. Rev.* 58: 686–706.

Y. Deng, J. Ren, G. Chen, G. Li, X. Wu, G. Wang, G. Gu, J. Li. 2017. Injectable in situ cross-linking chitosan-hyaluronic acid based hydrogels for abdominal tissue regeneration. *Sci. Rep.* 7: 2699–2711.

D. Ding, Q. Zhu. 2018. Recent advances of PLGA micro/nanoparticles for the delivery of biomacromolecular therapeutics. *Mater. Sci. Eng. C* 92: 1041–1060.

A, Dirksen, J.H. Dijkman, F. Madsen, B. Stoel, D.C. Hutchison, C.S. Ulrik, L.T. Skovgaard et al. 1999. A randomized clinical trial of alpha(1)-antitrypsin augmentation therapy. *Am. J. Respir. Crit. Care Med.* 160: 1468–1472.

O.C. Farokhzad, R. Langer. 2009. Impact of nanotechnology on drug delivery. *ACS Nano* 3: 16–20.

L. Fernández-garcía, N. Marí-buyé, J.A. Barios, R. Madurga, M. Elices, J. Pérez rigueiro, M. Ramos, G.V. Guinea, D. González-Nieto. 2016. Safety and tolerability of silk fibroin hydrogels implanted into the mouse brain. *Acta Biomater.* 45: 262–275.

S. Foster, C.L. Duvall, E.F. Crownover, A.S. Hoffman, P.S. Stayton. 2010. Intracellular delivery of a protein antigen with an endosomal-releasing polymer enhances CD8 T-cell production and prophylactic vaccine efficacy. *Bioconjug. Chem.* 21: 2205–2212.

S.Q. Gao, T. Maeda, K. Okano, K. Palczewski. 2012. A microparticle/hydrogel combination drug-delivery system for sustained release of retinoids. *Invest. Ophthalmol. Visual. Sci.* 53: 6314–6323.

M.A. García-Santana, J. Duconge, M.E. Sarmiento, M.E. Lanio-Ruíz, M.A. Becquer, L. Izquierdo, A. Acosta-Domínguez. 2006. Biodistribution of liposome-entrapped human gamma-globulin. *Biopharm. Drug Dispos.* 27: 275–283.

S. Giarra, C. Ierano, M. Biondi, M. Napolitano, V. Campani, R. Pacelli, S. Scala, G. De Rosa, L. Mayol. 2018. Engineering of thermoresponsive gels as a fake metastatic niche. *Carbohydr Polym.* 191(March): 112–8.

D.V Göddel, D.G. Kleid, F. Bolivar, H.L. Heyneker, D.G. Yansura, R. Crea, T. Hirose et al. 1979. Expression in Escherichia coli of chemically synthesized genes for human insulin. *Proc. Natl. Acad. Sci. U.S.A.* 76: 106–110.

J. Hajavi, M. Ebrahimian, M. Sankian, M.R. Khakzad, M. Hashemi. 2018. Optimization of PLGA formulation containing protein or peptide-based antigen: Recent advances. *J. Biomed. Mater. Res. A* 106: 2540–2551.

M. Hattori. 2005. Finishing the euchromatic sequence of the human genome. *Tanpakushitsu Kakusan Koso* 50: 162–168.

N. Henry, J. Clouet, C., Le Visage, P. Weiss, E. Gautron, D. Renard, T.Cordonnier et al. 2017. Silica nanofibers as a new drug delivery system: A study of the protein–silica interactions. *J. Mater. Chem. B* 5: 2908.

T.R. Hoare, D.S. Kohane. 2008. Hydrogels in drug delivery : Progress and challenges. *Polymer* 49: 1993–2007.

S. Honary, F. Zahir. 2013. Effect of zeta potential on the properties of nano-drug delivery systems: A review (Part 2). *Trop. J. Pharm. Res.* 12: 265–273.

D. Ibraheem, A. Elaissari, H. Fessi. 2014. Administration strategies for proteins and peptides. *Int. J. Pharm.* 477: 578–589.

M. Iqbal, N. Zafar, H. Fessi, A. Elaissari. 2015. Double emulsion solvent evaporation techniques used for drug encapsulation. *Int. J. Pharm.* 496: 173–190.
H. Jia, G. Zhu, B. Vugrinovich, W. Kataphinan, D.H. Reneker, P. Wang. 2002. Enzyme-carrying polymeric nanofibers prepared via electrospinning for use as unique biocatalysts. *Biotechnol. Prog.* 18: 1027–1032.
H. Jiang, Y. Hu, P. Zhao, Y. Li, K. Zhu. 2006. Modulation of protein release from biodegradable core-shell structured fibers prepared by coaxial electrospinning. *J. Biomed. Mater. Res. Part B Appl. Biomater.* 79: 50–57.
M. Karimi, H. Malekzad, H. Mirshekari, P.S. Zangabad, S.M. Moosavi Basri, F. Baniasadi, M.S. Aghdam, M.R. Hamblin. 2018. Plant protein-based hydrophobic fine and ultrafine carrier particles in drug delivery systems. *Crit. Rev. Biotechnol.* 38: 47–67.
E.G. Kim, K.M. Kim. 2015. Strategies and advancement in antibody-drug conjugate optimization for targeted cancer therapeutics. *Biomol. Ther.* 23: 493–509. DOI: 10.4062/biomolther.2015.116.
M. Kim, Y. Kim, K. Gwon, G. Tae. 2012. Modulation of cell adhesion of heparin-based hydrogel by efficient physisorption of adhesive proteins. *Macromol. Res.* 20: 271–276.
T.G. Kim, T.G. Park. 2006. Surface functionalized electrospun biodegradable. Nanofibers for immobilization of bioactive molecules. *Biotechnol. Prog.* 22: 1108–1113.
M. Kimura, K. Fukumoto, J. Watanabe, K. Ishihara. 2004. Hydrogen-bonding-driven spontaneous gelation of water-soluble phospholipid polymers in aqueous medium. *J. Biomater. Sci. Polym. Ed.* 15: 631–644.
D.S. Kohane. 2007. Microparticles and nanoparticles for drug delivery. *Biotechnol. Bioeng.* 96: 203–209.
J. Lam, N.F. Truong, F. Segura. 2014. Design of cell–matrix interactions in hyaluronic acid hydrogel scaffolds. *Acta Biomaterialia* 10: 1571.
R. Langer, J. Folkman. 1976. Polymers for the sustained release of proteins and other macromolecules. *Nature* 263: 797–800.
B. Leader, Q.J. Baca, D.E. Golan. 2008. Protein therapeutics: A summary and classification. *Nat. Rev. Drug Discov.* 7: 21–39.
B.K. Lee, Y. Yun, K. Park. 2016. PLA micro- and nano-particles. *Adv. Drug Deliv. Rev.* 107: 176–191.
E.J. Lee, S.A. Khan, J.K. Park, K.H. Lim. 2012. Studies on the characteristics of drug-loaded gelatin nanoparticles prepared by nanoprecipitation. *Bioprocess Biosyst. Eng.* 35: 297–307.
J.H. Lee, J. Kim, H.-R. Baek, K.M. Lee, J.-H. Seo, H.-K. Lee, A.-Y. Lee et al. 2014. Fabrication of an rhBMP-2 loaded porous β-TCP microsphere-hyaluronic acid-based powder gel composite and evaluation of implant osseointegration. *J. Mater. Sci. Mater. Med.* 25: 2141–2151.
J. Li, D.J. Mooney. 2016. Designing hydrogels for controlled drug delivery. *Nat. Rev. Mater.* 1: 1–18.
S. Liu, M. Qin, C. Hu, F. Wu, T. Jin, W. Cui, S. Liu, C. Fan, C. Hu. 2013. Tendon healing and anti-adhesion properties of electrospun fibrous membranes containing bFGF loaded nanoparticles. *Biomaterials* 34: 4690–4701.
S. Liu, F. Yin, M. Zhao, C. Zhou, J. Ren, Q. Huang, Z. Zhao et al. 2016. The homing and inhibiting effects of hNSCs-BMP4 on human glioma stem cells. *Oncotarget.* 7: 17920–17931.
S. Liu, Y. Liu, L. Jiang, Z. Li, S. Lee, C. Liu, J. Wang, J. Zhang. 2018. Recombinant human BMP-2 accelerates the migration of bone marrow mesenchymal stem cells via the CDC42/PAK1/LIMK1 pathway in vitro and in vivo. *Biomater. Sci.* 7: 362–372.
W. Liu, A. Ye, W. Liu, C. Liu, J. Han, H. Singh. 2015. Behaviour of liposomes loaded with bovine serum albumin during in vitro digestion. *Food Chem.* 175: 16–24.
P.L. Lu, J.Y. Lai, D.H.K. Ma, G.H. Hsiue. 2008. Carbodiimide cross-linked hyaluronic acid hydrogels as cell sheet delivery vehicles: Characterization and interaction with corneal endothelial cells. *J. Biomater. Sci. Polym. Ed.* 19: 1–18.

H.H. Luu, W.X. Song, X. Luo, D. Manning, J. Luo, Z.L. Deng, K.A. Sharff, A.G. Montag, R.C. Haydon, T.C. He. 2007. Distinct roles of bone morphogenetic proteins in osteogenic differentiation of mesenchymal stem cells. *J. Orthop. Res.* 25: 665–677.

H.K. Makadia, S.J. Siegel. 2011. Poly lactic-co-glycolic acid (PLGA) as biodegradable controlled drug delivery carrier. *Polymers* 3: 1377–1397.

Manavitehrani, A. Fathi, H. Badr, S. Daly, A. Negahi Shirazi, F. Dehghani. 2016. Biomedical applications of biodegradable polyesters. *Polymers* 8(1): 20.

S. Martins, B. Sarmento, D.C. Ferreira, E.B. Souto. 2007. Lipid-based colloidal carriers for peptide and protein delivery: Liposomes versus lipid nanoparticles. *Int. J. Nanomed.* 2: 595–607.

W.F. McKay, S.M. Peckham, J.M. Badura. 2007. A comprehensive clinical review of recombinant human bone morphogenetic protein-2 (INFUSE® Bone Graft). *Int. Orthop.* 31: 729–734.

E. Mirzaei B., A. Ramazani, M. Shafiee, M. Danaei. 2013. Studies on glutaraldehyde crosslinked chitosan hydrogel properties for drug delivery systems. *Int. J. Polym. Mater. Polym.* Biomater. 62: 605–611.

S. Mitragotri, P.A. Burke, R. Langer. 2014. Overcoming the challenges in administering biopharmaceuticals: Formulation and delivery strategies. *Nat. Rev. Drug Discov.* 13: 655–72.

B. Moon, J. Yoon, M. Kim, S.H. Lee, T. Choi, K.Y. Choi. 2009. Bone morphogenetic protein 4 stimulates neuronal differentiation of neuronal stem cells through the ERK pathway. *Exp. Mol. Med.* 41: 116–125.

Muheem, F. Shakeel, M.A. Jahangir, M. Anwar, N. Mallick, G.K. Jain, M.H. Warsi, F.J. Ahmad. 2014. A review on the strategies for oral delivery of proteins and peptides and their clinical perspectives. *Saudi Pharm. J.* 24: 413–428.

K. Ochi, G. Chen, T. Ushida, S. Gojo, K. Segawa, H. Tai, K. Ueno et al. 2003. Use of isolated mature osteoblasts in abundance acts as desired-shaped bone regeneration in combination with a modified poly-DL-lactic-co-glycolic acid (PLGA)-collagen sponge. *J. Cell. Physiol.* 194: 45–53.

E. Ozdemir-Kaynak, A.A. Qutub, O. Yesil-Celiktas. 2018. Advances in glioblastoma multiforme treatment: New models for nanoparticle therapy. *Front. Physiol.* 9: 1–14.

M. Pascoli, R. de Lima, L.F. Fraceto. 2018. Zein nanoparticles and strategies to improve colloidal stability: A mini-review. *Front. Chem.* 6: 1–5.

S. Patel, K. Kurpinski, R. Quigley, H. Gao, S. Li, M.-M. Poo, B.S. Hsiao, S. Patel. 2007. Bioactive nanofibers: Synergistic effects of nanotopography and chemical signaling on cell guidance. *Nano Lett.* 7: 2122–2128.

H. Peng, Y. Wu, Z. Duan, P. Ciborowski, J.C. Zheng. 2012. Proteolytic processing of SDF-1 α by matrix metalloproteinase-2 impairs CXCR4 signaling and reduces neural progenitor cell migration. *Protein Cell* 3: 875–882.

S. Piccirillo, B. Reynolds, N. Zanetti, G. Lamorte, E. Binda, G. Broggi, H. Brem et al. 2006. Bone morphogenetic proteins inhibit the tumorigenic potential of human brain tumour-initiating cells. *Nature* 444: 761–765.

D.S. Pisal, M.P. Kosloski, S. V. Balu-Iyler. 2011. Delivery of therapeutic proteins. *NIH Public Access* 99: 1–33.

S.D. Putney, P. A. Burke. 1998. Improving protein therapeutics with sustained-release formulations. *Nat. Biotechnol.* 16: 153–157.

H. Qi, P. Hu, J. Xu, A. Wang. 2006. Encapsulation of drug reservoirs in fibers by emulsion electrospinning: Morphology characterization and preliminary release assessment. *Biomacromolecules* 7: 2327–2330.

M. Rafat, F. Li, P. Fagerholm, M. Griffith, M. Rafat, R. Munger, T. Matsuura, M.A. Watsky, N.S. Lagali. 2008. PEG-stabilized carbodiimide crosslinked collagen–chitosan hydrogels for corneal tissue engineering. *Biomaterials* 29: 3960–3972.

Z. Ren, Y. Zhang, Y. Li, B. Xu, W. Liu. 2015. Hydrogen bonded and ionically crosslinked high strength hydrogels exhibiting Ca^{2+}-triggered shape memory properties and volume shrinkage for cell detachment. *J. Mater. Chem. B* 3: 6347–6354.

B. Richter, G. Neises 2005. "Human" insulin versus animal insulin in people with diabetes mellitus. *Cochrane Database Syst. Rev.* 1: CD003816.

C.C. Rider, B. Mulloy. 2017. Heparin, heparan sulphate and the TGF- Cytokinesuperfamily. *Molecules* 22: 1–11.

S.K. Sahoo, V. Labhasetwar. 2003. Nanotech approaches to drug delivery and imaging. *Drug Discovery Today* 8: 1112–1120.

B. Sarker, R. Singh, R. Silva, J.A. Roether, J. Kaschta, R. Detsch, D.W. Schubert, I. Cicha, A.R. Boccaccini. 2014. Evaluation of fibroblasts adhesion and proliferation on alginate-gelatin crosslinked hydrogel. *PLoS One* 9: 1–12.

S. Schmidt, D.V. Volodkin. 2013. Microparticulate biomolecules by mild CaCO3 templating. *J. Mater. Chem. B* 1: 1210–1218.

E.L. Schneider, J. Henise, R. Reid, G.W. Ashley, D. V. Santi. 2016. Hydrogel drug delivery system using self-cleaving covalent linkers for once-a-week administration of exenatide. *Bioconjug. Chem.* 27: 1210–1215.

C. Shen, Y. Li, H. Wang, Q. Meng. 2017. Mechanically strong interpenetrating network hydrogels for differential cellular adhesion. *RSC Adv.* 7: 18046–18053.

R. Singh, J.W. Lillard Jr. 2009. Nanoparticle-based targeted drug delivery. *Exp. Mol. Pathol.* 86: 215–223.

M. Sokolsky-Papkov, K. Agashi, A. Olaye, K. Shakesheff, A.J. Domb. 2007. Polymer carriers for drug delivery in tissue engineering. *Adv. Drug Deliv. Rev.* 59: 187–206.

R. Stupp, M. Hegi. 2007. Targeting brain-tumor stem cells. *Nat. Biotechnol.* 25: 193–194.

M. Sutter, J. Siepmann, W.E. Hennink, W. Jiskoot. 2007. Recombinant gelatin hydrogels for the sustained release of proteins. *J. Control. Release* 119: 301–312.

J. Swaminathan, C. Ehrhardt. 2012. Liposomal delivery of proteins and peptides. *Expert Opin. Drug Deliv.* 9: 1489–1503.

Szentivanyi, T. Chakradeo, H. Zernetsch, B. Glasmacher. 2011. Electrospun cellular microenvironments: Understanding controlled release and scaffold structure. *Adv. Drug Deliv. Rev.* 63: 209–220.

H. Tai, M.L. Mather, D. Howard, W. Wang, L.J. White, J.A. Crowe, S.P. Morgan et al. 2007. Control of pore size and structure of tissue engineering scaffolds produced by supercritical fluid processing. *Eur. Cells Mater.* 14: 64–77.

M. Tarhini, H. Greige-Gerges, A. Elaissari. 2017. Protein-based nanoparticles: From preparation to encapsulation of active molecules. *Int. J. Pharm.* 522: 172–197.

M. Tarhini, I. Benlyamani, S. Hamdani, G. Agusti, H. Fessi, H. Greige-Gerges, A. Bentaher, A. Elaissari. 2018. Protein-based nanoparticle preparation via nanoprecipitation method. *Materials* 11: 394.

M. Tarhini, A. Pizzoccaro, I. Benlyamani, C. Rebaud, H. Greige-Gerges, H. Fessi, A. Elaissari, A. Bentaher. 2020. Human serum albumin nanoparticles as nanovector carriers for proteins: Application to the antibacterial proteins "neutrophil elastase" and "secretory leukocyte protease inhibitor". *Int. J. Pharm.* 579: 119150.

S. Teixeira, L. Yang, P.J. Dijkstra, M.P. Ferraz, F.J. Monteiro. 2010. Heparinized hydroxyapatite/collagen three-dimensional scaffolds for tissue engineering. *J. Mater. Sci. Mater. Med.* 21: 2385–2392.

C. Thomasin, H.P. Merkle, B.A. Gander. 1997. Physico-chemical parameters governing protein microencapsulation into biodegradable polyesters by coacervation. *Int. J. Pharm.* 147: 173–186.

Z. Tian, W. Liu, G. Li. 2016. The microstructure and stability of collagen hydrogel cross-linked by glutaraldehyde. *Polym. Degrad. Stab.* 130: 264–270.

M.K. Tran, A. Swed, F. Boury. 2012. Preparation of polymeric particles in CO_2 medium using non-toxic solvents: Formulation and comparisons with a phase separation method. *Eur. J. Pharm. Biopharm.* 82: 498–507.

M.-K. Tran, A. Swed, B. Calvignac, K.-N. Dang, L.N. Hassani, T. Cordonnier, F. Boury. 2015. Preparation of polymeric particles in CO_2 medium using non-toxic solvents: Discussions on the mechanism of particle formation. *J. Mater. Chem. B* 3: 1573–1582.

Utani, M. Nomizu, H. Matsuura, K. Kato, T. Kobayashi, U. Takeda, S. Aota, P.K. Nielsen, H. Shinkai. 2001. A unique sequence of the laminin α3 G domain binds to heparin and promotes cell adhesion through syndecan-2 and –4. *J. Biol. Chem.* 276: 28779–28788.

S.R. Van Tomme, B.G. De Geest, K. Braeckmans, S.C. De Smedt, F. Siepmann, J. Siepmann, C.F. Van Nostrum, W.E. Hennink. 2005. Mobility of model proteins in hydrogels composed of oppositely charged dextran microspheres studied by protein release and fluorescence recovery after photobleaching. *J. Control. Release* 110: 67–78.

J.C. Venter, M.D. Adams, E.W Myers, P.W. Li, R.J. Mural, G.G Sutton, H.O. Smith et al. 2001. The sequence of the human genome. *Science* 291: 1304–1351. DOI: 10.1126/science.1058040.

Vescovi, R. Galli, B. Reynolds. 2006. Brain tumour stem cells. *Nat. Rev. Cancer* 6: 425–436.

G. Walsh. 2006. Biopharmaceutical benchmarks 2006. *Nat. Biotechnol.* 24: 769–776.

T.Y. Wang, K.A.F. Bruggeman, R.K. Sheean, B.J. Turner, D.R. Nisbet, C.L. Parish. 2014. Characterization of the stability and bio-functionality of tethered proteins on bioengineered scaffolds. Implications for stem cell biology and tissue repair. *J. Biol. Chem.* 289: 15044–15051.

G. Wei, Q. Jin, W. V. Giannobile, P.X. Ma. 2007. The enhancement of osteogenesis by nanofibrous scaffolds incorporating rhBMP-7 nanospheres. *Biomaterials* 28: 2087–2096.

C.Y. Wong, H. Al-Salami, C.R. Dass. 2018. Recent advancements in oral administration of insulin-loaded liposomal drug delivery systems for diabetes mellitus. *Int. J. Pharm.* 549: 201–217.

K.M. Woo, V.J. Chen, P.X. Ma. 2003. Nano-fibrous scaffolding architecture selectively enhances protein adsorption contributing to cell attachment. *J. Biomed. Mater. Res. Part A* 67: 531–537.

M.A. Wozniak, R. Desai, P.A. Solski, C.J. Der, P.J. Keely. 2003. ROCK-generated contractility regulates breast epithelial cell differentiation in response to the physical properties of a three-dimensional collagen matrix. *J. Cell Biol.* 163(3): 583–595.

H. Yamaguchi, M. Miyazaki. 2014. Refolding techniques for recovering biologically active recombinant proteins from inclusion bodies. *Biomolecules.* 4: 235–251.

W.W. Yang, E. Pierstorff. 2012. Reservoir-based polymer drug delivery systems. *J. Lab. Autom.* 17: (50–58.

P. Ye, Z.K. Xu, J. Wu, C. Innocent, P. Seta. 2006. Nanofibrous membranes containing reactive groups: Electrospinning from poly(acrylonitrile-co-maleic acid) for lipase immobilization. *Macromolecules* 39: 1041–1045.

H.S. Yoo, T.G. Kim, T.G. Park. 2009. Surface-functionalized electrospun nanofibers for tissue engineering and drug delivery. *Adv. Drug Deliv. Rev.* 61: 1033–1042.

J.J. Yoon, T.G. Park. 2001. Degradation behaviors of biodegradable macroporous scaffolds prepared by gas foaming of effervescent salts. *J. Biomed. Mater. Res.* 55: 401–408.

Y.-P. Yun, D.H. Yang, S.-W. Kim, K. Park, J.-Y. Ohe, B.-S. Lee, B.-J. Choi, S.E. Kim. 2014. Local delivery of recombinant human bone morphogenic protein-2 (rhBMP-2) from rhBMP-2/heparin complex fixed to a chitosan scaffold enhances osteoblast behavior. *J. Tissue Eng. Regener. Med.* 11: 163–170.

Y.Z. Zhang, X. Wang, Y. Feng, J. Li, C.T. Lim, S. Ramakrishna. 2006. Coaxial electrospinning of (fluorescein isothiocyanate-conjugated bovine serum albumin)-encapsulated poly(ε-caprolactone) nanofibers for sustained release. *Biomacromolecules* 7: 1049–1057.

Z. Zhao, Y. Li, M. Xie. 2015. Silk fibroin-based nanoparticles for drug delivery. *Int. J. Mol. Sci.* 16: 4880–4903.

12 Nanomedicines for the Treatment of Infectious Diseases

Formulation, Delivery and Commercialization Aspects

Admire Dube, Boitumelo Semete-Makokotlela, Bathabile Ramalapa, Jessica Reynolds and Frank Boury

CONTENTS

12.1 Introduction 272
12.2 The Need for Nanomedicines for Infectious Diseases 273
12.3 Nanoparticles as Drug Delivery Systems and Applications in Infectious Disease Treatment 275
12.3.1 Liposomes 275
12.3.2 Polymeric Nanoparticles 276
12.3.3 Solid Lipid Nanoparticles 277
12.3.4 Metallic Nanoparticles 277
12.3.5 Calcium Carbonate Particles 278
12.4 Medicines Development Models and Considerations for Commercialization of Nanomedicines for Infectious Diseases 278
12.4.1 Impact of IP on Commercialization Models for Medicines 279
12.4.2 Cost of Research and Development 280
12.5 New Commercialization Models 280
12.5.1 World Health Organization (WHO) Expert Working Group on Financing and Coordination (CEWG) 281
12.5.2 Product Development Partnership (PDP): Example of Medicine for Malaria Venture (MMV) 281
12.5.3 De-linking R&D Costs from Price of Medicines 282
12.6 Conclusion 283
References 283

12.1 INTRODUCTION

An infectious disease is defined as a disease that can be transmitted from person to person or an animal to a person or vice versa. Common modes of transmission include coughing, sneezing and exchange of body fluids. In a broad sense, what differentiates this definition of an infectious disease from a hereditary disease is that in the case of infectious diseases, a pathogen or causative organism is involved and that pathogen is transmitted or spread through humans, animals or vectors such as insects, while a hereditary disease is inherited from one's parents (Nash et al., 2015). Table 12.1 lists some of the common infectious diseases, the pathogen involved and the major transmitting agent.

Over 200 infectious diseases are known to man and outbreaks of infectious diseases are not uncommon. In December 2019, the coronavirus disease 2019 (COVID-19) caused by the severe acute respiratory syndrome coronavirus 2 (SARS-CoV-2) was identified in China and spread throughout the world through human contact to create a pandemic causing significant loss of lives and major social and economic disruptions. By April 2020, this disease had claimed close to 200,000 lives across the world (Dong et al., 2020). Another example of a major infectious disease outbreak is the Ebola epidemic in West Africa in 2014. This outbreak was the largest Ebola virus epidemic in history and claimed close to 30,000 lives before containment (Spengler et al., 2016). The disease was able to spread to parts of Europe and the United States. There have also been other deadly outbreaks in recent history, including the Zika virus outbreak emerging from South America, and the Avian flu outbreak emerging from Asia (Brasil et al., 2016; Xie et al.,2017). However, there are three infectious diseases responsible for significant morbidity and mortality around the world every year: HIV/AIDS, TB and malaria. These diseases will be the focus of this chapter.

HIV/AIDS, TB and malaria combined accounted for over 2.7 million deaths worldwide in 2018 (WHO, 2019a, b; 2020). HIV-1 is ranked globally as the deadliest

TABLE 12.1
List of Common Infectious Diseases, Pathogen Involved and Main Transmitting Agent

Disease	Pathogen	Transmitting agent
HIV/AIDS	Human immunodeficiency virus (HIV)	Humans
Tuberculosis (TB)	*Mycobacterium tuberculosis*	Humans
Malaria	*Plasmodium falciparum, Plasmodium vivax*	*Anopheles* mosquito
Dengue fever	Dengue virus 1–4	*Aedes* mosquito
Avian Influenza	H5N1, H1N1 etc.	Infected poultry/birds
COVID-19	SARS-CoV-2	Humans
Zika	Zika virus	*Aedes* mosquito
Sleeping sickness (Human African Trypanosomiasis)	Tsetse fly	*Trypanosoma brucei gambiense*
Ebola	Ebola virus	Humans

single most infectious agent, with *Mycobacterium tuberculosis* (*M. tb*) following a close second (WHO, 2020). HIV/AIDS is no longer a death sentence, but there is uncertainty concerning the access and sustainability of long-term treatment, especially in resource-limited settings. HIV-1 is one of the major co-infections in patients with TB, and this population is several times more likely to develop active TB disease than people without HIV-1 (WHO, 2019a). An estimated 1.7 billion people worldwide are infected with *M. tb* and do not show symptoms of active disease (latent infection) (WHO, 2019a). Latently infected persons are at risk of developing active TB during their lifetime and thus spreading the disease, especially when immunocompromised, as in the case of concurrent HIV-1 infection. Malaria is also among the world's deadliest infectious disease. Annually, over 1.5 million people die and an estimated 40% of the world's population is at risk. Over 90% of deaths occur in sub-Saharan Africa (Alonso et al., 2011; Fong, 2013). Children under 5 years of age and pregnant women are most at risk. On average, a child dies of malaria every 12 seconds and children surviving are at an increased risk of long-term neurological and cognitive disabilities (Idro et al., 2010; Snow et al., 2005). Over 100,000 newborn deaths each year are attributed to malaria in pregnancy (Desai et al., 2007). Consequently, malaria has a huge societal and economic impact globally, and in sub-Saharan Africa in particular.

12.2 THE NEED FOR NANOMEDICINES FOR INFECTIOUS DISEASES

Several drugs are available for the treatment of infectious diseases. The drugs are used alone, or more commonly, in combination, in what are known as "drug cocktails". Administration of drugs as cocktails assists in killing various life stages of the pathogen concurrently. For example, in the case of TB treatment, some drugs target persistent bacterium, while others may target the rapidly replicating bacterium (Janin, 2007). Cocktails also assist to reduce the generation of drug-resistant pathogen strains. Existing drugs are generally potent at killing the pathogen. Therefore, it is not so much a case of needing new drug compounds, but it is also a case of effective use of existing drugs. Examples of general limitations faced by the drugs include poor oral bioavailability, short plasma half-life, high plasma protein binding and poor penetration across the blood–brain barrier (BBB) (Kutscher et al., 2016; Dube et al., 2013). For most infectious diseases, the pathogen predominantly resides within the intracellular space, and this poses an additional barrier for the drug compound to penetrate (Armstead and Li, 2011). Other limitations include severe adverse effects, poor availability of patient-friendly dosage forms, e.g. dosage forms for paediatrics, which affects patient compliance towards treatment regimens (Sosnik and Carcaboso, 2014). Overcoming these challenges could improve treatment compliance, the manner in which the drug is taken and the efficacy of the drug (Sosnik et al., 2010) and improve treatment outcomes and reduce morbidity and mortality from these diseases. Improved drug delivery would be expected to work in conjugation with other public health measures to reduce transmission and deaths and move towards disease eradication, e.g. efforts to increase insecticide spraying in high mosquito burden areas, and improving awareness of the need for hand-washing. There have been extensive efforts to improve dosage forms for infectious diseases, and one notable example is the development of fixed dose combination tablets for TB and HIV therapy, with resultant improved patient compliance (Bangalore et al., 2007). However, infectious diseases continue to pose a global threat and their spread can be attributed to ineffectiveness of current treatment regimens, increased international travel and trade, migration and increasing antimicrobial resistance.

Vaccine development is a major focus in the development of therapies for infectious diseases (Røttingen et al., 2017). There is currently no approved vaccine for HIV or malaria, and the vaccine for TB, i.e. the *Bacillus Calmette–Guérin* (BCG) vaccine is generally ineffective. Recent phase III clinical trials on the most promising candidate malaria vaccine RTS, S/AS01E indicate limited long-term efficacy, with the vaccine providing 43.6% protection in the first year and zero protection by the fourth year (Olotu et al., 2013).

Nanomedicine has the potential to address the challenges faced by therapeutics for infectious diseases, for example reformulating drugs to provide effective therapies in patient-convenient dosage forms and regimens, across the range of therapeutic interventions (Andrade et al., 2013; Sosnik et al., 2010; Dube, 2019). Nanomedicine is a relatively new technology utilizing nanometre-scale particles to improve drug delivery, i.e. pharmacokinetic profiles, achieve organ, cell or pathogen targeting and reduction in drug toxicity and to improve diagnostic capability (Moghimi et al., 2005). Nanomedicine has already impacted other diseases, e.g. cancer, exemplified by the reformulation of doxorubicin to provide a potent, extended half-life therapy with reduced side effects (Bobo et al., 2016; Anselmo and Mitragotri, 2019). A review of US Food and Drug Administration (FDA) approved nanomedicines by Bobo et al. (2016) reported that out of 52 approved nanomedicines on the market, only 4 are intended for the treatment of an infectious disease (Bobo et al., 2016). The disease conditions targeted for treatment by these nanomedicines are fungal infections (i.e. AmBisome® and Abelcet®) and Hepatitis B and C (i.e. Pegasys® and PegIntron® for Hepatitis C treatment) (Bobo et al., 2016). In 2019, Anselmo and Mitragotri reported at least three nanomedicines that are currently in clinical trials for an infectious disease, i.e. a liposome formulation for both hepatitis A and hepatitis B vaccines, as well as a liposome-based gene therapy for prevention of Chikungunya virus from Moderna Therapeutics (Anselmo and Mitragotri, 2019). Juxtaposed against the global morbidity and mortality of infectious diseases, and the challenges faced by existing drugs, there is need to develop more nanomedicines for the treatment of infectious diseases. Nanomedicine has likelihood to do the same for infectious diseases, as it did for cancer, radically improving treatment outcomes using currently available drugs, saving lives and moving towards complete eradication of these diseases. The year 2020, saw heightened research into developing drugs and vaccines for COVID-19, and this included the application of nanoparticles, with some formulations reaching clinical trials (Chan, 2020; ClinicalTrials.Gov, n.d.; Hu et al., 2020). There are significant ongoing infectious disease drug development (Nordling, 2013) and nanomedicine research activities occurring on the African continent (Dube and Ebrahim, 2017; Saidi et al., 2018). Researchers in Africa are key in infectious disease research as the continent bears the greatest burden of infectious diseases, and researchers also have access to patient populations for clinical studies. However, it is hoped that more countries in the world will extensively engage in the development of nanomedicines for TB, HIV and malaria, as these diseases are of global concern. Some issues around "attractiveness" of development and commercialization of medicines for infectious diseases are discussed in Section 12.4.

12.3 NANOPARTICLES AS DRUG DELIVERY SYSTEMS AND APPLICATIONS IN INFECTIOUS DISEASE TREATMENT

A variety of organic and inorganic biomaterials have been developed as delivery systems, from the first liposomal system described in 1965 to more recent systems with virus- or bacteria-like properties (biomimetic systems) or with capabilities of stimulating therapeutic release and action in response to interactions with the surrounding environment. This section of the chapter will review some of the common nanoparticle types and their formulation and describe selected studies in which the nanoparticles were investigated for infectious disease therapy. Due to the broadness of the infectious disease field, the examples provided herein are derived from studies directed towards HIV, TB and malaria treatment. Due to the intracellular residence of infectious disease pathogens, the design of nanoparticles should facilitate entry into the intracellular space and potentially including the nucleus (Fig. 12.1).

12.3.1 LIPOSOMES

Liposomes are spherical vesicles consisting of phospholipid bilayers capability to entrap water-soluble drugs in the hydrophilic compartment and hydrophobic drugs in the lipid layers. They therefore present an opportunity to deliver drug cocktails of both hydrophilic and hydrophobic drugs. Drug delivery with liposomes has been widely investigated and has had numerous commercial applications, including in

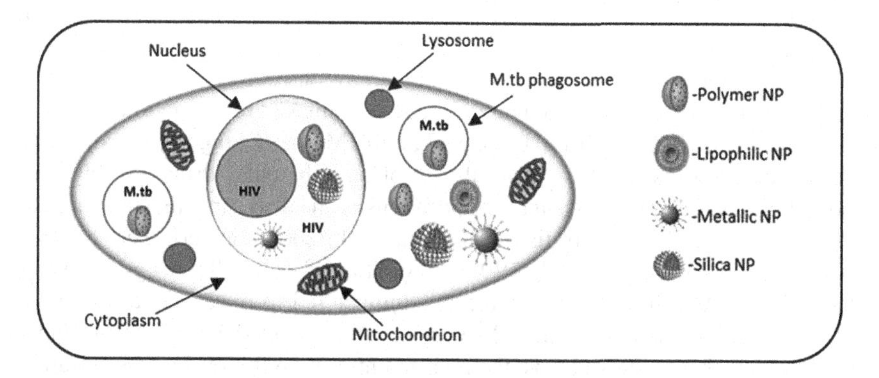

FIGURE 12.1 Schematic diagram illustrating the intracellular locality of *M. tb* and HIV pathogens, and the various types of nanoparticles. *M. tb* is typically contained within phagosomes, while HIV is located within the nucleus. In most cases, nanoparticles will need to penetrate the cellular host/intracellular space and/or nucleus to deliver therapeutic payload.

infectious diseases (Zazo et al., 2016). One such example is AmBisome® (liposomal Amphotericin B), approved by the FDA in 1997 for the treatment of fungal and protozoal infections. Liposomes have also been used to deliver latency activators to CD4+ T cells for the treatment of HIV (Kovochich et al., 2011). Kovochich et al. (2011) reported liposome-based co-delivery of nelfinavir and bryostatin-2 and consequent activation of latent virus and inhibition of virus spread. Mannose-decorated liposomes have been used by Chono et al. (2008) to achieve increased ciprofloxacin levels in macrophages and in plasma. Greco et al. (2012) constructed Janus-faced liposomes for TB treatment. The liposomes were constructed with external phosphatidylserine (induces phagocytic recognition and engulfment) and internal phosphatidic acid (promotes phagolysosome maturation). These liposomes could be taken up efficiently by macrophages leading to increased intracellular killing of *M. tb*.

12.3.2 Polymeric Nanoparticles

Biodegradable polymeric particles offer enhanced stability of drugs, biocompatibility with tissues and cells and controlled release of bioactives (Kumari et al., 2010). Various methods of synthesis have been developed leading to polymeric nanoparticles tailor-made according to the need of application and the drug to be encapsulated. Polyesters have been the most studied and well characterized of the synthetic biodegradable polymers and among them poly(ε-caprolactone) (PCL), poly(lactic acid) (PLA), poly(glycolic acid) (PGA) and their copolymer poly(lactic acid-*co*-glycolic acid) (PLGA) have received great attention due to better encapsulation, better controlled release and less toxicity (Kumari et al., 2010). PLGA has been the most successfully used and has received FDA approval in various drug delivery systems (Danhier et al. 2012). Natural polymers such as alginate, albumin and chitosan have also been used as drug delivery vehicles. Polymeric nanoparticles are typically coated with polyethylene glycol (PEG) to alter their distribution by enhancing their circulation time and increase the delivery of therapeutic molecules. PEG has the capability to minimize recognition of nanoparticles by plasma proteins and avoid uptake by macrophages for clearance (Semete et al., 2010a). Polymeric nanoparticles have been investigated for anti-TB and anti-HIV chemotherapy, designed to improve the pharmacokinetic profiles allowing for better dosage schedules to reduce cytotoxicity and side effects (Semete et al., 2010b; Dube et al., 2014; Makita-Chingombe et al., 2016). Early studies established that encapsulation of antitubercular drugs in polymeric nanoparticles could extend plasma concentrations of the drug, and also increase and extend drug residence time in tissue (Sharma et al., 2004; Gelperina et al., 2005; Pandey et al., 2003). Pandey et al. (2003) showed that when anti-TB drugs are encapsulated in PLGA nanoparticles and orally administered to mice, the drugs can be detected in plasma and tissues (lung, liver, spleen) for extended periods of time (up to 9–11 days) at concentrations above the required minimum inhibitory concentration. This is in contrast to free drug which was eliminated within 24 hours in plasma and 48 hours in tissue (Pandey et al., 2003). In a study utilizing *M. tb* infected guinea pigs, Ahmad et al. administered nebulized alginate nanoparticles at three doses that were spaced 15 days apart for 45 days, whereas free drugs were orally administered daily for 45 days. The study reported undetectable mycobacterial colony-forming units in the lungs and the spleen, suggesting the potential of nanoparticles to modify dosing regimens (Zahoor et al., 2005). The Gendelman group at the University of Nebraska has performed extensive studies to develop long-acting antiretrovirals encapsulated in nanoparticles (Gendelman et al., 2019; Hilaire et al., 2019; Ibrahim et al., 2019; Smith et al., 2019; Soni et al., 2019; Wang et al. 2020; Lin et al., 2018; McMillan et al., 2018; Sillman et al., 2018; Zhou et al., 2018; Guo et al., 2017; Singh et al., 2016; Edagwa et al., 2014). For example, a prodrug of lamivudine was encapsulated within polymeric nanoparticles producing a long-acting nanoformulation. Pharmacokinetic studies in rats demonstrated sustained drug levels in blood and tissues for 30 days (Smith et al., 2019). Recently, there has been a shift from utilizing polymeric nanoparticles to modulate intracellular drug pharmacokinetics (Tukulula et al., 2018) to utilizing the nanoparticles to also activate the innate immune system, i.e. immunotherapy for infectious diseases (Dube and Reynolds, 2016; Liu et al., 2016; Bekale et al.; 2018). Dube et al. (2014) synthesized a β-glucan functionalized chitosan–PLGA nanoparticles and demonstrated that these nanoparticles activated macrophages, i.e. resulted in a significant enhancement of reactive oxygen species and proinflammatory cytokines, including TNF-α production, compared to nanoparticles without β-glucan functionalization. This study, and that described by Greco et al., working with liposomes (Greco et al., 2012), demonstrates the potential of this immunotherapy approach towards eradication of intracellular pathogens.

12.3.3 Solid Lipid Nanoparticles

Solid lipid nanoparticles (SLN) are colloidal drug delivery systems consisting of a hydrophobic solid lipid core covered with a monolayer of phospholipid coating (zur Mühlen et al., 1998). A majority of SLNs are prepared by high-pressure homogenization either at temperatures above the melting point of the lipid (hot homogenization) or at cold temperature (cold homogenization) in the presence of a surfactant/stabilizer. Hydrophobic drugs are dissolved in the melted lipid during hot homogenization. Hydrophilic drugs are encapsulated using the cold homogenization technique to partitioning between the melted lipid and the water phase during hot homogenization (Müller et al., 2000). Within the antimicrobial field, the therapeutic potential of anti-TB drug loaded SLNs has been investigated (Pandey and Khuller, 2005). A SLN formulation was demonstrated to improve half-life of the antimalarial tafenoquine, and also mitigate drug toxicity against red blood cells (Melariri et al., 2015). Lipid-based nanoformulations of the antiretroviral rilpivirine have been shown to be long acting, extending the drug half-life and improving tissue distribution (Hilaire et al., 2019).

12.3.4 Metallic Nanoparticles

Metallic nanoparticles are generally synthesized by the chemical reduction of a chemical salt (gold – Au, silver – Ag, titanium, platinum) with a reducing agent and their characteristics being modified by control of different synthesis conditions such as temperature, pH, reduction time or reducing agent concentration (Mody et al., 2010). Ag nanoparticles have inherent antimicrobial properties and have found their main therapeutic application in the antimicrobial field (Wei et al., 2015). Au nanoparticles have been used to target antibacterial drugs which were linked to the particles through Au-S or Au-amino bonds (Zhao and Jiang, 2013; Grace and Pandian, 2007). Metallic nanoparticles are also potential antitubercular agents. Spherical Au and Ag nanoparticles were reported to display good antibacterial activity against BCG (Zhou et al., 2012). With regards to HIV, Ag nanoparticles have been shown to exert antiviral action against HIV (Lara et al., 2010). Recently, metal organic framework nanoparticles have been explored for targeted delivery of antitubercular drugs (Guo et al., 2019; Wyszogrodzka et al., 2018). Due to their safety and high drug loading capacity, these particles are promising next-generation drug delivery systems for infectious diseases (Wyszogrodzka et al., 2018). For the treatment of neuro-AIDS, magnetoelectric nanoparticles have been investigated for the delivery of antiretrovirals across the blood–brain barrier. Saiyed et al. (2010) demonstrated the delivery of azidothymidine across the BBB using iron oxide nanoparticle loaded liposomes (Saiyed et al., 2010). Ferric-cobalt nanoparticles were synthesized, having an siRNA bound to the surface to target Beclin-1 and were found to be able to Saiyid cross the BBB and attenuate the neurotoxic effects of HIV-1 infection (Rodriguez et al., 2017). An HIV theranostic comprised of rilpivirine ^{177}lutetium labelled bismuth sulfide nanorods demonstrated potent antiretroviral activities in cell-based assays and was shown to be a viable tool to monitor the distribution and transport of the antiretroviral in a mouse model (Kevadiya et al., 2020).

12.3.5 Calcium Carbonate Particles

In recent times calcium carbonate ($CaCO_3$) has gained increasing attention in drug delivery due to its enhanced biocompatibility and biodegradability in comparison to its counterparts (silica nanoparticles, calcium phosphates, carbon nanotubes, hydroxylapatites). $CaCO_3$ can be obtained by precipitation of aqueous calcium and carbonate solutions (liquid–liquid reactions) or carbonation of a calcium solution (gas–liquid reactions) (Zeynep et al., 2015). Increased interest in bioinspired materials and environment-friendly processes has recently led to the formulation of $CaCO_3$ particles using supercritical CO_2 ($ScCO_2$) technology. $ScCO_2$ is a highly efficient and versatile approach for the synthesis of $CaCO_3$ and offers optimal experimental conditions ideal for sensitive therapeutic compounds (Hassani et al., 2013). $CaCO_3$ particles have been studied for the pulmonary delivery of antibiotics to treat lung infections. Moreover, their size (1–5 μm) and their density provides them an excellent mass median aerodynamic diameter, compatible with their local administration as dry powders, a good penetration and retention in the lungs in the presence of airway narrowing (Tewes et al., 2016).

12.4 MEDICINES DEVELOPMENT MODELS AND CONSIDERATIONS FOR COMMERCIALIZATION OF NANOMEDICINES FOR INFECTIOUS DISEASES

A critical component of the research and innovation value chain is the translation of research outputs, and in the case of nanomedicines for infectious diseases, there is critical need for more of this technology to reach the patient. In this section, we discuss the various medicine development models that are available and the commercialization considerations that apply to development of medicines for infectious diseases.

The low number of new drug molecules approved for infectious diseases such as TB, malaria, trypanosomiasis, etc., is evidence that drug development for these therapeutic areas remains low priority and is generally considered non-lucrative by the multinational pharmaceutical companies (Pedrique et al., 2013). It is clear from R&D investment portfolios, that the majority of global spend on R&D goes towards therapeutic areas where there is a strong economic base. Thus, the classical commercialization route, management of intellectual property (IP) and R&D investment levels remain suboptimal for molecules targeting infectious diseases. Conventional medicines development models include, but are not limited to, the following:

- In-house R&D and commercialization of the molecules identified by the pharmaceutical company, investing from lead discovery through to completion of clinical trials.
- Acquisition of small biotechnology firms that emerge from university R&D where the molecules were discovered, followed by investment into the clinical trials by pharmaceutical companies.
- Management of IP such that it provides exclusivity for a specific period, resulting in monopolies and a pricing model that builds in the cost of R&D into the final product.

12.4.1 Impact of IP on Commercialization Models for Medicines

An important issue relating to biotech companies and the commercialization model to be followed is IP management and exploitation. Depending on the sector, various forms of IP are common, e.g. trade secrets, know-how, lead times, first mover advantage, etc. However, in the biotech and pharmaceutical sector, patents are an important financial asset. A patent provides a temporary monopoly to the owner in excluding others from using it and is seen as the largest asset of any biotech or pharmaceutical firm. This short-term monopoly enables the firm to sustain the economic value of technological knowledge and innovation to enable companies to refund their investment into other innovations, producing a high risk-return ratio. Patents are however granted for a limited period of time, mostly, 20 years. During this time, the patent holder may transact using the patent through licensing options (Külpmann, 2005). Different countries have patent laws which are a set of legal rules that govern the validity and infringement of patents across a wide variety of technologies (Külpmann, 2005). The increase in the number of participants in the IP landscape (producers and users) has resulted in patents becoming a very competitive business tool. Large corporations have progressively developed patent portfolios to strengthen their bargaining and retaliation power or to exercise patent strategies to delude competitors. This competitive climate has led to an emergence of patent-based business models which exploit the patent values. These business models tend to create more value since they are more effective and efficient at application and exploitation of patents. On the other hand, detrimental effects can be noted; particularly the litigation-based business models of patent trolls (Su et al., 2012) and the impact on access models for the products by consumers. These challenges have also been noted in the development of innovative medicines for infectious diseases, where the patenting strategy is not a viable approach for many philanthropic funders due to the impact they have on commercialization models and the cost of the eventual product. Therefore, the conventional models may have negative effects of slowing down and hampering innovation and patent creation in the infectious disease medicine development space.

12.4.2 Cost of Research and Development

It is well understood that R&D is expensive, primarily in the pharmaceutical sector compared to other technology companies. A recent study by DiMasi et al. (2016) provides the latest costs of drug development which amount to pre-tax out-of-pocket per approval at US$1.4 billion (2013 dollars) and pre-tax capitalized per approval is $2.5 billion (2013 dollars). The study also indicates that costs for compounds that were abandoned were linked to costs of approved compounds, thus resulting in the costs having to be recovered through drug pricing strategies (DiMasi et al., 2016). The reason that society has accepted this model is that there is a clear need for R&D into new medicines. However, such models have led to disparities in access, underutilization of medically important medicines and financial hardships for consumers, including payers and providers.

Furthermore, because of the characteristics (small, costly, negative cash flows, long time to develop a product, etc.) of biotech/small pharmaceutical companies, their financing is challenging. Most biotechnology companies also explore the avenue of seeking financing through partnering with larger firms such as multinational pharmaceutical companies. This has recently been the major business and funding strategy for biotech firms, where large pharmaceutical companies have cut their R&D budgets, and are seeking for close to commercial innovations to include into their pipeline (Schiff and Murray, 2004).

Therefore, the practice of rewarding pharmaceutical companies with time-limited marketing monopolies through IP rights management and exploitation can create other problems such as biases in R&D investment that favour therapeutic areas where there are clear economic returns, leading to inadequate investment in early stage of R&D.

These two critical aspects, i.e. IP rights and cost of R&D have a major impact on access to medicines and vaccines, and thus there is a need for new innovative approaches to ensure that those that need the medicines and vaccines can access these timely and cost-effectively.

12.5 NEW COMMERCIALIZATION MODELS

Any new model for the development and commercialization of therapeutics and diagnostics for infectious diseases would require that the model firstly implement innovative mechanisms to manage, transfer and exploit IP. Secondly, innovative financing models for R&D costs, which are typically built into the cost of medicines, are required. The critical third aspect would involve innovative mechanisms for access to medicines by patients in the regions most affected. The models described below address these principles to some extent.

12.5.1 World Health Organization (WHO) Expert Working Group on Financing and Coordination (CEWG)

The WHO CEWG demonstration projects were set up as a prototype model that will provide evidence on innovative mechanisms to fund and coordinate public health R&D to address unmet medical needs, especially of developing countries using unconventional mechanisms. Furthermore, the CEWG were established to contribute to further discussion on a sustainable global framework for improving access to health care (//www.dndi.org/2013/advocacy/who-cewg-process-identification-of-health-rd-demonstration-projects/). The main guiding principles of this model whose implementation, impact and return for the researchers and developers of the solutions are still to be evaluated include:

a. **Open knowledge and innovation:** This principle refers to the open use, generation and management of any IP emanating from joint projects. All partners should ensure that any IP, data, publications, etc. are openly shared. Collaborative approaches to addressing the specific challenge are encouraged within this model.
b. **Sustainable financing of the initiatives:** This principle states that members of the initiative, primarily member countries, should commit to securing the requisite funding. The source of funding is not restrictive, thus creating room for private public partnerships, and pooling of funds.
c. **Equitable access and de-linkage:** This principle makes a clear requirement for equitable access to the therapeutics or diagnostic tools developed. Integral to this principle is the commitment for production and supply at cost with a minimal margin, registration and availability in all endemic countries, and open licensing of all IP with a possibility of technology transfer. It is anticipated that the policy would facilitate de-linking R&D costs from the final price of the product (discussed further in the third model).
d. **Continuous new incentives**: This principle is in place to foster effective and efficient coordination mechanisms amongst existing organizations and multinational firms to ensure shared value across stakeholders.

12.5.2 Product Development Partnership (PDP): Example of Medicine for Malaria Venture (MMV)

MMV is a leading PDP in the field of antimalarial drug research and development, and it aims to reduce the burden of malaria in disease-endemic countries by discovering, developing and facilitating delivery of new, effective and affordable antimalarial drugs using innovative partnership and finding models (https://www.mmv.org/about-us). MMV coordinates and works with partners across the drug discovery value chain, spanning from research institutions, multinational pharmaceutical companies across various regions. This model has also been applied by the Global TB Alliance (https://www.tballiance.org/).

MMV and its partners manage a portfolio of 65 projects, the largest portfolio of antimalarial R&D, which includes nine new drugs in clinical development. Further details on these new molecules can be accessed on the MMV website (https://www.mmv.org/research-development/mmv-supported-projects) where up to date information is provided. Key to the MMV's success are the following factors:

First, coordination of large multi-country studies and a broad range of partners, which at the time of compilation of this work was about 400 pharmaceutical, academic and endemic country partners in more than 55 countries. Due to the socioeconomic challenges in the malaria endemic regions, MMV extends its support to work with distribution, pharmaceutical companies and country stakeholders to ensure efficient uptake that will ensure that the most appropriate medicines are available as quickly as possible.

The second key aspect to its success is its funding model. MMV receives sustained funding and support from government agencies, private foundation, international organizations, private individuals and corporate foundations. Of the total MMV funding, 40% comes from the Bill and Melinda Gates foundation and 14.5% from the United Kingdom Department of International Development and the rest from various other funders. These diverse funding streams are used to fund R&D as well as specific targeted access and delivery interventions that aim to make it easier for vulnerable population in endemic countries to access antimalarial products.

Through this model and since the inception of MMV in 1999, US$709 million has been invested into building this large portfolio, with six new antimalarial drug being brought into the market and distributed to those that need the drugs the most. MMV estimates that a minimum of US$430 million over 5 years would be required to sustain its work and outputs.

12.5.3 De-linking R&D Costs from Price of Medicines

As discussed in the first model, the management of IP makes the international trade agreements restrictive in that the exclusive rights regime tends to be very expensive. Thus, the WHO established a Consultative Expert Working Group on Research and Development: Financing and Coordination (CEWG), whose objective was to explore innovative approaches to manage innovation and access of new medicine innovations. One of the approaches considered is de-linking the cost of R&D from the price of the medicines. De-linkage is a concept that is strongly anchored in the WHO Global strategy and plan of action for public health, innovation and intellectual property and positioned within resolution WHA63.28. The difference with this approach is that it aims to:

- Eliminate monopolies on final product, fostered by IP exclusivities
- Implement product development with or without IP protections as long as IP rights are not implemented as the exclusive right to manufacture, sell or distribute products
- Decentralize systems for manufacturing, distribution and marketing
- Build-in incentives to reward investment in products that have the greatest impact on health outcomes
- Finance a wider range of R&D, including that of neglected diseases
- Foster the development and supply of knowledge as a public good

By embracing new policies that de-link the cost of R&D from product process, it is possible to achieve the following:

- Expand access to new innovative products
- Implement targeted R&D incentives
- Establishing a global framework that guides how every aspect of the R&D development value chain will be funded
- The new global norms would replace the current trade agreements that focus on higher prices and stronger product monopolies solely focused on incentives for private for-profit companies.

The main elements of the model, as much as they are still very nuanced and may appear complex, will enable an environment where access and innovation are longer competing objectives requiring trade-offs (Love, 2011).

12.6 CONCLUSION

Infectious diseases are a global health concern, and more international efforts need to be undertaken to research and develop nanomedicines for treatment of infectious diseases; in particular, nanomedicines for the three major killers, i.e. HIV, TB and malaria. Preclinical studies demonstrate the potential of nanoparticle drug delivery systems to address some of the drug delivery and treatment challenges of infectious diseases. There are several next-generation drug delivery systems and novel therapeutic approaches, which deserve to be explored further (in preclinical and clinical trials) to eventually reach the patient. Innovative commercialization models are in place for development of medicines for infectious diseases. These can be exploited, and possibly refined, to bring more nanomedicines for infectious diseases to the market.

REFERENCES

Alonso, Pedro L., Graham Brown, Myriam Arevalo-Herrera, Fred Binka, Chetan Chitnis, Frank Collins, Ogobara K Doumbo, Brian Greenwood, B Fenton Hall, and Myron M Levine. 2011. "A research agenda to underpin malaria eradication." *PLoS Medicine* 8, no. 1: e1000406.

Andrade, Fernanda, Diana Rafael, Mafalda Videira, Domingos Ferreira, Alejandro Sosnik, and Bruno Sarmento. 2013. "Nanotechnology and pulmonary delivery to overcome resistance in infectious diseases." *Advanced Drug Delivery Reviews* 65, no. 13–14: 1816–1827. DOI: 10.1016/j.addr.2013.07.020.

Anselmo, Aaron C., and Samir Mitragotri. 2019. "Nanoparticles in the clinic: An update." *Bioengineering & Translational Medicine* 4, no. 3: e10143. DOI: 10.1002/btm2.10143.

Armstead, Andrea L., and Bingyun Li. 2011. "Nanomedicine as an emerging approach against intracellular pathogens." *International Journal of Nanomedicine* 6: 3281–3293. DOI: 10.2147/IJN.S27285.

Bangalore, Sripal, Gayathri Kamalakkannan, Sanobar Parkar, and Franz H. Messerli. 2007. "Fixed-dose combinations improve medication compliance: A meta-analysis." *The American Journal of Medicine* 120, no. 8: 713–719. DOI: 10.1016/j.amjmed.2006.08.033.

Bekale, Raymonde B., Su-Mari Du Plessis, Nai-Jen Hsu, Jyoti R. Sharma, Samantha L. Sampson, Muazzam Jacobs, Mervin Meyer, Gene D. Morse, and Admire Dube. 2018. "Mycobacterium tuberculosis and interactions with the host immune system: Opportunities for nanoparticle based immunotherapeutics and vaccines." *Pharmaceutical Research* 36, no. 1: 8. DOI: 10.1007/s11095-018-2528-9.

Bobo, Daniel, Kye J. Robinson, Jiaul Islam, Kristofer J. Thurecht, and Simon R. Corrie. 2016. "Nanoparticle-based medicines: A review of FDA-approved materials and clinical trials to date." *Pharmaceutical Research* 33, no. 10: 2373–2387. DOI: 10.1007/s11095-016-1958-5.

Brasil, Patrícia, Guilherme Amaral Calvet, André Machado Siqueira, Mayumi Wakimoto, Patrícia Carvalho de Sequeira, Aline Nobre, Marcel de Souza Borges Quintana et al. 2016. "Zika virus outbreak in Rio de Janeiro, Brazil: Clinical characterization, epidemiological and virological aspects. *PLOS Neglected Tropical Diseases* 10, no. 4: e0004636. DOI: 10.1371/journal.pntd.0004636.

Chan, W. C. W. 2020. "Nano research for COVID-19." *ACS Nano* 14, no. 4: 3719–3720. DOI: 10.1021/acsnano.0c02540.

Chono, Sumio, Tomoharu Tanino, Toshinobu Seki, and Kazuhiro Morimoto. 2008. "Efficient drug targeting to rat alveolar macrophages by pulmonary administration of ciprofloxacin incorporated into mannosylated liposomes for treatment of respiratory intracellular parasitic infections." *Journal of Controlled Release* 127, no. 1: 50–58. DOI: 10.1016/j.jconrel.2007.12.011.

ClinicalTrials.Gov. 2020. "ClinicalTrials.Gov identifier: NCT04283461." Accessed: 25 April 2020. https://clinicaltrials.gov/ct2/show/NCT04283461

Danhier, Fabienne, Eduardo Ansorena, Joana M. Silva, Régis Coco, Aude Le Breton, and Véronique Préat. 2012. "PLGA-based nanoparticles: An overview of biomedical applications." *Journal of Controlled Release* 161, no. 2: 505–522. DOI: 10.1016/j.jconrel.2012.01.043.

Desai, Meghna, Feiko O. ter Kuile, François Nosten, Rose McGready, Kwame Asamoa, Bernard Brabin, and Robert D. Newman. 2007. "Epidemiology and burden of malaria in pregnancy." *The Lancet Infectious Diseases* 7, no. 2: 93–104. DOI: 10.1016/S1473-3099(07)70021-X.

DiMasi, Joseph A., Henry G. Grabowski, and Ronald W. Hansen. 2016. "Innovation in the pharmaceutical industry: New estimates of R&D costs." *Journal of Health Economics* 47: 20–33. DOI: 10.1016/j.jhealeco.2016.01.012.

Dong, E., H. Du, and L. Gardner. 2020. "An interactive web-based dashboard to track COVID-19 in real time." *The Lancet Infectious Diseases* 20, no. 5: 533-534. DOI: 10.1016/S1473-3099(20)30120-1.

Dube, Admire. 2019. "Nanomedicines for infectious diseases." *Pharmaceutical Research* 36, no. 4: 63. DOI: 10.1007/s11095-019-2603-x.

Dube, Admire, and Jessica L. Reynolds. 2016. "Modulation of innate immune responses using nanoparticles for infectious disease therapy." *Current Bionanotechnology* 2, no. 1: 60–65.

Dube, Admire, and Naushaad Ebrahim. 2017. "The nanomedicine landscape of South Africa." *Nanotechnology Reviews* 6, no. 4 (2017): 339–344.

Dube, Admire, Yolandy Lemmer, Rose Hayeshi, Mohammed Balogun, Philip Labuschagne, Hulda Swai, and Lonji Kalombo. 2013. "State of the art and future directions in nanomedicine for tuberculosis." *Expert Opinion on Drug Delivery* 10, no. 12: 1725–1734. DOI: 10.1517/17425247.2014.846905.

Dube, Admire, Jessica L. Reynolds, Wing-Cheung Law, Charles C. Maponga, Paras N. Prasad, and Gene D. Morse. 2014. "Multimodal nanoparticles that provide immunomodulation and intracellular drug delivery for infectious diseases." *Nanomedicine: Nanotechnology, biology and medicine* 10, no. 4: 831–838. DOI: 10.1016/j.nano.2013.11.012.

Edagwa, B. J., T. Zhou, J. M. McMillan, X. M. Liu, and H. E. Gendelman. 2014. "Development of HIV reservoir targeted long acting nanoformulated antiretroviral therapies." *Current Medicinal Chemistry* 21, no. 36: 4186–4198. DOI: 10.2174/0929867321666140826114135.

Fong, Ignatius W. 2013. "Challenges in the control and eradication of malaria." In *Challenges in Infectious Diseases*, 203–231. New York: Springer.

Gelperina, Svetlana, Kevin Kisich, Michael D Iseman, and Leonid Heifets. 2005. "The potential advantages of nanoparticle drug delivery systems in chemotherapy of tuberculosis." *American Journal of Respiratory and Critical Care Medicine* 172, no. 12: 1487–1490.

Gendelman, H. E., J. McMillan, A. N. Bade, B. Edagwa, and B. D. Kevadiya. 2019. "The promise of long-acting antiretroviral therapies: From need to manufacture." *Trends in Microbiology* 27, no. 7: 593–606. DOI: 10.1016/j.tim.2019.02.009.

Grace, A Nirmala, and K Pandian. 2007. "Antibacterial efficacy of aminoglycosidic antibiotics protected gold nanoparticles: A brief study." *Colloids and Surfaces A: Physicochemical and Engineering Aspects* 297, no. 1: 63–70.

Greco, Emanuela, Gianluca Quintiliani, Marilina B. Santucci, Annalucia Serafino, Anna Rita Ciccaglione, Cinzia Marcantonio, Massimiliano Papi et al. 2012. "Janus-faced liposomes enhance antimicrobial innate immune response in Mycobacterium tuberculosis infection." *Proceedings of the National Academy of Sciences* 109, no. 21: E1360–E1368. DOI: 10.1073/pnas.1200484109.

Guo, Ailin, Mikhail Durymanov, Anastasia Permyakova, Saad Sene, Christian Serre, and Joshua Reineke. 2019. "Metal organic framework (MOF) particles as potential bacteria-mimicking delivery systems for infectious diseases: Characterization and cellular internalization in alveolar macrophages." *Pharmaceutical Research* 36, no. 4: 53. DOI: 10.1007/s11095-019-2589-4.

Guo, D., T. Zhou, M. Araínga, D. Palandri, N. Gautam, T. Bronich, Y. Alnouti, J. McMillan, B. Edagwa, and H. E. Gendelman. 2017. "Creation of a long-acting nanoformulated 2′,3′-dideoxy-3′-thiacytidine." *Journal of Acquired Immune Deficiency Syndromes* 74, no. 3: e75–e83. DOI: 10.1097/qai.0000000000001170.

Hassani, Leila N., Francois Hindre, Thomas Beuvier, Brice Calvignac, Nolwenn Lautram, Alain Gibaud, and Frank Boury. 2013. "Lysozyme encapsulation into nanostructured CaCO3 microparticles using a supercritical CO2 process and comparison with the normal route." *Journal of Materials Chemistry B* 1, no. 32: 4011–4019. DOI: 10.1039/C3TB20467G.

Hilaire, James R., Aditya N. Bade, Brady Sillman, Nagsen Gautam, Jonathan Herskovitz, Bhagya Laxmi Dyavar Shetty, Melinda S. Wojtkiewicz et al. 2019. "Creation of a long-acting rilpivirine prodrug nanoformulation." *Journal of Controlled Release* 311–312: 201–211. DOI: 10.1016/j.jconrel.2019.09.001.

Hu, T. Y., M. Frieman, and J. Wolfram. 2020. "Insights from nanomedicine into chloroquine efficacy against COVID-19." *Nature Nanotechnology* 15, no. 4: 247–249. DOI: 10.1038/s41565-020-0674-9.

Ibrahim, I. M., A. N. Bade, Z. Lin, D. Soni, M. Wojtkiewicz, B. L. Dyavar Shetty, N. Gautam et al. 2019. "Synthesis and characterization of a long-acting emtricitabine prodrug nanoformulation." *International Journal of Nanomedicine* 14: 6231–6247. DOI: 10.2147/ijn.s215447.

Idro, R., K. Marsh, C. C. John, and C. R. J. Newton. 2010. "Cerebral malaria: Mechanisms of brain injury and strategies for improved neurocognitive outcome." *Pediatric Research* 68, no. 4: 267–274.

Janin, Yves L. 2007. "Antituberculosis drugs: Ten years of research." *Bioorganic & Medicinal Chemistry* 15, no. 7: 2479–2513. DOI: 10.1016/j.bmc.2007.01.030.

Kevadiya, Bhavesh D., Brendan Ottemann, Insiya Z. Mukadam, Laura Castellanos, Kristen Sikora, James R. Hilaire, Jatin Machhi et al. 2020. "Rod-shape theranostic nanoparticles facilitate antiretroviral drug biodistribution and activity in human immunodeficiency virus susceptible cells and tissues." *Theranostics* 10, no. 2: 630–656. DOI: 10.7150/thno.39847.

Kovochich, Michael, Matthew D. Marsden, and Jerome A. Zack. 2011. "Activation of latent HIV using drug-loaded nanoparticles." *PLoS ONE* 6, no. 4: e18270. DOI: 10.1371/journal.pone.0018270.

Külpmann, Mathias. 2005. "The economics of biotechnology." *Journal of Biosciences* 30, no. 2: 151–154. DOI: 10.1007/bf02703694.

Kumari, Avnesh, Sudesh Kumar Yadav, and Subhash C. Yadav. 2010. "Biodegradable polymeric nanoparticles based drug delivery systems." *Colloids and Surfaces B: Biointerfaces* 75, no. 1: 1–18. DOI: 10.1016/j.colsurfb.2009.09.001.

Kutscher, Hilliard L., Paras N. Prasad, Gene D. Morse, and Jessica L. Reynolds. 2016. "Emerging nanomedicine approaches to targeting HIV-1 and antiretroviral therapy." *Future Virology* 11, no. 2: 101–104. DOI: 10.2217/fvl.15.114.

Lara, Humberto H., Nilda V. Ayala-Nuñez, Liliana Ixtepan-Turrent, and Cristina Rodriguez-Padilla. 2010. "Mode of antiviral action of silver nanoparticles against HIV-1." *Journal of Nanobiotechnology* 8, no. 1: 1. DOI: 10.1186/1477-3155-8-1.

Lin, Z., N. Gautam, Y. Alnouti, J. McMillan, A. N. Bade, H. E. Gendelman, and B. Edagwa. 2018. "ProTide generated long-acting abacavir nanoformulations." *Chemical Communications* 54, no. 60: 8371–8374. DOI: 10.1039/c8cc04708a.

Liu, Jiaying, Pallab Pradhan, and Krishnendu Roy. 2016. "Synthetic polymeric nanoparticles for immunomodulation." In *Nanomaterials in Pharmacology*, 413–438. New York: Humana Press.

Love, J. 2011. De-linking R&D costs from product prices. WHO consultative expert working group (CEWG) on research and development: Financing and coordination, Geneva, Switzerland. Retrieved from http://www. who. int/phi/news/phi_cewg_1stmeet_10_KEI_submission_en. pdf.

Makita-Chingombe, Faithful, Hilliard L. Kutscher, Sara L. DiTursi, Gene D. Morse, and Charles C. Maponga. 2016. "Poly(lactic-co-glycolic) acid-chitosan dual loaded nanoparticles for antiretroviral nanoformulations." *Journal of Drug Delivery* 2016: 3810175. DOI: 10.1155/2016/3810175.

McMillan, J., A. Szlachetka, L. Slack, B. Sillman, B. Lamberty, B. Morsey, S. Callen et al. 2018. "Pharmacokinetics of a long-acting nanoformulated dolutegravir prodrug in rhesus macaques." *Antimicrobial Agents and Chemotherapy* 62, no. 1. doi: 10.1128/aac.01316-17.

Melariri, Paula, Lonji Kalombo, Patric Nkuna, Admire Dube, Rose Hayeshi, Benhards Ogutu, Liezl Gibhard et al. 2015. "Oral lipid-based nanoformulation of tafenoquine enhanced bioavailability and blood stage antimalarial efficacy and led to a reduction in human red blood cell loss in mice." *International Journal of Nanomedicine* 10: 1493–1503. DOI: 10.2147/IJN.S76317.

Mody, Vicky V., Rodney Siwale, Ajay Singh, and Hardik R. Mody. 2010. "Introduction to metallic nanoparticles." *Journal of Pharmacy and Bioallied Sciences* 2, no. 4: 282–289. DOI: 10.4103/0975-7406.72127.

Moghimi, S. Moein, A. Christy Hunter, and J. Clifford Murray. 2005. "Nanomedicine: current status and future prospects." *The FASEB Journal* 19, no. 3: 311–330. DOI: 10.1096/fj.04-2747rev.

Müller, Rainer H., Karsten Mäder, and Sven Gohla. 2000. "Solid lipid nanoparticles (SLN) for controlled drug delivery: A review of the state of the art." *European Journal of Pharmaceutics and Biopharmaceutics* 50, no. 1: 161–177. DOI: 10.1016/S0939-6411(00)00087-4.

Nash, Anthony A., Robert G. Dalziel, and J. Ross Fitzgerald. 2015. "Chapter 1: General principles." In *Mims' Pathogenesis of Infectious Disease*, 1–7. 6th ed. Boston: Academic Press.

Nordling, Linda. 2013. "Made in Africa." *Nature Medicine* 19, no. 7: 803–806. DOI: 10.1038/nm0713-803.

Olotu, Ally, Gregory Fegan, Juliana Wambua, George Nyangweso, Ken O. Awuondo, Amanda Leach, Marc Lievens et al. 2013. "Four-year efficacy of RTS,S/AS01E and its interaction with malaria exposure." *New England Journal of Medicine* 368, no. 12: 1111–1120. DOI: 10.1056/NEJMoa1207564.

Pandey, Rajesh, and G. K. Khuller. 2005. "Solid lipid particle-based inhalable sustained drug delivery system against experimental tuberculosis." *Tuberculosis* 85, no. 4: 227–234. DOI: 10.1016/j.tube.2004.11.003.

Pandey, Rajesh, A. Zahoor, Sadhna Sharma, and G. K. Khuller. 2003. "Nanoparticle encapsulated antitubercular drugs as a potential oral drug delivery system against murine tuberculosis." *Tuberculosis* 83, no. 6: 373–378. DOI: 10.1016/j.tube.2003.07.001.

Pedrique, Belen, Nathalie Strub-Wourgaft, Claudette Some, Piero Olliaro, Patrice Trouiller, Nathan Ford, Bernard Pécoul, and Jean-Hervé Bradol. 2013. "The drug and vaccine landscape for neglected diseases (2000–11): A systematic assessment." *The Lancet Global Health* 1, no. 6: e371–e379. DOI: 10.1016/S2214-109X(13)70078-0.

Rodriguez, M., A. Kaushik, J. Lapierre, S. M. Dever, N. El-Hage, and M. Nair. 2017. "Electromagnetic nano-particle bound beclin1 siRNA crosses the blood-brain barrier to attenuate the inflammatory effects of HIV-1 infection in vitro." *Journal of Neuroimmune Pharmacology* 12, no. 1: 120–132. DOI: 10.1007/s11481-016-9688-3.

Røttingen, John-Arne, Dimitrios Gouglas, Mark Feinberg, Stanley Plotkin, Krishnaswamy V. Raghavan, Andrew Witty, Ruxandra Draghia-Akli, Paul Stoffels, and Peter Piot. 2017. "New vaccines against epidemic infectious diseases." *New England Journal of Medicine* 376, no. 7: 610–613. DOI: 10.1056/NEJMp1613577.

Saidi, Trust, Jill Fortuin, and Tania S. Douglas. 2018. "Nanomedicine for drug delivery in South Africa: A protocol for systematic review." *Systematic Reviews* 7, no. 1: 154. DOI: 10.1186/s13643-018-0823-5.

Saiyed, Z. M., N. H. Gandhi, and M. P. Nair. 2010. "Magnetic nanoformulation of azidothymidine 5′-triphosphate for targeted delivery across the blood-brain barrier." *International Journal of Nanomedicine* 5: 157–166. DOI: 10.2147/ijn.s8905.

Schiff, Leora, and Fiona Murray. 2004. "Biotechnology financing dilemmas and the role of special purpose entities." *Nature Biotechnology* 22, no. 3: 271.

Semete, B., L. I. J. Booysen, L. Kalombo, J. D. Venter, L. Katata, B. Ramalapa, J. A. Verschoor, and H. Swai. 2010a. "In vivo uptake and acute immune response to orally administered chitosan and PEG coated PLGA nanoparticles." *Toxicology and Applied Pharmacology* 249, no. 2: 158–165. DOI: https://doi.org/10.1016/j.taap.2010.09.002.

Semete, Boitumelo, Laetitia Booysen, Yolandy Lemmer, Lonji Kalombo, Lebogang Katata, Jan Verschoor, and Hulda S. Swai. 2010b. "In vivo evaluation of the biodistribution and safety of PLGA nanoparticles as drug delivery systems." *Nanomedicine: Nanotechnology, Biology and Medicine* 6, no. 5: 662–671. DOI: 10.1016/j.nano.2010.02.002.

Sharma, Anjali, Rajesh Pandey, Sadhna Sharma, and G. K. Khuller. 2004. "Chemotherapeutic efficacy of poly (dl-lactide-co-glycolide) nanoparticle encapsulated antitubercular drugs at sub-therapeutic dose against experimental tuberculosis." *International Journal of Antimicrobial Agents* 24, no. 6: 599–604. DOI: 10.1016/j.ijantimicag.2004.07.010.

Sillman, B., A. N. Bade, P. K. Dash, B. Bhargavan, T. Kocher, S. Mathews, H. Su et al. 2018. "Creation of a long-acting nanoformulated dolutegravir." *Nature Communications* 9, no. 1: 443. DOI: 10.1038/s41467-018-02885-x.

Singh, D., J. McMillan, J. Hilaire, N. Gautam, D. Palandri, Y. Alnouti, H. E. Gendelman, and B. Edagwa. 2016. "Development and characterization of a long-acting nanoformulated abacavir prodrug." *Nanomedicine* 11, no. 15: 1913–1927. DOI: 10.2217/nnm-2016-0164.

Smith, Nathan, Aditya N. Bade, Dhruvkumar Soni, Nagsen Gautam, Yazen Alnouti, Jonathan Herskovitz, Ibrahim M. Ibrahim et al. 2019. "A long acting nanoformulated lamivudine ProTide." *Biomaterials* 223: 119476. DOI: 10.1016/j.biomaterials.2019.119476.

Snow, R. W., C. A. Guerra, A. M. Noor, H. Y. Myint, and S. I. Hay. 2005. "The global distribution of clinical episodes of Plasmodium falciparum malaria." *Nature* 434, no. 7030: 214–217. DOI: 10.1038/nature03342.

Soni, D., A. N. Bade, N. Gautam, J. Herskovitz, I. M. Ibrahim, N. Smith, M. S. Wojtkiewicz et al. 2019. "Synthesis of a long acting nanoformulated emtricitabine ProTide." *Biomaterials* 222: 119441. DOI: 10.1016/j.biomaterials.2019.119441.

Sosnik, Alejandro, and Angel M. Carcaboso. 2014. "Nanomedicines in the future of pediatric therapy." *Advanced Drug Delivery Reviews* 73: 140–161. DOI: 10.1016/j.addr.2014.05.004.

Sosnik, Alejandro, Ángel M Carcaboso, Romina J Glisoni, Marcela A Moretton, and Diego A Chiappetta. 2010. "New old challenges in tuberculosis: Potentially effective nanotechnologies in drug delivery." *Advanced Drug Delivery Reviews* 62, no. 4: 547–559.

Spengler, Jessica R, Elizabeth D Ervin, Jonathan S Towner, Pierre E Rollin, and Stuart T Nichol. 2016. "Perspectives on West Africa Ebola virus disease outbreak, 2013–2016." *Emerging Infectious Diseases* 22, no. 6: 956.

Su, Hsin-Ning, Carey Ming-Li Chen, and Pei-Chun Lee. 2012. "Patent litigation precaution method: Analyzing characteristics of US litigated and non-litigated patents from 1976 to 2010." *Scientometrics* 92, no. 1: 181–195.

Tewes, Frederic, Oliviero L. Gobbo, Carsten Ehrhardt, and Anne Marie Healy. 2016. "Amorphous calcium carbonate based-microparticles for peptide pulmonary delivery." *ACS Applied Materials & Interfaces* 8, no. 2: 1164–1175. DOI: 10.1021/acsami.5b09023.

Tukulula, Matshawandile, Luis Gouveia, Paulo Paixao, Rose Hayeshi, Brendon Naicker, and Admire Dube. 2018. "Functionalization of PLGA nanoparticles with 1,3-β-glucan enhances the intracellular pharmacokinetics of rifampicin in macrophages." *Pharmaceutical Research* 35, no. 6: 111. DOI: 10.1007/s11095-018-2391-8.

Wang, W., N. Smith, E. Makarov, Y. Sun, C. L. Gebhart, M. Ganesan, N. A. Osna, H. E. Gendelman, B. J. Edagwa, and L. Y. Poluektova. 2020. "A long-acting 3TC ProTide nanoformulation suppresses HBV replication in humanized mice." *Nanomedicine: Nanotechnology, Biology and Medicine,* 28: 102185. DOI: 10.1016/j.nano.2020.102185.

Wei, Liuya, Jingran Lu, Huizhong Xu, Atish Patel, Zhe-Sheng Chen, and Guofang Chen. 2015. "Silver nanoparticles: Synthesis, properties, and therapeutic applications." *Drug Discovery Today* 20, no. 5: 595–601. DOI: 10.1016/j.drudis.2014.11.014.

WHO (World Health Organization). 2019a. *Global Tuberculosis Report 2019*. Geneva: World Health Organization.

WHO. 2019b. *World Malaria Report 2019*. Geneva: World Health Organization.

WHO. 2020. "Global health observatory data." Accessed 08 February 2020. https://www.who.int/gho/hiv/en/.

Wyszogrodzka, Gabriela, Przemysław Dorożyński, Barbara Gil, Wieslaw J. Roth, Maciej Strzempek, Bartosz Marszałek, Władysław P. Węglarz, Elżbieta Menaszek, Weronika Strzempek, and Piotr Kulinowski. 2018. "Iron-based metal-organic frameworks as a theranostic carrier for local tuberculosis therapy." *Pharmaceutical Research* 35, no. 7: 144. DOI: 10.1007/s11095-018-2425-2.

Xie, Yubin, Xiaotong Luo, Zhihao He, Yueyuan Zheng, Zhixiang Zuo, Qi Zhao, Yanyan Miao, and Jian Ren. 2017. "VirusMap: A visualization database for the influenza A virus." *Journal of Genetics and Genomics* 44, no. 5: 281–284. DOI: 10.1016/j.jgg.2017.04.002.

Zahoor, A., Sadhna Sharma, and G. K. Khuller. 2005. "Inhalable alginate nanoparticles as antitubercular drug carriers against experimental tuberculosis." *International Journal of Antimicrobial Agents* 26, no. 4: 298–303. DOI: 10.1016/j.ijantimicag.2005.07.012.

Zazo, Hinojal, Clara I. Colino, and José M. Lanao. 2016. "Current applications of nanoparticles in infectious diseases." *Journal of Controlled Release* 224: 86–102. DOI: 10.1016/j.jconrel.2016.01.008.

Zeynep, Ergul Yilmaz, Debuigne Antoine, Calvignac Brice, Boury Frank, and Jerome Christine. 2015. "Double hydrophilic polyphosphoester containing copolymers as efficient templating agents for calcium carbonate microparticles." *Journal of Materials Chemistry B* 3, no. 36: 7227–7236. DOI: 10.1039/C5TB00887E.

Zhao, Yuyun, and Xingyu Jiang. 2013. "Multiple strategies to activate gold nanoparticles as antibiotics." *Nanoscale* 5, no. 18: 8340–8350. DOI: 10.1039/C3NR01990J.

Zhou, T., H. Su, P. Dash, Z. Lin, B. L. Dyavar Shetty, T. Kocher, A. Szlachetka et al. 2018. "Creation of a nanoformulated cabotegravir prodrug with improved antiretroviral profiles." *Biomaterials* 151: 53–65. DOI: 10.1016/j.biomaterials.2017.10.023.

Zhou, Yan, Ying Kong, Subrata Kundu, Jeffrey D. Cirillo, and Hong Liang. 2012. "Antibacterial activities of gold and silver nanoparticles against Escherichia coli and bacillus Calmette-Guérin." *Journal of Nanobiotechnology* 10, no. 1: 19. DOI: 10.1186/1477-3155-10-19.

zur Mühlen, Annette, Cora Schwarz, and Wolfgang Mehnert. 1998. "Solid lipid nanoparticles (SLN) for controlled drug delivery: Drug release and release mechanism." *European Journal of Pharmaceutics and Biopharmaceutics* 45, no. 2: 149–155. DOI: 10.1016/S0939-6411(97)00150-1.

13 Green-Synthesized Nanoparticles as Potential Sensors for Health Hazardous Compounds

Rachel Fanelwa Ajayi, Sphamandla Nqunqa, Yonela Mgwili, Siphokazi Tshoko, Nokwanda Ngema, Germana Lyimo, Tessia Rakgotho, Ndzumbululo Ndou, and Razia Adam

CONTENTS

13.1 Introduction 292
13.2 Effects of Hydrazine and Nitrobenzene on Human Health 292
13.3 Electrochemical Sensors 294
13.4 Green Method Nanoparticles 296
13.5 Methods Used in Green Synthesis of Nanoparticles 297
  13.5.1 Microwave Irradiation 297
  13.5.2 Ultrasound 297
  13.5.3 Photocatalysis 298
  13.5.4 Biotransformation 298
  13.5.5 Conventional Heating 298
13.6 Types of Green Method Synthesis Approaches and Sensor Application 299
  13.6.1 Plant-Mediated Synthesis of Nanoparticles 300
  13.6.2 *Other Green Synthesis Methods* 304
    13.6.2.1 Fungi-Mediated Synthesis of Nanoparticles 304
    13.6.2.2 Yeast-Mediated Synthesis of Nanoparticles 305
    13.6.2.3 Bacteria-Mediated Synthesis of Nanoparticles 305
References 305

13.1 INTRODUCTION

Nanomaterials are at the leading edge of the rapid development of nanotechnology. They have unique size-dependent properties which make them indispensable and superior in many areas of human activity (El-Rafie et al., 2012). Although physical and chemical methods could successfully produce pure, well-defined metal-based nanomaterials such as gold, zinc oxide and silver nanoparticles, these methods are quite expensive and potentially dangerous to the environment (Salata, 2004). Hence, the use of biological organisms such as plant extracts, microorganisms or plant biomass is preferable as alternatives to chemical and physical methods for the production of nanoparticles in an eco-friendly manner (Rao et al., 2013). Nanomaterials are known to have a long list of applicability that improves human life and the environment because of their optical and magnetic properties (Kotakadi et al., 2014). These materials are usually used to describe materials with one or more components that have at least one dimension in the range of 1–100 nm that includes nanotubes, nanoparticles, composite materials, nanofibers and nanostructured surfaces. The nanoparticles are usually a subset of nanomaterials currently well-defined by consensus as single particles with a diameter less than 100 nm (Borm et al., 2006).

Currently, the development of green processes for nanoparticles synthesis is developing into a central branch of nanotechnology. Most metal-based nanomaterials have drawn attention to scientists due to their extensive application in the development of new technologies in the areas of sensor development, material sciences, electronics and medicine at the nanoscale (Banu and Rathod, 2013). Nanocomposite materials containing nanoparticles such as silver and zinc oxide have also recently attracted a lot of attention because of their chemical stability, catalysis and good conductivity, and they are extensively used in construction materials and food industries as antibacterial agents in food packing (Bai et al., 2011). As a result, these nanoparticles have exceptional properties and are promising in new areas of research due to their size and highly active performance caused by their large surface areas yet; there are also some problems observed in the use of nanoparticles such as the high degree of agglomeration between nanoparticles (Haghparasti and Shahri, 2018). Moreover, expansion of green syntheses over chemical and physical methods is due to properties such as non-toxicity, the use of safe reagents, lack of need to use high pressure, energy, temperature and toxic chemicals and they are generally environment-friendly (Grodowska and Parczewski, 2010). Nanoparticles synthesized using green technology or biological methods have great stability, are diverse in nature and have proper dimensions since they are synthesized using a one-step method (Parveen et al., 2016).

Despite their ease of synthesis and benefits, the use of green method nanoparticles in the development of an electrochemical sensor for the detection of hazardous chemicals responsible for receding human health is less explored, particularly with regard to the detection of hydrazine and nitrobenzene.

13.2 EFFECTS OF HYDRAZINE AND NITROBENZENE ON HUMAN HEALTH

Hydrazine (Fig. 13.1) is a highly reductive and basic colourless inorganic liquid compound commonly used as a corrosion inhibitor in heating systems and in the

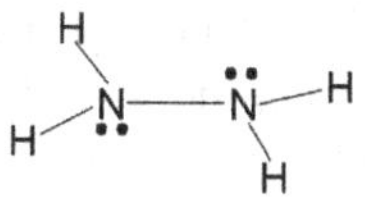

FIGURE 13.1 Structure of hydrazine.

propellant of fighter jets as well as in missiles and rockets board due to its high gas production and heat of combustion. Because of its elevated alkalinity and reducibility, hydrazine also serves as a significant manufacturing raw material and is commonly used as a synthetic precursor in the manufacturing of pharmaceuticals, blowing agents, pesticides and chemical dyes. Nonetheless, hydrazine is illustrious for its high toxicity and can cause serious hazards to human health (Lai et al., 2020; Wang et al., 2020). Hydrazine enters the body through osmosis, breathing and other means. The deadly properties of hydrazine on human health are irritation of eyes, temporary blindness, lung, liver, kidney and nervous system damage. Its high toxicity has led to a set minimum threshold limit value of 10 ppb in the environment. As such, hydrazine has been classified as a plausible human carcinogen by the United States Environmental Protection Agency (USEPA) and the European Commission's Scientific Committee on Occupational Exposure Limits (SCOEL). While this hazardous chemical is not endogenously created, certain medical remedies can be metabolized to form hydrazine in the human body and further compromise human health. Therefore, the selective and sensitive detection of trace hydrazine levels in biological and environmental systems has attracted great attention (Erdemira and Malkondub, 2020; Xingzong et al., 2020; Samanta et al., 2020).

One additional hazardous chemical to human health is nitrobenzene (Fig. 13.2), a toxic nitroaromatic compound used in the production of many commercially appropriate chemicals. Commercially, nitrobenzene and its derivatives are widely used in the manufacturing of dyes, explosives, perfumes, pesticides and pharmaceuticals. Conversely, the growth of these industries produces a large quantity of wastewater comprising of nitrobenzene, which brings toxicological inferences to human health such as anaemia, cancer and skin irritation (Li et al., 2020). Additionally, when inhaled, nitrobenzene is carcinogenic in humans and presents with symptoms such a cyanosis, dizziness, nausea, restlessness and vomiting. Other studies (e.g. Villegas et al., 2020) have reported additional symptoms such as a burning sensation in the mouth and throat, coordination disorders, a smell of bitter almonds in the exhaled air, signs of paralysis, tachycardia, a drop in blood pressure and unconsciousness. Despite its carcinogenic and toxic nature, the USEPA has certified nitrobenzene as group 2B carcinogen and has detailed a minimum threshold limit of 5 ppb for humans. Furthermore, nitrobenzene groundwater aquifer and soil pollution have become a serious issue resulting from its toxicity and potential harmful health

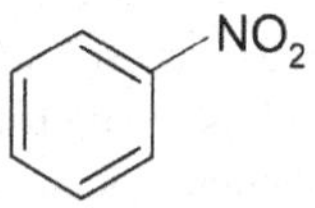

FIGURE 13.2 Structure of nitrobenzene.

impacts on humans. This is caused by the fact that these compounds are resistant to oxidative degradation resultant from the stability of the benzene rings and effects of the electron-withdrawing nitro groups making its quantification and detection a priority in most countries (Liu et al., 2020).

To date, various researchers (Penneman and Audrieth, 1948; Cigić and Prosen, 2009; Siddiquia et al., 2007; Hu et al., 2020; Sakthivel et al., 2020) have developed different analytical methods for the detection of hydrazine and nitrobenzene; however, electrochemical approaches with low detection limits are considered more selective, more sensitive and simpler to use as opposed to available spectrophotometric and chromatographic techniques. These devices are even more attractive when constructed using green method nanoparticles, particularly for health control and towards the detection of hazardous compounds such as hydrazine and nitrobenzene.

13.3 ELECTROCHEMICAL SENSORS

In summary, the development of analytical chemical testing methods demonstrates that electrochemical sensors signify the most promptly rising class of chemical sensors. This is attributed to their time-consuming processes, their expensive nature and their numerous challenges to perform rapid *in situ* analyses. Chemical sensors can be defined as devices capable of providing constant evidence about their locations and are able to provide certain responses related to quantities of specific chemical species (Stradiotto et al., 2003). These sensors are classified based on the property to be monitored or determined as their optical, mass, electrical or thermal, and are thus developed to detect and respond to an analyte of interest in either the liquid, gaseous or solid form. In their design, chemical sensors consist of a chemically selective layer responsible for isolating the analyte from its direct environment, and a transducer which transforms the desired response into a detectable signal. As opposed to thermal, optical and sensors, electrochemical sensors are especially attractive because of their notable sensitivity and selectivity, rapid response time, high portable field-based size, low cost and experimental simplicity (Wang et al., 2008; Tabrizi and Varkani, 2014). Electrochemical sensors have upgraded the presentation of conventional analytical methods through the elimination of expensive reagents and restrained preparation protocols and have delivered affordable analytical tools. Additionally, it is their portability, inexpensiveness and ease of operational analytical tools that make electrochemical sensors more advantageous over conventional analytical instruments. Thus, they offer abundant applications in food analysis, clinical diagnosis and environmental monitoring (Kimmel et al., 2011). The detection and monitoring of hazardous chemicals such as hydrazine and nitrobenzene have become a key focus to many researchers with the aim of controlling infections and the development of deadly diseases. The development of novel electrochemical sensors has found many applications in this field of study.

The principles of electrochemistry suggest that there is a relay of charge from an electrode surface to an alternative phase, which could either be a liquid or a solid sample. As this occurs, chemical changes occur at the electrode and the charge is steered through the bulk of the sample which can be moderated chemically and function as the basis of the sensing process. In most electrochemical sensor arrangements, biological elements are attached on electrode surfaces with the aim of creating a detailed interaction between the target compound and the biological element. The resultant electrical gesture from this interaction is proportional to the concentration of the target compound. In these sensors, the transducers comprise two or three electrodes, a silver chloride coated silver reference electrode, an auxiliary platinum wire electrode and a chemically stable working electrode (Zhu et al., 2015; Chen and Chatterjee, 2013). The biological element could be attached to a working electrode which may be glassy carbon electrodes, screen-printed electrodes, platinum electrodes, carbon paste electrodes, indium tin oxide electrodes and gold electrodes. Self-assembled monolayers are usually attached onto the electrode surfaces where they use their functional groups as supportive beds for the suitable immobilization of biological compounds. Currently, the application of nanomaterials such as metal oxides, gold nanoparticles, carbon nanotubes and fullerenes have played an integral part in the design of electrochemical sensors with the purpose of improving the resulting electrical signals to accomplish enhanced detection limits (Bakker and Telting-Diaz, 2002; Golichenari et al., 2019; Bandodkar and Wang, 2014). The basic structure of electrochemical sensors is illustrated in Fig. 13.3.

The most popular electrochemical sensors for the detection of hazardous compounds such as hydrazine and nitrobenzene are conductometric, amperometric and potentiometric sensors (Zhou et al., 2011). Amperometric sensors are leading and are more preferred in electrochemical sensing platforms, particularly when enzymes are used in the sensor construction because these techniques produce electroactive responses that can be easily detected by amperometry. In amperometry, a particular potential is applied to the working electrode against the reference electrode and the subsequent current is measured (Noah and Ndangili, 2019). Finally, in potentiometric sensing, very small currents are allowed; the potential difference between the working electrode and the reference electrode is determined without polarizing the electrochemical cell. The working electrode develops an adjustable potential

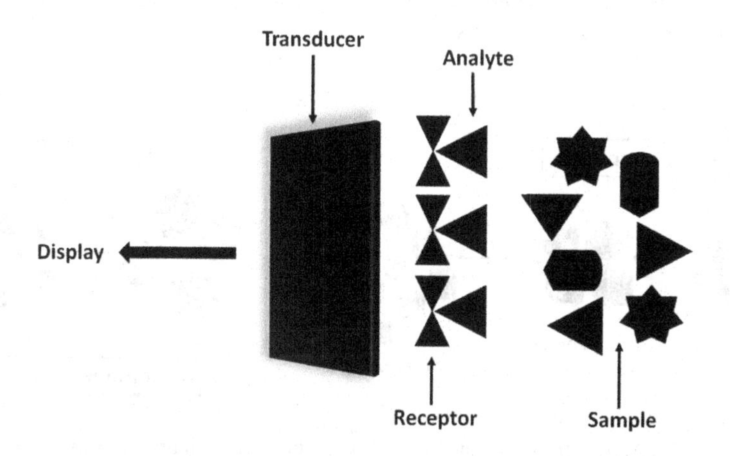

FIGURE 13.3 Basic structure of an electrochemical sensor.

depending on the action or the concentration of the analyte of interest. The change in potential is related to concentration in a logarithmic manner (Garzón et al., 2019).

The next section of this chapter will discuss how green method synthesized nanoparticles influences the operation of these sensors, especially those developed for the detection of hydrazine and nitrobenzene.

13.4 GREEN METHOD NANOPARTICLES

Nanotechnology has shown to be a rapid and successful growth since it allows known materials to be developed with different properties (Mazur, 2004). Decreasing the size of any material to nanoscale may result in the change in its intrinsic properties, thus making them suitable for various applications. As such, metal nanoparticles have been the subject of focused research in recent years mainly because of their unique optical, electronic, mechanical, magnetic and chemical properties that are significantly different from those of bulk materials (Kharissova et al., 2013). These special and unique properties could be attributed to their small sizes and large surface areas. Increased surface area contributes to their enhanced physical and chemical properties which are useful in different field cells (Moodley et al., 2018).

At present, there are two main approaches for the fabrication of nanostructures and the synthesis of nanomaterials: "bottom-up" and "top-down" approaches (Fig. 13.4). The top-down approach is a physical method known as microfabrication method where suitable bulk materials break down into fine particles by size reduction with various lithographic techniques such as milling, thermal/laser ablation, grinding and sputtering (Singh et al., 2016). However, the operative and acceptable method for nanoparticle preparation is the bottom-up approach where nanoparticles are synthesized using either biological or chemical methods through the self-assembling of atoms to new nuclei which grow into particles at the nanoscale (Ahmed et al., 2016). The chemical synthesis approach massively consumes expensive organic solvents, reducing agents and non-renewable solvents, which leads to environmental pollution. Henceforth, the biological paths of nanoparticle synthesis processes are developing as greener and novel strategies (Sathiyanarayanan et al., 2017).

The extraordinary accomplishments in the fields of nanoscience and nanotechnology have shown that nanomaterial-based electrochemical signal magnifications have

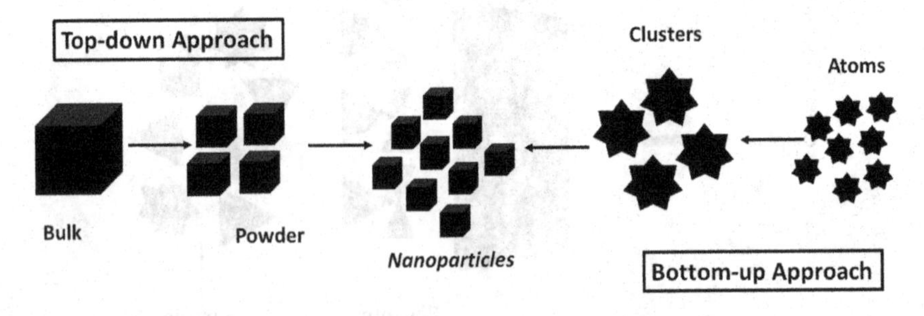

FIGURE 13.4 An illustration of the top-down and bottom-up approaches for the synthesis of nanoparticles.

countless possibilities of improving both selectivity and sensitivity in electrochemical sensing. Also, electrode materials play an imperative role in the development of sensing platforms for the detection of particles of interest through various analytical principles. Additionally, functional nanomaterials have the ability to intensify bio-recognition processes with precisely designed signal codes, leading to highly sensitive sensing. With different types of nanoparticles, green nanoparticles based sensors and biosensors strategies have shown countless potential in the enrichment of sensitivity and specificity detection of hazardous materials in developing countries (Parveen et al., 2016; El-Samadony et al., 2017).

13.5 METHODS USED IN GREEN SYNTHESIS OF NANOPARTICLES

13.5.1 Microwave Irradiation

Microwave irradiation (MI) method denotes a major invention in synthetic chemical methodologies: a dramatic change in the way green synthesis is performed (Grewal et al., 2013). This method holds great significance in the regime of nanoparticle synthesis as it controls high temperature enhancing the nucleation process by which nanoparticles are initially formed. MI forms nanoparticles that provide greater control over the shape and morphology of the nanostructures produced and have higher degrees of crystallinity (Parveen et al., 2016). This method consists of magnetic and electric fields, and thus it releases electromagnetic energy during synthesis. The electromagnetic energy performs as a non-ionizing radiation that causes rotation of the dipoles and molecular motions of ions but does not affect molecular structures. The internal heating in MI is much more homogeneous than classical heating (Majumder et al., 2013). Microwave irradiation is commonly used in agrochemical, pharmaceutical, and in the synthesis of organic materials (Díaz-Ortiz et al., 2018). Its principal interest concerns the saving of energy for heating by focusing efficient energy on the sample, enhancing product purities avoiding possible side reactions, the increase in product yields and dramatic reduction of reaction times and amount of solvents (Veitía and Ferroud, 2015).

13.5.2 Ultrasound

The ultrasound method has steadily been introduced as a green synthetic approach in the field of chemistry as it is safe for the environment. This method is known to be one of the most rapid techniques for the synthesis of metal nanoparticles due to its high pressure and temperature (Popov et al., 2015). These extreme conditions from ultrasound irradiation can possibly produce reactive free radical species followed by the production of metal nanoparticle in a short time. One of its greatest advantages is that it improves the yield, produces nanoparticles that are non-toxic, environment-friendly and simple (Faried et al., 2016). Currently, the ultrasonic-assisted synthesis of nanoparticles has been found to be affective in size distribution of the particles in a narrower range by thermal convection due to the penetration property of ultrasonic irradiation through solution, causing uniform activation energy for the reaction solution. The synthesis of nanoparticles by this method can also control the morphology and size distribution of the nanoparticles (Suib, 2013). This technique presently is widely used in environmental and fine chemistry.

13.5.3 Photocatalysis

Photocatalysis synthesis methods play a significant role in green processes since it provides an alternative to classical chemistry. It offers suitable tools for industrial reactions using cells and enzymes, which can be carried out under mild conditions, with a great control over chemo-, regio- and stereoselectivity by using appropriate enzymes and with the use of heavy metals (Hernaiz et al., 2010). Additionally, in fine chemistry, the activation by photocatalysis under visible light could be another method. This innovative approach is very attractive due to circumvents of the use of heavy metals. In this method, light can be considered as an ideal reagent which is environmentally friendly in green chemical synthesis. These days, the advancement of photo-redox catalysis originated from visible light is of real significance. The reaction of this method is usually carried out by photo-redox catalysts and organometallic complexes containing iridium and ruthenium (Grewal et al., 2013).

13.5.4 Biotransformation

Biotransformation is a widely used method in green chemistry, and it massively contributes to the development of chiral chemistry in aqueous medium merging the constraints enforced by the efficient synthesis with the constraints related in respect to the environment (Veitía and Ferroud, 2015). This method deals with the use of a biocatalyst for the mediation of a chemical reaction and for the synthesis of an organic chemical. Biotransformation is currently playing a crucial role in many industries, including animal feedstock, chiral drug formation and vitamin production. Even the use of microbes and enzyme for synthesis is expected to grow enormously since the industries are being forced by the public to shift toward "green chemistry", which uses safer and cleaner chemicals in their manufacturing processes and produces less toxic effluents (Doble et al., 2004).

13.5.5 Conventional Heating

Conventional heating (CH) is the most frequently used method to obtain green-synthesized nanoparticles, and it usually involves the use of oil baths or furnaces which heats the walls of the reactors by conduction or convection. In this particular method, the process initiates from the source where heat is generated and it then gets transferred from the external to the internal part of the material used through convection, conduction and radiation steadily (Jin et al., 2017). Conventional heating can be used to produce nanoparticles in enormous quantities with defined shape and sizes in a short period of time; however, it is also inefficient, outdated and complicated. In recent years, there has been rising concern in synthesis of nanoparticles which need to be environmentally friendly and should produce non-toxic waste products during the manufacturing process. Recent studies (Ezeonu et al., 2012; Liu et al., 2013) have started succeeding through benign synthesis techniques of a biological nature using biotechnological tools that are considered safe and ecologically sound for nanomaterial fabrication as a substitute to conventional chemical and physical methods. This has certainly given rise to the concept of green nanobiotechnology or green technology (Patra and Baek, 2014).

13.6 TYPES OF GREEN METHOD SYNTHESIS APPROACHES AND SENSOR APPLICATION

The development of eco-friendly technologies in material synthesis is considered an important ground to expand their biological applications. One way to achieve this goal is by using plants and microorganisms to synthesize the nanoparticles where they are mixed with metal salts. A variety of biological agents (plant extracts, bacterial and fungal derived compounds) may be used to convert metal salts to nanoparticles where they act as reducing, capping and stabilizing agents. The formation of these nanoparticles from plant or microbe extracts is influenced by parameters such as temperature, time of reaction and pH. The nature of biomolecules present in the plant extracts could be the most relevant factor in the bioprocess. Thus, the selected extract is important as it provides bioreduction agents in the synthesis such as phenolic compounds, terpenoids, flavonoids, alkaloids, polysaccharides, proteins, enzymes and amino acids (Carmon et al., 2017). Microbes in nature are capable of fabricating nanoparticles under ambient conditions without additional physical conditions or chemical agents. The particles generated by these processes have higher catalytic reactivity, greater specific surface area and an improved contact between the enzyme and metal salt in question due to the bacterial carrier matrix (Fang et al., 2019). In this case, nanoparticles are biosynthesized when the microorganisms grab target ions from their environment and then turn the metal ions into the element metal generated by the cell activities. The microorganisms' ability to synthesize inorganic materials is classified as intracellular or extracellular methods based on the formation of the nanoparticles. In intracellular process, the ions are transported into the microbial cell to form nanoparticles in the presence of enzymes. This process may however require additional steps such as ultrasound treatment or reactions with suitable detergents to release the synthesized nanoparticles (Nadagouda and Varma, 2008; Balakumaran et al., 2015; Nayak et al., 2015). In the extracellular process, the ions are reduced by proteins, enzymes and organic molecules in the medium or by the cell wall components. This process is found to be cheap, and it requires simpler downstream processing. The following section will focus on the use of the plant-based green-synthesized nanoparticles: gold, silver and platinum nanoparticles. Furthermore, this section will give various descriptions on the use of these nanoparticles in sensor development for the detection of the hazardous compounds: hydrazine and nitrobenzene. It should be noted that these two compounds are highly toxic and carcinogenic since they are readily absorbed through human skin. Prolonged exposure to these compounds results in the following combined health disorders: serious damage to the central nervous system, visual impairment, severe burns and lung irritation among others (Haigler and Spain, 1991; Chen and Wenhai, 2009; Martínez-Huitle et al., 2015). It is this reason that electrochemical sensors are required to promptly and effectively detect these compounds in order to prevent severe health effectives which have the potential of causing deadly diseases.

13.6.1 Plant-Mediated Synthesis of Nanoparticles

Plant extracts are the most common biological substrate used for the synthesis of nanoparticles because they are cost-effective, easily accessible and less toxic than microorganisms. These extracts contain secondary metabolites such alkaloids, amino acids, as polysaccharides, polyphenolic compounds and vitamins that act as reducing, stabilizing and capping agents. Different plant parts such as roots, leaves, seeds, stem and fruits have been utilized for the synthesis of nanoparticles (Darvishi et al., 2019; Bandaire et al., 2020). In sensor development, a small change in the size, morphology, surface nature or shape of the nanoparticles can have an effect on the sensing properties. In colorimetric sensors, aggregation of nanoparticles results in surface plasmon coupling, for example, nanoparticles that are close to each other combine to form a cluster resulting into colour changes after they combine (Kelly et al., 2003; Baetsen-Young et al., 2018; De Luca et al., 2018). In optical sensors, the surface plasmon resonance (SPR) absorbance will shift to higher wavelength (red shift) or lower wavelength (blue shift) and this shift depends on the concentration of the metal ions present in nanoparticles of a tested solution (Wang and Yu, 2013; Howes et al., 2014; Polavarapu et al., 2014).

In recent years, gold nanoparticles have been widely used in sensors since current methodologies of sensor development are not cost-effective and are extremely time-consuming. The unusual properties of gold nanoparticles make them excellent as a sensing tools in sensors such as SPR sensors, immunosensors (Kim and Lee, 2017), colorimetric (Zhou et al., 2012) and electrochemical sensors (Ye et al., 2018) to target hazardous compounds such as cancer cells, pesticides and water pollutants. The green method synthesis of gold nanoparticles is slowly gaining popularity amongst scientists due to the use of biodegradable compounds without compromising the application of the nanoparticles. Thus, this section will explore the use of plant-mediated silver, gold and platinum nanoparticles in the development of sensors for the detection of hydrazine and nitrobenzene.

A study by Karthik et al. (2016a) reports the use of *Cerasus serrulata* leaf extracts for the synthesis of gold nanoparticles where they were used as both stabilizing and reducing agents. The chemical constituents of *C. serrulata* leaf extracts used in the biosynthesis of the AuNPs were identified through gas chromatography–mass spectroscopy (GC-MS), while transmission electron microscopy (TEM) was used to confirm their spherical shape and to determine the approximate sizes of the nanoparticles to be 5–25 nm. The GC-MS revealed the main chemical constituents involved in the synthesis process to be 2,3,5-trimethyl decane, a-methylpropanoic acid, hydrocoumarin, butylhydroxytoluene and coumarin. This study explored the dual effect of the synthesized AuNPs; their antibacterial activity against gram-negative (*Escherichia coli*) and gram-positive (*Staphylococcus aureus*) bacteria as well and sensing application. The electrochemical properties of the green-synthesized AuNPs were studied using cyclic voltammetry (CV) on glassy carbon electrodes (GCE) and the modified electrodes were used as sensors for the detection of hydrazine. The developed sensor showed exhibited a wide linear range of 5 nM to 272 μM, a low detection limit 0.05 μM and presented good selectivity towards other hydrazine (Karthik et al., 2016a).

In another study by Emmanuel et al. (2014) involving the green synthesis of AuNPs, twig bark *Acacia nilotica* extracts extracted at room temperature were used in sensor construction for the determination of trace-level concentrations of nitrobenzene. Fourier transform infrared spectroscopy (FTIR) revealed that the phytochemicals present in the extracts are responsible for the stabilization and reduction of the AuNPs to be alkaloids, glycosides, tannins, saponins and terpenoids. The developed procedure revealed the biosynthesis of the starting material chloroauric acid to be within 10 minutes, an indication of an advanced reaction rate as compared to a majority if not all existing chemical synthesis methods. Differential pulse voltammetry (DPV) was used to electrochemically determine nitrobenzene using the green-synthesized AuNPs modified of GCE where excellent reduction responses towards nitrobenzene were observed. The sensor displayed a wide linear response range from 0.10 to 600 μM, a high sensitivity of 1.01 μA/(μM cm^2) and a low limit of detection of 0.01 μM. Excellent recovery results towards nitrobenzene were observed when the AuNPs-modified electrodes were subjected to various nitrobenzene-contaminated real water samples (Emmanuel et al., 2014).

Although there are reports on the application of green-synthesized gold nanoparticles towards the development of hydrazine and nitrobenzene sensors, this area continues to thrive. Table 13.1 lists some plant-based gold nanoparticles with potential applications in the development of sensors for the detection of hazardous compounds such as hydrazine and nitrobenzene.

Just like AuNPs, the synthesis of silver nanoparticles (AgNPs) is drawing more attention due to their applications in catalysis (Kumar et al., 2014): in antimicrobial application, in biomolecular detection and diagnostics, in microelectronics (Gittins et al., 2000), in sensing devices and towards the targeting of drugs (Sengupta et al., 2005). Some studies have also shown AgNPs in the medicinal field where they have been used as anti-inflammatory, antidiabetic and antioxidant agents as well as in cancer treatment and diagnosis (Chen et al., 2013). Several physical and chemical synthesis routes have been applied to produce AgNPs. However, there are drawbacks that come with these methods since they include the use of toxic precursor chemicals such as sodium borohydride, potassium nitrate, ethylene glycol, sodium dodecyl benzyl sulfate and polyvinyl pyrrolidone, thus generating toxic by-products (Roy et al., 2019). Since these nanoparticles are now applied to areas involving human contact, a need to develop environmentally friendly processes for nanoparticle synthesis is of great need. With the advancement of science, alternative synthesis routes which are eco-friendly, less costly, energy-efficient and non-toxic have been developed through green synthesis methods.

Sphere-like AgNPs were synthesized using *Camellia japonica* leaf extracts and were applied in both the remediation and detection of nitrobenzene. GC-MS

TABLE 13.1
A List of Green-Synthesized Gold Nanoparticles

Plant	Part used	References
Salvia officinalis, Lippia citriodora, Pelargonium graveolens and *Punica granatum*	Leaves	Rao et al. (2013)Elia et al. (2014)
Ephedra sinica	Plant	Park et al. (2019)
Leucosidea sericea	Leaves	Badeggi et al. (2020)
Macrotyloma uniflorum	Fruit	Aromal et al. (2012)
Rhizome	Plant	Mahakham et al. (2016)Singh et al. (2018)
Millettia pinnata	Leaves	Rajakumar et al. (2017)Kumar et al. (2019)
Indigofera tinctoria	Leaves	Vijayan et al. (2018)López-Millán et al. (2019)
Citrus limon, Citrus reticulata and *Citrus sinensis*	Fruit	Sujitha and Kannan (2012)Yang et al. (2019)
Genipa americana	Fruit	Ueda et al (1991)Kumar et al. (2016)
Mentha piperita	Leaves	Klekotko et al. (2015), Kuppusamy et al. (2016)

confirmed the main constituents of *Camellia japonica* involved in the reduction of silver ions to AgNPs are butyl tetradecyl ester, phthalic acid, 2-(pentyloxy)- ethylacetate, 2-methyl-propanoic acid, 4-azido-heptane, dibutyl phthalate and acetic acid. Amperometric studies using CV revealed that the electrochemical behaviour of the AgNPs modified on GCE were capable of reducing nitrobenzene. The sensor displayed a wide linear response range of 0.05–21 μM and 23–2,593 μM and a low limit of detection of 12 nM. Even in the presence of nitroaromatic-containing compounds and common metal ions, the sensors showed excellent selectivity towards the detection of nitrobenzene. The synthesized nanoparticles also revealed their photocatalytic and electrocatalytic properties towards the reduction of nitrobenzene (Karthik et al., 2017).

In a slightly different study by Karuppiah et al. (2015), an electrochemical sensor for nitrobenzene (NB) was developed using reduced graphene oxide (RGO) modified with green-synthesized Ag-NPs on GCE. *Justicia glauca* leaf extracts were used as stabilizing and reducing agents in the synthesis of AgNPs while the RGO–AgNPs composite used to modified the electrode surface was prescribed by a simple electrochemical reduction of AgNPs dispersed in a grapheme (GO) solution. Good efficiencies were produced by the modified electrodes and lower overpotentials for the electrocatalytic reduction of nitrobenzene were observed as opposed to individually modified electrodes with either AgNPs or RGO. The DPV data confirmed the nitrobenzene reduction peak currents to be linear over a concentration range of 0.5–900 μM. The detection limit of the sensor was established to be 0.26 μM, and the sensitivity of the sensor was determined to be 0.83 μA/(μM cm^2). Another encouraging aspect about this study was that the sensor showed good selectivity for nitrobenzene even in the presence of similar interfering compounds (Karuppiah et al., 2015).

TABLE 13.2
A List of Green-Synthesized Silver Nanoparticles with Potential Application in Electrochemical Sensors

Plant	Part used	References
Memecylon edule	Leaves	Elavazhagan and Arunachalam (2011)
Chrysopogon zizanioides	Leaves	Arunachalam and Annamalai (2013)
Camellia sinensis	Leaves	Vilchis-Nestor et al. (2008) Loo et al. (2012)
Ziziphora tenuior	Leaves	Sadeghi and Gholamhoseinpoor (2015)Sedaghat and Afshar (2016)
Ocimum sanctum	Leaves	Singal et al. (2011)Subba et al. (2013)
Coccinia grandis	Leaves	Arunachalam et al. (2012)Mala et al. (2017)
Eriobotrya japonica	Leaves	Rao and Tang (2017)Yu et al. (2019)
Tithonia diversifolia	Flower	Tran et al. (2013)Dada et al. (2018)
Gleichenia pectinata	Plant	Femi-Adepoju et al. (2019)
Citrullus lanatus	Fruit	Patra et al. (2016)Ndikau et al. (2017)

In a similar study, *Momordica charantia* leaf extracts were used in the development of a green and single-step process for the fabrication of Ag@AgCl nanoparticles. A highly novel method was developed illustrating fast synthesis rates of 1 minute for synthesis of Ag@AgCl NPs. The synthesized Ag@AgCl NPs revealed catalytic activities for the reduction of 2,4-dinitrophenyl hydrazine (DNPH) where the rate constant (k) for the reduction of 2,4-DNPH was found to be 0.05 min^{-1}. This study was the first of its kind to achieve the reduction of 2,4-dinitrophenyl hydrazine using Ag@AgCl nanoparticles as catalysts (Devi et al., 2016). Table 13.2 gives an illustration of some green method synthesized silver nanoparticles with potential sensor applications.

Another set of green method of synthesizing nanoparticles which continue to gain interest towards the detection of hazardous compounds is platinum nanoparticles. Platinum metals are known to be resistant to chemical outbreaks and corrosion and it possesses a high melting point and surface area. It is a resourceful catalytic agent in hydrogen storage, in direct methanol fuel cells and in the reduction of automobile pollution among others. On the other hand, various methodologies have been developed for the synthesis of platinum (Pt) nanoparticles (PtNPs) such as chemical precipitation, hydrothermal synthesis, sol process, sol–gel route and vapour deposition which unfortunately all come with restrictions such as high energy requirements, multistep processes and non-safety. Plant-mediated synthesis methods for the synthesis of PtNPs are the solution to remedy these issues since they are eco-friendly, low cost and simple. Thus far, very few reports are available for the synthesis of PtNPs using plant extracts (Jia et al., 2009; Nasrollahzadeh et al., 2016; Sun et al., 2014) and their use in the detection of hazardous compounds such as hydrazine and nitrobenzene.

The first reported study involving the detection of hydrazine involved the development of a sensor based on rapid, facile and eco-friendly *Quercus glauca* mediated synthesized PtNPs was reported by Karthik et al. (2016b). The resultant near spherical nanoparticles with sizes in the range of 5–15 nm were characterized using analytical and spectroscopic techniques, while CV was used to study the electrocatalytic performance of hydrazine using PtNPs-modified GCE in various water samples. A sharp peak was observed at very lower onset oxidation potentials coupled to a very lower detection limit. The green-synthesized PtNPs-modified GCE sensor showed excellent selectivity and a good sensitivity of 1.70 μA/(μM cm^2) (Karthik et al., 2016b). Unfortunately, the use of green method PtNPs for the detection of nitrobenzene is rare, while the synthesis of green method PtNPs is slowly gaining popularity.

One such study by Thirumurugan et al. (2016) is the synthesis of PtNPs using neem (*Azadirachda indica*) extracts which showed capabilities of reducing Pt^{4+} ion into PtNPs. The FTIR revealed the presence of functional groups such as alkanes, aliphatic amines and carbonyls believed to be the surface-active molecules responsible for the stabilization of the nanoparticles (Thirumurugan et al., 2016). Alshatwi et al. (2015) also reported the use of tea polyphenols (TPP) as both a reducing and surface-modifying agents in the synthesis of PtNPs resulting in flower-shaped well-dispersed 30–60-nm-sized TPP@Pt nanoparticles (Alshatwi et al., 2015). Similar studies with potential use in the sensor development for the detection of hydrazine and nitrobenzene are listed in Table 13.3.

13.6.2 *Other Green Synthesis Methods*

13.6.2.1 Fungi-Mediated Synthesis of Nanoparticles

Fungi are preferred microbes for the synthesis of nanoparticles because of their tolerance, better metal bioaccumulation ability, their affordability, their ease to handle in biomass and their ease in the scale-up process (Kalpana et al., 2018; Rajan et al., 2016). Fungi have the ability to secrete different enzymes that are used in various

TABLE 13.3
Examples of Green Method Synthesized Platinum Nanoparticles with Potential Application in Electrochemical Sensors

Plant	Part used	References
Pinus resinosa	Bark	Manikandan et al. (2016)
Ocimum sanctum	Leaves	Soundarrajan et al. (2012)Prabhu and Gajendran (2017)Pareek et al. (2017)
Doipyros kaki	Leaves	Sathishkumar et al. (2009)Zang et al. (2016)
Cacumen platycladi	Leaves	Zheng et al. (2013)Dobrucka (2019)
Dates	Plant	Al-Radadi (2019)
Bidens tripartitus	Leaves	Dobrucka (2016)
Prunus yedoensis	Gum	Velmurugan et al. (2016)Manikandan et al. (2016)

applications, which also play a major role in the synthesis of nanoparticles because of their metal bioaccumulation and toleration properties (Atalla et al., 2017).

13.6.2.2 Yeast-Mediated Synthesis of Nanoparticles

Yeast have the ability to synthesize nanoparticles in either intracellular or extracellular by reducing protein in enzymes. Synthesis of nanoparticle from yeast is a promising approach because they can be easily handled in the laboratory environment, they have abundant enzyme synthesis ability and they are able to rapidly grow without the need of complex nutrients. This is additional to their ability to grow at accelerated rates, therefore producing cultures and their storage in the laboratory is very practical (Moghaddam et al., 2017).

13.6.2.3 Bacteria-Mediated Synthesis of Nanoparticles

Literature (e.g. Prakash et al., 2010; Bandaire et al., 2020) indicate that extracellular synthesis involves the production and the release of protein and enzymes by bacteria that reduce metal ion and stabilize nanoparticles. Cyanobacteria are the most popular bacteria used for the synthesis of metal ion since they are *known* as largest groups of phototrophic bacteria with high potentials as excellent sources of fine chemicals and bioactive compounds, such as lipid-like compounds, amino acid derivatives, proteins and pigments. Bacteria such as *Bacillus cereus* (e.g. Hussein et al., 2009) and *Rhodococcus pyridinivorans* (e.g. Kundu et al., 2014) are commonly used for the synthesis of metal nanoparticles.

REFERENCES

Ahmed, S., Ahmad, M., Swami, B.L., and Ikram, S. 2016. Green synthesis of silver nanoparticles using Azadirachta indica aqueous leaf extract. *Journal of Radiation Research and Applied Sciences* 9: 1–7.

Al-Radadi, N.S. 2019. Green synthesis of platinum nanoparticles using Saudi's Dates extract and their usage on the cancer cell treatment. *Arabian Journal of Chemistry* 12: 330–349.

Alshatwi, A.A., Athinarayanan, J., and Vaiyapuri Subbarayan, P. 2015. Green synthesis of platinum nanoparticles that induce cell death and G2/M-phase cell cycle arrest in human cervical cancer cells. *Journal Material Science Materials in Medicine* 26: 1–9.

Aromal, S.A., Vidhu, V.K., and Philip, D. 2012. Green synthesis of well-dispersed gold nanoparticles using *Macrotyloma uniflorum. Spectrochimica Acta Part A: Molecular and Biomolecular Spectroscopy* 85: 99–104.

Arunachalam, K.D., and Annamalai, S.K. 2013. Chrysopogon zizanioides aqueous extract mediated synthesis, characterization of crystalline silver and gold nanoparticles for biomedical applications. *International Journal of Nanomedicine* 8: 2375–2384.

Arunachalam, R., Dhanasingh, S., Kalimuthu, B., Uthirappan, M., Rose, C., and Mandal, A.B. 2012. Phytosynthesis of silver nanoparticles using *Coccinia grandis* leaf extract and its application in the photocatalytic degradation. *Colloids and Surfaces B Biointerfaces* 1: 226–230.

Atalla, S.M.M., Mohamed, A.A., El-Gamal, N.G., and El-Shamy, A.R. 2017. Biosynthesis of zinc nanoparticles and its effect on enzymes production from *Chaetomium globosum* using different agricultural wastes. *Journal of Innovations in Pharmaceutical and Biological Sciences* 4: 40–45.

Badeggi, U.M., Ismail, E., Adeloye, A.O., Botha, S., Badmus, J.A., Marnewick, L.L., Cupido, C.N., and Hussein, A.A. 2020. Green synthesis of gold nanoparticles capped with procyanidins from *Leucosidea sericea* as potential antidiabetic and antioxidant agents. *Biomolecules* 10: 1–20.

Baetsen-Young, A.M., Vasher, M., Matta, L.L., Colgan, P., Alocilja, E.C., and Day, B., 2018. Direct colorimetric detection of unamplified pathogen DNA by dextrin-capped gold nanoparticles. *Biosensensors and Bioelectronics* 101: 29–36.

Bai, H., Li, C., and Shi, G. 2011. Functional composite materials based on chemically converted graphene. *Advanced Materials* 23: 1089–1115.

Bakker, E., and Telting-Diaz, M. 2002. Electrochemical Sensors. *Analytical Chemistry* 74: 2781–2800.

Balakumaran, M.D., Ramachandran, R., and Kalaichelvan. P.T. 2015. Exploitation of endophytic fungus, Guignardia mangiferae for extracellular synthesis of silver nanoparticles and their in vitro biological activities. *Microbiology Research* 178: 9–17.

Bandeira, M., Giovanela, M., Roesch-Ely, M., Devine, D.M., and Crespo, D.S. 2020. Green synthesis of zinc oxide nanoparticles: A review of the synthesis methodology and mechanism of formation. *Sustainable Chemistry and Pharmacy* 15: 100–223.

Bandodkar, A.J., and Wang, J. 2014. Non-invasive wearable electrochemical sensors: A review. *Trends in Biotechnology* 32: 363–371.

Banu, A., and Rathod, V. 2013. Biosynthesis of monodispersed silver nanoparticles and their activity against Mycobacterium tuberculosis. *International Journal of Biomedical Nanoscience and Nanotechnology* 3: 211–220.

Borm, P.J., Robbins, D., Haubold, S., Kuhlbusch, T., Fissan, H., Donaldson, K., Schins, R., Stone, V., Kreyling, W., Lademann, J., and Krutmann, J. 2006. The potential risks of nanomaterials: A review carried out for ECETOC. *Particle and Fibre Toxicology*, 3: 11.

Carmon, E.R., Benito, N., Plaza, T., and Recio-Sánchez, G. 2017. Green synthesis of silver nanoparticles by using leaf extracts from the endemic *Buddleja globosa* hope, *Green Chemistry Letters and Reviews* 10: 250–256.

Chen, B., and Wenhai, H. 2009. Effect of background electrolytes on the adsorption of nitroaromatic compounds onto bentonite. *Journal of Environmental Science* 21: 1044–1052.

Chen, C., and Chatterjee, S. 2013. Nanomaterials based electrochemical sensors for biomedical applications. *Chemical Society Review* 42: 5425.

Chen, S., Zhang, Q., Hou, Y., Zhang, J., and Liang, X. 2013. Nanomaterials in medicine and pharmaceuticals: Nanoscale materials developed with less toxicity and more efficacy. *European Journal of Nanomedicine* 5: 61–79.

Cigić, I.K., and Prosen, H. 2009. An overview of conventional and emerging analytical methods for the determination of mycotoxins. *International Journal of Molecular Sciences* 10: 62–115.

Dada, A.O., Inyinbor, A.A., Idu, E.I., Bello, O.M., Oluyori, A.P., Adelani-Akande, T.A., Okunola, A.A., and Dada, O. 2018. Effect of operational parameters, characterization and antibacterial studies of green synthesis of silver nanoparticles using *Tithonia diversifolia. PeerJ* 6: e5865.

Darvishi, E., Kahrizi, D., and Arkan, E. 2019. Comparison of different properties of zinc oxide nanoparticles synthesized by the green (using Juglans regia L. leaf extract) and chemical methods. *Journal of Molecular Liquids* 286: 110–831.

De Luca, G., Bonaccorsi, P., Trovato, V., Mancuso, A., Papalia, T., Pistone, A., Casaletto, M.P., Mezzi, A., Brunetti, B., and Minuti, L. 2018. Tripodal tris-disulfides as capping agents for a controlled mixed functionalization of gold nanoparticles. *New Journal of Chemistry* 42: 16436–16440.

Devi, T., Ahmaruzzaman, M., and Begum, S. 2016. A rapid, facile and green synthesis of Ag@AgCl nanoparticles for the effective reduction of 2,4-dinitrophenyl hydrazine. *New Journal of Chemistry* 40: 1497–1506.

Díaz-Ortiz, A., Prieto, P., and de la Hoz, A. 2018. A critical overview on the effect of microwave irradiation in organic synthesis. *The Chemical Record* 19: 85–97.

Doble, M., Kruthiventi, A.K., and Gaikar, V.G. 2004. *Biotransformations and Bioprocesses.* New York: Marcel Dekker.

Dobrucka, R. 2016. Synthesis and structural characteristic of platinum nanoparticles using herbal Bidens Tripartitus extract. *Journal of Inorganic Organometallic Polymers* 26: 219–225.

Dobrucka, R. 2019. Biofabrication of platinum nanoparticles using *Fumariae herba* extract and their catalytic properties. *Saudi Journal of Biological Sciences* 26: 31–37.

Elavazhagan, T., and Arunachalam, K.D. 2011. *Memecylon edule* leaf extract mediated green synthesis of silver and gold nanoparticles. *International Journal of Nanomedicine* 6: 1265–1278.

Elia, P., Zach, R., Hazan, S., Kolusheva, S., Porat, Z., and Zeiri, Y. 2014. Green synthesis of gold nanoparticles using plant extracts as reducing agents. *International Journal of Nanomedicine* 9: 4007–4021.

El-Rafie, M.H., Shaheen, T.I., Mohamed, A.A., and Hebeish, A. 2012. Biosynthesis and applications of silver nanoparticles onto cotton fabrics. *Carbohydrate Polymers* 90: 915–920.

El-Samadony, H., Althani, A., Tageldin, M.A., and Hassan, M.E. 2017. Azzazy nanodiagnostics for tuberculosis detection. *Expert Review of Molecular Diagnostics* 17: 427–443.

Emmanuel, R., Karuppiaha, C., Chena, S., Palanisamya, S., Padmavathyc, S., and Prakash, P. 2014. Green synthesis of gold nanoparticles for trace level detection of a hazardous pollutant (nitrobenzene) causing Methemoglobinaemia. *Journal of Hazardous Materials* 279: 117–124.

Erdemira, S., and Malkondub, S. 2020. A colorimetric and fluorometric probe for hydrazine through subsequent ring-opening and closing reactions: Its environmental applications. *Microchemical Journal* 152: 104375.

Ezeonu, C.S., Tagbo, R., Anike, E.N., Oje, O.A., and Onwurah, I.N.E. 2012. Biotechnological tools for environmental sustainability: Prospects and challenges for environments in Nigeria: A standard review. *Biotechnology Research International* 2012: 450802.

Fang, X., Wang, Y., Wang, Z., Jiang, Z., and Dong, M. 2019. Microorganism assisted synthesized nanoparticles for catalytic applications. *Energies* 12: 2–21.

Faried, M., Shameli, K., Miyake, M., Zakaria, Z., Hara, H., Khairudin, N.A., and Etemadi, M. 2016. Ultrasound-assisted in the synthesis of silver nanoparticles using sodium alginate mediated by green method. *Digest Journal of Nanomaterials and Biostructures* 11: 547–552.

Femi-Adepoju, A.G., Dada, A.O., Otun, K.O., Adepoju, A.O., and Fatoba, O.P. 2019. Green synthesis of silver nanoparticles using terrestrial fern (*Gleichenia Pectinata* (Willd.) C. Presl.): Characterization and antimicrobial studies. *Heliyon* 5: e01543.

Garzón, V., Pinacho, D.G., Bustos, R., Garzón, G., and Bustamante, S. 2019. Optical biosensors for therapeutic drug monitoring. *Biosensors* 9: 132.

Gittins, D.I., Bethell, D., Nichols, R., and Schiffrin, D.J. 2000. Diode-like electron transfer across nanostructured films containing a redox ligand. *Journal of Materials Chemistry* 10: 79–83.

Golichenari, B., Nosrati, R., Farokhi-Fard, A., Maleki, M.F., Hayat, S.M.G., Ghazvini, K., Vaziri, F., and Behravan, J. 2019. Electrochemical-based biosensors for detection of Mycobacterium tuberculosis and tuberculosis biomarkers. *Critical Reviews in Biotechnology* 39: 1–20.

Grewal, A.S., Kumar, K., Redhu, S., and Bhardwaj, S. 2013. Microwave assisted synthesis: A green chemistry approach. *International Research Journal of Pharmaceutical and Applied Sciences* 3: 278–285.

Grodowska, K., and Parczewski, A. 2010. Organic solvents in the pharmaceutical industry. *Acta Poloniae Pharmaceutica* 67: 3–12.

Haghparasti, Z., and Shahri, M.M. 2018. Green synthesis of water-soluble nontoxic inorganic polymer nanocomposites containing silver nanoparticles using white tea extract and assessment of their in vitro antioxidant and cytotoxicity activities. *Materials Science and Engineering: C* 87: 139–148.

Haigler, B.E., and Spain, J.C. 1991. Biotransformation of nitrobenzene by bacteria containing toluene degradative pathways. *Applied Environmental Microbiology* 57: 3156–3162.

Hernaiz, M.J., Alcantara, A.R., Garcia, J.I., and Sinisterra, J.V. 2010. Applied biotransformations in green solvents. *Chemistry-A European Journal* 16: 9422–9437.

Howes, P.D., Chandrawati, R., and Stevens, M.M. 2014. Colloidal nanoparticles as advanced biological sensors *Science* 346: 1247390.

Hu, Z., Yang, T., Liu, J., Zhang, Z., and Feng, G. 2020. Preparation and application of a highly sensitive conjugated polymer-copper (II) composite fluorescent sensor for detecting hydrazine in aqueous solution. Talanta 207: 120203.

Hussein, M.Z., Azmin, W.H., Mustafa, M., and Yahaya, A. 2009. *Bacillus cereus* as a biotemplating agent for the synthesis of zinc oxide with raspberry- and plate-like structures. *Journal of Inorganic Biochemistry* 108: 1145–1160.

Jia, L., Zhang, Q., Li, Q., and Song, H. 2009. The biosynthesis of palladium nanoparticles by antioxidants in *Gardenia jasminoides* Ellis: Long lifetime nanocatalysts for p-nitrotoluene hydrogenation. *Nanotechnology* 20: 385601.

Jin, S., Guo, C., Lu, Y., Zhang, R., Wang, Z., and Jin, M. 2017. Comparison of microwave and conventional heating methods in carbonization of polyacrylonitrile-based stabilized fibers at different temperature measured by an in-situ process temperature control ring. *Polymer Degradation and Stability* 140: 32–41.

Kalpana, V.N., Kataru, B.A.S., Sravani, N., Vigneshwari, T., Panneerselvam, A., and Rajeswari, V.D. 2018. Biosynthesis of zinc oxide nanoparticles using culture filtrates of *Aspergillus niger*: Antimicrobial textiles and dye degradation studies. *OpenNano* 3: 48–55.

Karthik, M., Govindasamy, M., Chen, S., Cheng, Y., Muthukrishnan, B., Padmavathy, S., and Elangovan, S. 2017. Biosynthesis of silver nanoparticles by using *Camellia japonica* leaf extract for the electrocatalytic reduction of nitrobenzene and photocatalytic degradation of Eosin-Y. *Journal of Photochemistry & Photobiology, B: Biology* 170: 164–172.

Karthik, R., Chen, S., Elangovan, A., Muthukrishnan, P., Shanmugam, R., and Lou, B. 2016a. Phyto-mediated biogenic synthesis of gold nanoparticles using *Cerasus serrulata* and its utility in detecting hydrazine, microbial activity and DFT studies. *Journal of Colloid and Interface Science* 468: 163–175.

Karthik, R., Sasikumar, R., and Chen, S. Govindasamy, M., Vinoth Kumar, J. and Muthuraj, V., 2016b. Green synthesis of platinum nanoparticles using *Quercus Glauca* extract and its electrochemical oxidation of hydrazine in water samples. *International Journal of Electrochemical Science* 11: 8245–8255.

Karuppiah, C., Muthupandi, K., Chen, S., Ali, M., Palanisamy, S., Rajan, A., Prakash, P., Al-Hemaid, F., and Lou, B. 2015. Green synthesized silver nanoparticles decorated on reduced graphene oxide for enhanced electrochemical sensing of nitrobenzene in wastewater samples. *RSC Advances* 5: 31139–31146.

Kelly, K.L., Coronado, E., Zhao, L.L., and Schatz, G.C. 2003. The optical properties of metal nanoparticles: The influence of size, shape, and dielectric environment. *Journal of Physical Chemistry* 107: 668–677.

Kharissova, O.V., Dias, H.R., Kharisov, B.I., Pérez, B.O., and Pérez, V.M. 2013. The greener synthesis of nanoparticles. *Trends in Biotechnology* 31: 240–248.

Kim, S., and Lee, H.J. 2017. Gold nanostar enhanced surface plasmon resonance detection of an antibiotic at attomolar concentrations via an aptamer-antibody sandwich assay. *Analytical Chemistry* 89: 6624–6630.

Kimmel, D.W., LeBlanc, G., Meschievitz, M.E., and Cliffel, D.E. 2011. Electrochemical sensors and biosensors. *Analytical Chemistry* 84: 685–707.

Klekotko, M., Matczyszyn, K., Siednienko, J., Olesiak-Banska, J., Pawlik, K., and Samoc, M. 2015. Bio-mediated synthesis, characterization and cytotoxicity of gold nanoparticles. *Physical Chemistry Chemical Physics* 17: 29014–29019.

Kotakadi, V.S., Gaddam, S.A., Rao, Y.S., Prasad, T.N.V.K.V., Reddy, A.V., and Gopal, D.S. 2014. Biofabrication of silver nanoparticles using *Andrographis paniculata. European Journal of Medicinal Chemistry* 73: 135–140.

Kumar, B., Smita, K., Cumbal, L., Debut, A., and Pathak, R.N. 2014. Sonochemical synthesis of silver nanoparticles using starch: A comparison. *Bioinorganic Chemistry and Applications* 2014: 784268.

Kumar, B., Smita, K., Cumbal, L., Camacho, J., Hernández-Gallegos, E., de Guadalupe Chávez-López, M., Grijalva, M., and Andrade. K. 2016. One pot phytosynthesis of gold nanoparticles using *Genipa americana* fruit extract and its biological applications. *Materials Science & Engineering C-Materials for Biological Applications* 62: 725–731.

Kumar, G., Ghosh, M., and Pandey, D,M. 2019. Method development for optimised green synthesis of gold nanoparticles from *Millettia pinnata* and their activity in non-small cell lung cancer cell lines. *IET Nanobiotechnology* 13: 626–633.

Kundu, D., Hazra, C., Chatterjee, A., Chauhari, A., and Mishra, S. 2014. Extracellular biosynthesis of zinc oxide nanoparticles using *Rhodococcus pyridinivorans* NT2: Multifunctional textile finishing, biosafety evaluation and in vitro drug delivery in colon carcinoma. *Journal of Photochemistry and Photobiology B: Biology* 140: 194–204.

Kuppusamy, P., Yusoff, M.M., Maniam, G.P., and Govindan, N. 2016. Biosynthesis of metallic nanoparticles using plant derivatives and their new avenues in pharmacological applications: An updated report. *Saudi Pharmaceutical Journal* 24: 473–484.

Lai Q, Si, S., Qin, T., Li, B., Wu, H., Liu, B., Xu, H., and Zhao, C. 2020. A novel red-emissive probe for colorimetric and ratiometric detection of hydrazine and its application in plant imaging. *Sensors and Actuators B: Chemical* 307: 127640.

Li, X., Zhang, X., Xu, Y., and Yu, P. 2020. Removal of nitrobenzene from aqueous solution by using modified magnetic diatomite. *Separation and Purification Technology* 242: 116792.

Liu, G., Dong, B., Zhou, J., Li, J., Jin, R., and Wang, J. 2020. Facilitated bioreduction of nitrobenzene by lignite acting as low-cost and efficient electron shuttle. *Chemosphere* 248: 125978.

Liu, W., Yuan, J.S., and Stewart Jr C.N. 2013. Advanced genetic tools for plant biotechnology. *Nature Reviews Genetics* 14: 781–793.

Loo, Y.Y., Chieng, B.W., Nishibuchi, M., and Radu, S. 2012. Synthesis of silver nanoparticles by using tea leaf extract from *Camellia sinensis. International Journal of Nanomedicine* 7: 4263–4267.

López-Millán, A., Del Toro-Sánchez, C.Z., Ramos-Enríquez, J.R., Carrillo-Torres, R.C., Zavala-Rivera, P., Esquivel, R., Álvarez-Ramos, E., Moreno-Corral, R., Guzmán Zamudio, R., and Lucero-Acuña, A. Biosynthesis of gold and silver nanoparticles using *Parkinsonia florida* leaf extract and antimicrobial activity of silver nanoparticles. *Materials Research Express* 6: 9.

Mahakham, W., Theerakulpisut, P., Maensiri, S., Phumying, S., and Sarmah, A.K. 2016. Environmentally benign synthesis of phytochemicals-capped gold nanoparticles as nanopriming agent for promoting maize seed germination. *Science of the Total Environement* 573: 1089–1102.

Majumder, A., Gupta, R., and Jain, A. 2013. Microwave-assisted synthesis of nitrogen-containing heterocycles. *Green Chemistry Letters and Reviews* 6: 151–182.

Mala, M., Hannah, H.A., and Jeya, J.G. 2017. Silver nanoparticles synthesis using *Coccinia grandis* (L.) Voigt and *Momordica charantia* L., its characterization and biological screening. *Journal of Bionanoscience* 11: 504–513.

Manikandan, V., Velmurugan, P., Park, J.H., Lovanh, N., Seo, S.K., Jayanthi, P., Park, Y.J., Cho, M., and Oh, B.T. 2016. Synthesis and antimicrobial activity of palladium nanoparticles from *Prunus xyedoensis* leaf extract. *Materials Letters* 185: 335–338.

Martínez-Huitle, C.A., Rodrigo, M.A., Sireis, I., and Scialdone, O. 2015. Single and coupled electrochemical processes and reactors for the abatement of organic water pollutants: A critical review. *Chemical Reviews* 115: 13362–13407.

Mazur M. 2004. Electrochemically prepared silver nanoflakes and nanowires *Electrochemistry Communications* 6: 400–403.

Moghaddam, A.B., Moniri, M., Azizi, S., Rahim, R.A., Arif, A.B., Saad, W.Z., Namvar, F., Navaderi, M., and Mohamad, R. 2017. Biosynthesis of ZnO nanoparticles by a new *Pichia kudriavzevii* yeast strain and evaluation of their antimicrobial and antioxidant Activities. *Molecules* 22: 872.

Moodley, J.S., Naidu Krishna, S.B, Pillay, K, Govender, S., and Govender, P. 2018. Green synthesis of silver nanoparticles from *Moringa oleifera* leaf extracts and its antimicrobial potential. *Advances in Natural Sciences: Nanoscience and Nanotechnology* 9: 015011.

Nadagouda, M.N., and Varma, R.S. 2008. Green synthesis of silver and palladium nanoparticles at room temperature using coffee and tea extract. *Green Chemistry* 10: 859–862.

Nasrollahzadeh, M., Sajadi, S.M., Rostami-Vartooni, A., Alizadeh, M., and Bagherzadeh, M. 2016. Green synthesis of the Pd nanoparticles supported on reduced graphene oxide using barberry fruit extract and its application as a recyclable and heterogeneous catalyst for the reduction of nitroarenes. *Journal of Colloid Interface Science* 466: 360–368.

Nayak, D., Pradhan, S., Ashe, S., Rauta, P.R., and Nayak. B. 2015. Biologically synthesized silver nanoparticles from three diverse family of plant extracts and their anticancer activity against epidermoid A431 carcinoma. *Journal of Colloid Interface Science* 457: 329–338.

Ndikau, M., Noah, N.M., Andala, D.M., and Masika, E. 2017. Green synthesis and characterization of silver nanoparticles using *Citrullus lanatus* fruit rind extract. *International Journal of Analytical Chemistry* 2017: 8108504.

Noah, N.M., and Ndangili, P.M. 2019. Current trends of nanobiosensors for point-of-care diagnostics. *Journal of Analytical Methods in Chemistry* 2019: 2179718.

Pareek, V., Bhargava, A., Gupta, R., Jain, N., and Panwar, J. 2017. Synthesis and applications of noble metal nanoparticles: A review. *Advanced Science, Engineering and Medicine* 9: 527–544.

Park, S.Y., Hye Yi, E., Kim, Y., and Park, G. 2019. Anti-neuroinflammatory effects of *Ephedra sinica* Stapf extract-capped gold nanoparticles in microglia. *International Journal of Nanomedicine* 14: 2861–2877.

Parveen, M., Ahmad, F., Malla, A.M., and Azaz, S. 2016. Microwave-assisted green synthesis of silver nanoparticles from *Fraxinus excelsior* leaf extract and its antioxidant assay. *Applied Nanoscience* 6: 267–276.

Patra, J.K., and Baek, K.H. 2014. Green nanobiotechnology: Factors affecting synthesis and characterization techniques. *Journal of Nanomaterials* 2014: 219.

Patra, J.K., Das, G., and Baek, K.H. 2016. Phyto-mediated biosynthesis of silver nanoparticles using the rind extract of watermelon (*Citrullus lanatus*) under photo-catalyzed condition and investigation of its antibacterial, anticandidal and antioxidant efficacy. *Journal of Photochemistry and Photobiology B* 161: 200–210.

Penneman, R. A., and Audrieth L. F. 1948. Quantitative determination of hydrazine. *Analytical Chemistry* 20: 1058–1061.

Polavarapu, J., Pérez-Juste, H., Qi, X., and Liz-Marzán, L. M. 2014. Optical sensing of biological, chemical and ionic species through aggregation of plasmonic nanoparticles. *The Journal of Materials Chemistry C* 2: 7460.

Popov, V., Hinkov, I., Diankov, S., Karsheva, M., and Handzhiyski, Y. 2015. Ultrasound-assisted green synthesis of silver nanoparticles and their incorporation in antibacterial cellulose packaging. *Green Processing and Synthesis* 4: 125–131.

Prabhu, N., and Gajendran, T. 2017. Green synthesis of noble metal of platinum nanoparticles from *Ocimum sanctum* (Tulsi) plant extracts. *IOSR Journal of Biotechnology and Biochemistry* 3: 107–112.

Prakash, S.A., Sharma, S., Ahmad, N., Ghosh, A., and Sinha, P. 2010. Bacteria mediated extracellular synthesis of metallic nanoparticles. *International Research Journal of Biotechnology* 1: 071–079.

Rajakumar, G., Gomathi, T., Thiruvengadam, M., Rajeswari, V.D., Kalpana, V.N., and Chung, M. 2017. Evaluation of anti-cholinesterase, antibacterial and cytotoxic activities of green synthesized silver nanoparticles using from *Millettia pinnata* flower extract. *Microbial Pathogenesis* 103: 123–128.

Rajan, A., Cherian, E., and Baskar, G. 2016. Biosynthesis of zinc oxide nanoparticles using *Aspergillus fumigatus* JCF and its antibacterial activity. *International Journal of Modern Science and Technology* 1: 52–57.

Rao, A., Mahajan, K., Bankar, A., Srikanth, R., RaviKumar, A., Gosavic, S., and Zinjarde, S. 2013. Facile synthesis of size-tunable gold nanoparticles by pomegranate (*Punica granatum*) leaf extract: Applications in arsenate sensing. *Materials Research Bulletin* 48: 1166–1173.

Rao, B., and Tang, R. 2017. Green synthesis of silver nanoparticles with antibacterial activities using aqueous *Eriobotrya japonica* leaf extract. *Advances in Natural Sciences: Nanoscience and Nanotechnology* 8: 015014.

Rao, Y.S., Kotakadi, V.S., Prasad, T.N.V.K.V., Reddy, A.V., and Gopal, D.S. 2013. Green synthesis and spectral characterization of silver nanoparticles from *Lakshmi tulasi* (*Ocimum sanctum*) leaf extract. *Spectrochimica Acta Part A: Molecular and Biomolecular Spectroscopy* 103: 156–159.

Roy, A., Bulut, O., Some, S., Kumar. A., and Yilmaz, D. 2019. Green synthesis of silver nanoparticles: Biomolecule-nanoparticle organizations targeting antimicrobial activity. *RSC Advances* 9: 2673–2702.

Sadeghi, B., and Gholamhoseinpoor, F. 2015. A study on the stability and green synthesis of silver nanoparticles using *Ziziphora tenuior* (Zt) extract at room temperature. *Spectrochimica Acta Part A: Molecular and Biomolecular Spectroscopy* 134: 310–315.

Sakthivel, R., Palanisamy, S., Chen, S., Ramaraj, S., Velusamy, V., Yi-Fan, P., Hall, J.M., and Ramaraj, S.K. 2017. A robust nitrobenzene electrochemical sensor based on chitin hydrogel entrapped graphite composite. *Journal of the Taiwan Institute of Chemical Engineers* 80: 663–668.

Salata, O.V. 2004. Applications of nanoparticles in biology and medicine. *Journal of Nanobiotechnology* 2: 3.

Samanta, S.K., Maiti, K., Ali, S.S., Guria, U.N., Ghosh, A., Datta, P., and Mahapatr, A.K. 2020. A solvent directed D-π-A fluorescent chemodosimeter for selective detection of hazardous hydrazine in real water sample and living cell. *Dyes and Pigments* 173: 107997.

Sathishkumar, M., Sneha, K., and Yun, Y.S. 2009. Palladium nanocrystals synthesis using *Curcuma longa longa* extract. *International Journal of Material Science* 4:11–17.

Sathiyanarayanan, G., Dineshkumar, K., and Yang, Y.H. 2017. Microbial exopolysaccharide mediated synthesis and stabilization of metal nanoparticles. *Critical reviews in microbiology* 43: 731–752.

Sedaghat, S., Afshar, P. 2016. Green biosynthesis of silver nanoparticles using *Ziziphora tenuior* L water extract. *Journal of Applied Chemical Research* 10: 103–109.

Sengupta, S., Eavarone, D, Capila, I., Zhao, G., Watson, N., Kiziltepe, T., and Sasisekharan, R. 2005. Temporal targeting of tumour cells and neovasculature with a nanoscale delivery system. *Nature* 436: 568–572.

Siddiquia, M.R., AlOthman, Z.A., and Rahman, N. 2007. Analytical techniques in pharmaceutical analysis: A review. *Arabian Journal of Chemistry* 10: S1409–S1421.

Singh, P., Pandit, S., Beshay, M., Mokkapati, V.R.S.S., Garnaes, J., Olsson, E.M., Sultan, A. et al. 2018. Anti-biofilm effects of gold and silver nanoparticles synthesized by the *Rhodiola rosea* rhizome extracts. *Artificial Cells Nanomedicine and Biotechnology* 46: S886–S899.

Singh, R., Nawale, L., Arkile, M., Wadhwani, S., Shedbalkar, U., Chopade, S., Sarkar, D., and Chopade, B.A. 2016. Phytogenic silver, gold, and bimetallic nanoparticles as novel antitubercular agents. *International Journal of Nanomedicine* 11: 1889.

Singhal, G., Bhavesh, R., Kasariya, K., Sharma, A.R., and Singh, R.P. 2011. Biosynthesis of silver nanoparticles using *Ocimum sanctum* (Tulsi) leaf extract and screening its antimicrobial activity. *Journal of Nanoparticle Research* 13: 2981–2988.

Soundarrajan, C., Sankari, A., Dhandapani, P., Maruthamuthu, S., Ravichandran, S., Sozhan. G., and Palaniswamy, N. 2012. Rapid biological synthesis of platinum nanoparticles using *Ocimum sanctum* for water electrolysis applications. *Bioprocess Biosystems Engineering* 35: 827–833.

Stradiotto, N.R., Yamanaka, H., and Zanoni, M.V.B. 2003. Electrochemical sensors: A powerful tool in analytical chemistry. *Journal of the Brazilian Chemical Society* 14: 159–173.

Subba, R.Y., Kotakadi, V.S., Prasad, T.N., Reddy, A.V., and Sai Gopal, D.V. 2013. Green synthesis and spectral characterization of silver nanoparticles from *Lakshmi tulasi* (*Ocimum sanctum*) leaf extract. *Spectrochimica Acta Part A: Molecular and Biomolecular Spectroscopy* 103: 156–199.

Suib, S.L., 2013. *New and Future Developments in Catalysis: Catalysis by Nanoparticles.* Connecticut, USA: Elsevier.

Sujitha, M.V., and Kannan, S. 2012. Green synthesis of gold nanoparticles using Citrus fruits (*Citrus limon*, *Citrus reticulata* and *Citrus sinensis*) aqueous extract and its characterization. *Spectrochimica Acta Part A: Molecular and Biomolecular Spectroscopy* 102: 15–23.

Sun, D., Zhang, G., Huang, J., Wang, H., and Li, Q. 2014. Plant-mediated fabrication and surface enhanced Raman property of flower-like Au@Pd nanoparticles. *Materials* 7: 1360–1369.

Tabrizi, M.A., and Varkani, J.N. 2014. Green synthesis of reduced graphene oxide decorated with gold nanoparticles and its glucose sensing application. *Sensors and Actuators B*, 202: 475–482.

Thirumurugan, A., Aswitha, A., Kiruthika, C., Nagarajan, S., and Nancy Christy, A. 2016. Green synthesis of platinum nanoparticles using *Azadirachta indica*: An eco-friendly approach. *Materials Letters* 170: 175–178.

Tran, T.T.T., Vu, T.T.H., and Nguyen, T.H. 2013. Biosynthesis of silver nanoparticles using *Tithonia diversifolia* leaf extract and their antimicrobial activity. *Materials Letters* 105: 220–223.

Ueda, S., Iwahashi, Y., and Tokuda, H. 1991. Production of anti-tumor-promoting iridoid glucosides in *Genipa americana* and its cell cultures. *Journal of Natural Products* 54: 1677–1680.

Veitía, M.S.I., and Ferroud, C. 2015. New activation methods used in green chemistry for the synthesis of high added value molecules. *International Journal of Energy and Environmental Engineering* 6: 37–46.

Velmurugan, P., Shim, J., Kim, K., and Oh, B. 2016. Prunus *x yedoensis* tree gum mediated synthesis of platinum nanoparticles with antifungal activity against phytopathogens. *Materials Letters* 174: 61–65.

Vijayan, R., Joseph, S., and Mathew, B. 2018. *Indigofera tinctoria* leaf extract mediated green synthesis of silver and gold nanoparticles and assessment of their anticancer, antimicrobial, antioxidant and catalytic properties. *Artificial Cells Nanomedical Biotechnology* 46: 861–871.

Vilchis-Nestor, A.R., Sánchez-Mendieta, V., Camacho-López, M.A., Gómez-Espinosa, R.M., Camacho-López, M.A., and Arenas-Alatorre, J.A. Solventless synthesis and optical properties of Au and Ag nanoparticles using *Camellia sinensis* extract. *Materials Letters* 62: 3103–3105.

Villegas, V.A.R., Ramı́rez, J.I.D.L., Guevara, H.E., Sicairos, S.P., Ayala, L.A.H.A., and Sanchez, L.B. 2020. Synthesis and characterization of magnetite nanoparticles for photocatalysis of nitrobenzene. *Journal of Saudi Chemical Society* 24: 223–235.

Wang C., and Yu C. 2013. Detection of chemical pollutants in water using gold nanoparticles as sensors: A review. *Reviews in Analytical Chemistry* 32: 1–14.

Wang, M., Wang, X., Xueyan Li, X., Yang, Z., Guo, Z., Zhang, J., Ma, M., and Wei, C. 2020. A coumarin-fused 'off-on' fluorescent probe for highly selective detection of hydrazine. *Spectrochimica Acta Part A: Molecular and Biomolecular Spectroscopy* 230: 118075.

Wang, Y., Xu, H., Zhang, J., and Li, G. 2008. Electrochemical sensors for clinic analysis. *Sensors* 8: 2043–2081.

Xingzong, J., Zhen, L., Mingqin, S., Sili, Y., Xiaoyang, Z., Yongle, Z., and Linxi, H. 2020. A fluorescence "turn-on" sensor for detecting hydrazine in environment. *Microchemical Journal* 152: 104376.

Yang, B., Qi, F., Tan, J., Yu, T., and Qu, C. 2019. Study of green synthesis of ultrasmall gold nanoparticles using *Citrus sinensis* peel. *Applied Sciences* 9: 2423.

Ye, L., Zhao, G., and Dou, W. 2018. An electrochemical immunoassay for *Escherichia coli* O157:H7 using double functionalized Au@Pt/SiO2 nanocomposites and immune magnetic nanoparticles. *Talanta* 182: 354–362.

Yu, C., Tang, J., Liu, X., Ren, X., Zhen, M., and Wang. L. 2019. Green biosynthesis of silver nanoparticles using *Eriobotrya japonica* (Thunb.) leaf extract for reductive catalysis. *Materials* 12: 189.

Zhang, Z., Suo, Y., He, Y., Li, G., Hu, G., and Zheng, Y. 2016. Selective hydrogenation of ortho-chloronitrobenzene over biosynthesized ruthenium–platinum bimetallic nanocatalysts. *Industrial and Engineering Chemistry Research* 55: 7061–7068.

Zheng, B., Kong, T., Jing, X., Odoom-Wubah, T., Li, X., Sun, D., Lu, F., Zheng, Y., Huang, J., and Li, Q. 2013. Plant-mediated synthesis of platinum nanoparticles and its bioreductive mechanism. *Journal of Colloid and Interface Science* 396: 138–145.

Zhou, L., He, X., He, D., Wang, K., and Qin, D. 2011. Biosensing technologies for mycobacterium tuberculosis detection: Status and new developments. *Clinical and Developmental Immunology* 2011: 193963.

Zhou, Y., Kong, Y., Kundu, S., Cirillo, J. D., and Liang, H. 2012. Antibacterial activities of gold and silver nanoparticles against *Escherichia coli* and Bacillus Calmette-Guerin. *Journal of Nanobiotechnology* 10: 1–9.

Zhu, C., Yang, G., Li, H., Du, D., and Lin, Y. 2015. Electrochemical sensors and biosensors based on nanomaterials and nanostructures. *Analytical Chemistry* 87: 230–249.

Index

A

Active pharmaceutical ingredients (APIs), 5
Aluminium hydroxide nanoparticle, 93–94
Amino acids, 130
Anodization, 47
Antidiabetic activity, 32
Antidiabetic agents, 35–37
Antidiabetic medicinal plant, 33
Antidiabetic plant extracts, 35
Antifungal therapy, 46–47
Antimicrobial activity, 220
Antimicrobial agents, 190
Antimicrobial resistance, 4, 126–127
Antimicrobial therapeutics, 189–191

B

Beer-Lambert theory, 91
Biofabrication techniques, 151
Biofilm formation, 135
Biofilms, 135
Biogenic silver particles, 151
Biosafety level 4 (BSL-4), 72
Biotransformation, 298
Blood-brain barrier (BBB), 18

C

Calcium carbonate particles, 278
Cancer theranostics nanoparticles, 18
Candida albicans biofilm, 55
Candida spp.
  metallic nanoparticles, 49–50
  zinc oxide NPS, 50, 52–55
Carbon-based nanomaterials
  carbon nanotubes, 172
  fullerenes, 171–172
Carbon nanotubes, 172
Chemical vapour deposition (CVD), 47
Chemiosmosis process, 127
Chemotherapy treatment, 152
Chitosan nanocapsules (CS-NC), 94
Chitosan nanoparticles, 151
Communicable diseases (CDs), 4
Community or community-associated MRSA (CA-MRSA), 188
Conventional antidiabetic drugs, 31
Conventional cancer treatments, 17
Conventional drug delivery system, 34
Conventional heating (CH), 298–299
Copper nanoparticles, 92
Copper oxide nanoparticle (CuO-NP), 93
COVID-19, 185, 272
Cytotoxicity, 21

D

Dalton's lymphoma ascites (DLA), 22
Dendrimers, 173–175
Denture stomatitis, 45, 52
Diabetes mellitus (DM), 31
  antidiabetic agents, 35–37
  antidiabetic plant extracts, 35
  vaccine and insulin delivery gene and cell therapies, 34
3-(4, 5-dim ethyl thiaz ol-2-yl)-2,5-di pheny l-2H-tetrazolium bromide (MTT), 19
Directly Observed Treatment Short (DOTS) programme, 129
Drug delivery systems
  calcium carbonate particles, 278
  liposomes, 275–276
  metallic nanoparticles, 277–278
  polymeric nanoparticles, 276–277
  solid lipid nanoparticles, 277
Drug resistance, 126

E

Ebola virus disease (EVD)
  nanotechnology-based approach, 72–73
  nanotechnology platform
    disinfection and textile applications, 73–74
    inorganic nanoparticles, 70
    lipid-based nanoparticles, 68–69
    polymer-based nanoparticles, 69–70
    vaccines, 70–71
  pathogenesis, 66–67
Echinocandins, 47
Electrochemical sensors, 294–296
Electron beam lithography, 47
Electron microscopy, 72
Enzyme-linked immunosorbent assay (ELISA), 72
Epidemiological data, 150
Extensively drug-resistant tuberculosis (XDR-TB), 125, 127–129, 132, 135–136
Extracellular polymeric substances (EPS), 135

F

Flavonoids, 130
FTIR spectroscopy, 89–90
Fullerenes, 171–172

G

Gold nanoparticles (AuNPs), 93, 138
Allium cepa, 92
metal nanocrystals, 72
quantum dots, 72
stabilization of, 92
theranostics and optical imaging, 23
Gold-palladium nanoparticls, 92–93
Green bionanotechnology, 18
Green gold nanoparticles (GAuNPs)
antibacterial effect of, 210
antimicrobial effect of, 212
antimicrobial potential of, 210
Au-WAs, 212
Catharanthus roseus (CR) and *Carica papaya* (CP) leaf extracts, 212
FTIR spectra, 211
green extracellular biosynthetic method, 211
green-synthesized GAuNPs, 211
isotopic GAuNPs, 210
XRD patterns, 209
Zizyphus mauritiana extracts, 211
Green method nanoparticles, 296–297
Green nanoparticles
copper nanoparticles CuNPs, 214–215
GZnO-NPs, 213, 214
iron NPs, 217
mechanism of actions, 205
medicinal plants, 199–201
MgO nanoparticles (MgO-NPs), 219
nickel oxide nanoparticles (NiO-NPs), 216
palladium nanoparticles, 217–218
selenium nanoparticles (SeNPs), 219–220
titanium dioxide (TiO_2) and copper (Cu) NPs, 214
titanium NPs (TiO-NPs), 215–216
Green nanotechnology, 47–48
Green silver nanoparticles (GAgNPS), 203–204
Anthemis atropatana extract, 204
CSE-AgNPs, 205
HR-TEM analysis, 193
medicinal plants extract, 194–196
Murraya koenigii, 207
Green synthesis
anticancer mechanism of action, 8
antimicrobial mechanism of action, 7–8
bacteria-mediated synthesis, 305
biotransformation, 298
carbon nanotubes, 172
commercialization aspects, 10
communicable diseases, 9
conventional heating (CH), 298–299
cost-effectiveness, 6
of dendrimers, 175
fungi-mediated synthesis, 304–305
green nanoparticles, sources of, 48–49
microwave irradiation, 297
multidrug resistance *S. aureus*
GC-MS analysis, 222
Hydrastis canadensis L, 223
Quercus infectoria G, 224
Rhodomyrtus tomentosa leaf ethanolic extract, 223–224
ursolic and oleanolic acids, 222
nanomedicine innovations, 9–10
non-communicable diseases, 8–9
photocatalysis synthesis methods, 298
plant extraction, 49
plant-mediated synthesis
detection of hydrazine, 304
electrochemical properties, 301
gold nanoparticles, 300, 302
nitrobenzene (NB), 302
platinum nanoparticles, 303
silver nanoparticles, 303
silver nanoparticles (AgNPs), 301
surface plasmon resonance (SPR), 300
ultrasound method, 297–298
yeast-mediated synthesis, 305

H

HIV-1, 273
Hydrazine
corrosion inhibitor, 292
detection of, 294
structure of, 293

I

Infectious diseases
blood-brain barrier (BBB), 273
COVID-19, 274
drug delivery systems
calcium carbonate particles, 278
liposomes, 275–276
metallic nanoparticles, 277–278
polymeric nanoparticles, 276–277
solid lipid nanoparticles, 277
HIV, 274
TB, 274
vaccine development, 274
Innovative technologies, 18
Inorganic nanoparticles, 32, 67, 70, 126
amino acids, 131

biofilm formation, 135
biofilms, 135
"down-up" approach, 132
extracellular polymeric substances (EPS), 135
flavonoids, 130
gallium, 132
gold nanoparticles (AuNPs), 138
hydrophilic functional groups, 136
labelled inorganic nanoparticles, 138–139
metal ions, 131
nanoparticle synthesis protocol, 131
pharmaceutical agents, 138
pH value, 132
phytochemicals-coated nanoparticles, 137
plant metabolites, 130
TB-infected macrophages, 137
"top-down" approach, 132
XDR-TB, 125, 127–129, 132, 135–136
Inorganic nanotechnology, 129–130
Ion and plasma etching, 47

K

Kaposi's sarcoma, 17

L

Laser pyrolysis, 47
Levofloxacin, 153
Lipid-based nanoparticles, 68–69
Lipid nanocapsules (LNP), 94
Liposomes, 173, 250, 275–276

M

Malaria, 273
aetiology of infection, 104
clinical research, 114–115
computational techniques and tools, 113
discovery of, 104
fatality of, 104
history of, 103
management of, 105
nanomedicine
handling uncertainties, 112–113
informatics and bioinformatics, 111–112
nanoparticles, management of
clinical trials, 108
conventional insecticides, 110
conventional therapeutics, 107
green-synthesized AuNPs, 110
herbal medicine and nanomedicine, 107
multidrug resistance, 107
silver nanoparticles (AgNP), 108
therapeutic agents, 108
Zornia diphylla leaf extract, 108
pharmacological and therapeutic agents, 105
plant-based biogenic nanoparticles, 115–116
preventive treatment, 104
In silico modelling, 113
traditional antimalarials, 105
treatment
chemotherapeutic treatment of, 105
pharmacotherapeutic agents for, 106
verification and validation, 114
Metallic nanoparticles, 6, 192, 277–278
antidiabetic activity, 38–39
antimicrobial activity, 220
applications of, 183–184
biological synthesis of, 169–171
Candida spp., 49–50
culture conditions, 184
S. aureus and mechanism of action, 169
Metal oxide nanoparticles (MO-NPs)
biological synthesis of, 170–171
S. aureus and mechanism of action, 170
Methicillin-resistant *S. aureus* (MRSA), 196
MgO nanoparticles (MgO-NPs), 219
Microorganisms, 48
Milling techniques, 47
Minimal bactericidal concentration (MBC), 94
Minimal inhibitory concentration (MIC), 94
Multidrug-resistant bacteria (MDR), 10, 184
Multidrug-resistant TB (MDR-TB), 127
Multidrug-resistant tuberculosis, 159–160
Multilamellar vaccine particle system (MVPS), 71
Murraya koenigii, 207
Mycobacterium tuberculosis (Mtb), 150
characterization of, 89–90
DNA deletions, 82
genome sequence of, 86
history of, 83
laboratory diagnosis, 85–86
phytonanotechnology, 153–155
plant-based nanoparticles
lipid peroxidation, 88–89
nanoparticles, 87–88
sugar leakages, 88
transmission and pathogenesis, 83–85
treatment of, 86–87

N

Nanoantibiotics, 129
Nanocomposite materials, 292
Nanoemulsion-based nanocapsules, 94
Nanoemulsion drug delivery systems, 68
Nanomedicine
cancer diagnosis, 18
for infectious diseases
blood-brain barrier (BBB), 273

COVID-19, 274
HIV, 274
TB, 274
vaccine development, 274
malaria
handling uncertainties, 112–113
informatics and bioinformatics, 111–112
medicines development models and commercialization
IP management and exploitation, 279–280
product development partnership (PDP), 281–282
research and development, 280, 282–283
WHO CEWG demonstration projects, 281
research and development (R&D), 4
Nanoparticle-based delivery systems, 34
Nanoparticles (NPs), 126
bacteria-mediated synthesis, 305
bioengineered plant-based nanoparticles, 91–93
biological barriers, 18
biological methods, 182
biowastes, bioproducts and biomolecules, 182
characteristics, 182
fungi-mediated synthesis, 304–305
green nanoparticles, 49
green synthesis (*see also* Green synthesis)
activation phase, 132
growth stage, 132
termination stage, 132
morphological characterizations, 90
morphological parameters, 32
optical characterization, 91
organic and inorganic NPs, 32
particle size and surface area characterization, 90
plant-based metal NPs, 184
production, 3
structural characterizations, 90
structural elucidation, of biogenic particles, 89–90
yeast-mediated synthesis, 305
Nanoscale drug delivery systems (nano-DDS), 129
Nanostructures delivery systems
polymeric nanoparticles (PNPs), 94–95
protein nanoparticles, 95
solid-lipid nanoparticles, 95–96
Nanotechnology
advantage of, 5
diabetes mellitus (DM) (*see* Diabetes mellitus (DM))
and EVD (*see* Ebola virus disease (EVD))
green nanotechnology, 47–48
history of, 4–5
Nanotheranostics approach, 18
Nitrobenzene
detection of, 294
structure of, 293
uses of, 293
Non-communicable diseases (NCDs), 4
Non-Hodgkin lymphomas, 17
Non-toxic edible and medicinal plants, 184

O

Oral candidiasis
denture stomatitis, 45
principles, 46
risk factors, 46
Organic carbon-based nanomaterials
dendrimers, 173–174
liposomes, 173
Organic nanoparticles, 32, 67

P

Palladium nanoparticles, 93, 217–218
Pancreatic cancer cell line (PANC-1), 21
Photocatalysis synthesis methods, 298
Photolithography, 47
Physiochemical properties, 139–140
Phytochemicals-coated nanoparticles, 137
Phytonanotechnology, 18
anticancer activity, 19–20
cancer nanotheranostics, 23, 24
mechanism of action for cancer treatment, 20–22
metallic nanoparticles, 19
Pinus merkusii nanoparticle, 94
Plant-based biogenic nanoparticles, 115–116
Plant-based synthesis, 4, 6
Plant crude extracts, 3
Plant-derived antimicrobials, 190
Plant extract components, 48
Plant-mediated synthesis method, 48
Plant metabolites, 130
Plant phytochemicals
alkaloids and saponins, 226
chemical interactions, 224
G-NPs, formation and stabilization, 225
Plants extract synthesis, 150
Plasmodium falciparum, 106
Plasmodium vivax, 106
Polymer-based nanoparticles, 69–70
Polymer-based protein delivery systems
fibrous scaffolds
BMP-7-loaded PLGA nanoparticles, 263
bovine serum albumin (BSA), 261
chemical immobilization, 263
functionalized polystyrene (PS) nanofibers, 263

heparin-functionalized fibrous scaffolds, 262
immobilization of, 262
scaffold preparation technique, 260
hydrogels, 258–259
micro/nanoparticles, 256–258
porous scaffolds, 259–260
Polymeric nanoparticles (PNPs), 94–95, 276–277
Polymethylmethacrylate (PMMA), 50
Polythene glycol, 151
Proof-of-concept, 250
Protein nanoparticles, 95
Protein therapeutics
advantages of, 252
bone scaffolds, 253
bone tissue engineering, 253
brain tumour glioblastoma (GBM), 253
clinical applications, 252
HrBMP4, 253
limitations and challenges, 254–256
pharmacological activity, 251
recombinant production, 252
Pulmonary drug delivery (PDD), 151

R

Research and development (R&D), 4
Reticuloendothelial system (RES), 95
Reverse-transcrip-tase polymerase chain reaction (RT-PCR), 72
Ribonucleoprotein (RNP) complex, 67
Rifampicin, 151

S

Scanning electron microscope (SEM), 19, 90
Selenium nanoparticles (SeNPs), 219–220
Severe acute respiratory syndrome coronavirus 2, 272
Silver nanoparticles
antifungal and anti-microbial activities, 91
AuNP photosensitizer (PS), 50
biosynthesis of, 37
multidrug-resistant microorganism, 91
and *Pterocarpus marsupium*, 37
Syzygium cumini, 91
Sol-gel method, 47, 93
Solid-lipid nanoparticles, 95–96, 277
Staphylococcus aureus
dendrimers, 173–174
liposomes, 173
metal oxide nanoparticles (MO-NPs), 170
methicillin-resistant *S. aureus* (PRSA), 188
multidrug-resistant bacteria (MDR), 185–186
penicillin-resistant *S. aureus* (PRSA), 187–188
vancomycin intermediate resistant *S. aureus*, 188–189
vancomycin-resistant *S. aureus*, 188–189
Synthesized NPs, 192

T

T-cell immunoglobulin, 66
TEM, 90
Titanium dioxide (TiO_2) nanoparticles, 93
Traditional medicines, 34
Transfer ribonucleic acid (tRNA), 141
Tripolyphosphate (TPP), 94
Tuberculosis (TB)
conventional chemotherapy of, 151–152
drug-resistant tuberculosis
management of, 155–156
synthetic drugs, 157–159
nanotechnology, 152
nanotechnology in treatment of, 152–153
pathogenesis and iypes, 151
therapeutic management of, 150
transmission of, 151
Type II diabetes mellitus (T2DM), 33

U

Ultrasound method, 297–298
Untreated control cells, 201
UV-vis absorption spectroscopy, 89

V

Vector-borne diseases, 4
VISA, 186

X

X-ray diffraction (XRD) analysis, 89

Z

Zaire species of EBOV (ZEBOV), 69
Zinc oxide nanoparticles (ZnO-NPs), 54
antimicrobial activity of, 54
biomedical applications, 50
and *Candida* Species, 53
gram-positive and gram-negative bacteria, 50
plant-mediated synthesis of, 53
PMMA mechanical properties, 50, 52

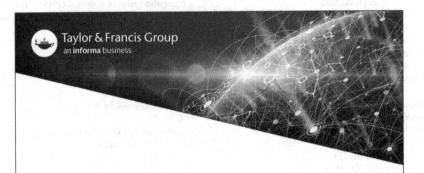

Printed in the United States
By Bookmasters